THE TRAGIC SENSE OF LIFE

Ernst Haeckel (*seated*) and his assistant Nikolai Miklucho on the way to the Canary Islands in 1866. Haeckel had just visited Darwin in the village of Downe. (Courtesy of Ernst-Haeckel-Haus, Jena.)

THE TRAGIC SENSE OF LIFE

Ernst Haeckel and the Struggle over Evolutionary Thought

ROBERT J. RICHARDS

THE UNIVERSITY OF CHICAGO PRESS
CHICAGO AND LONDON

The University of Chicago Press, Chicago 60637
The University of Chicago Press, Ltd., London

Paperback edition 2009
Printed in the United States of America

18 17 16 15 14 13 12 11 10 09 3 4 5 6

ISBN-13: 978-0-226-71214-7 (cloth)
ISBN-13: 978-0-226-71216-1 (paper)
ISBN-10: 0-226-71214-1 (cloth)
ISBN-10: 0-226-71216-8 (paper)

Library of Congress Cataloging-in-Publication Data

Richards, Robert J. (Robert John)
The tragic sense of life: Ernst Haeckel and the struggle over evolutionary thought
p. cm.
Includes bibliographical references and index.
ISBN-13: 978-0-226-71214-7 (cloth : alk. paper)
ISBN-10: 0-226-71214-1 (cloth : alk. paper)
1. Haeckel, Ernst Heinrich Phillipp August, 1834–1919. 2. Biologists—Germany—Biography. 3. Zoologists—Germany—Biography. 4. Evolution (Biology)—History. I. Title.
QH31.H2R53 2008
570.92—dc22
[B]

2007039155

♾ The paper used in this publication meets the minimum requirements of the American National Standard for Information Sciences—Permanence of Paper for Printed Library Materials, ANSI Z39.48-1992.

FOR MY COLLEAGUES

AND STUDENTS AT

THE UNIVERSITY OF CHICAGO

CONTENTS

List of Illustrations *xi*

Preface *xvii*

1. Introduction *1*

The Tragic Source of the Anti-Religious Character of Evolutionary Theory *13*

2. Formation of a Romantic Biologist *19*

Early Student Years *20*
University Years *26*
Habilitation and Engagement *49*

3. Research in Italy and Conversion to Darwinism *55*

Friendship with Allmers and Temptations of the Bohemian Life *57*
Radiolarians and the Darwinian Explanation *63*
Appendix: Haeckel's *Challenger* Investigations *75*

4. Triumph and Tragedy at Jena *79*

Habilitation and Teaching *80*
Friendship with Gegenbaur *84*
For Love of Anna *90*
The Defender of Darwin *94*
Tragedy in Jena *104*

5. Evolutionary Morphology in the Darwinian Mode *113*

Haeckel's *Generelle Morphologie der Organismen* *118*
Haeckel's Darwinism *135*
Reaction to Haeckel's *Generelle Morphologie* *162*
Conclusion *165*
Appendix: Haeckel's Letter to Darwin *168*

6. Travel to England and the Canary Islands: Experimental Justification of Evolution *171*

Visit to England and Meeting with Darwin *172*
Travel to the Canary Islands *176*
Research on Siphonophores *180*
Entwickelungsmechanik *189*
A Polymorphous Sponge: The Analytical Evidence for Darwinian Theory *195*
Conclusion: A Naturalist Voyaging *213*

7. The Popular Presentation of Evolution *217*

Haeckel's *Natural History of Creation* *223*
Conclusion: Evolutionary Theory and Racism *269*

8. The Rage of the Critics *277*

Critical Objections and Charges of Fraud *278*
Haeckel's Responses to His Critics *296*
The Epistemology of Photograph and Fact: Renewed Charges of Fraud *303*
The Munich Confrontation with Virchow: Science vs. Socialism *312*
Conclusion *329*

9. The Religious Response to Evolutionism: Ants, Embryos, and Jesuits *343*

Haeckel's Journey to the Tropics: The Footprint of Religion *344*
"Science Has Nothing to Do with Christ"—Darwin *350*
Erich Wasmann, a Jesuit Evolutionist *356*
The Keplerbund vs. the Monistenbund *371*
The Response of the Forty-six *382*
Conclusion *383*

10. Love in a Time of War 391

At Long Last Love 391
The World Puzzles 398
The Consolations of Love 403
Second Journey to the Tropics—Java and Sumatra 405
Growth in Love and Despair 413
Lear on the Heath 419
The Great War 425

11. Conclusion: The Tragic Sense of Ernst Haeckel 439

Early Assessments of Haeckel Outside of Germany 440
Haeckel in the English-Speaking World at Midcentury 442
Haeckel Scholarship in Germany (1900–Present) 444
The Contemporary Evaluation: Haeckel and the Nazis Again 448
The Tragedy of Haeckel's Life and Science 453

Appendix 1: A Brief History of Morphology 455

Johann Wolfgang von Goethe (1749–1832) 456
Karl Friedrich Burdach (1776–1847) 461
Lorenz Oken (1779–1851) 464
Friedrich Tiedemann (1781–1861) 466
Carl Gustav Carus (1789–1869) 470
Heinrich Georg Bronn (1800–1862) 474
Karl Ernst von Baer (1792–1876) 478
Richard Owen (1804–1892) 481
Charles Darwin (1809–1882) 484

Appendix 2: The Moral Grammar of Narratives in the History of Biology—the Case of Haeckel and Nazi Biology 489

Introduction: Scientific History 489
The Temporal and Causal Grammar of Narrative History 492
The Moral Grammar of Narrative History 497
The Case of Ernst Haeckel 500
The Moral Indictment of Haeckel 502
Nazi Race Hygienists and Their Use of Haeckelian Ideas 504

The Judgment of Historical Responsibility 505
The Reaction of Contemporary Historians 506
Principles of Moral Judgment 509
Conclusion 512

Bibliography 513

Index 541

ILLUSTRATIONS

Frontispiece, Ernst Haeckel and Nikolai Miklucho
1.1 Embryos from two stages of development, from Moore's *Before We Are Born* (1989) *5*
1.2 Isadora Duncan *12*
2.1 Ernst Haeckel with his parents *18*
2.2 Alexander von Humboldt *21*
2.3 Frontispiece of Matthias Schleiden's *Die Pflanze und ihr Leben* (1848) *23*
2.4 Members of the medical faculty at Würzburg *28*
2.5 Johann Wolfgang von Goethe *37*
2.6 Johannes Peter Müller *41*
2.7 Ernst Haeckel in 1858 *43*
2.8 Anna Sethe *52*
3.1 Hermann Allmers *58*
3.2 Ernst Haeckel in 1860 *60*
3.3 Micrographs of radiolarians *66*
3.4 *Heliosphaera,* from Haeckel's *Die Radiolarien* (1862) *73*
3.5 HMS *Challenger* *76*
4.1 Main university building of Friedrich-Schiller-Universtät *81*
4.2 Carl Gegenbaur *85*
4.3 Charles Darwin in 1860 *95*
4.4 Rudolf Virchow *103*
4.5 Anna Sethe Haeckel and Ernst Haeckel *105*
4.6 *Mitrocoma Annae,* from Haeckel's *System der Medusen* (1879) *110*
5.1 Ernst Haeckel and companions on trip to Helgoland, 1865 *118*

5.2 Life cycle of a medusa *131*

5.3 Stem-tree of plants, protists, and animals, from Haeckel's *Generelle Morphologie der Organismen* (1866) *139*

5.4 Stem-tree of lineal progenitors of man, from Haeckel's *Anthropogenie* (1874) *141*

5.5 Karl Ernst von Baer, about age eighty *150*

5.6 Nauplius *153*

5.7 Bronn's tree of systematic relationships *159*

5.8 Tree of Indo-German languages, from August Schleicher's *Darwinsche Theorie und die Sprachwissenschaft* (1863) *160*

5.9 Schleicher's scheme for illustrating morphology and descent of languages *160*

5.10 Stem-tree of the vertebrates, from Haeckel's *Generelle Morphologie der Organismen* (1866) *161*

6.1 Charles Darwin in 1874 *175*

6.2 Nikolai Miklucho *180*

6.3 Freshwater hydra, from Haeckel's *Arbeitsteilung in Natur und Menschenleben* (1868, 1910) *182*

6.4 Hydrozoan colony, from Haeckel's *Arbeitsteilung in Natur und Menschenleben* (1868, 1910) *183*

6.5 Siphonophore, from Haeckel's *Arbeitsteilung in Natur und Menschenleben* (1868, 1910) *184*

6.6 Experiments on siphonophore larvae, from Haeckel's *Zur Entwickelungsgeschichte der Siphonophoren* (1869) *187*

6.7 Calcareous sponge, from Haeckel's *Die Kalkschwämme* (1872) *198*

6.8 *Guancha blanca*, from Miklucho's "Beiträge zur Kenntniss der Spongien I" (1868) *200*

6.9 Cell cleavage and formation of the blastula, from Haeckel's "Die Gastrula und die Eifurchung der Thiere" (1875) *204*

6.10 Gastrula of several phyla of animals, from Haeckel's "Die Gastraea-Theorie" (1874) *205*

6.11 *Ascetta primordialis*, from Haeckel's *Die Kalkschwämme* (1872) *207*

6.12 Letter from Haeckel to Darwin (1873) *209*

7.1 Agnes Haeckel *219*

7.2 Ernst Haeckel in 1868 *220*

7.3 Frontispiece of Haeckel's *Natürliche Schöfungsgeschichte* (1868) *225*

7.4 Twelve human species, from Haeckel's *Natürliche Schöpfungsgeschichte* (1870) *226*

7.5 Haeckel's illustration of the biogenetic law (1868) *235*

7.6 Dog embryo, from Bischoff's *Entwicklungsgeschichte des Hunde-Eies* (1845) *236*

7.7 Human embryo, from Ecker's *Icones physiologicae* (1851–59) *237*

7.8 Human embryo, from Kölliker's *Entwicklungsgeschichte des Menschen und der höheren Thiere* (1861) *238*

7.9 Human embryos, from Ecker's *Icones physiologicae* (1851–59) and Haeckel's *Natürliche Schöpfungsgeschichte* (1868) *239*

7.10 Illustration of the biogenetic law, from Haeckel's *Anthropogenie* (1874) *240*

7.11 Eggs of human, ape, and dog, from Haeckel's *Natürliche Schöpfungsgeschichte* (1868) *241*

7.12 Sandal embryos, from Haeckel's *Natürliche Schöpfungsgeschichte* (1868) *242*

7.13 Stem-tree of the human species, from Haeckel's *Natürliche Schöpfungsgeschichte* (1868) *245*

7.14 Stem-tree of the human species, from Haeckel's *Natürliche Schöpfungsgeschichte* (1870–79) *247*

7.15 Stem-tree of the human species, from Haeckel's *Natürliche Schöpfungsgeschichte* (1889–1920) *249*

7.16 Map of human dispersal *251*

7.17 *Pithecanthropus alalus,* ape-man without speech *254*

8.1 Wilhelm His *280*

8.2 Human sandal embryo, from Haeckel's *Anthropogenie* (1874) *287*

8.3 Human sandal embryo with primitive streak, from Graf von Spee's "Beobachtungen an einer menschlichen Keimscheibe" (1889) *288*

8.4 Human sandal embryo with primitive streak, from Haeckel's *Anthropogenie* (1903) *289*

8.5 Human embryo with allantois, from Haeckel's *Anthropogenie* (1874) *290*

8.6 Alexander Goette *292*

8.7 Albert von Kölliker *294*

8.8 Richardson's embryos compared with Haeckel's *305*

8.9 Altered photos of Richardson's embryos *307*

8.10 Normal table of embryos, from His's *Anatomie menschlicher Embryonen* (1880–85) *309*

8.11 Emil Du Bois-Reymond *316*

8.12 Carl Nägeli *317*

8.13 Rudolf Virchow *319*

8.14 Hôtel de Ville, Paris *320*
8.15 Barricades and cannons on a Paris street *321*
8.16 Illustration of the biogenetic law, from Haeckel's *Anthropogenie* (1903) *335*
8.17 Echidna embryos, from Semon's *Zoologische Forschungsreisen* (1893–) *337*
8.18 Illustration of embryonic similarity of human and dog, from Darwin's *Descent of Man* (1871) *339*
8.19 Illustration of von Baer's law that organisms develop from a general to a more specific morphology during ontogeny, from Gilbert's *Developmental Biology* (1985–97) *340*
9.1 Ernst Haeckel on the way to Ceylon *342*
9.2 Haeckel's house, Villa Medusa *350*
9.3 Market square in Jena *359*
9.4 Erich Wasmann as seminarian *362*
9.5 Two species of the "guests of ants" *366*
9.6 Father Erich Wasmann, S.J. *369*
9.7 Meeting of the Monistenbund *373*
9.8 Eberhard Dennert *374*
9.9 Embryos illustrating the biogenetic law, from Haeckel's *Das Menschen-Problem* (1907) *378*
9.10 Ape embryos at comparable stages, from Selenka's *Menschenaffen* (1903) and Haeckel's *Menschen-Problem* (1907) *379*
9.11 Human embryos at comparable stages, from His's *Anatomie menschlicher Embryonen* (1880–85) and Haeckel's *Menschen-Problem* (1907) *380*
9.12 Comparison of ape skeletons with the human skeleton, from Haeckel's *Menschen-Problem* (1907) *381*
9.13 John Wendell Bailey *389*
10.1 Frida von Uslar-Gleichen and Bernhard von Uslar-Gleichen *394*
10.2 Agnes Haeckel *397*
10.3 René Binet's Porte Monumentale *408*
10.4 Discomedusa *Rhopilema Frida* *411*
10.5 Hamburg German-American liner *Kiautschou* *412*
10.6 Haeckel's letter to Frida *414*
10.7 Frida von Uslar-Gleichen *417*
10.8 Haeckel's illustration of the "Apotheosis of Evolutionary Thought" *418*

10.9 Isadora Duncan with Haeckel *421*
10.10 Phyletic Museum, Jena *422*
10.11 Else Meyer, Ernst Haeckel, and Walter Haeckel *424*
10.12 French trenches at Verdun *428*
10.13 German postcard with "Argonnerwald-Lied" *430*
10.14 Haeckel in his study, 1914 *433*
10.15 Haeckel's grave marker *437*
App. 1.1 Johann Wolfgang von Goethe *457*
App. 1.2 Karl Friedrich Burdach *462*
App. 1.3 Lorenz Oken *465*
App. 1.4 Friedrich Tiedemann *468*
App. 1.5 Carl Gustav Carus *470*
App. 1.6 Vertebrate archetype, from Carus's *Von den Ur-theilen des Knochen- und Schalengerüstes* (1828) *471*
App. 1.7 Ideal vertebra, from Carus's *Von den Ur-theilen des Knochen- und Schalengerüstes* (1828) *473*
App. 1.8 Heinrich Georg Bronn *475*
App. 1.9 Karl Ernst von Baer *480*
App. 1.10 Richard Owen *482*
App. 1.11 Vertebrate archetype, from Owen's *On the Nature of Limbs* (1849) *483*
App. 1.12 Vertebrate limbs of mole and bat *484*
App. 1.13 Charles Darwin *486*
App. 2.1 Plaque in the main university building at Jena *507*

Plates follow p. 172

Plate 1 Radiolaria of the subfamily Eucyrtidium, from Haeckel's *Radiolarien* (1862)
Plate 2 Alexander von Humboldt and Aimé Bonpland
Plate 3 Haeckel's portrait of the *Nationalversammlung der Vögel*
Plate 4 Haeckel's watercolor of his study on Capri
Plate 5 Discomedusa *Desmonema Annasethe*, from Haeckel's *System der Medusen* (1879)
Plate 6 *Physophora magnifica*, from Haeckel's *Zur Entwickelungsgeschichte der Siphonophoren* (1869)
Plate 7 Scene from Haeckel's *Arabische Korallen* (1876)
Plate 8 Haeckel's landscape of the highlands of Java

PREFACE

The nineteenth century was an age of enlightened science and romantic adventure. The age rippled with individuals of outsize talents. Johann Wolfgang von Goethe, the great German poet-scientist, joined aesthetic considerations with analytical observations to engage in two great scientific pursuits, a recalcitrant study of optics and an innovative construction of morphology. The former foundered on the rocks of his poetic genius, but the latter gave birth to a new discipline that became integral to biology. Alexander von Humboldt, a dashing disciple of Goethe, sailed to the New World in 1799 and spent five years exploring the jungles and social character of South and Central America. The intellectual results of his quest elevated him to the very summit of European science and culture. His travels became the inspiration for that other great romantic adventure, Charles Darwin's journey on HMS *Beagle.* Darwin's theory of evolution by natural selection transformed the thought of the period as had no other scientific accomplishment before or since. The last part of the nineteenth century was dominated in theoretical physics and experimental physiology by the polymath Hermann von Helmholtz, an individual who vied with Goethe for cultural hegemony. And at the very end of the century, Sigmund Freud completed his *Interpretation of Dreams,* which would become an icon of modernist science during the first half of the twentieth century, competing with Einstein's discoveries in broad intellectual significance, if not scientific import.

Another individual of comparable stature in his own time and with a reverberating impact on ours was Ernst Haeckel, Darwin's great champion in Germany. His name is not as well known as some of the others I have mentioned, but virtually everyone is aware of the principle he made famous: the biogenetic law that ontogeny recapitulates phylogeny—that is, that the embryo of a contemporary species goes through the same morphological changes in its development as its ancestors had in their evolutionary descent. More people at the turn of the century were carried to evolutionary theory on the torrent of his publications than through any other source, including Darwin's own writings. He waged war against orthodoxies of every sort and is largely responsible for fomenting the struggle between evolutionary science and religion that still stirs our social and political life. Like Goethe and Humboldt, whom he revered, his science was transported by deep currents of aesthetic inspiration. He was a gifted artist who illustrated all of his own works, making them accessible to a wider audience and a target for conservative opponents. Despite the maelstrom of controversy that engulfed his work, few individuals, except perhaps Darwin and Helmholtz, garnered from contemporaries more notable prizes, honorary degrees, and prestigious accolades. Though today the term "genius" has been debased and regarded as suspect, if it means startling creativity, tireless industry, and deep artistic talent, it should not be denied to Haeckel. His scientific ideas rebounded on Darwin, especially regarding human evolution. Helmholtz supported him and Freud made recapitulation a central doctrine of psychoanalysis. Casting one's historical vision lower, to the area of his special expertise, marine invertebrate biology, one still finds more creatures—radiolaria, medusae, siphonophores, sponges—having their species designation bearing his name than that of any other investigator.

In our time, this thinker of extraordinary depth, scope, and influence has yet been cast into the Mephistophelean role, one of a sinister individual whose science was meretricious and intent malign. Some contemporary scholars have accused him of fraud and—even worse—of not being a real Darwinian. Others have linked him with Nazi racism, though he died a decade and a half before Hitler came to power. There is little doubt that Haeckel was a man of contradictions and a personality of magnetic proportions—with one pole pulling the best biological students to his little redoubt in Jena and the other repulsing the orthodox all over the world. His energy and combativeness derived, I believe, from the tragedy that haunted him most of his days. That searing experience explains, at least in part, both his pulsing creativity and his incessant struggles. For any historian or philosopher of biology, Haeckel offers an irresistible subject of investigation.

My own interest in the man began some time ago. I first briefly visited Jena and Ernst-Haeckel-Haus, the repository of Haeckel's manuscripts, during those oppressive East German times. Some of the scholars I met at the

Institut für Geschichte der Medizin und der Naturwissenschaften, also located in Haeckel-Haus, inspired confidence that there would be better days. I returned to Jena when the promise began to be realized in January and February 1990, shortly after the fall of the Berlin Wall. I became acquainted with the director of the institute at the time, who was later revealed to be a high level Stasi, and with the archivist of the institute, Erika Krauße. Good socialist that she was, Krauße remained cautiously protective, during that uncertain period, of the very rich archive—thousands of letters, mostly to Haeckel, and the stacks of his manuscripts, paintings, and drawings as well as memorabilia of various sorts. More recently I have come to know individuals who have turned that archive into an open scholarly source, and I am deeply indebted to them for their help with materials under their custody. Beyond scholarship, however, Olaf Breidbach, the present director of Ernst-Haeckel-Haus, and Uwe Hoßfeld, a coworker with incomparable knowledge of German evolutionary biology, have become good friends. Mario Di Gregorio, another frequent visitor to Haeckel-Haus, has shared my interest in, if not my perspective on, the course of Haeckel's career; and I have learned much from him.

I began writing this book in 1994 but put it away after composing a few chapters. In attempting to prepare the ground for the study, I indulged in considerable research and reading about the earlier period of German Romanticism and was ineluctably and happily pulled back to that extraordinary time. This new departure yielded a book in 2002 under the title *The Romantic Conception of Life: Science and Philosophy in the Age of Goethe.* After its publication, I returned to Haeckel. In 2004–2005 I enjoyed the support of the National Science Foundation and the John Simon Guggenheim Memorial Foundation, which enabled me essentially to complete the present study, which might be regarded as a companion to that prior volume.

Some parts of this project have previously appeared in *Annals of the History and Philosophy of Biology; The University of Chicago Record; The Many Faces of Evolution in Europe, 1860–1914,* edited by Mary Kemperink and Patrick Dassen; and *Darwinian Heresies,* edited by Abigail Lustig, Michael Ruse, and Robert J. Richards. All translations, except as otherwise noted, are my own.

No scholar works alone, especially if he or she has ambitions to move beneath encrusted thought and to reevaluate the career of a multifaceted individual about whom influential judgments have long been confidently rendered. Old friends, as well as new acquaintances, have scrutinized my manuscript and tried to mend some of my ways. Lorraine Daston, Garth

Nelson, and Christopher Starr made important recommendations regarding various chapters. Christopher DiTeresi, Uwe Hoßfeld, Lynn Nyhart, Alessandro Pajewski, Trevor Pearce, Andrew Reynolds, and Cecelia Watson had the patience to read through the entire manuscript. The deep knowledge of these scholars ranged from the history of science to contemporary biology, from the logic of argument to the logic of the comma. I am deeply grateful for their aid. Erin DeWitt, with sure eye and steady hand, rendered my text smoother and more consistent than I could ever have managed.

My more indirect debt has been to colleagues and students at the University of Chicago. Their voracious and unrelenting intellectual appetites do not tolerate pabulum or mediocre fare. I know that many of my confections have not gone down easily with them. And while I may not have always met their demands, I am constantly reminded of and inspired by their standards. My wife, Barbara, has provided all that one could desire, and more need not be said.

CHAPTER ONE

Introduction

In late winter of 1864, Charles Darwin received two folio volumes on radiolarians, a group of one-celled marine organisms that secreted skeletons of silica having unusual geometries. The author, the young German biologist Ernst Haeckel, had himself drawn the figures for the extraordinary copper-etched illustrations that filled the second volume.[1] The gothic beauty of the plates astonished Darwin (see, for instance, plate 1), but he must also have been drawn to passages that applied his theory to construct the descent relations of these little-known creatures. He replied to Haeckel that the volumes "were the most magnificent works which I have ever seen, & I am proud to possess a copy from the author."[2] A few days later, emboldened by his own initiative in contacting the famous scientist, Haeckel sent Darwin a newspaper clipping that described a meeting of the Society of German Natural Scientists and Physicians at Stettin, which had occurred the previous autumn. The article gave an extended and laudatory account of Haeckel's lecture defending Darwin's theory.[3] Darwin

1. Ernst Haeckel, *Die Radiolarien. (Rhizopoda Radiaria). Eine Monographie*, 2 vols. (Berlin: Georg Reimer, 1862).

2. Darwin to Haeckel (3 March 1864), in the Correspondence of Ernst Haeckel, in the Haeckel Papers, Institut für Geschichte der Medizin, Naturwissenschaft und Technik, Ernst-Haeckel-Haus, Friedrich-Schiller-Universität, Jena. The letter has recently been published in *The Correspondence of Charles Darwin*, vol. 12: *1864*, ed. Frederick Burkhardt et al. (Cambridge: Cambridge University Press, 2001), 61. For a calendar of Haeckel's correspondence, see *Haeckel-Korrespondenz: Übersicht über den Briefbestand des Ernst-Haeckel-Archivs*, ed. Uwe Hoßfeld and Olaf Breidbach (Berlin: Verlag für Wissenschaft und Bildung, 2005).

3. "Vorträge Ernst Haeckels," *Stettiner Zeitung*, no. 439, 20 September 1863. The author began: "The first speaker [Haeckel] stepped up to the podium and delivered to rapt attention a lecture on Darwin's theory of creation. The lecture captivated the auditorium because of

immediately replied in a second letter: "I am delighted that so distinguished a naturalist should confirm & expound my views; and I can clearly see that you are one of the few who clearly understands Natural Selection."[4] Darwin recognized in the young Haeckel a biologist of exquisite aesthetic sense and impressive research ability and, moreover, a thinker who obviously appreciated his theory.

Haeckel would become the foremost champion of Darwinism not only in Germany but throughout the world. Prior to the First World War, more people learned of evolutionary theory through his voluminous publications than through any other source. His *Natürliche Schöpfungsgeschichte* (Natural history of creation, 1868) went through twelve German editions (1868–1920) and appeared in two English translations as *The History of Creation.* Erik Nordenskiöld, in the first decades of the twentieth century, judged it "the chief source of the world's knowledge of Darwinism."[5] The crumbling detritus of this synthetic work can still be found scattered along the shelves of most used-book stores. *Die Welträthsel* (The world puzzles, 1899), which placed evolutionary ideas in a broader philosophical and social context, sold over forty thousand copies in the first year of its publication and well over fifteen times that during the next quarter century—and this just in the German editions.[6] (By contrast, during the three decades between 1859 and 1890, Darwin's *Origin of Species* sold only some thirty-nine thousand copies in the six English editions.)[7] By 1912 *Die Welträthsel* had been translated, according to Haeckel's own meticulous tabulations, into twenty-four languages, including Armenian, Chinese, Hebrew, Sanskrit, and Esperanto.[8] The young Mohandas Gandhi had requested permission

---

its illuminatingly clear presentation and extremely elegant form." The author then gave an extensive précis of the contents of the entire lecture. He concluded by reporting that "a huge applause followed this exciting lecture."

4. Darwin to Haeckel (9 March 1864), in the Haeckel Correspondence, Haeckel-Haus, Jena; *Correspondence of Charles Darwin,* 12:63.

5. Erik Nordenskiöld, *The History of Biology: A Survey* (1920–24), trans. Leonard Eyre, 2nd ed. (New York: Tudor, 1936), 515.

6. See the introduction to a modern edition of Haeckel's *Die Welträtsel,* ed. Olof Klohr (Berlin: Akademie, 1961), vii–viii. See also Erika Krauße, "Wege zum Bestseller, Haeckels Werk im Lichte der Verlegerkorrespondenz: Die Korrespondenz mit Emil Strauss," in *Der Brief als wissenschaftshistorische Quelle,* ed. Erika Krauße (Berlin: Verlag für Wissenschaft und Bildung, 2005), 145–70 (publication details on 165–66).

7. See the introduction to *The Origin of Species by Charles Darwin: A Variorum Text,* ed. Morse Peckham (Philadelphia: University of Pennsylvania Press, 1959), 24.

8. Haeckel's charting is in an unnumbered document in the Haeckel Papers, Haeckel-Haus, Jena.

to render it into Gujarati; he believed it the scientific antidote to the deadly wars of religion plaguing India.[9]

Haeckel achieved many other popular successes and, as well, produced more than twenty large technical monographs on various aspects of systematic biology and evolutionary history. His studies of radiolarians, medusae, sponges, and siphonophores remain standard references today. These works not only informed a public; they drew to Haeckel's small university in Jena the largest share of Europe's great biologists of the next generation, among whom were the "golden" brothers Richard and Oscar Hertwig, Anton Dohrn, Hermann Fol, Eduard Strasburger, Vladimir Kovalevsky, Nikolai Miklucho-Maclay, Arnold Lang, Richard Semon, Wilhelm Roux, and Hans Driesch. Haeckel's influence stretched far into succeeding generations of biologists. Ernst Mayr, one of the architects of the modern synthesis of genetics and Darwinism in the 1940s, confessed that Haeckel's books introduced him to the attractive dangers of evolutionary theory.[10] Richard Goldschmidt, the great Berlin geneticist who migrated to Berkeley under the treacherous shadow of the Nazis in the 1930s, later recalled the revelatory impact reading Haeckel had made on his adolescent self:

> I found Haeckel's history of creation one day and read it with burning eyes and soul. It seemed that all problems of heaven and earth were solved simply and convincingly; there was an answer to every question which troubled the young mind. Evolution was the key to everything and could replace all the beliefs and creeds which one was discarding. There were no creation, no God, no heaven and hell, only evolution and the wonderful law of recapitulation which demonstrated the fact of evolution to the most stubborn believer in creation.[11]

Haeckel gave currency to the idea of the "missing link" between apes and man; and in the early 1890s, Eugène Dubois, inspired by Haeckel's ideas, actually found its remains where the great evolutionist had predicted, in

9. Joseph McCabe to Haeckel (July 1909), in the Haeckel Correspondence, Haeckel-Haus, Jena. McCabe, Haeckel's English translator, met Gandhi in London. In his book *Ethical Religion,* which was originally published as articles in early 1907, Gandhi looked to the evolutionary account of morality as demonstrating its ubiquity in nature and its supreme value. See Mahatma Gandhi, *Ethical Religion,* trans. A. Rama Lyer, 2nd ed. (Madras: S. Ganesan, 1922), 49–56.

10. Ernst Mayr, personal communication, 1995.

11. Richard Goldschmidt, *Portraits from Memory: Recollections of a Zoologist* (Seattle: University of Washington Press, 1956), 35.

the Dutch East Indies.[12] Haeckel formulated the concept of ecology; identified thousands of new animal species; established an entire kingdom of creatures, the Protista; worked out the complicated reproductive cycles of many marine invertebrates; identified the cell nucleus as the carrier of hereditary material; described the process of gastrulation; and performed experiments and devised theories in embryology that set the stage for the groundbreaking research of his students Roux and Driesch. His "biogenetic law"—that is, that ontogeny recapitulates phylogeny[13]—dominated biological research for some fifty years, serving as a research tool that joined new areas into a common field for the application of evolutionary theory. The "law," rendered in sepia tones, can still be found nostalgically connecting contemporary embryology texts to their history (figs. 1.1 and 8.18).[14]

Haeckel, however, has not been well loved—or, more to the point, well understood—by historians of science. E. S. Russell, whose judgment may usually be trusted, regarded Haeckel's principal theoretical work, *Generelle Morphologie der Organismen* (General morphology of organisms, 1866), as "representative not so much of Darwinian as of pre-Darwinian thought." "It was," he declared, "a medley of dogmatic materialism, idealistic morphology, and evolutionary theory."[15] Gavin De Beer, a leading embryologist of the first half of the twentieth century, blamed Haeckel for putting embryology in "a mental strait-jacket which has had lamentable

12. Haeckel speculated that the transition from ape to man via *Pithecanthropus alalus* (ape-man without speech) took place in the area of Borneo, Sumatra, and Java. Inspired by Haeckel, Eugène Dubois searched these regions while stationed there as a physician in the Dutch army. Amazingly, in 1890 and 1891, he discovered in Java the remains of what became known as *Homo erectus*, certainly the best candidate for the missing link. See Eugène Dubois, *Pithecanthropus erectus, eine menschenähnliche Übergangsform aus Java* (Batavia: Landesdruckerei, 1894); and "Pithecanthropus Erectus—A Form from the Ancestral Stock of Mankind," *Annual Report, Smithsonian Institution* (1898): 445–59.

13. Specifically the principle states that the developing embryo of an advanced species passes through the morphological stages of its more primitive evolutionary ancestors—that, for instance, the human embryo begins as a one-celled creature, just as our progenitor presumably did hundreds of millions of years ago, and then passes through stages similar to that of an early invertebrate, of a primitive vertebrate (e.g., a fish), of a primate, and finally of a human being.

14. Richardson and Keuck have listed about a dozen text books from the 1980s to the present that have used Haeckel's embryo illustrations. See Michael Richardson and Gerhard Keuck, "Haeckel's ABC of Evolution and Development," *Biological Review* 77 (2002): 495–528; the list is on 515.

15. E. S. Russell, *Form and Function: A Contribution to the History of Animal Morphology* (1916; repr., Chicago: University of Chicago Press, 1982), 247–48.

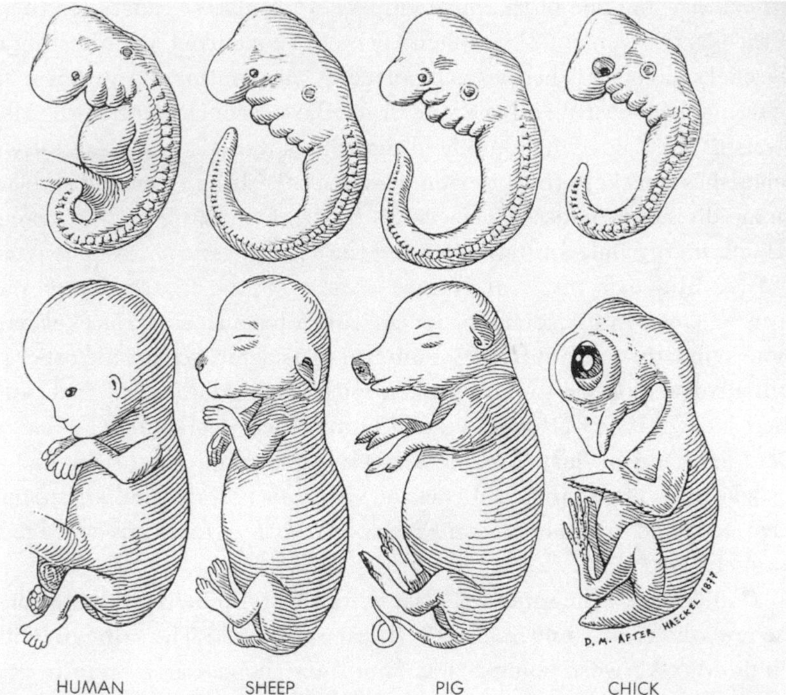
D. M. AFTER HAECKEL 1877
HUMAN
SHEEP
PIG
CHICK
Figure 6–16. Drawings of embryos of four species showing how their early characteristics are similar. By the eighth week, human embryos have distinctive characteristics.
70

Fig. 1.1. Depiction of different embryos at two stages of development "after Haeckel." (From Keith Moore's *Before We Are Born,* 1989.)

effects on biological progress."[16] Peter Bowler endorses these evaluations and further judges that the biogenetic law "illustrates the non-Darwinian character of Haeckel's evolutionism."[17] Bowler believes Haeckel's theory of evolution ideologically posited a linear and progressive trajectory toward man. Haeckel, he assumes, did not take seriously Darwin's conception of branching descent. Daniel Gasman has argued that Haeckel's "social Dar-

16. G. R. De Beer, *Embryos and Ancestors* (Oxford: Clarendon Press, 1940), 97.

17. Peter Bowler, *The Non-Darwinian Revolution* (Baltimore: Johns Hopkins University Press, 1988), 83–84. I have argued, on the contrary, that the recapitulational thesis forms the heart of Darwin's own theory of evolution. See Robert J. Richards, *The Meaning of Evolution: The Morphological Construction and Ideological Reconstruction of Darwin's Theory* (Chicago: University of Chicago Press, 1992), 91–166. See also the exchange in Peter Bowler, "A Bridge Too Far," *Biology and Philosophy* 8 (1993): 98–102; and Robert J. Richards, "Ideology and the History of Science," *Biology and Philosophy* 8 (1993): 103–8.

winism became one of the most important formative causes for the rise of the Nazi movement."[18] Stephen Jay Gould concurred, maintaining that Haeckel's biological theories, supported by an "irrational mysticism" and a penchant for casting all into inevitable laws, "contributed to the rise of Nazism." Like Bowler, Gould held that the biogenetic law essentially distinguishes Haeckel's thought from Darwin's.[19] Adrian Desmond and James Moore divine the causes of Haeckel's mode of thinking in "his evangelical upbringing and admiration for Goethe's pantheistic philosophy [which] had led him to a mystical Nature-worship at the University of Würzburg."[20] German historians of recent times have treated Haeckel hardly more sympathetically. Jürgen Sandmann considers Haeckel and other Darwinists of the period to have broken with the humanitarian tradition by their biologizing of ethics.[21] Peter Zigman, Jutta Kolkenbrock-Netz, and Gerd Rehkämper—just to name a few other German historians and philosophers who have analyzed Haeckel's various theories and arguments—have rendered judgments comparable to their American and English counterparts.[22]

Could this be the same scientist whom Darwin believed to be "one of the few who clearly understands Natural Selection"? The same individual whom Max Verworn eulogized as "not only the last great hero from the

18. Daniel Gasman, *The Scientific Origins of National Socialism: Social Darwinism in Ernst Haeckel and the German Monist League* (New York: Science History Publications, 1971), xxii. See also Daniel Gasman, *Haeckel's Monism and the Birth of Fascist Ideology* (New York: Peter Lang, 1998).

19. Stephen Jay Gould, *Ontogeny and Phylogeny* (Cambridge, MA: Harvard University Press, 1977), 77–81.

20. Adrian Desmond and James Moore, *Darwin: The Life of a Tormented Evolutionist* (New York: Norton, 1991), 538–39.

21. Jürgen Sandmann, *Der Bruch mit der humanitären Tradition: Die Biologisierung der Ethik bei Ernst Haeckel und anderen Darwinisten seiner Zeit* (Stuttgart: Gustav Fischer, 1990). See also his "Ernst Haeckels Entwicklungslehre als Teil seiner biologistischen Weltanschauung," in *Die Rezeption von Evolutionstheorien im 19. Jahrhundert*, ed. Eve-Marie Engels (Frankfurt: Suhrkamp, 1995).

22. See Peter Zigman, "Ernst Haeckel und Rudolf Virchow: Der Streit um den Charakter der Wissenschaft in der Auseinandersetzung um den Darwinismus," *Medizin-Historisches Journal* 35 (2000), 263–302; Jutta Kolkenbrock-Netz, "Wissenschaft als nationaler Mythos: Anmerkungen zur Haeckel-Virchow-Kontroverse auf der 50. Jahresversammlung deutscher Naturforscher und Ärzte in München (1877)," in *Nationale Mythen und Symbole in der zweiten Hälfte des 19. Jahrhunderts*, ed. Jürgen Link and Wulf Wülfing (Stuttgart: Kolett-Cotta, 1991), 212–36; and Gerd Rehkämper, "Zur frühen Rezeption von Darwins Selektionstheorie und deren Folgen für die vergleichende Morphologie heute," *Sudhoffs Archiv* 81 (1997): 171–92. Uwe Hoßfeld offers a quite different perspective in "Haeckelrezeption im Spannungsfeld von Monismus, Sozialdarwinismus und Nationalsozialismus," *History and Philosophy of the Life Sciences* 21 (1999): 195–213.

classical era of Darwinism, but one of the greatest research naturalists of all times and as well a great and honorable man"?[23]

Ernst Haeckel was a man of parts. It is not surprising that assessments of him should collide. I believe, however, that Darwin and Verworn, his colleagues, exhibited a more reliable sense of the man. This is not to suggest, though, that other of his contemporaries would not have agreed with the evaluations made by the historians I have cited. The philosophers, especially the neo-Kantians, were particularly enraged. Erich Adickes at Kiel dismissed *Die Welträthsel* as "pseudo-philosophy."[24] The great Berlin philosopher Friedrich Paulsen erupted in molten anger at the book and released a flood of searing invectives that would have smothered the relatively cooler judgments of the historians mentioned above. He wrote:

> I have read this book with burning shame, with shame over the condition of general education and philosophic education of our people. That such a book was possible, that it could be written, printed, bought, read, wondered at, believed in by a people that produced a Kant, a Goethe, a Schopenhauer—that is painfully sad.[25]

The Swiss zoologist Ludwig Rütimeyer stumbled across one of Haeckel's more crucial lapses of judgment and instigated a charge of scientific dishonesty that would hound him for decades.[26] And, of course, Haeckel's continued baiting of the preachers evoked from them an enraged howl of warning about "the depth of degradation and despair into which the teaching of Haeckel will plunge mankind."[27] Contemporary creationists and those advocating intelligent design have heeded the warning; they have ignited thousands of websites in an electronic auto-da-fé in which Ernst Haeckel's reputation is sacrificed to appease an angry God.

23. Max Verworn, "Ernst Haeckel," *Zeitschrift für allgemeine Physiologie* 19 (1921): i. Verworn was a student of Haeckel and later professor of physiology at Göttingen, director of the Physiological Institute at Bonn, and editor of *Zeitschrift für allgemeine Physiologie.*

24. Erich Adickes, "The Philosophical Literature of Germany in the Years 1899 and 1900," *Philosophical Review* 10 (1901): 386–416; see especially 404–7.

25. Friedrich Paulsen, *Philosophia militans: Gegen Klerikalismus und Naturalismus* (Berlin: Reuther & Reichard, 1901), 187.

26. Ludwig Rütimeyer, Review of "Ueber die Entstehung und den Stammbaum des Menschengeschlechts" and *Natürliche Schöpfungsgeschichte,* by Ernst Haeckel, *Archiv für Anthropologie* 3 (1868): 301–2. I will discuss the charges below.

27. R. F. Horton, "Ernst Haeckel's 'Riddle of the Universe,'" *Christian World Pulpit* 63 (10 June 1903): 353.

Haeckel's evolutionary convictions, fused together by the deep fires of his combative passions, kept the human questions of evolution ever burning before the public, European and American, through the last half of the nineteenth century and well into ours. The controverted implications of evolutionary theory for human life—for man's nature, for ethics, and for religion—would not have the same urgency they still hold today had Haeckel not written.

The measure of Haeckel is usually taken, I believe, using a one-dimensional scale. His acute scientific intelligence moved through many diverse areas of inquiry—morphology, paleontology, embryology, anatomy, systematics, marine biology, and his newly defined fields of phylogeny, ecology, and chorology (biogeography)[28]—and to all of these he made important contributions. But more significantly, through a deft construction of evolutionary processes, he reshaped these several disciplines into an integrated whole, which arched up as a sign of the times and a portent for the advancement of biological science. He anchored this evolutionary synthesis in novel and powerful demonstrations of the simple truth of the descent and modification of species. Haeckel supplied exactly what the critics of Darwin demanded, namely, a way to transform a *possible* history of life into the *actual* history of life on this planet. Certainly he merited Darwin's accolade and was, I believe, the English scientist's authentic intellectual heir. But Haeckel, needless to say, was not Darwin. His accomplishments must be understood as occupying a different scientific, social, and psychological terrain, through which passed a singular intellectual current that flowed powerfully even into the second half of the nineteenth century, namely, Romanticism.[29]

Both by intellectual persuasion and temperament, Haeckel was a Romantic. His ideas pulsed to the rhythms orchestrated by Johann Wolfgang von Goethe, Alexander von Humboldt, and Matthias Jakob Schleiden.

28. Haeckel was notorious for formulating jaw-breaking terms to define new or reconceived areas of research—"phylogeny," "ontogeny," "gastrulation," and "ecology" being those that have stuck the tightest to contemporary theory. He defined ecology as "the entire science of the relationships of the organism to its surrounding external world, wherein we understand all 'existence-relationships' in the wider sense." Chorology was the "entire science of spatial dispersion of organisms, of their geographical and topographical spread over the earth's surface." Haeckel conceived chorology as part biogeography and part the morphology of populations (much in the manner of Alexander von Humboldt). See Ernst Haeckel, *Generelle Morphologie der Organismen,* 2 vols. (Berlin: Georg Reimer, 1866): 2:286–87.

29. For a discussion of the ways the Romantic movement shaped biological thought in the first half of the nineteenth century, see Robert J. Richards, *The Romantic Conception of Life: Science and Philosophy in the Age of Goethe* (Chicago: University of Chicago Press, 2002).

They, and other similarly disposed figures from the first half of the century, inspired Haeckel in the construction of his evolutionary morphology. They had proposed that archetypal unities ramified through the wild diversity of the plant and animal kingdoms. Such Ur-types focused consideration on the whole of the creature in order to explain the features of its individual parts. When the theory of the archetype became historicized in evolutionary theory, it yielded the biogenetic law, the lever by which Haeckel attempted to lift biological science to a new plane of understanding. The Romantic thinkers to whom Haeckel owed much regarded nature as displaying the attributes of the God now in hiding; for them, and Haeckel as well, it was *Deus sive natura*—God and nature were one. This metaphysical persuasion required that the sterile mechanisms described by low-grade Newtonians be replaced by a fecund nature from whose creative depths greatly disparate forms could arise. Nature, under their conception, feigned no indifference to moral concerns or to beauty. Darwin himself, as I have shown elsewhere, shared this Romantic conception of nature.[30] These earlier Romantic scientists insisted that the understanding of organic forms, whether manifested in the individual or in the population, required not only theoretic consideration but aesthetic evaluation as well. The artistic features of organic forms had to be included in the proper assessment of their development and function; and for this purpose, Haeckel's talent with the artist's brush served him no less than his dexterity with the scientist's microscope. And just as Goethe sought the concrete realization of his theory of types in an aesthetically imagined primitive plant, the *Urpflanze,* so Haeckel pictured a polymorphous organism—a perverse sponge artfully conceived—that seemed to bring an ideal evolutionary theory into actual history.

Haeckel's Romanticism reached down to the inmost feelings of his being; and so to comprehend his scientific achievement, we must also probe his character. The strategy of causally linking the theories of a scientist not only to the ideas supplied by predecessors and contemporaries but also to the deeper forces of the self is born of a historiographic conviction, one given firm expression by Miguel de Unamuno, author of an earlier *Tragic Sense of Life*. In his *Del sentimiento trágico de la vida* (1913), he objected:

> In most of the histories of philosophy that I know, philosophic systems are presented to us as if growing out of one another spontaneously, and their authors, the philosophers, appear only as mere pretexts. The inner

30. Ibid., epilogue.

> biography of the philosophers, of the men who philosophized, occupies a secondary place. And yet it is precisely this inner biography that explains for us most things.[31]

The historical explanation of a scientist's ideas requires as well, I believe, a descent to that inner self, without neglecting, of course, the force of evidence and the compulsion of logic.

In this book I wish to explain why Haeckel adopted Darwinian theory and why that theory came to have, in his rendering, the special features it did. I will account for his initial acceptance of evolution, in large part, by showing how his own research became illuminated and inspired by his reading of Darwin's *Origin of Species.* Of course, many other biologists read Darwin in the 1860s but did not come away evolutionists—quite the contrary. The task, therefore, must be further to situate his reading in the context of the intimate experiences and profound beliefs that allowed Darwin's message to become in Haeckel's case virtually a religious calling, which he followed throughout the rest of his life.

Haeckel first read *The Origin of Species* immediately after research on a class of animals providing evidence that bespoke species transmutation; but, again, such evidence would bear fruit only in a mind prepared by certain other fertile conceptions—in Haeckel's case prominently among them were those Romantic notions I have mentioned, as well as the traditions of morphology in which he was schooled. Ideas will have causal efficacy because of their logical and semantic character. But this can hardly be enough. Logic and meaningful fit of ideas have potency only if invested with it by the person. To adapt Novalis's adage, logic and semantics bake no bread. Only when the fire is struck from below, in the depths of personality, will the logical and causal relations of ideas become solidified: the relations of ideas are human relations. Ideas that are logically or semantically fit to be cause and effect of one another must yet be brought into proximity and charged with causal energy through hopes and fears, desires and sufferings. Without the infusions of personality, ideas floating through the mind of a scientist will remain limp and anemic, poor effete creatures that evanesce away. Haeckel's ideas had martial force. So the study of his scientific ideas, their origin and trajectory, must be grounded in his character formation—in his *Bildung,* the Romantics would say—and in the enlarged

31. Miguel de Unamuno, *Tragic Sense of Life,* trans. J. E. Crawford Flitch (London: Macmillan, 1921), 2. Unamuno offers a clue, I believe, for the solution to the puzzle of Ernst Haeckel, a matter discussed briefly at the end of this chapter and in chapter 11.

passions of the man, in a deep need to find the truth about the world, especially a truth that would mitigate the overwhelming tragedy that touched virtually all of his work in evolutionary theory.

In the following chapters, then, I will trace the unfolding of Haeckel's thought, especially its Romantic connections, as it reaches up to the great synthesis of his early career, his *Generelle Morphologie der Organismen.* This work, born in despair, formed the trunk whence sprang the many branches of his later science. In order to appreciate the resolving power of Haeckel's theory, I will treat in some detail his great monographs on various marine organisms that appeared in the decade and a half surrounding his *Generelle Morphologie.* Those monographs, while still known to the relevant specialist in marine biology, remain forbidding waters to most others. Yet these volumes reveal his remarkable abilities as a research scientist and display the singular discoveries by which Darwinian theory achieved concrete realization. Indeed, Haeckel's empirical accomplishments in his vast studies of marine fauna provide counterweight to the presumption of many contemporary historians that his evolutionary theory fled sound science to reside in a speculative land of gothic dreams. Haeckel's research, richly detailed and technically sophisticated even to modern eyes, reached back, admittedly, through theoretical and aesthetical attachments to the works of Goethe, Humboldt, and Schleiden. Yet this only indicates, as I will argue, that Romanticism had features attractive and fecund enough to seduce thoroughly modern science.

Haeckel did not remain hidden behind the researcher's microscope. Because of a great personal tragedy, he took on Darwinian theory as a kind of theological doctrine, recasting it as the foundation for his "religion of monism." He preached this doctrine from a number of venues—the popular book, the vituperative essay, the revivalist lecture. These works brought him the admiration of a liberal, emancipated public during the last part of the nineteenth century and allowed him to cultivate relationships with such political, intellectual, and artistic luminaries as Edward Aveling (consort of Karl Marx's daughter and translator of *Das Kapital*), David Friedrich Strauss (theologian and iconoclastic author of the *Life of Jesus*), Ernst Mach (positivist and physicist at Vienna), and Isadora Duncan (freelover and dancer).

After his extraordinary empirical accomplishments of the 1860s and 1870s, Haeckel fought one battle after another, right through the First World War, against the enemies of his Romantic evolutionism, that is, his passionately applied Darwinism. The heated controversies in which he became engaged reflect, from a particular perspective, the course of evo-

Fig. 1.2. Isadora Duncan (1877–1927): "My writing table at Phillips Ruhe. I look upon your lovely picture. Yours in friendship, Isadora Duncan, July 1904." (Courtesy of Ernst-Haeckel-Haus, Jena.)

lutionary theory from the second half of the nineteenth century through the first part of the twentieth. These controversies concerned internal disputes of evolutionists as well as external conflicts with religious enemies. The politics of evolution even spilled over into Haeckel's efforts to enlist scientists to ward off the coming war that would devastate Europe. I will sketch these battles and thereby offer one portrait of the course of evolutionary theory during the period. I will also attempt to develop several themes of more historiographic concern, namely: the rhetorical structure of disputes in science, the role of graphic representation in the explanation and demonstration of particular theories, and the justification for making ethical evaluations of historical figures—this latter will occupy the second appendix.

Haeckel's greatest sin in the eyes of many historians and philosophers is that he was not Darwin. But not even Darwin was Darwin, at least as he is usually depicted in contrast to Haeckel. This study will, I hope, make it more difficult both to dismiss Haeckel's scientific accomplishments as anti-Darwinian and to denigrate his character as meretricious. I also hope that this book will expose those Romantic roots of evolutionary theory that have made it bloom with such diverting and sweetly compelling ideas.

The Tragic Source of the Anti-Religious Character of Evolutionary Theory

Had Charles Darwin or Ernst Haeckel not lived, I believe that in due course a theory of evolution by natural selection would have been formulated—Alfred Russel Wallace, after all, came very close to beating Darwin to the punch, though it may have been a punch not many people would have felt, initially at least. But in Germany prior to 1859, there were several biologists of prominence who had advanced one or another version of a theory of descent with modification; for some, the modifications were wrought by Lamarckian devices, for others by the divine hand. During the first half of the century, the evidence accumulated: the fossil evidence, the biogeographical evidence, the anatomical evidence, the embryological evidence, the practical evidence from breeders—all of these avenues led in the same direction. Moreover, though many different devices had been proposed to explain transmutation, the seeming analytic clarity of the principle of natural selection and the persuasive model of artificial selection could be expected, even without the *Origin of Species*, to reveal the power of the selective device, elevating it to become a leading contender for the position of chief causal source of species alteration. It is certainly not unreasonable to suppose, absent Darwin, that both of these ideas—descent with modification and natural selection—would have rather quickly become dominant in biological science during the latter part of the century. Why would they become dominant? Well, because, as the best evidence we have shows, they conform to features of the natural world.[32] How else to explain the rapid spread of evolutionary theory in radically different political cultures, eth-

32. There are certain Kantian problems with the concept of "the natural world" that need not be explored at this juncture.

nic domains, and religious orientations in the last part of the nineteenth century—from social conservatives to liberal Marxists, from western Europeans to eastern Asians, from militant atheists to militant Jesuits?

So I reject the so-called contingency thesis proposed by several sociologists and historians of science.[33] The thesis itself cannot, I think, even be coherently expressed. The notion seems to be something like this: major features of science—the experimental method, for instance—need not have come to characterize a successful *modern science;* rather those features resulted simply from a collocation of chance historical events that introduced and sustained them; and thus the development of an equally effective modern science could have occurred without the techniques of empirical experiment. If the contours of Robert Boyle's experimental profile, like Cleopatra's nose, had a different shape, then modern science would have developed in a dramatically different way—perhaps along the lines of a Hobbesian metaphysics. Yet in this scenario, which has been proffered by some contemporary historians, the contingency thesis cannot be intelligibly expressed. It cannot be intelligibly expressed because by "modern science" we mean that interconnected set of laws established by experimental procedures.[34] No doubt, it might possibly have occurred that the Black Death was more lethal to European populations than was historically the case and that virtually the entire intellectual community was obliterated. One could imagine—though with some difficulty—that the saved remnants reverted to doctrinaire superstition that became fanatically entrenched, so that its system came to dominate what subsequently

33. Hacking discusses the various formulations and implications of the contingency thesis. See Ian Hacking, *The Social Construction of What?* (Cambridge, MA: Harvard University Press, 1999), especially 63–99. While Hacking thinks the thesis not exactly clear, he agrees with it in a limited fashion.

34. Shapin and Schaffer have argued for the contingency thesis in their historical analysis of the controversy between Thomas Hobbes, whom they take to reject experimental methods to establish the fundamental elements of science, and Robert Boyle, whom they represent as advancing those methods. See Steven Shapin and Simon Schaffer, *Leviathan and the Air-Pump* (Princeton, NJ: Princeton University Press, 1985). They say: "Our goal is to break down the aura of self-evidence surrounding the experimental way of producing knowledge. . . . [W]e want to show that there was nothing self-evident or inevitable about the series of historical judgments in that context [of the Hobbes-Boyle debate] which yielded a natural philosophical consensus in favour of the experimental programme. Given other circumstances bearing upon that philosophical community, Hobbes's views might have found a different reception" (13). Shapin and Shaffer further contend that the victory of Boylean experimentalism in the history of early modern science was inextricably intertwined with his political and religious ideology—a quite contingent matter—and that this connection was a principal factor in the success of his program (80–109).

passed for intellectual thought. But simply said, that would not be science. It makes no sense to say that modern science could have developed quite nicely without modern (experimental) science. I do not think the thesis could be rationally expressed if one focused on modern biology and held that it only contingently featured evolutionary theory. As Theodosius Dobzhansky famously observed, nothing in biology makes sense except in light of evolution. Thus again, without this major feature—evolutionary theory—one could not have the development of "modern biology."

Well, these may seem like the niggling semantic objections of a paleo-positivist. I do believe, nonetheless, they go quite deep. Yet for my purposes in this history, it is not crucial that the reader accept these analytical objections to the contingency thesis. Indeed, I want to argue for an attenuated version of the thesis, a version that, I think, can be coherently stated. This version considers certain non-essential aspects of modern evolutionary theory, namely, its materialistic and anti-religious features. These, I believe, are contingent cultural traits of the modern theory. As I have attempted to show elsewhere, many of the early proponents of Darwinian theory were both spiritualists—that is, they accepted a nonmaterialistic metaphysics—and believers—that is, they integrated their scientific views with a definite, or sometimes an indefinite, theology.[35] Asa Gray, William James, and Conwy Lloyd Morgan are just a few prominent examples of advocates of evolutionary theory who nevertheless rejected a stony, desiccated materialism.

During the late nineteenth and through the twentieth century, however, the cultural representation of the evolutionary doctrine took on a different cast: evolutionary theory became popularly understood as materialistic and a-theistic, if not atheistic. I believe this cultural understanding is principally due to the tremendous impact and polarizing influence of Ernst Haeckel. Had Haeckel not lived, evolutionary theory would have turned a less strident face to the general public. At least, the antagonism with religion would not have been so severe. It was Haeckel's formulations that, as I will maintain, created the texture of modern evolutionary theory as a cultural product. My thesis is even more specific, namely: had Haeckel not suffered the tragic events that caused him to dismiss orthodox religion as unmitigated superstition and to advance a militant monistic philosophy, his own version of Darwinian theory would have lost its markedly hostile

35. See Robert J. Richards, *Darwin and the Emergence of Evolutionary Theories of Mind and Behavior* (Chicago: University of Chicago Press, 1987), 331–408.

features and these features would not have bled over to the face turned toward the public.

⁂

Miguel de Unamuno, in his *Del sentimiento trágico de la vida*, explored what he took to be the soul-splitting experience of Western intellectuals, their tragic sense of life. He depicted the struggles of a skeptical reason, especially in philosophy and science, as courageously insisting that human striving is mortal, that its efforts end in the grave; yet such reasoning cannot, he thought, overcome the vital desire for life, for transcendence.[36] Ernst Haeckel experienced the passion for transcendence through a love that lifted him to ecstasy and then crushed him in despair. This experience invaded his insistently rational attitudes, even transforming his science into a means for escaping the grasping hand of mortality. My overarching argument will be that Haeckel's science and his legacy for modern evolutionary theory display the features they do because of his tragic sense of life.

36. I will return to consider Unamuno's thesis in relation to Haeckel's accomplishments in the conclusion to this book.

Fig. 2.1. Ernst Haeckel (1834–1919), with his parents, Charlotte (1799–1889) and Karl Gottlob Haeckel (1781–1871). (Courtesy of Ernst-Haeckel-Haus, Jena.)

CHAPTER TWO

Formation of a Romantic Biologist

"I am decidedly a 'Leptoderm,' that is, 'thin-skinned,' and thus have experienced much more suffering and, also, more intense joy than the run of men."[1] Even a slight acquaintance with Haeckel's life would confirm what might appear a rather Romantic self-appraisal. He believed he owed his mercurial emotions to his mother, with whom, during his early years, he had a strong bond. From his father, he thought he had inherited an ever-curious intelligence, which constantly labored to restrain—not always successfully—his volatile impulses. Whether these traits burgeoned in the blood or in an exceedingly warm and encouraging home environment, Haeckel had full opportunity to cultivate them during his early years and

1. Ernst Haeckel, from a brief biographical note, quoted by Erika Krauße, *Ernst Haeckel* (Leipzig: Teubner, 1984), 10. Most of my information about Haeckel's early school days comes from three sources. The first is his own "Biographische Notizen," in the Haeckel Papers, Institut für Geschichte der Medizin, Naturwissenschaft und Technik, Ernst-Haeckel-Haus, Friedrich-Schiller-Universität, Jena; these notes were apparently used for a biographical sketch edited by Heinrich Schmidt, in *Ernst Haeckel: Gemeinverständliche Werke*, 6 vols. (Leipzig: Alfred Kröner, 1924), 1:ix–xxxi. The second source is Wilhelm Bölsche's *Ernst Haeckel: Ein Lebensbild* (Berlin: Georg Bondi, 1909). Bölsche had interviewed Haeckel's aunt Bertha Sethe, younger sister of Haeckel's mother, about Haeckel's early days; and Haeckel himself reviewed the manuscript. Finally, there is the biography prepared by Heinrich Schmidt for the celebration of Haeckel's eightieth birthday in 1914: "Was Wir Ernst Haeckel Verdanken," in *Was Wir Ernst Haeckel Verdanken: Ein Buch der Verehrung und Dankbarkeit*, ed. Heinrich Schmidt, 2 vols. (Leipzig: Unesma, 1914), 1:7–194; Haeckel reviewed this biography as well. Also useful are Walther May, *Ernst Haeckel: Versuch einer Chronik seines Lebens und Wirkens* (Leipzig: Johann Ambrosius Barth, 1909); Peter Klemm, *Ernst Haeckel: Der Ketzer von Jena* (Leipzig: Urania, 1966); and Krauße's little book, mentioned above. More recently, Mario Di Gregorio has published a comprehensive account of Haeckel's life in *From Here to Eternity: Ernst Haeckel and Scientific Faith* (Göttingen: Vandenhoeck & Ruprecht, 2005). I have learned a great deal from Di Gregorio's work.

in the more demanding but immensely fertile university life he later spent at Würzburg and Berlin.

Early Student Years

Ernst Heinrich Philipp August Haeckel was born 16 February 1834 in Potsdam, where his father, Karl (1781–1871), a jurist, served as privy counselor to the Prussian court. He was the second and last child of Charlotte (1799–1889), whose own father, Christoph Sethe (1767–1855), had been a jurist who won fame during the Napoleonic occupation when he faced down the French authority with the declaration that a bullet for him would also pierce the law. His older brother, Karl (1824–1897), would follow grandfather and father into the legal profession. The family moved in the year after Haeckel's birth to Merseburg, where the father assumed ministerial responsibility for schools and ecclesiastical affairs.

During his seventeen years in the small capital of Saxony, the young Haeckel enjoyed a rather solitary but intellectually full life. His mother nurtured him on classic German poetry, especially that of her favorite, Friedrich Schiller, while his father discussed with him the nature philosophy of Goethe and the religious views of Friedrich Schleiermacher, who had been an intimate of the family and even presided over the baptism of Haeckel's aunt Bertha. Karl Haeckel had a keen interest in geology and foreign vistas, which undoubtedly led his son to treasure the travel literature of Alexander von Humboldt and Charles Darwin. The boy devoured their books, which set the deep root of a lasting desire for adventure in exotic lands. His judicial heritage may also have fostered a lingering impulse to bring legal clarity, through the promulgation of numerous laws, into what he later perceived as ill-ordered biological disciplines.

The Influence of Goethe, Humboldt, Darwin, and Schleiden

From the books of his early years, Haeckel recalled in particular Alexander von Humboldt's *Ansichten der Natur* (Views of nature, 1808), Matthias Jakob Schleiden's *Die Pflanze und ihr Leben* (The plant and its life, 1848), and Charles Darwin's *Naturwissenschaftliche Reisen* (Natural scientific journeys, 1844—Darwin's *Beagle Voyage*). He thought these books determined the course of his professional life.[2] Humboldt (1769–1859) sketched a

2. Haeckel, "Biographische Notizen," 3, in the Haeckel Papers, Haeckel-Haus, Jena. Haeckel also related to his biographer Wilhelm Bölsche that in gymnasium his three favorite

Fig. 2.2. Alexander von Humboldt (1769–1859). Portrait (1806) by F. C. Weitsch; painted shortly after Humboldt's return from his five-year voyage (1799–1804) to the Americas. (Courtesy of Preussischer Kulturbesitz, Berlin.)

theory of the topography of the plant environment—cultivating ideas from Goethe (1749–1832)—according to which similar vegetative forms might be found distributed throughout the globe along corresponding latitudes and altitudes. Humboldt suggested that vital forces, perhaps unknown chemical interactions, accounted for the phenomena of life.[3] Schleiden

books were those of Humboldt, Schleiden, and Darwin. See Haeckel to Bölsche (4 November 1899), in *Ernst Haeckel–Wilhelm Bölsche: Briefwechsel 1887–1919*, ed. Rosemarie Nöthlich (Berlin: Verlag für Wissenschaft und Bildung, 2002), 110–11.

3. Humboldt's notions about the environmental morphology of geographical regions are sketched in "Ideen zu einer Physiognomik der Gewächse," in *Ansichten der Natur, mit wissenschaftliche Erläuterungen*, 3rd ed. (1808; repr., Stuttgart: Cotta'schen Buchhandlung, 1871). Haeckel would later develop Humboldt's suggestions into "ecology" and "chorology," both disciplines he first formulated (see note 28 of the previous chapter).

(1804–1881), a theist for whom nature reached up toward Divinity, yet put aside the biblical account of creation and sketched a naturalistic transformation theory. He proposed that simple organisms had developed in the sea through physical influences, which no longer obtained. He supposed that chemical forces, especially active in tropical climates, had transformed those simple organisms into species whose remnants were preserved in fossils and whose descendants now, in different form, populated the earth.[4] Both Humboldt and Schleiden insisted that the proper evaluation of nature required aesthetic as well as theoretic judgment, and their own poetic descriptions overflowed the banks of their scientific narratives, carrying along the boy's lively imagination. Later in this chapter, I will provide a more detailed analysis of the conceptual foundations of Humboldt's and Schleiden's natural historical and aesthetic positions.

Darwin's (1809–1882) exciting account of his *Beagle* voyage whetted the young Haeckel's taste for exotic travel.[5] The Englishman suggested that an interest in botanical and zoological nature portended journeys to strange lands, where adventure awaited in every jungle clearing. When Haeckel began the study of medicine, thoughts of sailing away to tropical islands would continually divert his fantasies away from the profession for which he prepared. And when he later read Darwin's *Entstehung der Arten* (*Origin of Species*), his childhood memories of the exciting vistas opened by that congenial Englishman would encourage him to take further steps along the path his scientific colleagues attempted to close off.[6]

The young, introverted boy presumably had few philosophic considerations in mind when whiling away hours with these treasured books. He was simply attracted to the delightful botanical descriptions and exciting foreign vistas depicted on their pages. Humboldt's *Ansichten* and Darwin's *Reisen* were replete with exotic tales of adventure in steamy jungles and of encounters with South American Indians and Spanish gauchos, stories

4. M. J. Schleiden, *Die Pflanze und ihr Leben: Populäre Vorträge* (Leipzig: Wilhelm Engelmann, 1848), 257–84. Rupke discusses several thinkers—Hermann Burmeister, Heinrich Georg Bronn, and Carl Vogt—who, like Schleiden, developed theories of autochthonous generation; that is, theories that proposed a spontaneous generation of primordial germs that gave rise to species. See Nicolaas Rupke, "Neither Creation nor Evolution: The Third Way in Mid-Nineteenth-Century Thinking about the Origin of Species," *Annals of the History and Philosophy of Biology* 10 (2005): 143–72.

5. Charles Darwin, *Naturwissenschaftliche Reisen*, trans. Ernst Dieffenbach, 2 parts (Braunschweig: Vieweg, 1844).

6. Haeckel remembered the charm of Darwin's many tales when he later recommended the *Reisen* to the audiences for his lectures. See, for example, Ernst Haeckel, *Natürliche Schöpfungsgeschichte* (Berlin: Georg Reimer, 1868), 106. This latter book arose from stenographic notes of lectures he gave in Jena during the winter term of 1868.

Fig. 2.3. Frontispiece of Matthias Jakob Schleiden's *Die Pflanze und ihr Leben* (1848). (From the author's collection.)

that could not fail to stimulate a reclusive boy's dreams. And Schleiden's colorful descriptions of plants piqued the youngster's growing enthusiasm for botany and the aesthetic joys of nature, attitudes that were initially nurtured in his mother's gardens and in the instruction he received from his much-beloved tutor Karl Gude. Gude led him through an elementary systematics of Linnaeus (1707–1778), which would have driven to tears most any other twelve-year-old. To balance the stolid Scandinavian, his tutor also introduced him to the writings of Lorenz Oken (1779–1851), whose inspired descriptions would sometimes fall off into speculative fantasy.[7] Gude encouraged the young Haeckel to build a herbarium, which, as he recalled, "would be ordered now according to this system, now according to that."[8]

In the Merseburg Dom-Gymnasium, which he entered in 1843, Haeckel

7. For a discussion of Oken's various theories, see Robert J. Richards, *The Romantic Conception of Life: Science and Philosophy in the Age of Goethe* (Chicago: University of Chicago Press, 2002), 492–502. See also the first appendix of the present volume.

8. Haeckel, "Biographische Notizen," 2, in the Haeckel Papers, Haeckel-Haus, Jena.

was fortunate enough to have had a teacher, Otto Gandtner (later of the University of Berlin), who continued to cultivate the young boy's interest in nature study and introduced him to the elements of chemistry. The school's curriculum, though, remained rooted in the classical humanistic tradition, which Haeckel would later condemn for narrowness but which, nevertheless, had residual effects on his attitudes and tastes. The director of the Gymnasium, Ferdinand Wiek, revered Goethe and would read long passages of the master to his classes. Haeckel thought this experience, along with the influence of his father, made him especially receptive to a Goethean monism, which supplied the metaphysical foundation for his later evolutionary theory. Both his schooling and the liberal political ideals of his father planted in Haeckel the seeds of a deep passion for the culture of Germany and a desire for German national unity, though under a new social dispensation.

Striving for German Unity

That goal of political union was initially realized in the civil war of 1866, when Chancellor Otto von Bismarck (1815–1898) and his general Graf Helmuth von Moltke (1800–1891)—aided greatly by the new needle guns of the infantry—broke the Austrian alliance and brought several of the German states under the Prussian king.[9] By the early 1870s, Bismarck forced the remaining German lands into union. Prior to that aggressive political resolution, Germany existed principally as a cultural entity. Through the eighteenth century, it consisted of a multitude of sovereign territories, ruled over by kings, princes, knights, and ecclesiastical nobles, and of numerous free cities (such as Frankfurt and Munich), all of which owed nominal allegiance to the Austrian Holy Roman Emperor.

At the Congress of Vienna, convened to make final settlement of the Napoleonic wars, the allies—Britain, Russia, Austria, and Prussia—sought not only to contain France but to resolve boundary disputes, while yet preserving much of the old social and political order. The Austrian Klemens Wenzel Prince von Metternich (1773–1859), the presiding genius with an exquisite sense of balance, skillfully reproduced in the new German Confederation (1815) a modern, functional equivalent of the old empire. He balanced off the

9. For a fuller account of Bismarck's efforts at unification of the Germanies, see James J. Sheehan, *German Liberalism in the Nineteenth Century* (Chicago: University of Chicago Press, 1978), and *German History, 1770–1866* (Oxford: Clarendon Press, 1989); Gordon A. Craig, *Germany, 1866–1945* (Oxford: Oxford University Press, 1980); and William Carr, *A History of Germany, 1815–1990*, 4th ed. (London: Arnold, 1991).

danger of France with an enlarged Prussia at her border. Prussia had to cede its Polish lands to Russia but received in return from defeated France the Rhineland, the Duchy of Westphalia, and upper Saxony. This trade eliminated the largest fraction of non-Germans from Prussia while doubling its total population. The thirty-nine states of the Confederation retained their sovereignty but were held in modest check by the ministrations of the Federal Diet, which was controlled by the most powerful members of the Confederation, the Hohenzollerns in Prussia and the Hapsburgs in Austria.

Though Napoleon's armies initially employed muskets to win the hearts and minds of the German peoples, republican ideology proved ultimately more successful. Those enlightenment conceptions mingled with a burgeoning Romanticism that emphasized both the spiritual need of the individual freely to construct the self (an idea at the root of Schelling's idealism) and the destiny of that self in an ancient and fabled fatherland (the mythical history that Herder and Fichte limned). This roiling stream of often conflicting individual and social ideals initially cleared the way for modest reforms within the Prussian kingdom: for instance, those wrought by the imperial noble Baron Karl vom und zum Stein (1757–1831), adviser to the king, who abolished serfdom and reorganized municipal governments; or by Prince Karl August von Hardenberg (1750–1822), the chancellor who, though of a conservative bent, urged the adoption of democratic principles of economic and social freedom within the monarchy; or by Wilhelm von Humboldt (1767–1835), Alexander's brother, who stimulated a pedagogical renaissance with the foundation of the University of Berlin. In the wake of these promising beginnings, liberals like Karl Haeckel cultivated hope of national union, in which the petty princes and nobles would be politically expunged and the country ruled by an enlightened parliamentary government under a single sovereign. Certainly the desire that Germany's richly diversified culture might be incorporated within a single nation, that freedom of intellectual inquiry and expression might be guaranteed, that political representation might become a reality—these fervent republican hopes came to life in Karl Haeckel's son and continued as a deep passion throughout his years.

In 1849, however, the aspirations of the liberals crashed upon the failure of the Frankfurt Parliament. Periodically through the 1830s and 1840s, students in the new *Burschenschaften* (the first of these fraternities was founded at Jena), professors of liberal persuasion, and the rising educated middle class agitated for greater political freedoms and recognition of rights, all to be specified in written constitutions of the various German lands. These movements usually evoked repressive responses from aristocratic rulers, which only made the desire for freedom keener. Unrest reached a crisis pitch in the mid-1840s, when an economic depression caused widespread social misery. The shrillest retort, though, came in February 1848, when the French king Louis-Philippe (1773–1850) fled Paris for England, under the nom de guerre "Mr. Smith," and the second Republic was born. Europe so dangerously smoldered with the heat of revolutionary passion that the new Prussian king, Frederick William IV (1795–1861), employing a co-optive strategy, declared his solidarity with the liberals and called for a new German nation having a parliament and constitution.

The Frankfurt Parliament (Die Deutsche Nationalversammlung), sanctioned by the king, met in 1848 to devise a constitution and settle the boundaries of the proposed new nation. However, the moderate majority grew wary of the radical members, which sought to prosecute a nationalistic war over Danish territorial claims in Schleswig-Holstein. Radicalized workers launched mass protests over the failure of the Parliament to defend German sovereignty; their efforts brought out the Prussian and Austrian troops, who fired on the demonstrators. The Parliament quickly lost the faith of the workers. And when Frederick William refused what he regarded as the tinsel crown of the new nation—expressing a fearful disdain for the rabble and a renewed nostalgia for the empire—the Parliament shuddered as its main supports gave way. The meetings of the representatives in the Paulskirke during 1848 and 1849 initially had fired the imagination of German liberals and nationalists. Even the sixteen-year-old Ernst Haeckel felt some reflected glow—at least his large painting of the *Parliament of Birds* (*Nationalversammlung der Vögel*), done in 1850 (see plate 3), suggests an inchoate patriotic sentiment, banked by scientific interest, that would finally ignite during his university years and continue to burn hotly throughout his days. But unlike the parliament of birds, the parliament of the incipient nation collapsed, and so the hopes of German liberals took flight. Shortly thereafter, in 1851, Karl Haeckel retired from government service.

University Years

Medical School at Würzburg

Upon completion of his *Abitur* examination in March 1852, Haeckel prepared to matriculate at Jena, where he intended, with reluctance, to enroll in medicine. The real attraction of Jena was Schleiden, who lectured there on botany. The young Haeckel—he was eighteen—had momentarily, however, to shelve his plans after suffering a knee injury while hunting wild plants during the Easter holiday. He convalesced at his parents' home in Berlin and matriculated at the university there in the summer term. A friend of the family, the great botanist Alexander Braun (1805–1877), who was only slightly less famous than Schleiden, lectured at this time in Berlin. Initially, Haeckel followed Braun's lectures with enthusiasm; but he quickly grew disappointed with their elementary character. During this Berlin sojourn, Haeckel's father began to worry about his son's professional trajectory. Karl Haeckel, being the solid, middle-class professional that he was, did not believe the future glowed bright with promise for another botanist. His son simply had to take medicine more seriously.

In late August 1852, respecting the counsel of his father, Haeckel entered the University of Würzburg, having probably the best medical faculty in Germany at the time (see fig. 2.4). Students at the university—some six hundred in 1852—came from all over the German lands to study with such luminaries as Albert von Kölliker (1817–1905) and Rudolf Virchow (1821–1902). Kölliker taught histology and introduced Haeckel to what would quickly flower into a sweet delight—at least for one so disposed—namely, microscopic study. The professor's just-published *Handbuch der Gewebelehre des Menschen* (Guide to the doctrine of human tissues, 1852) became the student's vade mecum.

Virchow, however, was the star of the faculty. He excited a frisson of danger in the active imaginations of students because of his history of radical politics. As *Privatdozent* under Johannes Müller (1801–1858), he had become a leader in the medical reform movement and a political activist of the left. In 1848, when revolution broke out in Berlin, he not only cared for the wounded at the Charité hospital; he actually mounted the barricades, pistol in hand.[10] Müller, who was rector in that fateful year, feared his more radical students would burn down the university. After the collapse of the constitutional movement in 1849, Virchow momentarily escaped the

10. Virchow wrote his father on 19 March 1848, late in the evening: "From this moment [the afternoon of 18 March] the revolution began. Everything screamed betrayal and revenge. In a few hours the whole of Berlin was under barricades and whoever could get a gun armed himself.... The number of wounded and dead cannot be estimated at this time. In the Charité, we had 52 wounded and 11 dead from the civil guard, 24 dead lie in the river-side church and at least as many at the palace.... My part in the uprising was relatively insignificant. I helped build a few barricades; but since I could only get a pistol, I wasn't of much use because most of the soldiers were too far away to shoot." See Rudolf Virchow, *Briefe an seine Eltern, 1839 bis 1864*, ed. Marie Rabl, 2nd ed. (Leipzig: Wilhelm Engelmann, 1907), 134–37.

Fig. 2.4. Members of the medical faculty at Würzburg. *Back row from left:* Rudolf Virchow (pathological anatomy), Albert von Kölliker (histology); *first row:* J. J. von Scherer (chemistry), Franz Kiwisch von Rotterau (gynecology), Franz Rinecker (pharmacology). Photo taken in 1850. (Courtesy of Ernst-Haeckel-Haus, Jena.)

maelstrom of politics when he married and removed himself to Würzburg. In the medical arena, his ideas concerning the cellular basis of life and disease proved just as revolutionary; and his reputation for deep research and academic controversy ensured his lectures would be jammed. His electrifying talent as a scientist drew Haeckel to his classes, but his insulated and cool personality kept the two from ever becoming very close—quite in contrast to Haeckel's relationship with Kölliker, with whom he would strongly disagree intellectually but would remain on the warmest personal terms throughout their years. Virchow and Haeckel would later interact in proper professional ways until, that is, the famous scientist started preaching the dangers of evolutionary theory for untutored minds. In 1877, in the wake of Haeckel's urging that modern science, especially evolutionary theory, be introduced into the lower school curriculum, Virchow protested. He admonished his colleagues not to press for evolutionary theory to be taught in the German schools, since, as he argued, it lacked scientific evidence, was an affront to religion, and smoothed the way to socialism (discussed in chapter 8). Haeckel's sulfuric reaction to this politically tinged attack on Darwinian science undoubtedly released a force building since his student days.[11]

Haeckel did not take naturally to the idea of medical school and its likely consequence, clinical practice. Shortly after he arrived at Würzburg, he suffered what was to be a recurring feeling of revulsion about becoming a physician. He felt a disgust for illness and disease that he could never quite overcome. In his early months at Würzburg, this distaste was carried on waves of homesickness. He wrote his parents of his doubts, a message that would be reiterated throughout his time in medical school:

> I will straightaway tell you openly and fully that the study of medicine has never caused me so much pain as now. I have now the strong conviction, which many wiser men than I have already had, that I can never become a practicing physician or even study medicine.[12]

Two cords seemed to have kept Haeckel tethered to medical school, nevertheless: a tempered passion for the kind of fundamental science he experienced with Kölliker and Virchow; and a strategy for utilizing medicine to achieve the scientific vocation he envisioned from his reading of Humboldt and Darwin. Under the affable tutelage of Kölliker, he grew to love precise work in histology, especially since he had a talent with the microscope. He could simultaneously peer with one eye through the lens and with the other draw in exquisite detail the minute structures of tissues. "Vivant cellulae! Vivat Microscopia!" he exulted to his father at Christmas of 1853. But it was Virchow's lectures during his second year that confirmed a resolve, made to his father, to stick with medicine. He provided his father a description of the arresting experience:

11. German government officials were preparing pedagogical reforms for the lower schools, reforms in which natural science would play a more important role than it had. At the meeting of the Society of German Natural Scientists and Physicians at Munich in 1877, Virchow cautioned his colleagues not to push for evolutionary theory to become part of the curriculum. He thought only the secure facts of science ought to be represented; and evolutionary theory had no real empirical support—especially those aspects that Haeckel insisted upon, namely, spontaneous generation and the descent of man. See Virchow's talk given at the Versammlung Deutscher Naturforscher and Ärtze at Munich, September 1877, and reprinted as *Freiheit der Wissenschaft im modernen Staat* (Berlin: Wiegandt, Hempel & Parey, 1877). See also Haeckel's reply in *Freie Wissenschaft und freie Lehre: Eine Entgegnung auf Rudolf Virchow's Münchener Rede über "Die Freiheit der Wissenschaft im modernen Staat"* (Stuttgart: Schweizerbart'sche, 1878). I will return to this controversy in chapter 8.

12. Haeckel to his parents (1 November 1852), in *Entwicklungsgeschichete einer Jugend: Briefe an die Eltern, 1852–1856*, ed. Heinrich Schmidt (Leipzig: K. F. Koehler, 1921), 6.

> Virchow's lectures are rather difficult, but extraordinarily beautiful. I have never before seen such a pregnant concision, a compressed power, a tight consistency, a sharp logic, and yet the most insightful descriptions and compelling liveliness as are here united in lectures. Though, if one does not bring to the lectures an intense concentration and a good philosophical and general culture, it is very difficult to follow him and to get ahold of the thread that he so beautifully draws through everything; a clear understanding will be taxed considerably by a mass of dark, quickly moving expressions, learned allusions, and a generous use of foreign terms, which are often very superfluous.[13]

Haeckel obviously tempered his adolescent admiration for Virchow's powerful and enigmatic intellect with some suspicion of his rhetorical displays.

Haeckel filled his days with lectures and exercises—during his first term, for example: osteology with Heinrich Müller (1820–1864) four times a week; August Schenk's (1815–1891) course on cryptogamic plants twice a week; Kölliker's human anatomy for two hours every day; and drawing anatomical structures or attending a private tutorial in microscopy with Franz Leydig (1821–1908; at the time, a *Privatdozent* and one in whom Haeckel would find a soul mate). During his second year, he added lectures and tutorials with Virchow. He experienced at Würzburg the highest level of laboratory biology and clinical medicine achievable at the time. In the late evening, though, he turned to science in a different key. He always preserved an hour or so for reading Humboldt.[14] Through his first two years at medical school, Haeckel went through all of the major works of Humboldt, as well as a great many of the minor ones, and, of course, he always had Goethe by his side.[15]

The Aesthetic Science of Humboldt, Kant, Schelling, and Goethe

It is worth spending a few minutes on Humboldt's aesthetic approach to science, since something comparable underlay Haeckel's conception of na-

13. Haeckel to his parents (16 November 1853), in ibid., 80.

14. Haeckel outlined his schedule for his parents in a letter of 6 November 1852, in ibid., 9.

15. Haeckel's reading can be determined from the list of books he had with him at Würzburg, as preserved in MS no. 398, in the Haeckel Papers, Haeckel-Haus, Jena. The catalog records 161 books. Letters to his parents (e.g., 17 February 1854, *Briefe an die Eltern*, 100–101) indicate his reading from the university library.

ture and gently guided his later morphological and evolutionary considerations. Humboldt's theories lead, as well, back to Goethe, without whom Haeckel's science is unimaginable. Both Humboldt's and Goethe's views about the relation between science and art had their source in Kant (and perhaps Schelling), about whom a few words must also be expended. These intellectual connections likely encouraged Haeckel as a young docent at Jena to attend Kuno Fischer's (1824–1907) lectures on Kant, lectures that he greatly admired.[16]

Alexander von Humboldt's *Ansichten der Natur* (Views of nature, 1808), in which Haeckel delighted as a young student, comprises a medley of essays designed to convey vivid impressions of nature, of the kind he himself experienced during his five-year journey through South, Central, and North America.[17] Humboldt's extended description of his trip, *Voyage aux régions equinoxiales du nouveau continent, fait en 1799–1804* (Travel to the equinoctial regions of the new continent, made from 1799–1804, published 1807–35),[18] had inspired many famous naturalists, including Haeckel and the young Charles Darwin. During his own voyage to South America, Darwin wrote to his mentor John Stevens Henslow: "I formerly admired Humboldt, I now almost adore him; he alone gives any notion of the feelings which are raised in the mind on first entering the Tropics."[19] Humboldt had designedly attempted to so elevate the minds of his readers. He believed that the unity of form underlying the diverse profusion of life,

16. During the summer term of 1862, while he lectured on anatomy, Haeckel also attended Kuno Fischer's lectures on Kant four times a week; and he read the first *Critique*, which, as he explained to his parents, was "for the natural researcher of the highest importance." He wrote his father that he was spending four to six hours a day on Kant. See Haeckel's letters to his fiancée (9 and 17 May 1862) and to his parents (5 and 24 June 1862), in Ernst Haeckel, *Himmelhoch Jauchzend: Erinnerungen und Briefe der Liebe*, ed. Heinrich Schmidt (Dresden: Carl Reissner, 1927), 281, 284, 287, 295.

17. Haeckel likely read the second edition of Humboldt's book. See Alexander von Humboldt, *Ansichten der Natur, mit wissenschaftliche Erläuterungen*, 2nd ed., 2 vols. (Stuttgart: Cotta'schen Buchhandlung, 1826).

18. Alexander von Humboldt and Aimé Bonpland, *Voyage aux régions equinoxiales du nouveau continent, fait en 1799–1804*, 29 vols. (Paris: F. Schoell, 1807–35).

19. Darwin to J. S. Henslow (18 May 1832), in *Correspondence of Charles Darwin*, vol. 1: *1821–1836*, ed. Frederick Burkhardt et al. (Cambridge: Cambridge University Press, 1985), 237. Humboldt's narrative became the filter through which a good portion of Darwin's own perceptions passed, as his sister Susan detected in her brother's journals: "I thought in the first part (of this last journal) that you had, probably from reading so much of Humboldt, got his phraseology & occasionally made use of the kind of flowery french expressions which he uses, instead of your own simple straight forward & far more agreeable style. I have no doubt you have without perceiving it got to embody your ideas in his poetical language." See Caroline Darwin to Charles Darwin (28 October 1833), in ibid., 345.

which the tropics especially exhibited, could be expressed in biogeographical calculations, with which even his casual essays bulged. Fat numbers alone, though, could not adequately portray the face of nature—only the art of narrative, the poetry of description, could convey to discriminating sensibilities her active, vital features. Behind Humboldt's declarations about the obligation of the naturalist to convey a certain feeling for nature lay the epistemological and metaphysical structures erected by Kant, Schelling, and Goethe.[20]

In the *Kritik der Urteilskraft* (Critique of the power of judgment, 1790), Kant maintains that only teleological judgment can capture what seems to be the designed and purposive features of nature. When we try to understand nature in Newtonian terms, we bring causes and their effects under determinate and necessary quantitative laws: say, for instance, the laws regulating the refraction of light on to the retina of the eye as it passes through the cornea, lens, and vitreous. What a system of such laws cannot render intelligible, however, is the coordination of causes, the specific arrangement of the various media of the eye for the overall end, the production of clear vision. Rather we can comprehend the interactive relationships of the elements of the eye only under the assumption that they have been organized for the purpose of achieving focused images. And vision itself must be conceived as a means to a further end, the well-being of the organism. A healthy organism, of course, will have rebounding causal effects on the operations of the eye and its parts. It is in this sense, Kant holds, that an organism teleologically conceived is one in which the parts are construed to be mutually cause and effect of each other.

Ultimately we make understandable the structures and functions of organisms *as if* they were the causal product of "an *intellectus archetypus*,"[21] an intellect that operates according to certain archetypal ideas. Moreover, we must also assume that such purposeful intelligibility was designed with us in mind, designed so that we might comprehend the complex web of natural organization. Indeed, the naturalist cannot even begin investigation, except under the assumption that the apparently contingent aspects of nature, her separate causal strands, will ultimately yield an intelligibly harmonious weave. Kant argues, however, that such reflective judgments, which leads us along a teleological path from organs and organ-

20. I have discussed the aesthetic interpretation of scientific judgment in my *Romantic Conception of Life*.

21. Immanuel Kant, *Kritik der Urteilskraft*, A346–47, B350–51, in *Kants gesammelte Schriften*, ed. Akademie der Wissenschaften, 23 vols. (Berlin: Gruyter, 1902–), 5:408.

isms to nature at large and finally to the Divine, does not have the same logical status as those determinate judgments that express the a priori categories constituting our scientific understanding of nature, the judgments formulating mechanistic laws. We cannot suppose that those reflective, teleological considerations reveal designs actually embedded in empirical reality. Judgments of purpose cannot, in Kant's terms, be constitutive but only regulative, suggesting avenues of investigation for the discovery of mechanistic principles but not providing a demonstrative means to vault the phenomenal world into the supersensible world.

Kant's analysis of teleological judgment comes in the second part of the third *Critique*. In the first part, he discusses the structure of aesthetic judgment, which he maintains is, at a certain level, isomorphic with that of teleological judgment. Judgments of beauty assume that the various sensuous elements of an aesthetic object, whether artificial or natural, insofar as they portend a cognitive unity, have the purpose of producing a feeling of aesthetic delight in the perceiver, a delight that results from the play of imagination as it attempts to harmonize with the understanding. Hence, when nature becomes the object of consideration, reflective judgments about the orchestrated interaction of various causes ought to have an aesthetic as well as a theoretical dimension. At least this is the conclusion toward which Kant's devoted follower, Friedrich Schelling, drove.

Schelling maintains, in parts 5 and 6 of his *System des transscendentalen Idealismus* (1800), that the creation of beautiful objects by an artistic genius supplies the model by which to understand nature's productions of organic beings.[22] The self constitutes within the sphere of conscious perception natural objects that display a purposive structure. This self, standing behind, as it were, our consciousness of the world, draws upon unconscious resources, whence even the objective givenness of nature could be deduced (in a kind of Kantian transcendental deduction). For us to understand this process, according to Schelling, we need a perspicuous model, which the actions of the artistic genius furnish. For genius brings forth from a reservoir of unconscious natural talent the productive inspiration realized in his conscious craft. The aesthetic product thus becomes the very standard for understanding nature's fecund creations. But for Schelling, artistic action serves not merely as a model—for the self is that very natural genius that constructs extrinsic beings in its objective perception of

22. Friedrich W. J. Schelling, *System des transscendentalen Idealismus*, with an introduction by Walter Schulz (1800; repr., Hamburg: Felix Meiner, 1962), 275–98 [607–29 of the *Gesamtausgabe*].

the world. Hence, at bottom, artistic perception and scientific perception are one.

The epistemology and aesthetics of Kant and Schelling came to Haeckel through several sources: as a beginning professor at Jena, he read Kant with Kuno Fischer, rector of the university and great expositor of both Kant and Schelling (see chapter 4). But for the young medical student, the swirl of these conceptions came from Alexander von Humboldt, whose various treatises yielded up Kantian and Schellingean ideas through the interstices of exacting scientific examinations.[23] The exciting works of Humboldt frequently pulled the student away from more tedious medical tracts.

In his *Voyage*, in his *Ansichten*, and especially in his famous *Kosmos*, Humboldt attempted to formulate and plait together a great many empirical laws—those characterizing astronomy, chemistry, physics, geology, botany, and zoology. He believed that the principles of those several disciplines touching on the phenomena of life all harmoniously articulated with one another and thus demonstrated that "a common, lawful, and eternal bond runs through all of living nature."[24] The task of the natural scientist, then, was to reveal this harmony of laws that produced a unified whole, to work through the vast and wondrous diversity of nature to discover the underlying forms. The harmony of nature—a *cosmos*, according to Humboldt—was discovered to both our reason and poetic imagination. In the case of biology, the natural scientist had to consider not only the quantitative principles of plant morphology and biogeography (upon which Humboldt especially focused) but also the aesthetic features of the living environment. For aesthetic judgment was no less important for human understanding than mechanistic determination. "Descriptions of nature," he observed in a Kantian vein,

> can be sharply delimited and scientifically exact, without being evacuated of the vivifying breath of imagination. The poetic character must

23. Humboldt had studied Kant with his brother Wilhelm when they were students at Göttingen; at Jena, he became an acquaintance of Schiller, who was a champion of Kant's aesthetic doctrine; and when he returned from his voyage to the Americas, he studied Schelling's *Ideen zu einer Philosophie der Natur* and *System des transscendentalen Idealismus*, which embodied many residual Kantian notions, but which made teleological principles a part of transcendentally constituted nature. Haeckel, as mentioned above, studied Kant with Kuno Fischer at Jena; Fischer also wrote a comprehensive study of Schelling, his *Friedrich Wilhelm Joseph Schelling*, 2 vols.; vol. 6 of *Geschichte der neuern Philosophie* (Heidelberg: Carl Winter's Universitätsbuchhandlung, 1872–77).

24. Alexander von Humboldt, *Kosmos: Entwurf einer physischen Weltbeschreibung*, 5 vols. (Stuttgart: Gotta'scher, 1845–58), 1:9.

> derive from the intuited connection between the sensuous and the intellectual, from the feeling of the vastness, and of the mutual limitation and unity of living nature.[25]

This same basic premise, that teleological judgments and aesthetic judgments about living nature have the same structure and aim—that they deliver to comprehension the unity and diversity of nature but portend the sublime—this premise was of Kantian origin but likely of more immediate Goethean derivation. It had been a subject of some conversation between Goethe and Humboldt during the many years of their friendship.[26]

The young Goethe had studied Kant's *Kritik der reinen Vernunft* (Critique of pure reason, 1781), but, like many readers, he came away benumbed. The *Kritik der Urteilskraft,* which he read just as he finished his *Metamorphose der Pflanzen* (Metamorphosis of plants, 1790), yet captured his admiration, if not his complete understanding. Goethe's notion that the several parts of a plant exemplified variations on a fundamental type, that of the ideal leaf—this notion closely conformed to Kant's conception of an *archetype,* the unifying structure that lies beneath the variety of organisms of a given kind.[27] Kant had argued that the biological researcher had initially to comprehend the archetype—the design of the organism—in order to appreciate the ways in which individual parts were related to

25. Ibid., 2:74.

26. In 1794 the two brothers Humboldt came to reside in Jena. There Wilhelm formed a close friendship with Schiller and Alexander with Goethe. Alexander stimulated Goethe to develop and put his osteological ideas on paper. See Johann Wolfgang von Goethe, *Tag- und Jahreshefte* (1794), in *Johann Wolfgang von Goethe Werke, Hamburger Ausgabe,* 14 vols. (München: Deutscher Taschenbuch, 1988), 10:441; and notes to "Erster Entwurf einer allgemeinen Einleitung in die vergleichende Anatomie, ausgehend von der Osteologie" (1795), in ibid., 13:591–92. Goethe indicated to Eckermann that he and Humboldt, over the thirty years of their friendship, spoke on all topics of science and literature. See Johann Peter Eckermann, *Gespräche mit Goethe in den letzten Jahren seines Lebens,* 3rd ed. (Berlin: Aufbau, 1987), 161. Alexander von Humboldt's understanding of Kant was, of course, fostered by his many conversations with Schiller, Fichte, and Schelling while at Jena. When Humboldt returned from his five-year voyage to the Americas, he immediately immersed himself in Schelling's philosophy of nature. See the exchange of letters between Schelling and Humboldt, as quoted in Karl Bruhns, *Life of Alexander von Humboldt,* trans. J. and C. Lassell, 2 vols. (London: Longmans, Green, 1873), 1:202–4.

27. See Johann Wolfgang von Goethe, *Italienische Reise,* in *Werke,* 11:375: "In order to facilitate further understanding, I would like briefly to say this: It occurred to me that the true Proteus lay hidden, a structure that could conceal itself in all forms and reveal itself as well. It was that very organ of the plant that we usually speak of as the leaf. Forwards and backwards, the plant is simply only a leaf, inseparably united so craftily with the seed that one is not able to think of one without the other. To grasp such a concept, to hold it, and to find it in nature is a task that drops us in a painfully sweet situation."

each other and were arranged for the welfare of the whole creature. And for Kant, the archetype pointed mutely to the transcendently divine mind that harbored it, an *intellectus archetypus.* Goethe accepted this kind of teleology, especially insofar as his Spinozistic monism—the view that mind and matter express two aspects of the same underlying *Urstoff*—allowed him to attribute the causal potency of archetypes, not to a transcendent Creator but to an immanent source of creation, to *Deus sive natura.*[28] Thus nature herself, in Goethe's view, was a creative font that showered diversity, though along unified trajectories.[29] This fundamental view would underlay all of Haeckel's work in science.

While traveling in sun-soaked southern Italy (1786–88), Goethe became convinced that the wild fecundity of living nature lay grounded in a transcendent unity, a *real* unity that might break through the shadowy metaphysical restraints that Kant would cast over living nature. Goethe searched through the gardens and fields of Sicily for the *Urpflanze,* the primal plant—the perfect embodiment of the ideal type. He never found his plant but soon came to detect an even more fundamental unity grounding the forms of all plants: their various parts—stem, leaves, petals, seeds—were the multiple expressions of a fundamental form, that of the ideal leaf. In the 1820s, Goethe would extend his conception of a primal structure to the animal kingdom. He came to argue, for instance, that the vertebrate skull, as well as the whole skeleton, comprised a metamorphosed series of the elemental unit, the vertebra.[30] Both the plant and the animal thus could be understood as modeled on archetypes, ideals that lay in the bosom of nature and exhibited creative power.

In Goethe's interpretation of Kant, which favored his own deep inclinations, the unity in diversity expressed by the teleological structure of nature could also be perceived through poetic imagination. Science and

28. Goethe's monistic belief that "matter can never exist and be effective without mind, nor mind without matter" became a foundation for Haeckel's own monism. See Goethe, "Erläuterung zu dem Aphoristischen Aufsatz 'Die Natur,'" in *Werke,* 13:48. See also below for a discussion of Haeckel's debt to Goethe's monistic philosophy.

29. After citing Kant's remarks about the *intellectus archetypus,* Goethe muses in his essay "Anschauende Urteilskraft": "So it is likely the same in the intellectual sphere, that we make ourselves worthy, through an intuition of the eternal creativity of nature, of participating mentally in her productivity." See Goethe, *Werke,* 13:30–31.

30. Goethe published his vertebral theory of the skull only in the second number of his *Zur Morphologie* (1820). See Johann Wolfgang von Goethe, *Die Schriften zur Naturwissenschaft,* 1st division, vol. 9: *Morphologische Hefte,* ed. Dorothea Kuhn (Weimar: Böhlaus Nachfolger, 1954), 185. Lorenz Oken (1779–1851) claimed also to have made the discovery of the vertebral nature of the skull, and so published in 1807. He and Goethe had a bitter priority dispute concerning the discovery. See my discussion in *The Romantic Conception of Life,* chap. 11.

Fig. 2.5. Johann Wolfgang von Goethe (1749–1832). Chalk portrait (1791) by Johann Heinrich Lips. (Courtesy Archiv für Kunst und Geschichte.)

poetry ultimately had the same functions: to reveal the harmonious relation of parts to the whole organism, to manifest the details of structure as they contributed to the perfection of life, and, thereby, to lead us to appreciate and participate in the infinite creativity of nature. Haeckel's consumption of great quantities of Humboldt and Goethe during his medical school years caused his own ideas to pulse with their conceptions of science and art. These Romantic rhythms would sustain him throughout his life. (See the first appendix for further discussion of Goethe's morphology.)

The Research Ideal

Goethean and Humboldtian ideas fueled Haeckel's own natural propensities toward the solitary life. But while in medical school, he was hardly an isolated figure. He had good friends among his classmates, with whom he learned to lift a draught. His friends, though, were aware of his tendencies toward study and solitude. They sought to loosen him up a bit. On one occasion they enticed him to attend a masked ball held in Würzburg. When he got there, he was astonished to have a mysterious young woman—at the time he knew only two women, both wives of professors!—come up and chide him for not socializing more. Delighted, he asked her to write her name on a slip of paper. "Mysterious" was the name she wrote, and then she vanished. Haeckel suspected that his friends had put her up to it. This and like experiences perhaps brought him to a certain resolution: the next year he took dancing lessons.[31] But in those moments of the adolescent's deep reflections and inevitable anxieties, he found great consolation in the Romantics' traditional resources—nature and poetry. After having dinner with a friend or alone, he often stole out into the countryside to savor the delights of nature settling into evening. Or in the twilight of his darkening room, he would light a candle and pull down his Schiller, Goethe, or perhaps read from a translation of Shakespeare—a favorite of the Romantics. He expressed his feelings about these activities to his parents:

> I can't tell you what joy the pleasure of nature provides me, whether nature be smiling beautifully or overcast and gloomy. I feel that all my troubles, which I suffer from during the day, are immediately lifted from me. It is as if the peace of God and of Nature, which I otherwise so vainly seek, suddenly entered my heart. What the consideration of world history and the general fate of men is for you, dear Father, the general and special contemplation of nature, perhaps even more so, is for me. . . . Again I find another great pleasure and consolation in poetry. Recently I have learned rightly to treasure this. Poetry raises a man above the dust and worry of everyday life and banishes evil thoughts.[32]

Though he often felt he had two souls dwelling in one breast—that of the "loving man," who feels deeply and kindles his passions with nature and poetry, and that of the "scientific man," who splashes cold reason on the emotions to achieve objective understanding—he yet conceived of a way to temper these disjoint inclinations. This was through a Humboldtian vision of the researcher who works in exotic lands and occasionally attends to the medical needs of the natives. He used this image to fortify

31. See his letters to his parents (20 March 1845 and 19 November 1855), in *Briefe an die Eltern*, 107, 167.

32. Haeckel to his parents (27 November 1852), in ibid., 19.

his efforts at medicine, which he never loved. It was an adolescent dream, but one that, remarkably, would materialize in a few years. He wrote his parents to describe his plan:

> I would like this fervent wish, nothing more nothing less, to come true, whose realization I dream of day and night, really a dream I have had since I was a child, namely, of a great trip into the tropics, thus something not really new but rather old now . . . to stay in some tropical land (in Brazil, Madagascar, Borneo, or some other such place), where I can sit in some primeval forest with my wife (that is, my inseparable microscope) and, insofar as my bodily powers allow, to anatomize and microscopize animals and plants, to collect all sorts of zoological, botanical, and geographical knowledge, so that this material will allow me to accomplish something coherent. . . . This dream, this beautiful golden castle in the air, satisfies my intention in every way. Namely, it shows me the fixed goal toward which I must steer; it mirrors for me the reality of my most desired wish; it spurs me on to perfect myself in every way possible in the beloved sciences; and it forces me morally to stay the path toward the hated medicine. In all of these respects, especially the last, this beautiful dream can be useful to me, even if it should not be realized, as I fear will be the case.[33]

Perhaps no experience confirmed Haeckel in his goal of biological (as opposed to medical) research more than his new relationship with the most famous physiologist and zoologist of his day, Johannes Müller. In the spring of 1854, Haeckel decided to take his summer term in Berlin. Away from provincial Würzburg, he would indulge himself in this "metropolis of intellect" and, of course, visit with his parents and relatives. He would also have opportunity to study with the renowned Müller.[34]

During the summer term at Berlin, Haeckel attended Müller's lectures on comparative anatomy and physiology, and those of his former teacher Alexander Braun, who then was occupied with generational alteration in plants, a topic that certainly engaged Haeckel more than the elementary subjects he tolerated with Braun on his first stint at Berlin. The decisive experience with Müller, though, came during the summer vacation.

At the end of August 1854, Haeckel and his friend Adolph de la Valette St. George (1831–1910) decided to travel to Helgoland (two islands in the

33. Haeckel to his parents (17 February 1854), in ibid., 101.

34. Haeckel to his parents (25 March 1854), in ibid., 109.

North Sea, west of Schleswig-Holstein). They planned to meet other student friends there for collecting seaweeds and rather desultory anatomical study—all to be refreshed by a good deal of sea bathing. Likely Müller's stories of collecting off the islands, along with other tourist delights, inspired them to go. On the way they passed through the port city of Hamburg, whose shops carried exotic wares from all over the globe and whose streets could hardly contain the crowds of sailors, tourists, peddlers, and citizens of all stations and dress. The harbor itself displayed to the entranced students a tangled forest of masts and rigging from ships that plied the seas of the world. After a harrowing passage on a new three-masted iron steamer during a great gale, Haeckel and Valette disembarked on the principal island of Helgoland in the late afternoon of 17 August. They settled into a routine of sea bathing at 6:00 a.m. and collecting and dissecting during the rest of the day. It was a revealing experience for Haeckel, as he indicated to his parents: "You cannot believe what new things I see and learn here every day; it exceeds by far my most exaggerated expectations and hopes. Everything that I studied for years in books, I see here suddenly with my own eyes, as if I were cast under a spell, and each hour, which brings me surprises and instruction, prepares wonderful memories for the future."[35]

Rather unexpectedly, Johannes Müller and his son Max arrived in Helgoland for two weeks of research on echinoderms (starfish, sea urchins, etc.). Müller immediately invited Haeckel and Valette to accompany his son and him on their fishing and research expedition. The friendship of his revered teacher and the marvel of the invertebrates they brought up for study each day irrevocably altered the course of Haeckel's research interests, from botany to marine invertebrate zoology, a transition sealed with the publication the next year of his maiden research article in Müller's *Archiv*.[36] Much later, in 1905, during a series of confrontational lectures given in Berlin, Haeckel recalled the magical period he spent with Müller in Berlin and on that wonderful trip to Helgoland. He conjured up a memory, undoubtedly bent against the winds of the intervening half century, of a question he put to Müller as they brought up extraordinary sea creatures:

> As we fished together in the boat and captured the beautiful medusae, I asked how then the astonishing generational alternation of these creatures was to be explained. Whether or not the medusae, out of whose

35. Haeckel to his parents (30 August 1854), in ibid., 122.

36. Ernst Haeckel, "Über die Eier der Scomberesoces," *Archiv für Anatomie und Physiologie* (1855): 23–32.

Fig. 2.6. Johannes Peter Müller (1801–1858). (Courtesy of the Smithsonian Institution.)

> eggs polyps today develop, thus originally themselves arose out of more simply organized polyps? To this rather forward question I heard the resigned answer: "Yes, we are faced with a great riddle! We know nothing of the origin of species!"[37]

Whether the memory remained green after all those years is impossible to say. But it is true that the phenomenon of generational alternation in invertebrates led Haeckel to his formulation of and confidence in the biogenetic law, an essential principle of his and Darwin's evolutionary theory.[38]

37. Ernst Haeckel, *Der Kampf um den Entwickelungs-Gedanken: Drei Vorträge, gehalten am 14, 16, und 19 April 1905 im Saale der Sing-Akademie zu Berlin* (Berlin: Georg Reimer, 1905), 24.

38. I have argued that the recapitulation principle—that the embryo of a species passes through morphological stages characteristic of the species forms of its evolutionary history—is authentically Darwinian. See my *Meaning of Evolution: The Morphological Construction and Ideological Reconstruction of Darwin's Theory* (Chicago: University of Chicago Press, 1992), 91–166. See also chapter 5 below.

Had Haeckel interacted with Müller so intimately a year later, he might well not have followed his course in life. In the autumn of 1855, Müller and several students had journeyed to the coast of Norway for marine research. On the return, their ship was rammed by another vessel in the dark and sank. One of the students drowned, and Müller and another student survived only by a very long swim through frigid waters in total darkness. The experience so affected the fifty-four-year-old professor that he would never again board even a light fishing boat. All his further research would be done from catches that local fishermen brought to dry land. This was a practice that Haeckel believed led to certain deficiencies in Müller's later work, since many of the specimens his mentor investigated would be damaged by the locals' rough handling of them.[39]

Haeckel extended his stay in Berlin through the winter semester of 1854–55 but returned to Würzburg the following spring. He spent the summer term of 1855 in clinical training and in the fall would commence with the actual treatment of patients. During the summer, though, he also found time to take a short course in the dissection of invertebrates offered by two *Privatdozenten,* Franz Leydig and Carl Gegenbaur (1826–1903), both of whom worked with Kölliker. Haeckel's clinical experience was usually confined to the poor and destitute of Würzburg, and the cases with which he dealt—in children, for example, horrible worms, rickets, scrofula, and eye diseases—did little to stimulate his appetite for the practice of medicine. The only part he really enjoyed was the postmortem anatomies, of which there seemed to be no short supply.

His salvation during this period lay in the tutelage of Virchow, who encouraged the young student in pathological anatomy. Virchow oversaw Haeckel's next two publications, which embroiled the apprentice in a controversy with his mentor's opponents.[40] "But how sweet to be attacked in defense of Virchow," he wrote his parents.[41] After a successful comparative anatomy exam, Haeckel became Virchow's assistant for the summer of 1856 and harbored the hope that the great man would take him along in the autumn to the University of Berlin, to which the renowned scientist had been called. But during that summer, Haeckel again began to despise the clinical practice of medicine and longed to be able to pursue what he thought

39. Haeckel recounted this episode in *Die Radiolarien (Rhizopoda Radiaria). Eine Monographie,* 2 vols. (Berlin: Georg Reimer, 1862): 1:17.

40. Ernst Haeckel, "Zwei medizinische Abhandlungen aus Würzburg: I. Über die Beziehungen des Typhus zur Tuberkulose; II. Fibroid des Uterus," *Wiener medizinische Wochenschrift* 6 (1856): 1–5, 17–20, 97–101.

41. Haeckel to his parents (8 June 1856), in *Briefe an die Eltern,* 184–88.

Fig. 2.7. Ernst Haeckel in 1858, when he passed his medical exams. (Courtesy of Ernst-Haeckel-Haus, Jena.)

his true vocation—biological research. Moreover, though his relationship with Virchow was cordial, the cool and reserved character of the professor ill complemented the passionate and excitable nature of the student.

After the tedious summer weeks of clinical work, Haeckel was invited by Kölliker to travel with him to Nice for collection and study of invertebrates. He rejoiced at the opportunity, made good with the help of some 150 *Reichstaler* from his father. On the seductive French Riviera, the company met Müller, and the whole experience convinced the young scientist that he had entered paradise. But the bliss of biology gave way again to dreaded medicine, and in the winter semester of 1856–57, Haeckel retreated to Berlin to prepare his medical dissertation, which he wrote under the guidance of Leydig. His study was on the histology of river crabs (*De telis quibusdam Astaci fluviatilis*), a subject conveniently ambiguous of disciplinary direction.[42] He received his medical doctorate in March 1857 and then felt compelled to spend the summer in Vienna for further clinical study, to prepare for the state medical exam, which, after more anxious preparation in Berlin during the winter semester, he passed the following March.

During his medical education, Haeckel became ever more passionate about his vocation: not that of a physician but that of a biological researcher, one whose ideal was formed in the exacting microscopical work done under the guidance of Kölliker and Virchow but whose deeply rooted inclinations were nourished by the kind of science practiced by Humboldt and Goethe. And like these latter paragons, Haeckel transferred an early religious enthusiasm on to nature. This transference was made easy by the religious attitudes instilled in him by his parents, who themselves adopted the ideals of the Romantic theologian Friedrich Schleiermacher (1768–1834).[43]

Religion vs. Science

Schleiermacher relocated the source of religion in the subjective awareness of a particular kind of feeling.[44] As he maintained in his famous opus *Der christliche Glaube* (2nd ed., 1835), the piety at the foundation of religion is "that we are conscious of ourselves as being utterly dependent, or, which is the same thing, as in relation to God."[45] For Schleiermacher, religion consisted principally in this feeling of complete dependence. Theological dogmas that described the attributes of God, the nature of the world, and the relations of man to God and the world—such dogmas, according to

42. Ernst Haeckel, *De telis quibusdam Astaci fluviatilis* (Berlin: Schade, 1857). A principal thesis that he argued in the dissertation was that "Formatio cellularum libera, et physiologica et pathologica, haud minus quam generatio animalium et planatarum spontanea rejicienda est." (The formation of free cells, either physiologically or pathologically, is to be rejected no less than the spontaneous generation of animals and plants.) The thesis suggests both Haeckel's adherence to Virchow's conviction that "every cell comes from a cell" and his belief in the stability of species. Academic dissertations, though, rarely reveal troubled convictions or differences with one's professors.

43. Schleiermacher, a close friend of Friedrich Schlegel, was a member of the circle of "early Romantics" that migrated between Jena and Berlin.

44. I have discussed the Romantic character of Schleiermacher's theological convictions and his proposals for science in my *Romantic Conception of Life*, 97–105.

45. Friedrich Schleiermacher, *Der christliche Glaube*, 2nd ed., 2 vols. (1835; repr., Berlin: De Gruyter, 1960), 1: sec. 4, p. 23.

Schleiermacher, only reflected certain aspects of the subject's feelings of dependence. Thus, what appeared to be objective propositions referring to external reality were only "conceptions of Christian pious affections expressed in speech."[46] This kind of feeling of dependence and reverence in the presence of a greater power could rather easily be transferred onto a nature that displayed a commanding force and sublimity, which yet might yield up certain secrets to the microscope but which would trail off again into mysterious depths.

The economic conviction that religion has as its source an internal feeling, which theological dogmas only symbolize, had to be affronted by the beliefs and practices of the Catholic Church, especially in the full flower of its Bavarian enthusiasms. During Haeckel's residence in Würzburg, he had ample opportunity to experience the spectacle of Roman observance and to contrast it with the simpler preaching he found in the Evangelical Church. He wrote his parents often enough of the extravagant Feast Day celebrations, with "idolatrous processions," in which would march "bishops and other high spirituals, bedecked resplendently in gold and purple, after whom would come the violet robed canons, no less well nourished, splendidly fat, and continuously snorting snuff."[47] But the mysteries of Catholicism attracted as much as they repelled. During the second term of his first year, Haeckel attended a mission preached by the Jesuits at the cathedral in Würzburg. He found the sermons of a young priest full of "eloquence, fire, and expression," as well as philosophically acute and intelligent. This he knew to be a Jesuitical trick, for "they first introduce only plausible matters, and then gradually they descend further and more particularly into their doctrines."[48] Haeckel's respectful detestation of the Jesuits would deepen just after the turn of the century during his debates with the Jesuit naturalist Father Erich Wasmann, who, because he adopted a version of evolutionary theory, epitomized for Haeckel the archetype of co-opting Jesuitical sophistry. After his encounters with Wasmann, he described as "Jesuit" any religious objector to his ideas.

Haeckel stoked his anti-Catholicism during trips to Italy in the late summer of 1855 and again in the summer of the next year. In the ripe atmosphere of the boot, he discovered a repulsive decadence that forever infected his imagination with the fear of Ultramontanism, a fear that in fu-

46. Ibid., 1: sec. 15, p. 105.

47. Haeckel to his parents (1 June 1853), in *Briefe an die Eltern*, 58.

48. Haeckel to his parents (17 February 1853), in ibid., 41.

ture years would fester as the political power of the Catholic Center party grew in Germany. The image of the tentacles of the papacy spreading out from Rome up toward Bavaria and into Prussia, strangling science and culture, seizing political power, and devouring a naive populace—this vision became for Haeckel the emblem of the dangers and pernicious influence of organized religion. He claimed to his parents, just before he left Würzburg, that it was just this "superstitious formalism and wholly un-Christian homage to images, this clerical rule and Marian cult of Catholicism," that helped drive him to the conclusion that "each person can and must form his own religion according to his own individual character, and thus to the truth of Schiller's words: 'Man depicts himself in his gods.' "[49]

Another force that helped shape Haeckel's religious abreactions during this period was the expressed materialism of close student friends and respected professors. He became unnerved after a particularly long discussion with his companion of many hours in the lab, Otto Beckmann (1832–1860), a beloved friend whose promising career abruptly ended with his unexpected death in Göttingen. Beckmann professed an extreme materialism and a disdain for religion, while yet remaining large in intellectual capacity, admirable in moral bearing, and strong in personal character. Haeckel thought his friend's attitudes must have resulted from the daily exposure to the "false, sham religion of the Catholic Church."[50] And then there was Virchow.

Haeckel inhaled Virchow's theory of the organism as a confederation of cells, a conception that later furnished the young scientist an experimental foundation for his proposals about the nature of the evolutionary individual. Virchow's cellular theory appeared at one level to be decidedly materialistic. In his lectures at Würzburg, he stressed that the more one studied the phenomena of life, the more "one becomes convinced that the variable life of a higher organism derives from the activities of its particular parts [its cells]. One finally comes to the conclusion, when making general comparisons, that such a complex organism can better be compared

49. Haeckel to his parents (10 February 1856), in ibid., 188–89. The line comes from Schiller's inaugural essay as an academic historian at Jena. Schiller describes the progressive development of Western civilization and culture, and contrasts it with the low state of wild savages (which mirrors an earlier developmental stage of Western civilization). He portrays the deities of these less advanced groups as reflections of human impulses: "There the piously simple-minded cast themselves down before some ridiculous fetish, and here before some abominable monster. Man depicts himself in his gods." See Friedrich Schiller, "Was Heisst und zu welchem Ende studiert man Universalgeschichte?" in *Sämtliche Werke*, ed. Jost Perfahl, 6th ed., 5 vols. (Düsseldorf: Artemis & Winkler, 1997), 4:705–20; quotation from 709.

50. Haeckel to his parents (17 June 1855), in *Briefe an die Eltern*, 146.

to a society than an individual, to a state than a citizen."[51] This reduction of life, say in a creature such as man, to the life of its smallest vital parts, the cells, did militate against an older vitalism, one that postulated a soul or *Lebenskraft* as the organizing force governing the whole creature.[52] And this opposition obviously struck home with Haeckel, at least he so conveyed it to his parents:

> Virchow is through and through a matter-of-fact person, a rationalist, and a materialist. He regards life as the sum of functions of the particular organs, which differ from one another materially, chemically, and anatomically. The entire living body reduces thus to a sum of particular centers of life, whose specific activities are bound up with the properties of their elementary parts, ultimately the properties of cells, out of which the whole body is constructed. Thus the activity of the soul [is to be regarded as] the inherent property of the living nerve cells, movement the result of the formation of the muscle cells, etc.[53]

Though Haeckel thought that this materialistic rationalism "springs from the very essence of Virchow," he nonetheless perceived it also as generally characteristic of the bulk of natural scientists of Germany.[54]

51. These remarks come from Virchow's course of lectures in general pathological anatomy, given at Würzburg in the winter semester of 1855–56, two years after Haeckel took the course. The lectures were recorded by a student, Emil Kugler. See Rudolf Virchow, *Die Vorlesungen Rudolf Virchows über Allgemeine Pathologische Anatomie aus dem Wintersemester 1855/56 in Würzburg,* recorded by Emil Kugler (Jena: Gustav Fischer, 1930), 24.

52. Virchow was certainly not original in rejecting the idea of a *Lebenskraft* governing the organism and in replacing it with the conception of a statelike confederation of units. Johann Christian Reil (1759–1813), the great medical anatomist at Halle, had argued a half century earlier that the processes of life were due to the particular powers (*Kräfte*) of matter as articulated by the laws of chemical affinity. He held that each organ of the body, down to the smallest fiber, remained independent of the others, though in causal interaction with them to maintain the whole body. "The animal body," he concludes, "is thus like a great republic that arises out of several parts." See Johann Christian Reil, "Von der Lebenskraft," *Archiv für die Physiologie* 1 (1796): 8–162; quotation from 105.

53. Haeckel to his parents (16 November 1853), in *Briefe an die Eltern,* 81.

54. Ibid. Virchow was not a materialist in any conventional sense of the term—though a few years later, M. J. Schleiden would also characterize him as a materialist (a dispute I will discuss in the next chapter). The life of the organism, in Virchow's conception, was reducible to the life of its cells. But the cells manifested life in an irreducible form. He argued that two forces operated in the cell: a molecular, physical force and a sui generis vital force. This latter could be called, despite its unfavorable associations, a "*Lebenskraft.*" If one could demonstrate a spontaneous generation of cells, then the life force might be dispensed with. But, according to Virchow, such had not been shown. Therefore, "we are not justified in concluding otherwise than that life finds its own particular property in the continuation of a movement that occurs

Haeckel fought to reconcile the scientific materialism that governed the attitudes of the professionals he most admired with his still-simmering religious piety. The compromise into which he entered at the end of his Würzburg years echoes the familiar doctrines of Schleiermacher. He wrote to his aunt Bertha, who had been baptized by the theologian, of the perplexity that darkened his last student days:

> There is a point which has bothered me, and the more light and truth I attempt to find concerning it, the darker and more confused it seems. It is the relationship of our modern natural science, whose enthusiastic disciple I am proud to be called, to Christianity on the one side and materialism on the other. The further that research presses, the clearer and simpler universal natural laws become, finally reducing to mechanistic relationships and ultimately to mathematical formulas (which indeed is the highest goal of the organic natural sciences), and the nearer comes the thought and the greater comes the temptation to seek the final ground of all things in such mechanistic, blind, unconscious, and exceptionless natural law and to draw there from all the consequences that modern materialism has drawn. . . . Yet one comes to a point . . . in which it is useless to seek a way out and one must simply remain stationary since our limited human understanding cannot go further. It is this point, where knowledge ceases and faith, which the materialists would completely reject and like to remove, begins. And indeed, it is this faith—which is perfected in Christianity and has found there its truest expression—that is the only anchor of salvation for the soul searching after other consolations and other satisfactions.[55]

Haeckel seems to have found momentary comfort in the Schleiermacherian view that there are two spheres, one of knowledge and science, the other of faith and religion. And if faith ultimately reflects, as he shortly thereafter wrote to his parents, the "individual's own character," then religion, shorn of dogmatic formulations, might lie down with science.[56] After

in the material. In the assumption of a similar movement joined to similar matter, I argue generally in a vitalistic direction [vitalistischen Richtungen]." See Virchow, *Vorlesungen über Allgemeine Pathologische Anatomie*, 30.

55. Haeckel to his aunt Bertha (1 February 1856), in *Briefe an die Eltern*, 177–78.

56. Haeckel's conception of two independent spheres of concern became a common-enough resolution to the conflict between science and religion during the nineteenth century. Virchow, in his dispute with Schleiden (see chapter 4), would advance the same conception. Andrew Dixon White, in his *History of the Warfare of Science with Theology in Christendom*, 2 vols. (New York: D. Appleton, 1896), would give the idea currency at the end of the century.

the two greatest events in his life, during the early 1860s, Haeckel would slide away even from this uneasy settlement. His God gradually slipped into the guise fashioned by Goethe and Spinoza—namely, Nature.

Habilitation and Engagement

After passing his state medical examinations in March 1858, Haeckel laid plans for the prosecution of his true vocation, research science. He arranged with Johannes Müller to conduct his habilitation research at Berlin—the habilitation, with its required monograph, was a requisite for an academic position. During this period, though, Müller suffered from the deepest of depressions, which led him to an ultimate solution. He took his own life with an overdose of opium—at least that was what Haeckel suspected.[57] Haeckel was devastated, not simply because of a lost opportunity, but because he truly revered and loved the man.

Haeckel's academic ambitions brightened when another Müller protégé, Carl Gegenbaur, his friend from Würzburg, invited him to visit Jena, where Gegenbaur had become ordinary professor of anatomy in the medical faculty.[58] During the visit in May 1858, Gegenbaur offered intimations of support and more straightforwardly asked Haeckel if he would care to travel to Messina with him in October. To Haeckel it seemed a dream materialized, and he quickly said yes. The dream began to dissolve, how-

---

Kleeberg artfully describes the liberal Protestant culture that formed Haeckel's general view of religion. See Bernhard Kleeberg, *Theophysis: Ernst Haeckels Philosophie des Naturganzen* (Weimar: Hermann Böhlau, 2005), 32–38.

57. While in his late sixties, Haeckel became enamored of a beautiful young woman over thirty years his junior, Frida von Uslar-Gleichen. He thought the relationship doomed, and in their voluminous correspondence he would often pour out his despair. In his letter to her of 11 January 1900, he mentioned that he often thought of suicide, and that his "great, highly revered master, Johannes Müller, ended his nervous condition (accompanied by sleeplessness) with morphine in April 1858." See *Das ungelöste Welträtsel: Frida von Uslar-Gleichen und Ernst Haeckel, Briefe und Tagebücher 1898–1900,* ed. Norbert Elsner, 3 vols. (Göttingen: Wallstein, 2000), 1:390. Gottfried Koller suggests that any overdose of morphine would have been accidental. See his *Das Leben des Biologen Johannes Müller, 1801–1858* (Stuttgart: Wissenschaftliche Verlagsgesellschaft, 1958), 234–36.

58. Gebenbaur was called to Jena as extraordinary professor (roughly the equivalent of an American associate professor) in 1855. Earlier, in 1851, he had met Johannes Müller, who persuaded him to do research in Helgoland; the next year, with Heinrich Müller and Kölliker, Gegenbaur traveled to Messina, where he became confirmed in his interests in marine invertebrates. He habilitated at the end of the winter term of 1853–54 with a monograph on generational alteration and reproduction in medusae and polyps. See Carl Gegenbaur, *Erlebtes und Erstrebtes* (Leipzig: Wilhelm Engelmann, 1901), 57–64, 87; and Georg Uschmann, *Geschichte der Zoologie under der zoologischen Anstalten in Jena 1779–1919* (Jena: Gustav Fischer, 1959), 28–29.

ever, when Gegenbaur and Moritz Seebeck (1805–1884), the curator of the university, took him aside to offer the advice of wisdom and age, that he should not even think about marriage lest his scientific career sink before being properly launched. That evening, with obviously troubled conscience, Haeckel sat down to write of this conversation to Anna Sethe, his first cousin and the woman to whom he had become secretly engaged.[59]

Two days after Müller's burial on 28 April 1858, Haeckel had pledged his troth to Anna Sethe. He first met his cousin at the wedding of his brother, Karl, and Anna's sister Hermine. Anna's father was the brother of Haeckel's mother, Charlotte Sethe. In his diary for 21 September 1852, when he was eighteen and she seventeen, he penned: "Celebration at Karl's wedding. Anna Sethe as an elf! Dancing. I knew how but couldn't dance and sat (as usual when others are having fun) in a melancholy mood by myself in the back of the room."[60] Haeckel would see Anna from time to time at various family gatherings. In 1856 she came with Haeckel's parents to visit him in Würzburg. After the death of her father, she and her mother moved to Berlin in 1857, during the time Haeckel spent working on his dissertation. Through the next year their relationship flowered, and in precipitous passion at the time of Müller's death, he asked her to marry him. It was only two months later that Gegenbaur and Seebeck offered their peremptory advice, which was often repeated by friends and relatives to whom he revealed his secret.[61]

The difficulties of managing both marriage and a career—a career that had not even really begun—agitated Haeckel through the summer of 1858 and beyond. But simultaneously he came to perceive Anna as the lodestar of his life—even more, as an all-consuming love that gave meaning to his work and, it is no exaggeration, to the entire universe. She was in many ways the young, long-haired, blond, blue-eyed scientist's female double, either in blood or in his own imagination, as his description for a friend suggests:

> A true German child of the forest, with blue eyes and blond hair and a lively natural intelligence, a clear understanding, and a budding imagi-

59. Haeckel to Anna Sethe (25 May 1858), in *Himmelhoch Jauchzend: Erinnerungen und Briefe der Liebe*, ed. Heinrich Schmidt (Dresden: Carl Reissner, 1927), 19. This volume contains Haeckel's letters to Anna from spring 1858 to fall 1862.

60. The passage from Haeckel's diary is quoted by Schmidt in the introduction to Haeckel's letters to Anna, ibid., 6.

61. Haeckel mentioned to Anna these several warnings. See his letters to Anna (9 April and 27 September 1858), in ibid., 64–65, 76.

> nation. She puts no stock in the so-called higher and finer world, for which I hold her even higher since she was brought up in it. She is rather a completely unspoiled, pure, natural person.[62]

Haeckel's letters to Anna over the period of their courtship express three intertwined themes: his love for her; his hopes of landing a professorship, which would allow them to marry; and his exuberant and irrepressible attachment to nature, an emotion that at times seems to rival that for her. But through this period, the latter themes gradually become submerged in an overflow of desire for Anna. "How our souls have already so closely and strongly grown together," he exclaimed to her in August, "so that absolutely nothing can separate them and so that every thought and every action are able to be realized only with and in the 'other ego.'" He thought of her love as a kind of salvation, a lifeline that would pull him back from the dark abyss of materialism toward which he felt himself dragged by his science. "When I press through from this gloomy, hopeless realm of reason to the light of hope and belief—which remains yet a puzzle to me—it will only be through your love, my best, only Anna."[63]

Their growing love pressed them to reveal officially what by midsummer most of their friends knew already; and so on 14 September 1858, in Anna's new family home in Heringsdorf (north of Berlin on the Baltic), they announced and celebrated their engagement. Two weeks later Haeckel wrote to his fiancée from Berlin, recalling with febrile delight their Sunday morning walk on the day of their festivities.

> My gay, frisky roe trotted by my side, happy and free over rocks and roots, slipping through thorns and thickets. [They sat down on the green moss bank] and your sighing breath, your warm cheek on mine announced to me at every blissful second that sweet unspeakable happiness that I held in my arms, close and sure, so that I might never, never lose it. Then we lay on my good old plaid, placed on the natural bed of the forest, upholstered with dry beech leaves, sloping down on the side, at the foot of two old boughs carved out for us, and we peered through the thousand smaller and larger holes between the round, green leaves up into the deep blue cloudless sky, whose bright sun so wonderfully shown on the happy pair as if it rejoiced with them. O Anna, those were moments I will never, never forget, moments of the greatest

62. Haeckel to a friend (14 September 1858), in ibid., 67.
63. Haeckel to Anna Sethe (22 August 1858), in ibid., 54.

Fig. 2.8. Anna Sethe (1835–1864), who became engaged to Haeckel in 1858. (Courtesy of Ernst-Haeckel-Haus, Jena.)

human happiness, the most happy because the individual himself is completely forgotten; he removes himself purely and completely from the dirty, spotted veil of a suffering personality in which he is wrapped, and lifts himself up and beyond into a full and pure intuition of the other in the joy of an absolute giving to the other. One forgets heaven and heart, past and future, and lives purely and completely in the present. Here Faust himself could exclaim, "Tarry a while, you are so beautiful," so he might secure the moment which sadly only too quickly dissolves.[64]

64. Haeckel to Anna Sethe (26 September 1858), in ibid., 72–73.

In the August prior to their engagement, Haeckel had traveled again to Jena, invited by Gegenbaur, for the celebration of the three hundredth anniversary of the university. Gegenbaur had also invited several other young scientists, who would soon make their marks: Julius Victor Carus (1823–1903), from Leipzig, the zoologist who would translate Darwin's and Huxley's works into German; Max Schultze (1825–1874), from Halle, the comparative anatomist whom Haeckel particularly liked and who would make important suggestions for the further development of cell theory; and Carl Nägeli (1817–1891), from Munich, who had already made significant contributions to plant anatomy and would work on cell theory and genetics. Several of the visitors crammed into Gegenbaur's small house: Dr. Schultze slept in the living room, Dr. Carus in the pantry, and Haeckel, "the Doctor of little," in the utility room.[65] The celebration of the university affected Haeckel deeply—this small court of learning, surrounded by rolling hills and haunted forests, had housed at the turn of the last century Schiller (for whom it is today named), Novalis, Fichte, Schelling, Oken, Hegel, the brothers Schlegel, and on frequent occasion Goethe himself. Sentiments of romance and freedom enlivened the very air Haeckel inspired, and he rejoiced in its "fresh natural spirit of liberalism."[66] The only unsettling experience he had was the realization that Gegenbaur would not likely travel to Italy after all.

Haeckel decided he had to make the trip nonetheless, even if he had to go it alone. It would be an excursion not simply to secure a subject for his *Habilitationsschrift,* but also one of *Bildung,* of intellectual and personal formation. He planned to spend the spring of 1859 in Florence and Rome studying art, the summer in Naples, where he would begin his marine research, and finish in Palermo and Messina in the winter. As a version of the kind of trip he always dreamed of, he expected his travel would "re-form and give rebirth to my whole outlook on life."[67] And so it did.

65. Haeckel to Anna Sethe (22 August 1858), in ibid., 37.
66. Ibid.
67. Ibid., 56.

CHAPTER THREE

Research in Italy and Conversion to Darwinism

Like Goethe—and many other Germans before him—Haeckel looked to Italy as a warm, vivifying balm for a cold, desiccated soul. He hoped it would produce a rebirth, though any keen observer of his life at this stage would have judged him hardly in need of resuscitation. Italy offered more a romantic enticement than a soul-saving escape.

On 28 January 1859 Haeckel made his move. He left Berlin, traveled back to Würzburg to collect materials and equipment, and then went on to Luzerne and Genoa. On 6 February he arrived in the artistic heart of Italy, Florence. For Haeckel, though, the heart beat dull and weak. He intended to study and copy the masterpieces that seemed to hang from every wall of the city. Quickly, however, he grew weary of the incessantly repeated themes—biblical images clung heavily to every surface. And then there were the countless Madonnas: Mary as a child, the Annunciation, the Birth, the Domestic model, the Grieving Mother, and now as an Italian, a Frenchwoman, a German, or a Spaniard, and each depicted in the garments of every century. The art was too religious, too Catholic, too much for Haeckel's liberal Protestant sensibilities.[1] In mid-February he traveled to Pisa for relief. Again he was surrounded by artful Virgins.

The Eternal City, which he reached on 23 February, seemed even more heavily caked with the cloying oils of southern religious sentiment. But worse yet, almost daily the streets of the ancient city were choked off with religious processions in celebration of one of the innumerable saints of the

1. One kind of artistry he did admire in Florence, though, was the craft of Giovanni Battista Amici, a famous microscope maker. Haeckel obtained a powerful instrument (1:1000) with a water-immersion objective. See Olaf Breidbach, "Einführung," in *Ernst Haeckel: Kunstformen aus dem Meer* (Munich: Prestel, 2005), 13.

Roman calendar. He saw cardinals from this or that cathedral riding in their gilded coaches and displaying to the poor of the city scarlet robes bedecked in jewels. He wrote to Anna that "had I not already during the last years—through a study of nature, pressing into her depths and finest parts—discarded the Christianity of the theologians, here in Rome I would surely become a pagan."[2]

Beneath the facade of the citadel ruled by "the pope with his band of Christian barbarians," Haeckel found the ancient city of Virgil, Horace, and Cicero. In the moonlight, he would walk through the ruins of that ghostly civilization and conjure up the shade of Goethe, who had lingered along the same paths during his own Italian journey in the 1780s.[3] But unlike Goethe—who could delight in the pomp of Catholicism, the craft of the Jesuits, and the decadence of the streets—Haeckel felt suffocated. He left Rome on 28 March and traveled to Naples, where he had to get to the chief business that brought him to Italy, biological research.

Naples was no joy. He had barely adequate accommodations, with constant noise from the streets. In the spring the weather was foul—frequent rain interrupted by oppressive heat and the unremitting winds of the sirocco out of North Africa. Nor did the Neapolitans elevate his measure of humankind: "The dishonesty, superficiality, thoughtlessness, the swindling selfishness overreaches all the usual bounds and for a true German this is all doubly painful."[4] Anna diagnosed his unhappiness in Naples as a consequence of his loss of religious faith. Haeckel agreed with this analysis but protested that even if he were ten times as unhappy, he could "never again accept an arbitrary dogma." "The fruit of the tree of knowledge," he wrote his Eve, "is worth the loss of Paradise."[5]

Despite his discomforts, Haeckel settled for almost six months in Naples, until mid-September. After he had arranged for a modestly regular and reliable supply of catch from local fishermen, he spent most of his day—roughly from 9:00 a.m. to 5:30 p.m.—examining and describing the various invertebrates that piled up on his table. But he had no direction in his research, and many a creature easily slipped through the gaps in his knowledge. He began to despair of ever becoming master of the field and of

2. Haeckel to Anna Sethe (28 February 1859), in Ernst Haeckel, *Italienfahrt: Briefe an die Braut, 1859–1860,* ed. Heinrich Schmidt (Leipzig: K. F. Koehler, 1921), 8.

3. Haeckel described his experiences in Rome to Anna Sethe (28 February, 1 March, and 15 March 1859), in ibid., 8–9, 14.

4. Haeckel to Anna Sethe (18 April 1859), in ibid., 28.

5. Haeckel to Anna Sethe (29 May 1859), in ibid., 65.

discovering something significant, which did not bode well for attaining an academic position and marrying Anna. Despite her constant efforts to cheer him, the lines of *Faust* came liquid to his pen: "I am plagued by no scruple or doubt, nor do I fear hell or the devil; yet all joy has been ripped from me, and I do not imagine I can know anything aright or teach anything to better men and convert them."[6]

Friendship with Allmers and Temptations of the Bohemian Life

On 17 June, no longer able to stomach the city, Haeckel took palate and easel and slipped across the bay of Naples to the beautiful island of Ischia. Under a sunny sky and surrounded by mountains and small forests, ripe for sketching or botanizing, Haeckel's mood shifted to contentment and then to something like happiness. But what made the trip more than a relief from tedium and frustration was his meeting there with the poet and painter Hermann Allmers (1821–1902), who would become his lifelong friend. Haeckel found in Allmers the odd complement. The poet was fourteen years older, gnomelike in appearance, and possessed of a "colossal Bedouin nose"[7]—the opposite of the tall, golden, and strikingly handsome young scientist. The contrasting but mutually attracting qualities reached down to the souls of each, as Haeckel reported to Anna:

> Allmers is above all a poet. He sees the whole of life, with all its light and shadowy sides, only from the beautiful, misty perspective of poetry, and so constitutes in this idealism a stark contrast to my natural-scientific realism, which strives to discard this misty, yet so very beautiful, gown and to view reality generally in its naked truth.[8]

These complements of talent and attitude—running over a deeper sexual feeling—supplemented the more repressed inclinations of each: Allmers could botanize with exactitude, and Haeckel often very happily would lose himself "in the misty distances of a dreamy poetry."[9] Haeckel wrote Anna that Allmers "has struck a responsive cord in me, has awakened feeling

6. Haeckel to Anna Sethe (9 May 1859), in ibid., 49–50.

7. See Haeckel to Allmers (14 May 1860), in *Haeckel und Allmers: Die Geschichte einer Freundschaft in Briefen der Freunde*, ed. Rudolph Koop (Bremen: Arthur Geist, 1941), 46.

8. Haeckel to Anna Sethe (1 August 1859), in *Briefe an die Braut*, 79.

9. Haeckel to Anna Sethe (1 August 1859), in ibid., 80.

Fig. 3.1. Hermann Allmers (1821–1902). Photo from 1860. (Courtesy of Ernst-Haeckel-Haus, Jena.)

and effort that I believed had already completely died; and, in a certain sense, he has given me back to myself."[10]

During that short week on Ischia (17 to 25 June) their friendship was sealed, as they tramped across the island, searching out plants, sketching, and thoroughly enjoying one another's company. On the third day out, as Allmers recalled, they came across the remains of an ancient Roman bath that enclosed one of the many thermal pools on the island. In view of the extraordinary landscape, they quickly shed all of their clothes and "naked, as a pair of truly natural men, plunged into the hot, flowing muck."[11] The two new friends returned to Naples together, planning other excursions

10. Haeckel to Anna Sethe (25 June 1859), in ibid., 69.

11. Allmers's reminiscence was communicated to Wilhelm Breitenbach, who used it in a sketch of Haeckel's life done for his seventieth birthday. See Wilhelm Breitenbach, *Ernst Haeckel, Ein Bild seines Lebens und seiner Arbeit* (Odenkirchen: Dr. W. Breitenbach, 1904), 22.

along the way. They sought vistas to paint, occasions to poetize, groves to botanize, and adventures to remember. And memorable, indeed, was their climb of Vesuvius (southeast of Naples) in July—a too vivid re-creation of the similar effort of Goethe and his painter-friend Johann Heinrich Tischbein three-quarters of a century before.[12]

They had planned to reach the summit before sunset. Allmers, however, was hardly up to the effort, even though Haeckel carried his pack. With Allmers lagging behind, the sun suddenly dropped below the horizon, and darkness fell quickly on everything in their path. They lost contact with each other. Haeckel thought the only way to go was up, since there was still faint light toward the top. Any attempt to turn back would pitch them into blackness and acute danger, since the night hid the fumaroles and the glassy-sharp ravines of solidified lava. At various intervals incandescent molten rock appeared through the cracks running along the often blazingly hot crust; the light served only as a frightening warning. All the way up, Haeckel kept calling to Allmers, whose replies grew fainter, until the last distant call "I can't go on." They lost contact altogether. Haeckel finally reached a spot near the crater's edge and kept shouting for Allmers. As he wrote to his parents, "I couldn't hurry to help him, since every step back was impossible; so we both had to stumble upward as the only sure way." Then after four hours of Haeckel's calling and whistling without any answer, Allmers suddenly crawled out of the dark. "Saved, we fell into one another's arms."[13] In later years they would often remind each other of this adventure. This and other such intimate experiences forged a friendship that slowly drew Haeckel away from steady work in biology.

In August they sailed to Capri, where they would spend the month leading a bohemian life of wandering through the countryside, bathing, and painting. Several times they had taken a boat out in the afternoon to the Blue Grotto, a rock formation just off the island, where phosphorescent organisms illuminated the waters with vivid color. Haeckel decided that they could best see the phenomenon in the dark, and so they rowed out under starry skies. When they got to the entrance of the circular formation, they could not quite steer the boat through, so Haeckel doffed his clothes, grabbed the rope at the bow, and swam in, singing songs from Goethe and Heine to ward off the sprits stirring in that place.[14] Capri seemed to

12. I describe the climb of Goethe and Tischbein in my *Romantic Conception of Life: Science and Philosophy in the Age of Goethe* (Chicago: University of Chicago Press, 2002), 394.

13. Haeckel to his parents (17 August 1859), in *Briefe an die Braut*, 92–93.

14. Haeckel to Anna Sethe (27 August 1859), in ibid., 95–96.

Fig. 3.2. Ernst Haeckel in 1860. (Courtesy of Ernst-Haeckel-Haus, Jena.)

Haeckel the realization of the dreams of his youth, dreams arising out of reading *Robinson Crusoe,* Goethe's *Italienische Reise,* and Humboldt's and Darwin's travel books, even if this Italian island melted into a glow the hardships described in that earlier literature. With the beauty of the island, the companionship of the other artists there, and the deepening friendship with Allmers, Haeckel was tempted to abandon his so far fruitless research and spend his days in landscape painting—his great delight—and his nights in dancing the tarantella, as he had the night of their departure from Capri.[15] What restrained his inclination was that he recognized his talent with watercolors was somewhat less than his aspirations; and, of course, it was obvious that the life of the bohemian did not pay very well—certainly not enough to support a wife, his Anna, to whom he felt ever closer the longer he remained away.

During that month on Capri, Haeckel cultivated his painterly eye for the glories of nature, while Allmers came to regard the robust, golden youth himself as one of those glories. Their talents fitted smoothly together and

15. Haeckel relates this to Breitenbach. See Breitenbach, *Ernst Haeckel, Ein Bild,* 25.

their politics were consonant. They both lauded Giuseppe Garibaldi's efforts to wrest Italy away from Austria and unite the country—a model for Germany, they thought. Their feelings adjusted harmoniously one with the other. Sometime after their return from Italy, Allmers composed a long poem for Haeckel that suggests the depth of his affection. It began:

Do you still think of that summer night—
Which I can never forget—the tide
That carried us to Ischia's sight?
How the ship did so quietly glide,
As the silence spread unbounded
Ineffably solemn and sublime,
While heavens and sea were surrounded
In the beauty of starlight so fine?
How the magic glow danced on the swell,
When the rudder stirred the water's bed
And how distant lava would tell
Of Vesuvius flowing dusty red?
We were still strangers to each other—
We spoke but a moment at the start—
And yet soon each like a brother
Revealed the depths of his heart.
We talked of things we might love and trust
And how it was before we came together,
What kind fate had allotted us,
What evils we had to weather;
And for many hours we spoke but truth
Of parents, homeland, and joys of youth.[16]

Haeckel's feelings for Allmers were no less profound, though in a different register. The only cloud that cooled momentarily the deep warmth of their

16. Allmers to Haeckel (undated), in *Haeckel und Allmers*, 17: "Gedenkst du noch der Sommernacht, / Mir kommt sie nimmer aus dem Sinn, / Die uns nach Ischia gebracht?— / Wie schwamm die Barke still dahin, / Wie war's so lautlos weit und breit, / Unnennbar feierlich und hehr, / In sterndurchstrahlter Herrlichkeit / Umfingen Himmel sich und Meer, / Und magish leuchtete die Flut, / Wenns Ruder leichte Wellen schuf, / Und drüben düsterrot die Glut / Der letzten Lava des Vesuv,— / Fremd waren wir einander noch— / Wir sahn uns ja kaum eine Stunde— / Und beide trieb's zu redden doch / Uns bald aus tiefstem Herzengrunde / Von allem, was uns lieb und wert, / Und wie's bisher mit us gekommen, / Was uns ein hold Geschick beschert, / Was uns ein Feindliches genommen, / Vom Elternhaus, vom Heimatland, / Von glückerfüllten Jungendtagen—."

bond was Allmers's incipient jealousy of Anna, which would slowly grow during the next several months.

The plan of Haeckel's itinerary now dictated that he leave for Messina, the Sicilian city where his revered teacher Müller had spent so many profitable days. Forty-eight hours after they returned from Capri, Haeckel and Allmers arrived in Messina, on 10 September 1859. They spent five weeks together traveling around the island by ship, wagon, mule, and foot. They climbed Mount Aetna, without the scare that Vesuvius had caused. Compared to Capri, which remained his "Italian Paradise," Sicily was disappointing in its quite ordinary flora and fauna. The forest had almost disappeared, and the cities had little to recommend them. Only ancient ruins offered some interest to the travelers. Haeckel found the Sicilians more to his liking than the Neapolitans, though only by a breath. "The Sicilians," he wrote Anna, "even if they are not comparably so depraved, so bereft of all virtue and honor as the completely bovine Neapolitans, they are, nonetheless, such a miserable group that a sensitive German conscience could never be reconciled to their superficial considerations and aspirations."[17]

In mid-October Allmers had to leave, and Haeckel at last turned to work. He justified to Anna and more especially to himself the time he had already expended as necessary for the development of his mind, of his character, and for the deepening of his appreciation of natural beauty. It was the sort of *Bildung* experienced by Goethe on his own Italian journey, and Haeckel hoped for a comparable result.[18] But now that Allmers had left, Haeckel's days took on a different rhythm. Typically he would begin at sunrise, when his assistant, Domenico Nina, awakened him for his morning sea bath. Returning from the harbor after his swim, he would stop by the fish market to inspect the early catch and then return to prepare for the day's work. At 8:00 a.m. he was brought breakfast, and after a quick check again at the market, he sat down at his microscope, describing and sketching the creature of the day. At 4:30 or 5:00 p.m., he would take a meal with friends and after a walk would return about 7:30 p.m. to go over the day's work and fill out his notes, perhaps do some reading, and retire about midnight. His companion at meals and in the evening walks, now that Allmers had left, was a Dr. Edmund von Bartels, a hypochondriacal and melancholy physician from Hamburg. Bartels rarely elevated Haeckel's spirits in the way Allmers had, and at the beginning of his serious work in Messina, Haeckel needed some uplifting words.

17. Haeckel to Anna Sethe (16 October 1859), in *Briefe an die Braut,* 112.

18. Haeckel to Anna Sethe (21 October 1859), in ibid., 116–17.

The flood of creatures that welled up in the seas around Messina—"the Eldorado of zoology," he called it—drove Haeckel to despair of seizing and reducing to actuality that great wealth of possibilities. Not only was he delivered of unusual species and genera, but of whole families, orders, and classes never before described, beautiful and astonishing animals—siphonophores, petropods, heteropods, radiolaria, medusae, and more. As the mountain buried him in its avalanche of goods, Haeckel pulled back into thoughts of the artist's existence, which promised "a rich, creative, colorful life of imagination, while that of the scientist offers a sober, cold, anatomical effort of reason that always soon leads to negation and skeptical dissolution, a reason that is oriented to a possible understanding of natural wonder that we can never comprehend." What kept him from casting off his plans—which now desiccated into that of "a repressed professor who in Jena or Freiberg or Tübingen or Königsberg or in some other small, petty university, every semester must take his one-and-a-half to three students and 'here and there, back and forth, lead them by the nose' "[19]—what constrained him on that gloomy professorial path was the image of the bright presence of Anna, who awaited at the end. Haeckel's despair at this juncture formed the negative image of his recent, glorious experience with Allmers and his desire for the distant Anna. But the bitter taste of research would quickly turn sweet as the topic for his *Habilitationsschrift* began to form in his mind.

Radiolarians and the Darwinian Explanation

At the end of November, with just a few months left for his research in Italy, Haeckel finally decided to focus on just one group of animals, the almost unknown radiolaria—a large class of one-celled marine organisms that secreted unusual skeletons of silica.[20] When he had traveled in the sum-

19. Haeckel to Anna Sethe (21 October 1859), in ibid., 118–19.

20. In 1836 and 1837, Christian Gottfried Ehrenberg (1795–1876) had described conglomerates of fossil protozoa, among which were, apparently, some radiolaria and perhaps Acantharia, distinguishable by the chemical composition of their skeletons, which in the fossilized state he described respectively as silica and flint. (Nonfossilized radiolaria and Acantharia skeletons we now know to be composed, respectively, of silica and strontium sulfate.) These remains were similar, he maintained, to certain living, freshwater siliceous protozoans (*Kiesel-Infusorien*). See Christian Ehrenberg, "Über das Massenverhältnis der jetz lebenden Kiesel-Infusorien und über ein neues Infusorien-Conglomerat als Polirschiefer von Jastraba in Ungarn," *Abhandlungen der Königlichen Akademie der Wissenschaften zu Berlin* (1836): 109–36. Since radiolaria are marine animals, likely Ehrenberg had observed living species of the class Heliozoa, which also have a silica skeleton but are freshwater. In 1847 Ehrenberg described fossilized silica conglomerates from Barbados. He called them "Polycystinen" and identified, on the basis of their skeletons, 282 species, arranged in 44 genera. See Christian Gottfried Eh-

mer of 1856 with Kölliker to Nice, they unexpectedly had met Johannes Müller, who had been collecting there. At that time Müller had been working on the radiolaria, and the great scientist returned to Saint-Tropez in 1857 to complete his research. Müller's short monograph on these animals was his final publication, appearing just after his death.[21] Haeckel had the foresight—or perhaps just the simple desire for remembrance—to bring the tract with him to Italy. During the course of his own research, the monograph became his "gospel," and he all but memorized it.[22] But Müller's work, it was clear, had been preliminary; and much remained for an ambitious researcher—especially to provide concrete meaning for that ever-nebulous claim of systematists that the several groups of organisms they treated were more closely or distantly *related.* When Haeckel produced his own monograph on the radiolaria—greater in length and breadth of con-

---

renberg, "Über die mikroskopischen kieselschaligen Polycystinen als mächtige Gebirgsmasse von Barbados," *Monatsbericht der Königlichen Akademie der Wissenschaften zu Berlin* (1847): 40–60. Thomas Henry Huxley, while serving on board HMS *Rattlesnake,* discovered what he thought to be a hitherto unknown zoophyte, which he called Thalassicolla (i.e., sea-jelly). Huxley skimmed connected masses of these one-celled creatures from the surface of the ocean. He noticed that glassy spicula would sometimes be found along the surface of a cell. See Thomas Henry Huxley, "Zoological Notes and Observations Made on Board H.M.S. Rattlesnake during the Years 1846–50" (1851), in *The Scientific Memoirs of Thomas Henry Huxley,* ed. M. Foster and E. Lankester, 4 vols. (London: Macmillan, 1898), 1:86–95. Huxley probably observed two related orders of the class of radiolaria now called Spumellaria—the Colloidea and the Beloidea. (Thalassicollida being a family of Colloidea). These orders either have imperfect skeletons or lack them entirely. Johannes Müller built upon the observations of Ehrenberg and Huxley in papers he read before the Berlin Academy of Sciences in 1855. In those papers, he confirmed Huxley's observations of Thalassicolla, and because of the associated spicula he suggested that they might be related to sponges, on the one hand (which also have siliceous spicula), and, on the other, to Ehernberg's Polycystina—Müller had found living specimens of these in waters off Messina in 1853. See his "Über Sphaerozoum und Thalassicolla" and "Über die im Hafen von Messina beobachteten Polycystinen," *Bericht über die Verhandlungen der Königlichen Preussischen Akademie der Wissenschaften zu Berlin* (1855): 229–54, 671–76. See also the next note.

21. Johannes Müller, "Über die Thalassicollen, Polycystinen und Acanthometren des Mittelmeeres," *Abhandlungen der Königlichen Akademie der Wissenschaften zu Berlin* (1858): 1–62. As with Huxley, Müller described the Thalassicolla to be without skeleton, or with skeleton only imperfectly represented (see previous note). The Polycystina, which Ehrenberg had identified in 1847, displayed the silica skeleton, and the Acantharia, which are now usually distinguished as a related class (both under the subphylum Sarcodina), also had a skeleton, but not of silica. Müller called them all by the common name "Rhizopoda radiaria" or "radiolaria" and regarded them as closely related to other Rhizopoda, such as the amoeba—a common judgment made today. Müller divided the radiolaria into two major groups, those living singly and those colonially. The Thalassicolla, Polycystina, and Acantharia lived separately, and the first two also had colonial forms, called respectively Sphaerozoum and Collosphaera. For the most detailed modern study of these creatures, see O. Roger Anderson, *Radiolaria* (New York: Springer, 1983).

22. Haeckel to Anna Sethe (29 February 1860), in *Briefe an die Braut,* 163.

sideration, more beautiful by far than that of his teacher—he dedicated it to Müller, so that natural piety linked Müller's tragic end with Haeckel's glorious beginning.

Haeckel wrote Anna to describe the creatures that would become his constant companions, though at one-thousandth to eight-hundredths of an inch in diameter, they were hardly companionable:

> The radiolaria are almost exclusively pelagic animals, that is, they only live swimming on the surface of the deep sea. . . . Their body consists of a hard and a soft part. The hard part is a siliceous skeleton, the soft is mostly a spherical, small, round capsule surrounded on all sides by an outcrop of many hundreds of exceptionally fine filaments, by which the animal moves and nourishes itself.[23]

Under his microscope, completely new radiolarian species began to appear, so that by the spring he was able to ship back to Berlin specimens of some 101 species never before described.[24] (With the dredging expedition of the *Challenger*, which traveled around the world in the 1870s, Haeckel added several thousand more radiolarian species to his catalog; see the appendix to this chapter.)

Shortly after returning to Berlin, at the end of April 1860, Haeckel arranged to work on his collection at the Berlin Zoological Museum, where he had earlier cultivated a circle of friends and patrons, including the director Wilhelm Peters (1815–1883) and the eminent Christian Ehrenberg (1795–1876), presiding secretary of the Berlin Academy of Sciences. Initially Haeckel prepared a report on his radiolarian work, which Peters presented to the academy.[25] The report carefully described the new species he had discovered and analyzed their internal structure, something never before done. His descriptions remain today the starting point for further explorations with the scanning electron microscope. (One might compare Haeckel's figures with recent micrographs, lest one assume that his imagination had fabricated the gothic structures he depicted: see his Eucyrtidium illustrations, plate 1, and micrographs of the same group, fig. 3.3.)

Haeckel determined the radiolaria to have a soft body consisting of a central capsule, with a minute inner vesicle (*Binnenblase*), and surrounded

23. Haeckel to Anna Sethe (29 February 1860), in ibid., 161–62.

24. Haeckel to Allmers (14 May 1860), in *Haeckel und Allmers*, 45.

25. Ernst Haeckel, "Über neue, lebende Radiolarien des Mittelmeeres," *Monatsberichte der Königlichen Preussische Akademie der Wissenschaften zu Berlin* (1860): 794–817, 835–45.

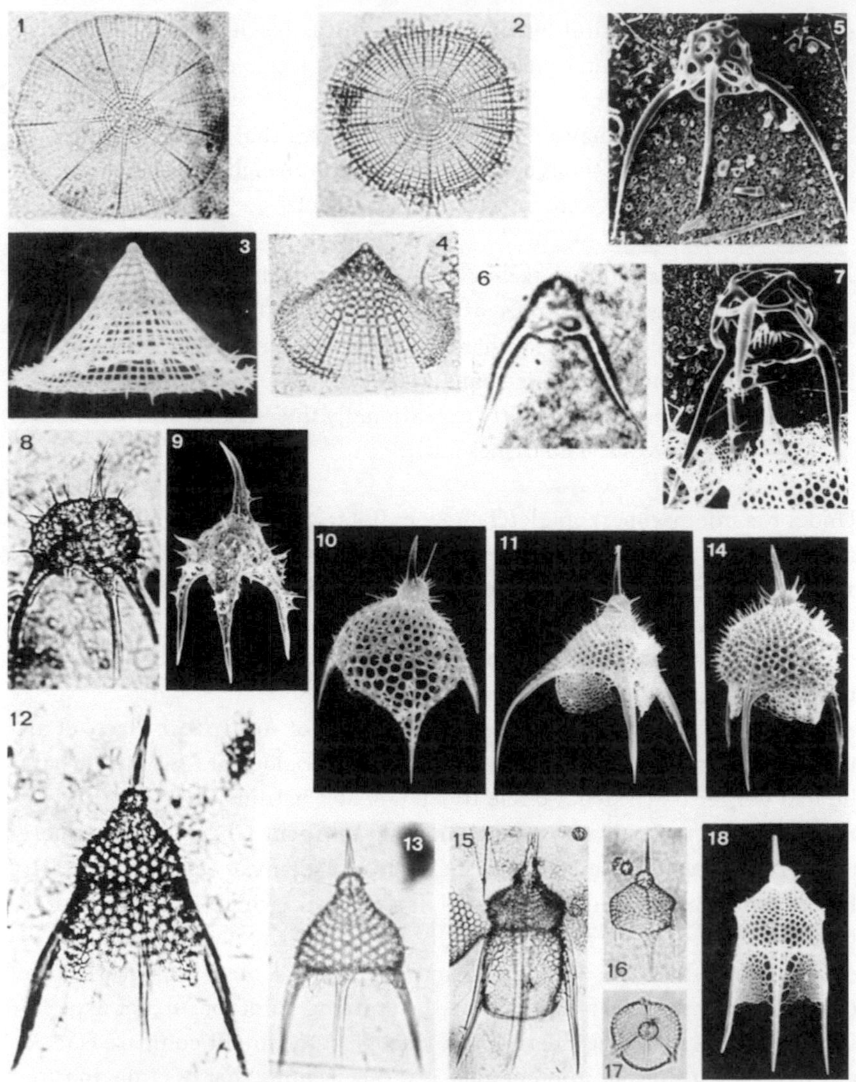

Fig. 3.3. Micrographs of the subfamilies Plectopyramidinae and Eucyrtidiinae. (From Kozo Takahashi and Susumu Honjo, *Radiolaria: Flux, Ecology, and Taxonomy in the Pacific and Atlantic*, 1991.)

by smaller vesicles (*Bläschen*), through which radiated a great number of stiff, threadlike pseudopodia.[26] Depending on the family, the skeleton either surrounded the central capsule (as with the solitary Polycystinae) or extended into the capsule (as with the Acanthometra and the colonial Polycystinae).[27] All of this was reiterated, with an elaboration of the systematics of the known species, in Haeckel's *Habilitationsschrift,* rendered into Latin and completed in 1861.[28]

Yet neither the readers of the academy report nor of the *Habilitationsschrift* would have been prepared for the large two-volume monograph Haeckel produced in 1862, his *Die Radiolarien (Rhizopoda Radiaria).* The first two exercises announced a scholar of competence and promise; the latter showed the promise already brilliantly fulfilled. The monograph—which so astonished Darwin and which would be awarded the prestigious Cothenius gold medal of the Leopold-Caroline Academy of German Natural Scientists (1863)—displayed through its over 570 pages of the first volume and the 35 copper plates of the second many extraordinary features. I will mention just a few of the more significant.

First of all, with his discoveries Haeckel increased by almost half the number of known species of radiolaria. Second, he provided the most careful description of the distinguishing characteristics of the skeletons and soft parts, including extraordinarily exact measurements. He employed, though, some rough models in this effort: he would stud a potato with rods to get the perspective correct, and then allow his painterly eye to take over.[29] The technique yielded not only amazingly precise but beautiful depictions. His discrimination of the central capsule and the associated smaller vesicles, as mentioned above, set the foundation for later anatomical research.[30] Third, in anticipation of the kind of chorological considerations he would develop in later work, he specified the various seas in which

26. These inner vesicles that Haeckel described are probably symbiotic organisms. See Breidbach's introduction, "Die allerreizendsten Tierchen: Haeckels Radiolarien-Atlas von 1862," to his beautiful reproduction of the atlas of Haeckel's *Die Radiolarien:* Ernst Haeckel, *Kunstformen aus dem Meer,* with an introduction by Olaf Breidbach (Munich: Prestel, 2005), 9.

27. Haeckel, "Über neue, lebende Radiolarien des Mittelmeeres," 795–97.

28. Ernst Haeckel, *De Rhizopodum finibus et ordinibus* (Berlin: Georg Reimer, 1861). The dissertation reappeared with few alterations as part 4 of his large monograph on the radiolarians. See Ernst Haeckel, *Die Radiolarien (Rhizopoda Radiaria). Eine Monographie,* 2 vols. (Berlin: Georg Reimer, 1862), 1:194–212.

29. Haeckel to Anna Sethe (14 August 1860), in Ernst Haeckel, *Himmelhoch Jauchzend: Erinnerungen und Briefe der Liebe,* ed. Heinrich Schmidt (Dresden: Carl Reissner, 1927), 133.

30. Haeckel, *Radiolarien,* 1:68–116.

a given species lived and the depths at which it could be found.[31] Fourth, and of considerable significance, he attempted to arrange his species into a "natural system" based on homology.[32] The two principal comparative axes for homological arrangement concerned the relation of the skeleton to the central capsule (either completely external to it or partly inside it) and the forms of the skeleton itself (or its absence). On this basis Haeckel distinguished, as they fell into patterns, some fifteen natural families.

Haeckel said he was inspired to attempt a natural system because of the extraordinary book he had read while preparing his specimens—*Über die Entstehung der Arten im Thier- und Pflanzen-Reich durch natürliche Züchtung; oder, Erhaltung der vervollkommneten Rassen in Kampfe um's Daseyn* by the English naturalist Charles Darwin.[33] Haeckel first looked into Heinrich Georg Bronn's (1800–1862) German translation of Darwin's *Origin of Species* while at the Berlin Museum in the summer of 1860, just after he had returned from Messina. Being an anti-authoritarian—in his later days to the point of dogmatism—Haeckel was probably enticed to read the new work because Ehrenberg and Peters both regarded it as a "completely mad book."[34] Though anti-authoritarian, Haeckel was not foolish; so it is not surprising that no mention of Darwin appeared in his academy report in the fall or in his *Habilitationsschrift*. It may be, however, that the full impact of the *Origin* had not struck home during the composition of those pieces. In November 1861, while laboring full bore on his monograph, he again opened up the *Origin* and, as he related to Anna, "buried" himself in it.[35] From that fertile womb he emerged newly born for Darwin's theory, and the zeal of his conviction never cooled through the later days.

What kept Haeckel's enthusiasm for evolutionary theory glowing was the special contribution he thought he could make, namely, to establish it empirically. He seems to have been especially provoked in this respect by

31. Ibid., 1:166–93.

32. Ibid., 1:213–40.

33. Charles Darwin, *Über die Entstehung der Arten im Thier- und Pflanzen-Reich durch natürliche Züchtung; oder, Erhaltung der vervollkommneten Rassen in Kampfe um's Daseyn* (based on 2nd English ed.), trans. Heinrich Georg Bronn (Stuttgart: Schweizerbart'sche, 1860).

34. Ilse Jahn, "Ernst Haeckel und die Berliner Zoologen," *Acta Historica Leopoldina* 16 (1985): 75.

35. Haeckel to Anna Sethe (4 November 1861), in *Himmelhoch Jauchzend*, 250. Haeckel's copy of Bronn's translation of Darwin's *Origin of Species* bears reading marks throughout. The copy is kept at Haeckel-Haus, Jena. Haeckel also kept a notebook as a kind of index of concepts of the *Origin*. Mario Di Gregorio gives an account of these jottings in his *From Here to Eternity: Ernst Haeckel and Scientific Faith* (Göttingen: Vandenhoeck & Ruprecht, 2005), 77–85.

Darwin's translator Bronn, a scientist with extensive knowledge of paleontology and morphology. In his rendering of the *Origin,* Bronn had added an epilogue in which he evaluated the merits of Darwin's theory. He voiced objections of varying weights, but his principal demur rested on the notion of variability of form: Darwin had postulated that multiple and very small changes in a variety constituted the initial stage of a new species form; however, the many small alterations, all of which had to produce an integrated structure, would be changing independently and at random, producing not a coherent type but only a confusion having no advantage over competitors.[36] His objections notwithstanding, Bronn had high praise for the ingenuity and stimulating character of Darwin's work. He also had some sympathy for the theory, which resembled the proto-evolutionary scheme he himself had advanced two years before Darwin published. According to Bronn's theory, however, integration of progressive traits into new species forms was shaped by the divine hand (see appendix 1). Despite his sympathy and admiration, he must have also sensed some real dangers in Darwin's doctrine, since he failed to translate the sentence that the Englishman had carefully dropped into the third-to-last paragraph of the book: "Light will be thrown on the origin of man and his history."[37] The danger merely threatened, however, since Darwin's hypothesis remained just that, a hypothesis, only a possible scenario of life's history. Bronn thus declared:

> We have therefore neither a positive demonstration of descent nor—from the fact that [after hundreds of generations] a variety can no longer be connected with its ancestral form [*Stamm-Form*]—do we have a negative demonstration that this species did not arise from that one. What might be the possibility of unlimited change is now and for a long time will remain an undemonstrated and, indeed, an uncontradicted hypothesis.[38]

Haeckel was not much troubled by Bronn's objections and, more importantly, he believed he could provide the required positive proof of descent. Through the next decade and a half, he cultivated the kind of evidence that

36. H. G. Bronn, "Schlusswort des Übersetzers," in Charles Darwin, *Über die Entstehung der Arten,* 503–4.

37. Charles Darwin, *On the Origin of Species* (London: Murray, 1859), 488.

38. Bronn, "Schlusswort des Übersetzers," 502.

he thought would empirically demonstrate Darwin's original conception, as well as lead to further important theoretical articulations. His appetite for this endeavor was first sharpened by his radiolarian work.

In *Die Radiolarien,* Haeckel boldly sided with the English scientist. He argued that the radiolaria provided the desired empirical support for the new theory of evolution, since the relatedness of species within families bespoke genealogy and the transitional species joining families seemed to confirm it.[39] In this light, Haeckel constructed a genealogical table that indicated, in part, the kind of descent relations these animals might actually express.[40] I say "in part," because the table had not abandoned principles employed in the older morphological tradition: it assumed, for example, the primitive form to be a sphere and derived subsidiary forms through a geometrical deformation and arrangement of the original type—a bit like Ptolemy reconstructing erratic planetary motion from the rotation of spheres (discussed further below). The Darwinian conception had not yet matured in Haeckel's thought. But the roots had found favorable soil.

Haeckel's adoption of Darwin's theory was facilitated by three features of his intellectual situation. First, the fact of several intermediate species forms between the major groups of radiolaria begged for an evolutionary interpretation. Second, Haeckel's still-revered teacher, Rudolf Virchow, had, in 1858, declared that the mechanistic view of life, which he believed the only scientific outlook, required the postulation of species transmutation.[41] Finally the morphological tradition in which Haeckel was schooled,

39. Haeckel, *Radiolarien,* 1:231–33.

40. Ibid., 1:234.The table depicted the Ur-organism as of the *Heliosphaera* type—namely, a radiolarian with a spherical form and symmetrically extended pseudopodia. Deformations of this original type—which Haeckel represents by Aulosphaera, Ethmosphaera, etc.—might then account, by reason of descent, for the families, subfamilies, and genera of the radiolaria.

41. Virchow had tentatively advanced the transmutation hypothesis in a lecture at the thirty-fourth meeting of the Society of German Natural Scientists and Physicians at Karlsruhe in September 1858. The lecture, entitled "Über die mechanische Auffassung des Lebens," was reprinted in Rudolf Virchow, *Vier Reden über Leben und Kranksein* (Berlin: Georg Reimer, 1862). The lecture in this latter printing sparked the ire of Matthias Schleiden, as I will discuss in the next chapter. The relevant passage concerning transmutation reads: "Our experience justifies us in not holding as an inviolable rule good for all time that species are unchangeable, which at present seems so certainly to be the case. For geology teaches us to recognize a certain progression in which one species follows upon another, the higher succeeding the lower; and though the experience of our time opposes this, I must recognize, it seems to me, as a requirement of science that we return again to the transmutability of species. The mechanistic theory of life will thus obtain real security by taking this path" (31). During his later years, Virchow felt less ready to endorse transmutationism as a viable hypothesis; and in his confrontation with his onetime student in 1877, he all but rejected the possibility of evolutionary transmutation. I will discuss the confrontation in chapter 8.

with its emphasis on homology, could easily be turned to evolutionary account, as Goethe had in the 1820s and Schleiden in the 1840s.[42]

Aside from the intellectually coercive evidence and the conceptual preparation that Haeckel had undergone, other more subjective, personal reasons may have inclined him to cast his lot with the new theory. In a long footnote to the section that displayed his empirical evidence for Darwin's theory, he referred to a clarion passage at the conclusion of the *Origin* in which the English scientist issued a call to all the up-and-coming young naturalists to judge his ideas without prejudice. The note indicates that one zealous young iconoclast heard the resounding message:

> I cannot let this opportunity pass without giving expression to the considerable astonishment I felt over Darwin's exciting theory about the origin of species. I am moved to do this even more because the German professionals have found this epoch-making work to be an unhappy presumption; they make this charge partly because they seem to misunderstand his theory completely. Darwin himself wished that his theory might be tested from every possible side and he looked "with confidence toward the young and striving naturalists who will be able to judge both sides of the question without partiality. Whoever is inclined to view species as changeable will, through the conscientious admission of his conviction, do a good service to science; only thereby can the mountain of prejudice under which this subject is buried be generally avoided." I share this view completely and believe for this reason that I must express my conviction that species are changeable and that organisms are really related genealogically. Though I have some reservations about extending Darwin's insight and hypothesis in every direction and about all his attempts to demonstrate his theory, yet I must admire in his work the first, earnest and scientific effort to explain all appearances of organic nature from one excellent, unitary viewpoint and his attempt to bring all sorts of inconceivable wonders under a conceivable law of nature. Perhaps there is in Darwin's theory, as the first effort of this sort, more error than truth. . . . The greatest confusion of the Darwinian theory lies probably herein, that it does not rest upon the origin of the Ur-organism—most probably a simple cell—whence all others have been developed. When Darwin assumes for this first species a special creative act, it seems of little consequence, and it

42. For a discussion of Goethe's evolutionary theory, see my *Romantic Conception of Life,* 476–86.

> seems to me not seriously meant. Aside from this and other confusions, Darwin's theory already has performed the immortal service of having brought the entire doctrine of relationships of organisms to sense and understanding. When one considers how every great reform, every strong advance has found a mighty opposition, the more he will oppose without caution the rooted prejudice and battle against the ruling dogma; so one will, indeed, not wonder that Darwin's ingenious theory has, instead of well-deserved recognition and test, found only attack and rebuff.[43]

Haeckel's support for Darwin's theory and his desire thereby to be accounted among the Darwinians would be reciprocated in the Englishman's own declaration, some years later, that most of his ideas about human evolution (in *The Descent of Man*) had been antecedently confirmed by Haeckel—so in this, and other respects, Darwin could be accounted a Haeckelian.[44]

In developing his argument for the reality of genealogical transformation, however, Haeckel invoked two apparently conflicting principles that led to different representations of the natural system that he claimed his research uncovered. One principle—which allowed him to organize his specimens into families, genera, and species—was that of progressive skeletization. The other principle stemmed from a Goethean aestheticized morphology.[45]

In the ordering of his thirty-five copper plates for the second volume of his monograph, Haeckel began with a species of Thalassicolla, which lacks a skeleton; plate 2 displays two aspects of *Aulacantha scolymantha,* which has some spicula, and *Thalassicolla zanclea* and *Thalassolampe margarodes,* both of which lack any hard parts; and plate 4 illustrates again *Aulacantha scolymantha,* along with forms (from different families and underfamilies) that have surrounding skeletons—namely, *Acanthodesmia prismatium,*

43. Haeckel, *Radiolarien,* 1:231–32n1.

44. In the preface to his *Descent of Man and Selection in Relation to Sex,* 2 vols. (London: Murray, 1871), 1:4, Darwin said of Haeckel's *Natürliche Schöpfungsgeschichte* (1868) that had it "appeared before my essay had been written, I should probably never have completed it. Almost all the conclusions at which I have arrived I find confirmed by this naturalist, whose knowledge on many points is much fuller than mine." Despite this avowal, to call Darwin a Haeckelian, as I have, will seem outrageous to some historians. I've tried to substantiate the sense in which this might be true in my *Meaning of Evolution: The Morphological Construction and Ideological Reconstruction of Darwin's Theory* (Chicago: University of Chicago Press, 1992), as well as in chapter 5, below.

45. See the first appendix for a description of Goethe's morphological ideas.

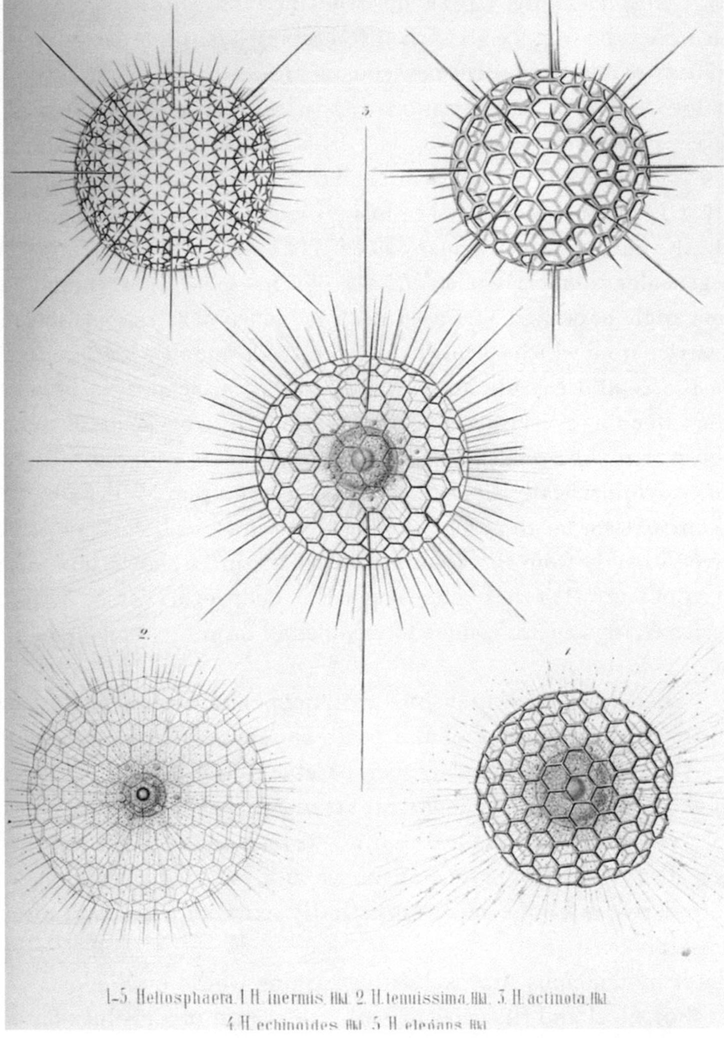

1-5. Heliosphaera. 1. H. inermis. Hkl. 2. H. tenuissima. Hkl. 3. H. actinota. Hkl.
4. H. echinoides. Hkl. 5. H. elegans. Hkl.

Fig. 3.4. The genus *Heliosphaera*. (From Ernst Haeckel, *Die Radiolarien*, 1862.)

Litharachnium tentorium, and *Eucyrtidium lagena*. All of this seemed quite consonant with a Darwinian paradigm of evolutionary progress. Yet Haeckel proposed another principle by which to understand the relationships among the morphological types. This principle harkens back to the older morphological tradition that I mentioned. He suggested that the Ur-type of the phylum, the one that might have given rise to others, was comparable to *Heliosphaera* (fig. 3.4). He hypothesized that this kind of organism was the archetype whence all of the fifteen families might be derived. Immediately after mentioning the new considerations that Darwin introduced into zoology and his own declaration of apostleship, Haeckel wrote:

> A continuous red thread passes through the entire series of these forms, so that I am already prepared to make the effort to represent graphically the connections and many-sided relations of all these forms in one genealogical table of relatedness. From this table all other possible forms might be derived. I see such an Ur-radiolarium as a simple spherical lattice from which spicula radially protrude and in whose internal area the central capsule floats. The lattice is suspended on pseudopodia extending everywhere. We actually find this Ur-type in the genus *Heliosphaera*. As a model, we can take *Heliosphaera actinota*, with its twenty symmetrically separated spicula (according to Müller's law).[46] Of course, I am far from maintaining that all the radiolaria must be derived directly from this form, only that it can be shown how, as a matter of fact, all of these extensively developed forms can be derived [*abgeleitet*] from such a common fundamental form.[47]

The idea that descent relationships might operate according to various mathematical deformations of the basic sphere was quite in the older Goethean tradition of morphology, comparable to Carus's derivation of the form of the vertebra from geometrical arrangements of the basic sphere.[48] Haeckel even suggested, as Goethe himself had, that once the archetype had been discovered through comparative analysis, the naturalist would be able to derive not only the forms actually existing, but also those that could possibly exist.

In later monographs, Haeckel's illustrations would more closely unite the morphological and the genealogical orders into one evolutionary tableau of systematic arrangement. In the *Challenger* volumes (see appendix

46. Müller's law was the principle formulated by Johannes Müller, in his "Über die Thalassicollen," 12. The law was quasi-empirical and quasi-mathematical. It stated that in the family of Acanthometriden, as well as in the genera *Haliomma, Actinomma, Heliosphaera,* and others, the largest number of spicula displayed would be twenty and that they would be symmetrically arranged around five equally spaced girdles or lines of latitude, with four evenly spaced spicula along each line.

47. Haeckel, *Radiolarien,* 1:233.

48. See the first appendix for a discussion of Carus's geometrical morphology.

to this chapter), he concluded that the Ur-type, the archetype and original organism from which all the radiolaria descended, was a spherical form with radiating pseudopodia but without a skeleton.[49]

The archetypal structures that Haeckel detected as the basic forms of different animal groups, those original forms of the progenitor organisms, could be comprehended only by the mind's eye—as Goethe had claimed earlier. But the essence of such forms could yet be rendered by the artistic hand. And this is why, for Haeckel and other biologists of the nineteenth century—and even today—artistic sensibility reveals what mechanical reproductions, like photographs, can only obscure. The dramatic and exotic beauty of Haeckel's illustrations would in future play decided roles in persuading his readers of the evolutionary theory that would stand ever more forcefully behind his art.

Appendix: Haeckel's *Challenger* Investigations

In the early 1870s, Charles Wyville Thomson (1830–1882), a naturalist at the University of Edinburgh, discussed with members of the Royal Society the possibility of an expedition to sound the oceans of the world to discover the chemical composition, temperatures, and depths of their various waters, as well as the character of their marine life. After some negotiations with the Royal Navy, a fighting ship, HMS *Challenger,* had most of its guns removed and replaced with dredging and other equipment needed to carry out the plan (fig.3.5). In December 1872 the three-masted ship, under Captain George Nares (1831–1915) and a crew of two hundred men, embarked on a three-and-a-half-year voyage of research. Thomson directed a team of some six scientists. The team included John Murray (1841–1914), who reported, for the first time, on the plateaus and deep ocean trenches

49. Gould launches a rather obtuse and tendentious complaint about Haeckel's radiolaria work. He suggests that Haeckel intentionally "improved" and "enhanced" his depictions to make them more symmetrical. See Stephen Jay Gould, "*Abscheulich!* (Atrocious!): Haeckel's Distortions Did Not Help Darwin," *Natural History* 109, no. 2 (2000): 42–49; quotations from 43. First, Haeckel did not portray all of his radiolaria as symmetrical, though some of them he did. But he was working at just the limits of microscopical acuity in 1861 and 1862. Undoubtedly he was swayed by the Goethean tradition. But as a matter of indisputable fact, many of the radiolarian species he depicted are indeed astoundingly symmetrical. See, for instance fig. 3.3, scanning electron micrographs by Kozo Takahashi. See his and Susumu Honjo's *Radiolaria: Flux, Ecology, and Taxonomy in the Pacific and Atlantic* (Woods Hole, MA: Woods Hole Oceanographic Institution, 1991). The exact forces creating the symmetrical forms of many of the radiolarian species remain unknown.

Fig. 3.5. HMS *Challenger.* (From C. Wyville Thomson, *Report on the Scientific Results of the Voyage of the Challenger,* 1878–95.)

of the Atlantic and Pacific—useful knowledge for the laying of telegraph cables in different regions of the world.

The ship traveled to the Canary Islands, down the mid-Atlantic, across to Brazil, and south along the coast, but then veered east to Africa, around the Cape of Good Hope, turned southeast toward the Antarctic, moved up toward the southern coast of Australia, across to New Zealand, cut north toward Fiji, then back west to the East Indies, sailed farther north to Japan, across the northern Pacific to mid-ocean, headed south through the Sandwich and Society islands, then east to the coast of Chile and down around the Horn, up the coast of Argentina, and finally sped northeast up the mid-Atlantic and back to England, returning in May 1876. Chemists, physicists, and marine biologists of international repute were contacted; those who accepted the offer were charged with the task of analytically describing the composition of the seas, the seabeds, and the various sorts of animals secured. During the late 1870s and early 1880s, the materials were distributed to the respective experts and they began their analyses.

Then from the early 1880s into the late 1890s, the *Challenger* Commission, under the direction of Thomson and then Murray, issued over fifty, very thick folio volumes of reports, thirty-two of which were devoted to zoology. Haeckel—because of his reputation and the systematic work he had done on radiolaria, medusae, siphonophores, and sponges—was asked to work on these creatures.

Haeckel's *Report on Radiolaria* took him the better part of a decade to finish. It described over four thousand species of radiolaria in 1,803 pages, which constituted two large folio volumes; a third volume of 140 plates completed the study.[50] As these numbers might suggest, Haeckel's analyses formed one of the largest single biological contributions to the *Challenger* research. For each species, he described the systematic relations, the morphology, the locality where taken (the latitude, longitude, and the nearest land), the abundance of the creatures, the depth and temperature of the waters, and the nature of the sea bottom.

In his *Report,* Haeckel formed the classificational system of radiolaria that is still generally in use. He remarked that though any effort at a natural system must rely on paleontology, comparative ontogeny, and comparative anatomy, in the case of radiolaria the first two remained obscure. His system, therefore, would have to be "a compromise between the natural and artificial systems."[51] So as much for convenience as evolutionary assumption, he divided the phylum of radiolaria into four classes: Spumellaria, Nassellaria, Acantharia, and Phaeodaria. He regarded the classes as natural stems for the somewhat more dubiously divided eight legions (or subclasses) and twenty orders.[52] He now slightly altered his views regarding the evolutionary progenitor of the entire phylum—or at least he thought the question open. He speculated that the simplest form might be the best candidate, namely, that of the Thalassicollidae.[53] This was a

50. Ernst Haeckel, *Report on Radiolaria,* vol. 18, parts 1 and 2 of *Report on the Scientific Results of the Voyage of H.M.S. Challenger during the Years 1873–1876,* prepared under the superintendence of the late Sir C. Wyville Thomson (London: Her Majesty's Stationery Office, 1887). See also Ernst Haeckel, *Report on the Deep-Sea Medusae dredged by H.M.S. Challenger,* vol. 14 of *Report on the Scientific Results of the Voyage of H.M.S. Challenger* (London: Longmans & Co., 1882); *Report on the Siphonorae collected by H.M.S. Challenger during the Years 1873–76,* vol. 28 of *Report on the Scientific Results of the Voyage of H.M.S. Challenger* (1888); and *Report on the Deep-Sea Keratosa collected by H.M.S. Challenger during the Years 1873–1876,* vol. 32 of *Report on the Scientific Results of the Voyage of H.M.S. Challenger* (1889).

51. Haeckel, *Report on Radiolaria,* part 1, p. ci.

52. These, roughly, are still used today, except for the class Acantharia, which is no longer regarded as a group of true radiolaria. See Anderson, *Radiolaria,* 7–21.

53. Haeckel, *Report on Radiolaria,* part 1, p. cv.

spherical form without an external skeleton, a form yet in keeping with the older morphological tradition.

Anyone slightly acquainted with Haeckel's *Challenger* volumes, or any of his other monographs on marine organisms, could not possibly entertain the idea that his evolutionary theories lacked extensive empirical foundation. Until quite recently, as O. Roger Anderson notes, Haeckel's work has "remained the major source of information on radiolarian diversity and taxonomy."[54]

54. Anderson, *Radiolaria,* 17. Haeckel's first major monograph, *Die Radiolarien,* served not only as the basis for his *Challenger* study, but it also led to further histological analyses conducted by his assistant Richard Hertwig, recorded in a monograph dedicated to his teacher. This is simply another testimony to the scientific fecundity of Haeckel's work. See Richard Hertwig, *Zur Histologie der Radiolarien* (Leipzig: Wilhelm Engelmann, 1876).

CHAPTER FOUR

Triumph and Tragedy at Jena

After his return from Italy in the spring of 1860, Haeckel faced conflicting pressures of varying intensities.[1] He ran into Virchow, who thought the Italian sojourn had altered his former student into a more identifiably German type, which caused Haeckel to muse whether his esteemed teacher referred merely to his blue eyes and sun-streaked blond hair or to some other, less definite quality. He was never quite sure of what Virchow meant. The weight of self-doubt increased when he read an obituary of Johannes Müller written by another former student of the master, Emil Du Bois-Reymond (1818–1896).[2] How could he even come close to the accomplishments of this genius, his revered teacher? Yet Haeckel felt he did have a real research project in the radiolarians; and the passion for publication, surging now to flood level and never easing even in his old age, propelled him forward on his book. Finally, against his need for uninterrupted time for writing, he had to weigh Gegenbaur's urgent invitation to serve as an assistant at Jena.[3] He had no choice. Jena, that warmhearted and energetic pulse of Romantic élan, could not be refused, especially since its embrace might also bring him into the arms of that other love, Anna. With demonlike energy and severe concentration, he set out to complete as much as he could on his monograph before teaching responsibilities might hobble his progress. In

1. He related these concerns to Anna. See his letters to Anna Sethe (1 June, 10 June, and 1 July 1860), in Ernst Haeckel, *Himmelhoch Jauchzend: Erinnerungen und Briefe der Liebe,* ed. Heinrich Schmidt (Dresden: Carl Reissner, 1927), 94, 96, 110.

2. Emil Du Bois-Reymond, "Gedächtnisrede auf Johannes Müller," *Abhandlungen der Königlichen Akademie der Wissenschaften zu Berlin* (1860): 25–190.

3. Carl Gegenbaur to Haeckel (13 August 1860), in the Correspondence of Ernst Haeckel, in the Haeckel Papers, Institut für Geschichte der Medizin, Naturwissenschaft und Technik, Ernst-Haeckel-Haus, Friedrich-Schiller-Universität, Jena.

August he decided to leave Berlin, where his friends provided only diverting enticements, and to move in with his brother, Karl, in Freidenwald (about twenty-five miles northwest of the city). Though isolated from colleagues, he did not deprive himself of what was becoming for him the very energy of life. Anna came to see him almost immediately, and she lingered long and palpably in his thoughts, as his letter just after her departure suggests: "I accompany this morning greeting with a kiss. That's what happens when one becomes so accustomed to kisses."[4] In the winter he left for Jena.

Habilitation and Teaching

Haeckel and Jena were made for each other. That birthplace of German Romanticism still echoed the symphilosophizing orchestrated by the Schlegels and Novalis, the idealistic reconstructions of the self executed by Fichte and Schelling, the classes on history tinged with poetry conducted by Schiller, and the analytically exact but synthetically wild lectures on biology given by Oken. The spirit of Goethe hovered over all, invading even the student *Kerker*—the jail room reserved for miscreant adolescents—which still held, as it does today, a charcoal likeness of the aged poet drawn by a talented inmate (who also sketched some of the famous professors of the 1820s strolling along, arm in arm, with well-known whores of the town).[5] Haeckel wrote Allmers in the spring of 1860: "You know me and you know Jena and you know, therefore, how I have been created for Jena."[6]

Haeckel's career at Jena, however, could not really begin until he had officially habilitated and had received an endorsement as *Privatdozent* from the medical faculty, where Gegenbaur, his supervisor, held an appointment. He glided through these tasks, since Gegenbaur had smoothed the way.[7] The report Haeckel had prepared for the Berlin Academy of Sciences,

4. Haeckel to Anna Sethe (9 August 1860), in *Himmelhoch Jauchzend*, 131.

5. The student *Kerker* is now connected with the modern medical school—a union of seventeenth-century Gothic with 1950s East German Stalinist style. In 1988 a professor of the medical school proudly showed me the room and the likenesses of Goethe and the roughly treated professors, observing wistfully: "We have the history, though you in the West have the science."

6. Haeckel to Hermann Allmers (1 June 1861), in *Haeckel und Allmers: Die Geschichte einer Freundschaft in Briefen der Freunde*, ed. Rudolph Koop (Bremen: Arthur Geist, 1941), 71.

7. Gegenbaur offered to the authorities a "not official report" on the course of Haeckel's activities. He urged his young protégé to work quickly on his habilitation. He indicated that not only he but Kuno Fischer, rector of the university, was "quite enthusiastic for the appointment." Gegenbaur to Haeckel (13 August and 28 December 1860), in the Haeckel

Fig. 4.1. The main university building in Jena (1905–8), now Friedrich-Schiller-Universität. (Photo by the author.)

presented the previous fall, served as the foundation for his *Habilitationsschrift,* the composition of which followed quickly enough. The venerable traditions of the university required that he render his tract into Latin and then defend it in a public *disputatio.* The intellectual fete occurred on 4 March 1861. He read to the assembly a précis of *De Rhizopodum finibus et ordinibus* (On the boundaries and systematics of Rhizopods) and then defended it against the prescribed two opponents.[8] The audience for this ancient academic ceremony consisted of the dean of the faculties and Haeckel's early idol Matthias Schleiden, the two opponents, the rector's representative, one friend, and a curious student who happened to wander in. The "swindle," as Haeckel called it, took fourteen minutes, exhausting the oral Latin of all concerned. The next day he had a more serious ordeal, a lecture to the medical faculty—"On the Vascular System of the Invertebrates." Gegenbaur advised that it should "not have anything too specialized" (*nichts gar zu sehr specielles*) and that he should keep it short.[9]

Correspondence, Haeckel-Haus, Jena. Gegenbaur managed Haeckel's early career with extraordinary care and in great detail.

8. See the end of the previous chapter for a discussion of this material.

9. Gegenbaur to Haeckel (2 February 1861), in the Haeckel Correspondence, Haeckel-Haus, Jena.

For the honor of the performance, he had to contribute five *Reichstaler* to the faculty coffers. Having performed these exercises, Haeckel habilitated and was given license to hold classes as *Privatdozent*, a kind of freelance teacher whom medical students would pay directly for instruction.[10] His life at the university had begun.

That life quickly distilled into two unremitting tasks: preparing courses to stay just ahead of his students and then reserving bits and pieces of the remaining hours for work on his book—a schedule quite familiar yet today to academics training to vault into tenure. Gegenbaur taught anatomy to the medical students and shared this task with his new assistant, but the assistant himself still had a lot to learn. So on Monday, Tuesday, Thursday, and Friday, Haeckel would arise at 4:30 a.m., take some coffee at 5:00 a.m., and from 5:30 a.m. until 1:30 p.m., he would prepare his lecture for the day. At two o'clock he would lunch with Gegenbaur and others at the local inn, the Schwarzer Bär (the nineteenth-century charm of which has recently been restored, with the addition of early twenty-first-century conveniences). The two friends would talk until midafternoon, when Haeckel would slip off to the newspaper room and read for a while. He might then go to the Zoological Museum to work for an hour arranging his notes. From 5:00 to 6:00 p.m., he would expel his morning's preparation in the appointed lecture. Thereafter he would cool down, walking with Gegenbaur or returning to the museum till about eight o'clock, at which time he would head home for a light meal and bed by ten.

Haeckel worked hard as a teacher, both because he loved the art and because, as he became increasingly aware, his courses forced him to extend and deepen his knowledge, thus aiding him in research.[11] During this first term on the other side of the desk, he had only eight students, of which five were paying and the other three, since they were too poor, attended gratis. Haeckel's effectiveness in the classroom can be measured by the steady increase in audience over the next several terms. Indeed, within a few years, when he began offering public lectures on Darwin's theory—graphically portrayed with large, dramatic illustrations by his own hand—he attracted not only significant numbers of students but faculty from various departments as well as townspeople. In the lecture hall, Haeckel became a pres-

10. Haeckel described the ritual defense of his thesis to Anna Sethe (4 March 1861), in *Himmelhoch Jauchzend*, 152–53, and to Allmers (9 March 1861), in *Haeckel und Allmers*, 63.

11. Haeckel made this observation in a letter to his father (1 August 1862), in *Himmelhoch Jauchzend*, 226.

ence. One of his former students, Max Fürbringer (1846–1920), the Heidelberg comparative zoologist, recalled his experience in 1866:

> And then he entered the auditorium. He proceeded to the podium, not with the thoughtful steps of the professor, but with the victorious rush of an Apollonian youth. He was a tall, slender, imposing figure, having a countenance that bespoke much thought and work, but not a hint of weakness, with great golden locks flowing from his large head, which itself evinced a great brain; and he had tremendous, blue eyes, blazing yet friendly—he was probably the most handsome man that I had ever seen up to that time, and he seemed to make that bright room even brighter. And then the lecture began, not in polished and well-wrought terms, but in a great gush, a showering illumination of new revelations. The phenomenon, the brilliance of thought, and the special form of his lectures first struck me, and then only later did the content hit me.[12]

Haeckel did not spend his early years at Jena completely consumed by teaching and research. He reserved time to take long hikes into the countryside around the city. These excursions into nature, which he often shared with Gegenbaur, supplied an elixir for his soul. But Haeckel flexed more than his contemplative muscles. Through the encouragement of a new acquaintance in the arts faculty, the linguist August Schleicher (1821–1868), he took up serious gymnastics. Schleicher was the leader of the *Turnverein*, the gymnastics group that organized exercises in running, jumping, bar work, and swimming. Haeckel practiced with the group twice a week and privately once a week with another friend in the medical school.[13] During the early 1860s, he became quite devoted to this exercise program. He wrote Allmers in early 1862 that when they next met he hoped his friend would be pleased and surprised at his "muscular development."[14] In the late summer of 1863, he attended a gymnastics festival at Leipzig, where he won a

12. Max Fürbringer, "Wie Ich Ernst Haeckel Kennen Lernte und mit Ihm Verkehrte und wie Er mein Führer in den grössten Stunden meines Lebens Wurde," in *Was Wir Ernst Haeckel Verdanken: Ein Buch der Verehrung und Dankbarkeit*, ed. Heinrich Schmidt, 2 vols. (Leipzig: Unesma, 1914), 2:336–37.

13. Haeckel described his gymnastics activities to Anna Sethe (1–15 June 1861), in *Himmelhoch Jauchzend*, 190.

14. Haeckel to Allmers (28 January 1862), in *Ernst Haeckel: Sein Leben, Denken und Wirken. Eine Schriftenfolge für seine zahlreichen Freunde und Anhänger*, ed. Victor Franz, 2 vols. (Jena: Wilhelm Gronau und W. Agricola, 1943–44), 2:29.

prize for a six-meter jump, something that made Anna quite proud of her "blond-headed boy."[15] At that event, Schleicher was not so lucky and came away with an injured arm.[16] Undoubtedly this Greek regimen, along with his constant travels, contributed to Haeckel's robust good health throughout his long life. His training had other consequences as well. He became good friends with Schleicher, who would suggest to him a conception that developed into the philosophical backbone of his evolutionary morphology, namely, monism. His friend also inspired him to represent genealogical relations via the treelike diagrams that came to dominate his expression and understanding of evolutionary descent. Additionally, Schleicher's ideas about language acquisition would play a crucial role in Haeckel's theory of human evolution. (I will return to these topics in chapters 5 and 7.)

Friendship with Gegenbaur

Haeckel's closest intellectual confidant in these early years was Carl Gegenbaur. Their friendship, which only faltered during the waning moments of the century, found its roots in a remarkably similar history.[17] Gegenbaur descended, like Haeckel, from a long line of government officials. During the Napoleonic Wars, his paternal grandfather, also named Carl, achieved some local fame because of courage displayed in anti-French agitation. The grandfather was interned and escaped execution only through his wife's intervention. On the mother's side, his heritage sprouted some small cultural branches. The grandfather, Jacob Roth, attended the University of Heidelberg and one of his sons, Joseph—a poet, painter, and world traveler—became Gegenbaur's favorite uncle. In a memoir, published in 1901, Gegenbaur reached back some sixty-five years to a memory, yet green, of his childhood self entranced by the buffalo hides and guns his uncle Joseph brought back from adventures on the plains of North America.

During his father's governmental posting to Würzburg, Carl Gegenbaur was born on 21 August 1826, the eldest of seven children, four of whom died in infancy. From 1834 to 1837, he attended the Latin school in Weis-

15. Anna Sethe Haeckel to Allmers (15 December 1863), in ibid., 2:38.

16. August Schleicher to Haeckel (5 September 1863), in the Haeckel Correspondence, Haeckel-Haus, Jena.

17. I have taken the following sketch of Gegenbaur's life from his autobiography *Erlebtes und Erstrebtes* (Leipzig: Wilhelm Engelmann, 1901). Also see Uwe Hoßfeld and Lennart Olsson, "The History of Comparative Anatomy in Jena—an Overview," *Theory in Biosciences* 122 (2003): 109–26; and Lynn Nyhart, "The Importance of the 'Gegenbaur School' for German Morphology," *Theory in Biosciences* 122 (2003): 162–73.

Fig. 4.2. Carl Gegenbaur (1826–1903). Photo from about 1860. (Courtesy of Ernst-Haeckel-Haus, Jena.)

senburg (just south of Würzburg). He thought the education he received in that Catholic secondary school had provided him a sound classical education. Contentment turned to anxiety in the stricter rule of the gymnasium in Arnstein (just north of Würzburg), the village to which his family moved in 1838. Against the Jesuits of his gymnasium, the boy found an ally in a local priest, who salted his religious convictions with an anti-Roman, anti-Jesuit attitude that likely contributed to the gradual withering, during his professorial days, of allegiance to formal Catholic doctrine. The chill of his scholastic regimen retreated in the warm after-school hours, when the young Gegenbaur would stroll with his mother through the countryside while she described the plants of the local area. As with Haeckel, botanical love blossomed under motherly cultivation. During vacations Gegenbaur's father often took him hunting. The son, however, did not relish blood sport but tolerated it as an accompaniment to the aesthetic experience of nature and as an opportunity to collect plants.

At nineteen, Gegenbaur entered the university at Würzburg with the faltering purpose of studying medicine. He quickly resolved himself to the goal of natural science, for which medicine formed only a means. At the university he became a follower of Haeckel's future mentors, Kölliker and Virchow. He received his degree in 1851, after he proposed a number of theses in medicine (e.g., "Omne vivum e cellulis"; "Nonnisi unum graviditatis signum certum")[18] and gave a general lecture (in German) on botany. In his memoir, as he examined the intellectual repertoire of that far-distant self, he conjured up from memory ideas concerning the instability of plant species that he felt anticipated Darwin's own proposals.[19] Looking back over that half century, he construed the eager student as formed for evolutionary theory; and, indeed, he came over to the new dispensation with as much conviction as Haeckel, though not with quite the same alacrity for public declaration.

Gegenbaur's transformation into an evolutionary morphologist advanced along a remarkably similar path to that of his younger colleague. During the summer of 1851, when he took a break from further medical work, he toured northern Germany, reaching what must have been a not-accidental destination, Berlin and Johannes Müller's Zoological Institute. The great scientist invited the young doctor to accompany him for research in Helgoland, which gave Gegenbaur a better vision of his own possibilities. Those possibilities were realized the next summer (1852) in a hajj to Messina with Kölliker and Heinrich Müller, another of his teachers at Würzburg. After the tutelage he received during the first several weeks, he continued on his own to seek out a habilitation project. During his eighteen months in Sicily, he fell in love with the country and its scientific and natural delights. When Etna erupted that fall, Gegenbaur, following the shades of Goethe and Humboldt, climbed the spewing mountain, even spending the night on its crest, sheltered from flying rocks.[20] During the next year he traveled extensively on the island, though he hardly neglected

18. Carl Gegenbaur, *Gesammelte Abhandlungen von Carl Gegenbaur*, ed. M. Fürbringer and H. Bluntschli, 3 vols. (Leipzig: Wilhelm Engelmann, 1912), 3:575. "All life derives from cells." "Only one indication of pregnancy is certain."

19. Gegenbaur, *Erlebtes und Erstrebtes*, 53.

20. Ibid., 65–66.

research. He eventually published over a dozen papers on his investigations of marine invertebrates.

Gegenbaur returned to Würzburg in the summer of 1853, where he continued work on the several publications that detailed his research in Messina. It was at this time he first met Haeckel, who was just beginning his second year of medical school. The younger student, by his later admission, had become entranced by Gegenbaur's tales of Italy and Sicily; these floated to the back of his mind, subtly coloring his future decisions about research travel. In the fall Gegenbaur completed his *Habilitationsschrift,* which discussed the alteration of generations and reproduction in medusae and polyps,[21] topics upon which Haeckel would also later meditate. Gegenbaur officially habilitated with Kölliker at the end of the winter term and became *Privatdozent.* Due to his many publications, Gegenbaur's reputation blossomed during the three terms he spent with Kölliker. The medical faculty at Jena, catching that southern breeze, smelled the sweet opportunity. They offered Gegenbaur a position as extraordinary professor of zoology. He realized his possibilities for professional growth at Würzburg were restricted, constrained as he was by Kölliker at the top and by Heinrich Müller and Franz Leydig pressing from either side. The natural beauty of the area surrounding Jena—the picturesque mountains, lush valleys, and thick forests—completely seduced him. And in the university itself, he discovered to his wonder men who had known Schiller and talked with Goethe.[22]

So even though his family remained in Würzburg, Gegenbaur accepted the invitation of a professorship at Jena and took up his new position in the summer of 1856. When Emil Huschke (1797–1858), the professor of anatomy and physiology in the medical faculty, died two years later, Gegenbaur was the natural candidate to replace him. Gegenbaur, a skilled anatomist, harbored little interest in physiology nor did he feel competent in that particular medical discipline. Moritz Seebeck, the civil administrator of the university, feared that if the faculty did not accommodate the young scientist, he would accept a call to Berlin, an invitation expected after Müller's death.[23] In Solomonic fashion, the university resolved the problem by hiring Albert von Bezold (1836–1868) to occupy a new position as extraordi-

21. Carl Gegenbaur, "Zur Lehre vom Generationswechsel und der Fortpflanzung bei Medusen und Polypen," *Verhandlungen der physikalisch-medicinischen Gesellschaft in Würzburg* 4 (1853): 154–221.

22. Gegenbaur, *Erlebtes und Erstrebtes,* 92.

23. Georg Uschmann, *Geschichte der Zoologie und der zoologischen Anstalten in Jena 1779–1919* (Jena: Gustav Fischer, 1959), 32.

nary professor of physiology, while Gegenbaur was advanced to ordinary professor of anatomy. In his autobiography, Gegenbaur proudly indicated that the professional division of anatomy from physiology—an act of some moment in German biological science—occurred in the summer of 1858, prior to their separation at Berlin after Müller died.[24]

Gegenbaur received Haeckel in the full flower of the younger naturalist's radiolarian enthusiasms and learned quickly of his assistant's plan to continue investigations of other marine invertebrates, now under the sign of descent theory. Undoubtedly Haeckel's interests, and the rich beginnings he had made, counseled Gegenbaur to shift his own research away from marine fauna, with which he had been exclusively concerned. He turned his attention to comparative work more fitting, perhaps, for someone ensconced in a medical school, namely, the vertebrates.[25] Gegenbaur probably read Darwin's *Origin of Species* in the Bronn translation sometime in 1860 or 1861. In 1864 he sent Darwin a copy of a newly published monograph, the first in a series, on the comparative anatomy of vertebrates.[26]

Passing strange, though, especially in light of his gift to Darwin, was Gegenbaur's reticence about the new theory. In none of the volumes of his monograph series (1864, 1865, 1872) did he refer to species descent.[27] He devoted the first volume, for instance, to a depiction of the carpus and tarsus bones in various families of reptiles, birds, and mammals (in man, the bones of wrist and ankle, respectively). He proclaimed that the general task of comparative anatomy was "to investigate whether and how the relationships found in the higher vertebrates were derivable [*ableitbar*] from the lower forms, whether certain common relationships of arrangement were fundamental to all the classes, and in what way the modifications were so disposed as to determine the arrangements distinguishing the particular

24. Gegenbaur, *Erlebtes und Erstrebtes,* 95. It is usually assumed that the separation of anatomy from physiology occurred first at Berlin—see, for example, Karl Rothschuh, *History of Physiology,* trans. Guenter Risse (Huntington, NY: Krieger, 1973), 153. Gegenbaur, mindful of the importance of this disciplinary event, maintained that the distinction belonged to Jena.

25. Gegenbaur's very first publication, done in collaboration when he was a young student at Würzburg, treated the skull of the axolotl (larval salamander). While a student, he also wrote two papers on sensory reception in mammals. The rest of his over forty research publications up to 1861 were devoted to invertebrates. In 1861, the time Haeckel arrived, he began issuing publications on vertebrates, a subject from which he hardly deviated thereafter.

26. Charles Darwin to Haeckel (8 October 1864), in the Darwin Papers, DAR 166.1, Special Collections, Cambridge University Library. Darwin asked Haeckel to thank Gegenbaur for the book.

27. Carl Gegenbaur, *Untersuchungen zur vergleichenden Anatomie der Wirbelthiere,* 3 vols. (Leipzig: Wilhelm Engelmann, 1864, 1865, 1872).

divisions of animals."[28] This conception, as so expressed, while it could easily be cashed out in evolutionary currency, was itself common enough coin to non-evolutionary comparative morphologists—Richard Owen, for example, whom Gegenbaur liberally cited. Only in 1870, in the second edition of his *Grundzüge der vergleichenden Anatomie* (Foundations of comparative anatomy), did Gegenbaur explicitly discuss Darwin's theory and forthrightly argue for an evolutionary grounding of morphology:

> From the standpoint of descent theory, the "relationship" of organisms has lost its metaphorical meaning. When we meet a demonstrable agreement of organization through precise comparison, this indicates an inherited trait stemming from a common origin. The task becomes to trace, step-by-step, the various paths the organ has followed by reason of acquired adaptation; it no longer suffices to derive each relationship from some remote similarity.[29]

Gegenbaur's efforts in morphology during the 1860s may have been illuminated by Darwin's theory, but, unlike Haeckel, he kept his light under a bushel.[30] After the appearance of Haeckel's *Natürliche Schöpfungsgeschichte* in 1868, it would have been difficult for a prominent morphologist not to take an explicit stand on evolutionary theory.

Like Haeckel, Gegenbaur was fully persuaded of the biogenetic law that ontogeny recapitulated phylogeny; he thought the principle could be quite helpful in establishing phylogenetic relationships. Nonetheless, in the actual practice of the researcher, investigations of the comparative anatomy of the adult vertebrate system, as opposed to the embryonic, would yield, in his view, more plentiful and reliable information about the systematic connections of organisms. During his later years at Heidelberg, whose call he accepted in 1873, Gegenbaur's influence in comparative anatomy would bring many promising students to his research benches—for instance, Max Fürbringer, Georg Ruge (1852–1919), Friedrich Maurer (1859–1936),

28. Ibid., 1:iv.

29. Carl Gegenbaur, *Grundzüge der vergleichenden Anatomie*, 2nd ed. (Leipzig: Wilhelm Engelmann, 1870), 19.

30. Nyhart, in her comprehensive and penetrating study of German morphology in the nineteenth century, suggests that in the 1860s Gegenbaur was explicitly advancing a program of evolutionary morphology. If so, it was done with extreme subtlety in his publications, unlike his colleague Haeckel. See Lynn Nyhart, *Biology Takes Form: Animal Morphology and the German Universities, 1800–1900* (Chicago: University of Chicago Press, 1995), 153–55.

Hermann Klaatsch (1863–1916), and Ernst Göppert (1866–1945). They would become known in the historiography of the period as the Gegenbaur school of comparative anatomy.[31]

During their years at Jena, the friendship between Gegenbaur and Haeckel grew slowly but steadily. Their mutual support in scientific interests began quite early and can be traced through their dedications of books to each other and the ever-deepening respect with which one would discuss the theories of the other.[32] From the beginning, Haeckel revered his colleague as a model of scientific industry and single-mindedness, but that made him even more sensitive to his own differently disposed nature. He parsed their respective attitudes to Anna after about five months at Jena:

> Gegenbaur is so thoroughly of a strict and pure scientifically minded character that this side of his personality—this terrible energetic, manifold, and almost absolute scientific effort—pushes all other human considerations into the background. . . . Thus he is deprived of all the wonderful pleasures of life, all the formative sides of a warm humane life, the sort of love life I have come to know with you, my dearest, the sort that completely overbalances my scientific endeavors and decisively pushes them into the background.[33]

If human love did not initially cement the relationship between these two men, then love of nature did. They took long hikes together into the Thuringian woods and mountains near Jena, allowing the richness of it all to flood over them. Haeckel thought Gegenbaur the only one of his colleagues who loved nature as much as he himself—though he suspected it was only nature that Gegenbaur loved.[34] In a short time, though, Haeckel would find a deeper well of humanity in Gegenbaur as they suffered a similar human tragedy.

For Love of Anna

Haeckel was driven by his desire for research and his desire for Anna. Both required a permanent professorial position, but that was by no means cer-

31. See ibid., 207–42.

32. Haeckel dedicated the first volume (1866) of his *Generelle Morphologie der Organismen* to Gegenbaur, who returned the favor in the third volume (1872) of his *Untersuchungen zur vergleichenden Anatomie der Wirbelthiere.*

33. Haeckel to Anna Sethe (15 June 1861), in *Himmelhoch Jauchzend,* 170–71.

34. Haeckel to Anna Sethe (30 August 1861), in ibid., 244.

tain at Jena. He did not wish to broach the topic of a professorship with Gegenbaur directly but silently hoped, as he confided to his fiancée, that his mentor would take the initiative without prompting.[35] In fact, Gegenbaur did. As Haeckel's work on his radiolarian book approached its end, Gegenbaur revealed to his young friend that Seebeck would appoint him extraordinary professor as soon as the book appeared. Haeckel would thus be relieved of depending on fluctuating student fees for his livelihood; and, of course, he and Anna could then be married.

At the beginning of March 1862—after a year and a half of continuous writing and research, topping almost a year of microscopic study of the very small subjects of his analysis—Haeckel was delivered by his publisher Reimer of two very large folio volumes, the first of which weighed in at over seven pounds and the second, the atlas, brimmed with extraordinarily precise and beautiful illustrations, many in color. He distributed three of the four advance copies of *Die Radiolarien* to Gegenbaur, Kuno Fischer (then rector), and Seebeck. In June Fischer told him that the four courts of the archdukedom had approved his appointment as extraordinary professor. He immediately wrote Anna to boast of his elevation as the "Archducal-Saxonish-Weimarish-Colburgish-Altenburgish-Meiningenish Extraordinary Professor."[36] He also learned that Gegenbaur would turn over to him the directorship of the small Zoological Museum, whose several collections of skeletons and animals preserved in spirits of wine—mammals, birds, amphibians, fish, mollusks, and over forty thousand insects—had all been desultorily added to the natural history cabinets that Goethe had originally brought together during the second decade of the century. Gegenbaur was glad to be relieved of the burden.[37]

With Haeckel's promotion came some financial security. He would receive one hundred *Taler* for teaching his seminars, fifty for his new rank, and another hundred for directing the Zoological Museum. In order for him to make it, though, his father still had to contribute two hundred *Taler* a year. But all together this was quite sufficient for a professor and his wife to live modestly. And living with Anna is what Haeckel most desired.

All the while Haeckel worked at Jena, he thought of his fiancée, if the

35. Haeckel to Anna Sethe (4 November 1861), in ibid., 248–49.

36. Haeckel to Anna Sethe (17 June 1862), in ibid., 281.

37. Georg Uschmann describes the origins and development of the museum in his *Geschichte der Zoologie und der zoologischen Anstalten in Jena 1779–1919*. In 1865, when Haeckel became ordinary professor of zoology, he received initial funds for establishing an Institute of Zoology. The building, however, materialized only much later; initially, the institute became incorporated into the Botanical Institute, occupying the first and third floors.

constant stream of letters to her provides any evidence of his obsession. These letters detail the terms of his love, not in the discreet numbers of a scientist but in the generous language of a man who has attained inner happiness against which a consuming passion pushes his emotions beyond reflective constraint. Typical is this letter penned in the late spring of 1861:

> The more I attain inner calm and clarity here through energetic external activity, as well as through lively mental exercise, and the more the peace of nature is drawn into my soul, the clearer it becomes to me what a great, inestimable, enviable happiness has bloomed in me during these last years in which I have possessed the loveliest, purest maiden soul and the most noble, most beautiful friendship, and these continue to mature into ever more blossoms and happy fruit. Love and Friendship! How happy they make me. I had earlier chosen science alone, but they promise me everything that science cannot give.[38]

With a position secure in Jena, Haeckel and Anna began to plan their late-summer wedding. The only slight blemish on their happiness occurred because Allmers could not (actually, would not) come to the celebration. He did send a landscape of southern Italy as a reminder to Haeckel of their "free and happy, sunny and blissful existence, on which I could not think back without giving over my heart to melancholy and rapture." He also included in his letter a sketch for a bathetic play. It was entitled *Polypa, the Radiolarian Sprite.* In the play Polypa entices a beautiful blond boy to her cavern and tempts him in dance; she flashes before his eyes "small crabs, then starfish, and finally the most delicate and interesting of her radiolarians until he succumbs." But just as she thinks she has him in her grasp and he has forgotten the past, he calls out: "O Anna, how you will enjoy all of this," and Polypa's magic dies in a trice. Polypa, however, is finally consoled about the impending marriage of the youth by the publication of the magnificent radiolarian work. With that "there is a great joy throughout the sea world, celebrated with dance and evolution."[39] What could Haeckel have made of this letter and its sentiments? What must Anna have thought or suspected? All suspicions, no doubt, disappeared in the glow of their festivities and honeymoon travels.

Ernst Haeckel and Anna Sethe were married in Berlin on 18 August 1862. Immediately that evening they took the night train to Dresden,

38. Haeckel to Anna Sethe (7 June 1861), in *Himmelhoch Jauchzend,* 187.

39. Allmers to Haeckel (10 August 1862), in *Haeckel und Allmers,* 85–88.

where they spent the first three days of their honeymoon visiting art galleries. On the twenty-first, they traveled through Hof, Bamberg, Nürnberg, Regensburg, finally reaching Passau, where they stopped over for two days. Then on 23 August, they traveled on to Salzburg, where they spent ten days walking through the neighboring alpine villages, climbing the mountains, and camping in the valleys. Haeckel took time, as the still dutiful son, to inform his parents of his new life with Anna: "With each day I am more in love with her, though I cannot believe any more love is possible. The liveliest interest in nature and art join us even closer together, more than pure attraction alone could."[40] On 7 September they left Salzburg for the Tyrol, where they enjoyed another two weeks hiking through the mountains and valleys, camping here and there, and resting in inns along the way. They traveled light, with Haeckel carrying all their extra clothing in a leather satchel on his back.[41] With Anna, whose joy in art and nature matched his own, Haeckel found, as he wrote Allmers later in March, "a completely different life."[42]

Anna shared Haeckel's passion for Darwin, at least in the way a lover might indulge the beloved's great obsession. She delighted in calling him, as he related to the master himself, "her German Darwin-man."[43] The name must have come easily, since she absorbed great quantities of evolutionary talk listening to her husband prepare for his many lectures on the subject. Indeed, his mission to instruct his students, colleagues, and the general public in the details of Darwinian theory began just as his honeymoon ended. During the winter term of 1862–63, he offered a public lecture series on evolutionary theory (in addition to three seminars: on zoology, histology, and osteology); and in March he lectured to the Weimar court on Darwin.[44] It was during the following September, however, that Haeckel's name became inextricably linked with Darwin's, at least in the mind of the larger German scientific community. The occasion was the famous meeting of the Society of German Natural Scientists and Physicians at Stettin in September 1863. It was the newspaper account of this meeting that Haeckel sent to Darwin (see chapter 1).[45]

40. Haeckel to his parents (26 August 1862), in *Himmelhoch Jauchzend*, 302–3.

41. Haeckel described his honeymoon trip to Allmers (23 December 1862), in *Haeckel und Allmers*, 30–32.

42. Haeckel to Allmers (29 March 1863), in ibid., 33.

43. Haeckel to Darwin (10 August 1864), in the Darwin Papers, Cambridge. Anna's term "Darwin-Mann" also means "Darwin-husband."

44. Haeckel to his parents (5 November 1862), in *Himmelhoch Jauchzend*, 307–8; and Haeckel to Allmers (29 March 1863), in *Haeckel und Allmers*, 33.

45. *Stettiner Zeitung*, no. 439 (20 September 1863); see also chap. 1.

The Defender of Darwin

The Society of German Natural Scientists and Physicians, founded by Lorenz Oken in 1822 at Leipzig, was the premiere organization of its type. It gathered annually in different cities throughout the Germanies with the express purpose of promoting social and intellectual exchange among scientific researchers in all disciplines. The thirty-eighth meeting, held on 17 to 23 September 1863 in the Prussian town of Stettin,[46] had drawn a large audience; well over five hundred members had officially registered and several times that had attended the plenary sessions as guests. Undoubtedly considerable interest was stimulated by word that Rudolf Virchow would take the occasion to defend his views against the charge, leveled by Matthias Schleiden, that his theories led to a spirit-killing materialism.[47] But the two thousand people who attended the first plenary session on 19 September testified to the general interest that Darwin's theory had aroused. Haeckel was singularly honored when asked to give that first lecture, "Ueber die Entwickelungstheorie Darwins" (On Darwin's evolutionary theory).[48]

Haeckel said he felt compelled to give an account of Darwin's theory in the brief hour he had—despite the inadequacies of time and medium—since the biological, geological, and philosophical worlds had been riven over the question of species. Daily the two parties, the progressive Darwinists and their conservative opponents, grew further apart and ever more ready to attack each other. Haeckel thus thought it necessary to bring before the eyes of all, professionals and laity alike, the leading features of evolutionary theory in a clear and calm fashion.

He granted that Darwin's fundamental proposal was hardly novel. Lamarck, Geoffroy Saint-Hilaire, and Lorenz Oken had already voiced the idea that plants and animals had undergone continuous change and that the later forms were related to the earlier genealogically. "The whole natural system of plants and animals," Haeckel allowed, "appears from this

46. The city, now Szczecin, lies on the Oder River entrance to the Baltic Sea, just on the Polish side of the border.

47. Schleiden made the charge a few months earlier in his "Über den Materialismus der neueren deutschen Naturwissenschaft, sein Wesen und seine Geschichte" (1863), reprinted in Matthias Jakob Schleiden, *Wissenschaftsphilosophische Schriften*, ed. Ulrich Charpa (Köln: Jürgen Dinter's Verlag für Philosophie, 1989), 265–308. The dispute is discussed below.

48. Ernst Haeckel, "Ueber die Entwickelungstheorie Darwins," in *Amtlicher Bericht über die acht und dreissigste Versammlung Deutscher Naturforscher und Ärzte in Stettin* (Stettin: Hessenland's Buchdruckerei, 1864), 17–30.

Fig. 4.3. Charles Darwin (1809–1882) in 1860. Photo taken by his son Erasmus. (Courtesy of the Archives of Gray Herbarium, Harvard University.)

perspective as a great stem-tree [*Stammbaum*], and so each genealogical table of relations can be represented intuitively in the form of a ramifying tree whose simple roots lie hidden in the past." This meant that "no species, perhaps with the exception of the first, has thus been independently created; rather, all derived in the course of immeasurable ages from several or a few primitive forms, which have, perhaps, spontaneously sprung up."[49]

Haeckel undoubtedly gave Oken too much credit as an evolutionist—he really was not.[50] But mention of the founder of the association then assembled was undoubtedly meant to link the Romantic *Naturphilosophie* of the previous generation with the new theory. The idea that spe-

49. Ibid., 20.

50. See Robert J. Richards, *The Meaning of Evolution: The Morphological Construction and Ideological Reconstruction of Darwin's Theory* (Chicago: University of Chicago Press, 1992), 39–42.

cies relations could be intuitively represented in a tree diagram may have been prompted by remarks of Darwin in the *Origin of Species*,[51] but more proximately Haeckel probably got the suggestion from two other sources: Heinrich Georg Bronn (see the appendix to this volume) and his friend the linguist August Schleicher. Schleicher in 1863 also had argued, in a little book dedicated to Haeckel, that languages could best be understood as having evolved in a Darwinian fashion and that this could be most perspicuously represented, as he in fact had, by a greatly ramified tree.[52] Haeckel himself would develop the art of tree diagrams a bit later. But one aspect of this early defense of Darwin he silently dropped from a later republication of his essay; this was the suggestion that the original form of life might have been "independently created."[53] Haeckel must have tossed in that line to anticipate possible objections to what might seem a thoroughly materialistic theory; later he became less careful about the sensibilities of opponents. During the next several years, he cultivated various theories of spontaneous generation of primitive life out of the *Urschleim*—a conviction growing from a more profound metaphysical belief, one initially planted by Spinoza and Goethe, that at a deep level nonlife and life, matter and mind, differently expressed the same underlying *stuff*.

Haeckel, of course, did not believe Darwin's theory to be simply derivative of earlier proposals. Darwin, he maintained, had purged the earlier *Naturphilosophische* views of errors and had established, through laws of inheritance and variability, a natural meaning for the homologies that animals displayed. Darwin's particular contribution to the theory of de-

51. Charles Darwin, *On the Origin of Species* (London: Murray, 1859), 129: "The affinities of all the beings of the same class have sometimes been represented by a great tree. I believe this simile largely speaks the truth. The green and budding twigs may represent existing species; and those produced during each former year may represent the long succession of extinct species. . . . The limbs divided into great branches, and these into lesser and lesser branches, were themselves once, when the tree was small, budding twigs; and this connection of the former and present buds by ramifying branches may well represent the classification of all extinct and living species in groups subordinate to groups."

52. August Schleicher, *Die Darwinsche Theorie und die Sprachwissenschaft* (Weimar: Hermann Böhlau, 1863). I will discuss Haeckel's tree diagrams, especially as represented in his later works, in the next chapter. For Schleicher's theory of the evolution of language, see Robert J. Richards, "The Linguistic Creation of Man: Charles Darwin, August Schleicher, Ernst Haeckel, and the Missing Link in Nineteenth-Century Evolutionary Theory," in *Experimenting in Tongues: Studies in Science and Language*, ed. Matthias Doerres (Stanford, CA: Stanford University Press, 2002), 21–48.

53. After the mid-1860s, Haeckel derided the idea of divine action in the natural world. So in the republication of his lecture on Darwin's theory, he dropped the suggestion that the first spark of life was independently ignited. See Ernst Haeckel, *Gesammelte populäre Vorträge aus dem Gebiete der Entwickelungslehre*, 2 vols. (Bonn: Emil Strauss, 1878), 1:1–29.

velopment, though, lay in the device brought to explain species change, namely, natural selection. Haeckel offered his audience a lucid account of the new idea:

> Now it is clear that in this struggle for existence those individuals of the same species will, on average, overcome and survive the others, if they have in their relationship any better position, or possess more strength to withstand attack, or have greater quickness to escape predators, or because of any organizational property have an advantage over the others. . . . A repetition of this process in the same species over many generations must have as a consequence a continuous perfecting [*Vervollkommnung*] of the species.[54]

Darwin's suggestion that natural selection was a perfecting agent obviously struck a resonant cord in Haeckel, and in his lecture he stressed this seemingly nonmaterialistic, virtually divine activity of nature:

> Without question, out of this general process, when considered on the whole and at large, there must necessarily follow a continuous, general alteration of the entire living world, a progressive metamorphosis [*progressive Metamorphose*], a progressive reformation and ennobling [*Veredelung*] of all organisms. The lower, less perfect [*unvollkommeneren*] forms will continually be eliminated; the higher, more perfect [*vollkommeneren*] will be preserved; and these latter will produce again a still greater number of yet more perfect forms through continuous variation and the origin of new species.[55]

In his lecture Haeckel had already anticipated Darwin's *Descent of Man:* he quickly applied the concept of natural selection to human cultural and social history. "We find essentially the same law of progress [*Gesetz des Fortschritts*]," he intoned, "to be operative in the historical development of human races. Thus in civil and social relationships again the same principle is at work—the struggle for existence and natural selection; this drives peoples irresistibly forward and progressively to higher levels of culture."[56]

When Darwin wrote Haeckel that his new friend was "one of the few

54. Haeckel, "Ueber die Entwickelungstheorie Darwins," 24.
55. Ibid., 26.
56. Ibid., 28.

who clearly understands natural selection,"[57] he was not simply flattering another potential ally. Haeckel well understood the way selection operated and what, at least in the Darwinian mode, its likely consequences would be—namely, a progressive development of ever more complex organisms.

This, however, is not the usual judgment made about Haeckel's understanding of Darwin's theory. Most historians believe that the progressivist note that Haeckel sounded was a false one, perhaps deceptively scaled by Bronn's German translation of the *Origin*. Bronn, after all, did frequently raise the pitch of the terms he rendered—"improvement," for instance, usually came out "*Vervollkommnung*," which might naturally be back-translated as "perfection." Add to this distinctively orchestrated translation Haeckel's own *Naturphilosophische* inclinations, and, it is often believed, we get that peculiar rendition of evolution that marks it as non-Darwinian—"non-Darwinian" because Darwin's theory, these historians believe, certainly was not progressivist in character.[58]

Darwin's notebooks, letters, and the plain text of the *Origin* testify otherwise. These documents show that he initially conceived of natural selection as producing progressive improvements in species.[59] The perfecting of species, moreover, would not be merely relative to local circumstances. He thought natural selection would supply ever more progressive types over time, so that, as he expressed it in the *Origin*, "the more recent forms must, on my theory, be higher than the more ancient; for each new species is formed by having had some advantage in the struggle for life over other

57. Darwin to Haeckel (9 March 1864), in the Haeckel Correspondence, Haeckel-Haus, Jena; *The Correspondence of Charles Darwin*, vol. 12: *1864*, ed. Frederick Burkhardt et al. (Cambridge: Cambridge University Press, 2001), 61. See chapter 1 for further quotations from Darwin's letter to Haeckel.

58. The idea that Darwin allowed only for relative progress—i.e., local improvements, which would be washed out with any migration or change of environment—dominates the current historical interpretation. See, for example, Peter Bowler, *The Non-Darwinian Revolution* (Baltimore: Johns Hopkins University Press, 1988), 84–90. See also Michael Weingarten,"Darwinismus und materialistischen Weltbild," in *Darwin und Darwinismus: Eine Ausstellung zur Kultur- und Naturgeschichte*, ed. Bodo-Michael Baumunk and Jürgen Riess (Berlin: Akademie, 1994), 80–81. Weingarten, referring to Haeckel's 1863 lecture, expresses an almost ineradicable opinion: "Progress [*Fortschritt*] is supposed to be a natural law that no social structures or institutions can abrogate. Haeckel believed that this progressive apologetic could be derived from Darwinian evolutionary theory; he failed to observe that Darwin, in contrast to Lyell, nowhere spoke of an absolute progress but only of a relative progress (in relation to occurrent environmental relationships)." In her meticulous study of the German morphological tradition, Lynn Nyhart also suggests that Bronn's translation of Darwin introduced the false supposition of progressive evolution. See her *Biology Takes Form*, 111–12.

59. I have discussed Darwin's progressivism at greater length in *The Meaning of Evolution*, 84–90.

and preceding forms."[60] Darwin thus urged that the progressive dynamic of natural selection would require Eocene fauna to succumb to modern types, just as Secondary fauna had to fall to Eocene, and Paleozoic to Secondary.[61] The idea that evolution by natural selection would yield progressively more advanced species—as human beings exemplified—remained Darwin's own belief; indeed, he thought such results "inevitable," or at least he so phrased it in the third and later editions of the *Origin.* Natural selection, he claimed, produces "improvements [that] *inevitably* lead to the gradual advancement of the organization of the greater number of living beings throughout the world."[62] So when Bronn has Darwin claim "da die Natürliche Züchtung nur durch und für das Gute eines jeden Wesens wirkt, so wird jede fernere körperliche und geistige Ausstattung desselben seine Vervollkommnung fördern,"[63] his translation faithfully captured Darwin's conviction about progressive development of species explicitly expressed in the *Origin:* "as natural selection works solely by and for the good of each being, all corporeal and mental endowments will tend to progress towards perfection."[64] (Bronn's translation—by omitting the word "tend"—does suggest that selection *will* produce an advance in perfection; a literal back-translation would be: "every successive corporeal and mental endowment will advance its [the being's] perfection." As mentioned above, this idea of *inevitable* progress does, nonetheless, represent Darwin's view.) Haeckel may have stressed the note of progress, but it was a note sounded first by Darwin himself and played out in any number of keys in the *Origin.*[65]

Haeckel concluded his defense of Darwin by mentioning the kind of

60. Darwin, *Origin of Species,* 337.

61. Ibid.

62. Charles Darwin, *The Origin of Species by Charles Darwin: A Variorum Text,* ed. Morse Peckham (Philadelphia: University of Pennsylvania Press, 1959), 221. The emphasis is mine.

63. Charles Darwin, *Über die Entstehung der Arten im Thier- und Pflanzen-Reich durch natürliche Züchtung; oder, Erhaltung der vervollkommneten Rassen in Kampfe um's Daseyn,* trans. Heinrich Georg Bronn (Stuttgart: Schweizerbart'sche Verlaghandlung, 1860), 494. This particular sentence, in both the German and English, remained unchanged through the subsequent editions of the *Origin.* Bronn did mislead in another fashion, but this in a nonprogressivist direction. He declined to translate a sentence at the end of the *Origin,* Darwin's remark that in view of his theory he could foresee that "Light will be thrown on the origin of man and his history" (*On the Origin of Species,* 488). This omission, however, did not prevent Haeckel from incorporating considerations of human evolution in his *Generelle Morphologie,* which I will treat in chapter 5.

64. Darwin, *Origin of Species,* 489.

65. A simple frequency count of the times the term "perfection" in one of its forms appears in the *Origin* reveals that, in the suspect sense, it occurs ninety-one times, or about once every five pages. This, of course, does not include semantic equivalents. See Paul Barrett, Donald

evidence for evolution that he thought the most persuasive—and a kind of evidence he would make central to his own later development of evolutionary theory, namely, "the threefold parallel between embryological, systematic, and paleontological development of the organism." Both Darwin and Haeckel thought this parallel, as Haeckel claimed, "the strongest proof of the truth of evolutionary theory."[66] Haeckel regretted he did not have time to explicate the parallel. Likely, however, his remarks were more commentary on the penultimate chapter of the *Origin of Species* than a carefully worked-out conception. The next year he would read Fritz Müller's *Für Darwin,* a book that developed the idea of embryological recapitulation in some detail. He recommended the book to Darwin, and he himself would elaborate the idea in his *Generelle Morphologie* in 1866.[67]

The *Stettiner Zeitung* for 20 September 1863 reported that Haeckel's lecture "captivated the auditorium because of its illuminatingly clear presentation and extremely elegant form." After an extensive précis of the lecture, the newspaper writer concluded with: "a huge applause followed this exciting lecture."[68] A majority of the audience apparently endorsed many of the ideas Haeckel expressed, though undoubtedly with varying levels of agreement. Two of the participants in the plenary sessions also engaged the question of evolution, as well as Haeckel's particular interpretation. Their conceptions, together with that of Haeckel, constitute what would become the three common types of evolutionary understanding in Germany during the last half of the nineteenth century.

Haeckel himself represents what I believe to be the authentic Darwinian strain of interpretation, a strain that, in a Romantic fashion, is progressivist and totalizing: all of nature, including human nature, develops temporally through more progressive forms, both in the microcosm of the individual and in the macrocosm of the phylum. The other two figures who

---

Weinshank, and Timothy Gottleber, eds., *A Concordance to Darwin's Origin of Species, First Edition* (Ithaca, NY: Cornell University Press, 1981), 553–56.

66. Haeckel, "Ueber die Entwickelungstheorie Darwins," 29. Darwin also claimed that recapitulation of phylogenetic forms in embryological development provided the strongest support for the general theory of evolution. In a letter to Asa Gray (10 September 1860), Darwin indicated that "embryology is to me by far the strongest single class of facts in favour of change of form." The letter is contained in *The Correspondence of Charles Darwin,* vol. 8: *1860,* ed. Frederick Burkhardt et al. (Cambridge: Cambridge University Press, 1993), 350. See my discussion of Darwin's theory of recapitulation in *The Meaning of Evolution,* 91–166.

67. Fritz Müller, *Für Darwin* (Leipzig: Wilhelm Engelmann, 1864). See also Haeckel to Darwin (26 October 1864), in the Haeckel Correspondence, Haeckel-Haus, Jena; and in *Correspondence of Charles Darwin,* 12:381. I will discuss Müller in the next chapter.

68. *Stettiner Zeitung,* no. 439 (20 September 1863).

addressed the question of evolution dissented from either the progressivist or the totalizing feature of Haeckel's construction.

The prominent geologist Georg Heinrich Otto Volger (1822–1897) from Frankfurt, speaking five days after Haeckel had opened the meeting, denied the developmental and directional character of evolution, which he believed Darwin's theory embraced and Haeckel's lecture faithfully defended.[69] Volger really offered nothing exceptional by way of opposition to the Darwinian version of transmutation, not even the arch condescension of the German professoriate—he "thanked" Haeckel for introducing, "with youthful conviction," the pertinent questions concerning Darwin's theory. Against the idea that one or a few original species had given rise to the rest through progressive advance, Volger maintained that new findings demonstrated many contemporary species had been around since the earliest times. He did allow that the fossil record revealed some changes in species, but certainly no progressive direction of those changes.[70] Haeckel had opportunity to respond to Volger and did so in a way that suggests the broader set of assumptions that made Darwin's theory so congenial to him.

Haeckel first rejoined that no one could really pronounce on evolutionary theory without a thorough understanding of anatomy and, especially, embryology. Here we have real knowledge that clearly shows progressive transformation and overrides our ignorance of the extremely fragmentary geological record. Moreover, the very nature of man as a progressive being suggests the truth of the theory:

> If one appeals to feeling, then this circular theory [i.e., Vogler's admission of nonprogressive alteration] leaves me little consolation, since the insight derived from the Darwinian theory of progressive development seems to correspond to the very nature of man. I am convinced that the history of human beings is only a product of the history of organisms of an earlier time; and even if we find in particular periods a retrogression, we cannot yet deny progress on the whole. I am convinced that this progress will not for long be constrained and that the whole history of organisms manifests the law of progress.[71]

69. Otto Volger, "Ueber die Darwin'sche Hypothese vom erdwissenschaftlichen Standpunkte aus," in *Amtlicher Bericht über die acht und dreissigste Versammlung Deutscher Naturforscher und Ärzte in Stettin* (Stettin: Hessenland's Buchdruckerei, 1864), 59–70.

70. Ibid., 68–69.

71. Ernst Haeckel, ["Response to Volger"], in *Amtlicher Bericht über die acht und dreissigste Versammlung Deutscher Naturforscher und Ärzte in Stettin* (Stettin: Hessenland's Buchdruckerei, 1864), 71.

The transcription of Haeckel's response indicates that with this peroration, the audience shouted "Bravo." Haeckel had made his mark as a defender of Darwin.

The second response to Haeckel's presentation came from his former teacher Rudolf Virchow, who rejected the totalizing claims of science, evolutionary or otherwise. Virchow inserted his animadversions within a lecture responding to the accusation of Schleiden that his natural scientific endeavors reeked of materialism. Schleiden had made the charge a few months earlier in his "Über den Materialismus der neueren deutschen Naturwissenschaft, sein Wesen und seine Geschichte" (On the materialism of recent German science, its character and history).[72] He objected to Virchow's assumption that man was not an absolute unity but only a confederation of smaller parts, of cells.[73] Virchow, in response, affirmed that science had demonstrated human beings to be composite: "the 'I' of today is no longer that of yesterday, and still less that of the day before; and we change even more in our bodily self." Any other position, he declared, must be derived from some transcendental source, which would have to spring from speculation and not natural science. Yet within Virchow's theory of the cell state lay a ticking consequence that would seem to vindicate Schleiden and force Virchow to a position that was cratered with metaphysical dangers. If the human body were a federation of independent units, what about the soul? Did each living cell have an animating principle, so that the human psyche, too, had to be regarded a multiplicity? To this last question, Haeckel would urge an affirmative answer. But this would come only in the wake of his confrontation with Virchow at the Munich meeting in 1877.[74]

Virchow, at the Stettin meeting, must have sensed the unhappy consequence toward which his theory of the cell state was driving him. To Schleiden's attack on the cell state, he defended the theory in its corporeal

72. Schleiden, "Über den Materialismus."

73. Virchow argued that animal and plant organisms were composed of cells held in confederation in his *Die Cellularpathologie in ihrer Begründung auf physiologische und pathologische Gewebelehre* (Berlin: Hirschwald, 1858), 12. Likely Virchow was indebted to Johann Christian Reil, who had argued in 1796 that the organs of the body, down to the smallest fiber, were independent of one another and existed only in extrinsic causal relationship. They functioned together, he maintained, like a "great Republic." See Johann Christian Reil, "Von der Lebenskraft," *Archiv für die Physiologie* 1 (1796): 8–162; see especially 105. I have discussed Reil's position in my "Rhapsodies on a Cat-Piano, or Johann Christian Reil and the Foundation of Romantic Psychiatry," *Critical Inquiry* 24 (1998): 700–736.

74. Haeckel drew out the implications of the cell-state theory in his monograph *Zellseelen und Seelenzellen*, 2nd ed. (1878; repr., Leipzig: Alfred Kröner, 1923).

Fig. 4.4. Rudolf Virchow (1821–1902) in the 1850s.
(Courtesy of the Smithsonian Institution.)

interpretation but then retreated to safer ground by acknowledging that science did not and could not bring all knowledge into its sphere of competence. Rational consciousness, he declared, would not yield to the methods of natural investigation. Rather, a transcendent, nonscientific arena protects "an independent soul, an independent mental power . . . upon which ground one's religious knowledge must be formulated, accordingly as it conforms to one's conscience and feeling." This, Virchow protested, is anything but materialism.[75]

Virchow's intellectual position seems dangerously vertiginous, bestrid-

75. See Rudolf Virchow, "Ueber den vermeintlichen Materialismus der heutigen Naturwissenschaft," in *Amtlicher Bericht über die acht und dreissigste Versammlung Deutscher Naturforscher und Ärzte in Stettin* (Stettin: Hessenland's Buchdruckerei, 1864), 35–42; quotations from 41–42.

ing as it does two gyrating spheres—that of natural science, which uncovers new empirical facts and rapidly forms theories to sustain the mechanistic construction of life, and that of religion, which is moved by currents of faith and deep feeling to sustain the nonmaterial. To maintain equilibrium between these two spheres becomes even more difficult in the presence of a new, electrifying scientific hypothesis that would jolt the very soul of man. Yet Virchow attempted it. In his plenary lecture, he endorsed Darwin's theory, which, as he reasonably interpolated, supposes a "transition from apes to man." The truth of this supposition, he maintained, had to be settled within the scientific realm, and neither religion nor the state should interfere in its resolution. To retain balance, though, the natural researcher had to recognize that not all knowledge on these questions may be attainable within science.[76] Virchow and many of his audience strained to keep the two spheres of knowledge separate, wishing to avoid a war of worlds. In short time, however, Haeckel would reject this irenic separation, and he would become, with death, a destroyer of worlds.

Tragedy in Jena

Anna had accompanied Haeckel to Stettin and undoubtedly gloried in the adulation he received. When they returned to Jena, the newly knighted champion plunged into further study of Darwinian theory. He wrote Allmers just before Christmas that "I am now convinced that a great future lies before this theory and that it will slowly but surely loose us from the bonds of a great and far-reaching prejudice. For this reason I shall dedicate my whole life and efforts to it."[77] Haeckel's devotion to evolutionary theory and his considerable teaching duties did not, however, prevent him and Anna from enjoying an active social life—at least of a sort familiar to academics in a small university town. So, for instance, they became accustomed to walking tours with Schleicher and his lively wife, who quickly became Anna's best friend; and their social circle widened to embrace the newly married Gegenbaur and his young spouse.[78] Haeckel's happiness during this period should have been rounded to a professional bliss by the great honor bestowed on him for his radiolarian monograph: in February

76. Ibid. Virchow, as he later liked to observe, had maintained, prior to Darwin, that evolution was a hypothesis suggested by many considerations of modern science. See the previous chapter.

77. Haeckel to Allmers (15 December 1863), in *Sein Leben, Denken und Wirken*, 2:36.

78. Haeckel to his parents (25 April 1863), in *Himmelhoch Jauchzend*, 309–10.

Fig. 4.5. Anna Sethe Haeckel and Ernst Haeckel about 1862. (Courtesy Ernst-Haeckel-Haus, Jena.)

he was to receive for his scientific achievement a diploma of merit and the Cothenius medal from the Leopold-Caroline Academy of German Natural Scientists. Carl Gustav Carus, onetime friend of Goethe and president of the academy, would confer the awards. The appointed day, however, turned black, and the magic circle was broken.

In late January 1864, Anna suffered a severe attack of pleurisy, which lasted through the first part of February. Haeckel became quite worried, but his wife seemed to recover. In mid-February she again became ill with severe abdominal pains. During the night of the fifteenth, her pain became acute, with great tenderness in the area of the liver (perhaps appendicitis).

She lost consciousness in the late morning of the sixteenth and died at three thirty that afternoon.[79] On that same day, Haeckel turned thirty years old and also received word that he had been awarded the Cothenius medal.[80]

Haeckel became mad with grief, falling unconscious and remaining in bed for some eight days in partial delirium. His parents were telegraphed, and they quickly came to care for him. They and his brother, Karl, kept watch, lest he take his own life, which they feared him quite capable of doing. He confessed to Allmers a month later: "Were my parents dead, I would soon follow. I am dead on the inside already and dead for everything. Life, nature, science have no appeal for me. How slowly the hours pass."[81] His parents arranged for their grieving son to travel to the Mediterranean, where once he had known happiness.

Gegenbaur wrote to convey the condolences of his wife and himself, hoping that the Italian heavens would gently carry away his terrible suffering. Undoubtedly fearing his friend might quit everything, he wanted to remind him that the work he had in Jena would act as a balm.[82] Shortly thereafter, as if their similar professional trajectory portended a common fate, Gegenbaur's own young wife died from childbed fever after the birth of their daughter, Emma (21 July 1864). The following Christmas, while staying with his parents in Berlin, Haeckel wrote a searing letter to his friend. In the letter he confided his continuing pain and depression, which now united them in the depths of their common sorrow. "Unmentionable suffering, inexpressible pain," he wrote, "has consumed us both today as the lights on the Christmas tree are lit and all families, old and young, prepare for the festivities and Christmas cheer." He told Gegenbaur he had that day walked past Anna's house in Berlin and recalled the years of their courtship, when he would pass by and she would throw down the key to him, and how after a wonderful evening, as he left, they would call to one another "felicissima notte." "All this," he said, "I have experienced, as a living fairy tale and yet so beautiful, so poetic as only one would expect to find in a book of fairy tales."[83]

Anna's loss marked Haeckel for life. A year later, on the anniversary of her death, his courage failed him and the tides of sorrow again washed over

79. There is an oral tradition at Haeckel-Haus that Anna died from the complications of a miscarriage, the details of which were kept from Haeckel.

80. Haeckel provided the details to Allmers (27 March 1864), in *Sein Leben, Denken und Wirken*, 2:41–42.

81. Haeckel to Allmers (27 March 1864), in ibid., 41.

82. Gegenbaur to Haeckel (26 April 1864), in the Haeckel Correspondence, Haeckel-Haus, Jena.

83. Haeckel to Gegenbaur (24 December 1864), in *Himmelhoch Jauchzend*, 322–24.

him, so that he "had to muster all his strength," he related to his parents, "not to be overcome by the bitterest and deepest pain." He had intended, he said, to spend the day with Gegenbaur, but as the time approached, he fled the city for the Thuringian mountains and woods, and then wandered over to the Schwarz Valley, where he had spent some of the happiest days of his life with Anna.[84] Even into his later years, on his birthday and the anniversary of her death, he could not work, could not eat, and often tempted himself with death. In 1899 he would write to a new Anna, a reincarnation of his love, that "Thursday, 16 February is my sixty-fifth birthday, for me the saddest anniversary of the year, since on this same day in 1864 I lost my most beloved and irreplaceable first wife. On this sad day, I am lost."[85] Even after thirty-five years, the blistering wound never healed.

Haeckel's scientific work, his perception of nature, and his metaphysical convictions—these all became transformed beyond their original proportions by the tragedy. Some hint of this transformation is provided in a letter to his parents, which he wrote from Nice, where they sent him in March of that fateful year to attempt a recovery.

> The last eight days have passed painfully. The Mediterranean, which I so love, has effected at least a part of the healing cure for which I hoped. I have become much quieter and begin to find myself in an unchanging pain, though I don't know how I shall bear it in the long run. . . . You conclude . . . that man is intended for a higher, godlike development, while I hold that from so deficient and contradictory a creation as man, a personal progressive development after death is not probable; more likely is a progressive development of the species on the whole, as Darwinian theory already has proposed it. . . . Mephisto has it right: "Everything that arises and has value comes to nothing."[86]

In a biographical note composed a decade later, Haeckel confirmed that the death of his wife "destroyed with one blow all the remains of my earlier dualistic worldview."[87]

With the extinction of love came emptiness, a void that quickly filled

84. Haeckel to his parents (18 February 1865), in ibid., 326–27.

85. Haeckel to Frieda von Uslar-Gleichen (14 February 1899), in *Das ungelöste Welträtsel: Frida von Uslar-Gleichen und Ernst Haeckel, Briefe und Tagebücher 1898–1900*, ed. Norbert Elsner, 3 vols. (Berlin: Wallstein, 2000), 1:128. The printed version has the date as 4 February, but the manuscript original (held in library of the Preussischer Kulturbesitz, Berlin) has 14 February.

86. Haeckel to his parents (21 March 1864), in *Himmelhoch Jauchzend*, 318–19.

87. See the excerpt printed in ibid., 333.

with the miasma of great stridency, bitterness, and ineluctable sadness, which not even friends like Schleicher could clear away.[88] Through this acid mist, Haeckel resolved to devote himself single-mindedly to a cause that might transcend individual fragility. He would incessantly push the Darwinian ideal and oppose it to those who refused to look at life, to look at death, face on: his own scientifically orthodox colleagues, who were mired in a useless past; and the religiously orthodox, who promised a deceptive future. After a period of recovery, Haeckel abandoned himself to an orgy of unrelenting work that yielded, after eighteen-hour days over twelve months, a mountainous two-volume monograph that laid out his fundamental ideas about evolution and morphology. The volcanic *Generelle Morphologie der Organismen* spewed fire and ash over the enemies of progress and radically altered the intellectual terrain in German biological science. The sulfuric passion in which his evolutionary ideas gushed out was propelled by his great pain, something he confessed to Darwin.[89] But so corrosive of orthodoxy and the orthodox was his monograph that T. H. Huxley, who sought to have it translated for an English audience, had to extract from Haeckel a promise to excise the polemics, the nasty asides, and the attacks—and this from Huxley, one of the most accomplished practitioners of polemical science.[90] Yet, not only did bitter despair fuse his ideas into a quick, cutting hardness, but there was another, quite opposite mood that more quietly breathed over his work, one of discovering in nature the beauty and solace lost in human love or, rather, of a transformation of nature through an

88. Schleicher's efforts to comfort Haeckel were considerable—from simple condolences to diverting discussions of his garden. He thought, however, that only the burden of "scientific activity" would bring him back to himself. The blow to Haeckel was also a blow to Schleicher, who regarded Haeckel as his only real friend in Jena. A bit later he got Haeckel to return to gymnastics practice. See Schleicher to Haeckel (5–6 April, 7 May 1864, 10 February 1865), in the Haeckel Correspondence, Haeckel-Haus, Jena.

89. Haeckel to Darwin (12 May 1867), in the Darwin Papers, DAR 166.1, Cambridge.

90. See the exchange of letters: Huxley to Haeckel (13 November 1868) and Haeckel to Huxley (18 November 1868), in "Der Briefwechsel zwischen Thomas Henry Huxley und Ernst Haeckel," ed. Georg Uschmann und Ilse Jahn, *Wissenschaftliche Zeitschrift der Friedrich-Schiller-Universität Jena, Mathematisch-Naturwissenschaftliche Reihe* 9 (1959–60): 7–33; relevant letters are on 19–20. Haeckel made a like promise to Darwin: Haeckel to Darwin (28 November 1868), in the Darwin Papers, DAR 166.1, Cambridge. Huxley was also worried about the technical aspects of Haeckel's *Generelle Morphologie,* and he had several recommendations about what to include and what to omit in preparing the work for translation. Many of the problems created by the technical character of the work were solved when Haeckel published in 1868 a series of more popular lectures that were based on the larger monograph. The decision was made to translate this instead, and the *Natürliche Schöpfungsgeschichte* appeared in English (1876) as *The History of Creation.* The polemical fire of the original was dampened but little in the translation.

apotheosis of such love. This alternate trajectory of Haeckel's thought can initially be gleaned from an experience during his convalescence at Nice.

While walking along the shore, lost in his grief, he gazed idly upon a medusa, of a species unknown to him, floating near the surface of a tidal pool (fig. 4.6). The creature seems to have been transformed before his eyes into something quite different. Later in 1879, in his giant two-volume *System der Medusen,* he recounted the experience in the fine print of his systematic description of the organism, a creature he named *Mitrocoma Annae*—Anna's headband. Any reader who chanced to fall upon this passage, buried as it is amongst technical descriptions of the over six hundred species of medusae cataloged, would certainly have been startled by its very personal character:

> *Mitrocoma Annae* belongs to the most charming and delicate of all the medusae. It was first observed by me in April 1864, in the Bay of Villafranca near Nice. . . . The movement of this wonderful Eucopide offered a magical view, and I enjoyed several happy hours watching the play of her tentacles, which hang like blond hair-ornaments from the rim of the delicate umbrella-cap and which with the softest movement would roll up into thick short spirals. . . . I name this species, the princess of the Eucopiden, as a memorial to my unforgettable true wife, Anna Sethe. If I have succeeded, during my earthly pilgrimage in accomplishing something for natural science and humanity, I owe the greatest part to the ennobling influence of this gifted wife, who was torn from me through sudden death in 1864.[91]

Haeckel wrote this about his "unforgettable true wife" in 1879, while married to his apparently forgettable second wife, Agnes.

Several years after he had the transforming experience at Nice, he discovered another medusa, which he thought even lovelier, and hence this as well had to embody the spirit of his "true, unforgettable wife Anna Sethe." He named it *Desmonema Annasethe* (plate 5).[92] Perhaps because of

91. Ernst Haeckel, *Das System der Medusen,* 2 vols. (Jena: Gustav Fischer, 1879), 1:189. Shortly after his return from Nice, Haeckel provided a brief account of *Mitrocoma Annae* in a paper describing several new species of medusae. See Ernst Haeckel, "Beschreibung neuer craspedoter Medusen aus dem Golfe von Nizza," *Jenaische Zeitschrift für Naturwissenschaft* 1 (1864): 325–42. In his description here, he did not mention the occasion for his naming of the species, but he did remark on its charm and its "blond hair-ornaments that hang from the cap" (333).

92. Haeckel, *System der Medusen,* 1:526–27.

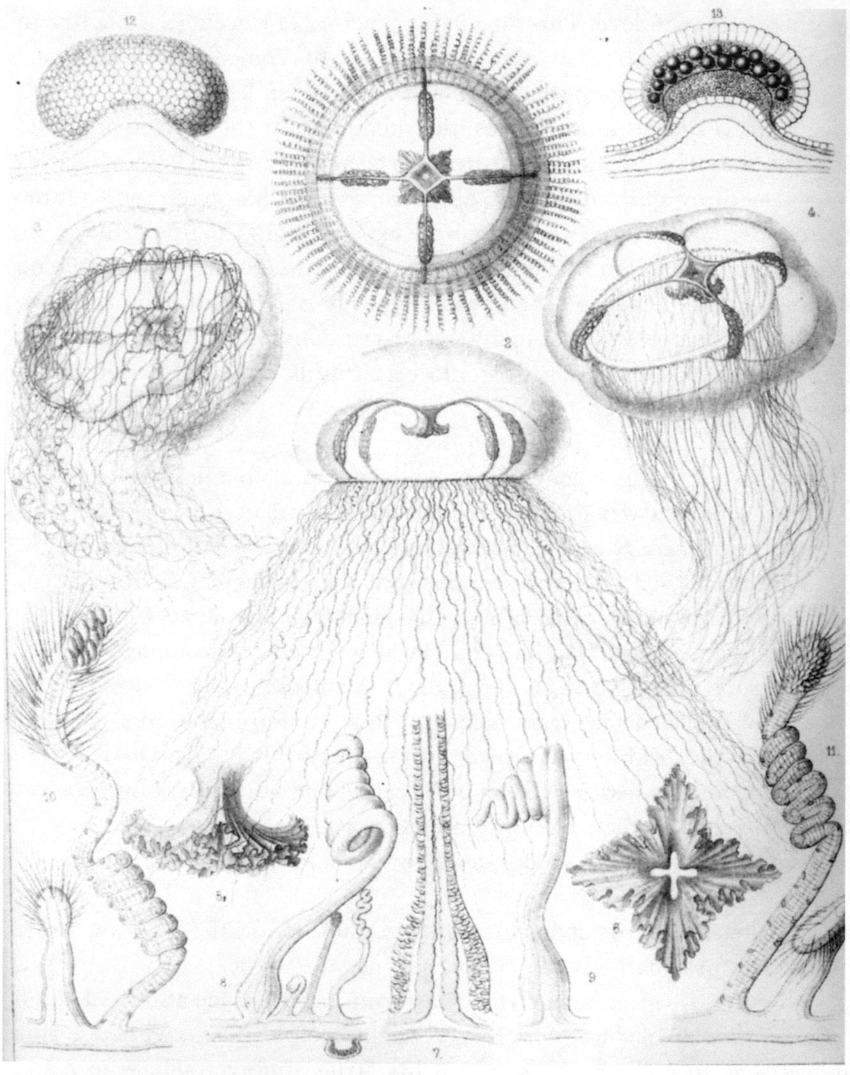

Fig. 4.6. *Mitrocoma Annae.* (From Haeckel, *System der Medusen,* 1879.)

these associations of medusae with his first wife, when, in 1882, he built a house in Jena, he decorated it with frescoes of medusae and called it Villa Medusa. Goethe and Humboldt believed, adapting ideas from Kant's third *Critique,* that aesthetic judgment complemented scientific understanding; each in its own mode captured the laws of nature, the principles according to which nature exhibited a unity underlying an ever-astonishing variety. With Haeckel, aesthetic judgment would be fused with Darwinian understanding through a love now lifted beyond the individual. The *Generelle Morphologie* would exhibit fundamental features of this new union, both in the bitter polemics—the other side of love—against the scientifically benighted and religiously stupefied, and in the metaphysical effort to absorb the individual into the whole, each life into *Deus sive natura* that would preserve it eternally. Monistic metaphysics, which would be voiced in the concluding chapter of the *Generelle Morphologie,* would be the substitute for traditional religion, a metaphysics that made no false promises of personal survival but that revealed a preservation of a different order. In the words of Goethe, which Haeckel chose as the initial epigram for his book:

> There is in nature an eternal life, becoming, and movement. She alters herself eternally, and is never still. She has no conception of stasis, and can only curse it. She is strong, her step is measured, her laws unalterable. She has thought and constantly reflects—but not as a human being, but as nature. She appears to everyone in a particular form. She hides herself in a thousand names and terms, and is always the same.[93]

For Haeckel, love fled and hid her face among sea creatures.

93. Johann Wolfgang von Goethe as quoted by Ernst Haeckel, *Generelle Morphologie der Organismen,* 2 vols. (Berlin: Georg Reimer, 1866), iv. There is some dispute today about whether these lines are actually by Goethe. The passage occurs in a handwritten journal—the so-called *Tiefurt Journal*—and was recorded at the end of 1782 or beginning of 1783. At the time, Goethe suggested it was by someone else (perhaps the young Swiss theologian Georg Christoph Tobler [1757–1812]); later, however, he said he simply couldn't remember. No one doubts, however, that it expresses Goethe's own view of nature. See Johann Wolfgang Goethe, *Die Natur,* in *Sämtliche Werke nach Epochen seines Schaffens* (Münchner Ausgabe), ed. Karl Richter et al., 21 vols. (Munich: Carl Hanser, 1985–98), 2.2:479.

CHAPTER FIVE

Evolutionary Morphology in the Darwinian Mode

Haeckel returned from the Mediterranean to rooms that briefly brightened with memories of earlier happiness but then quickly filled with shadows of his present sorrow. In quiet despair, he wrote Allmers a letter that ended with an adaptation of a stanza of one of his friend's poems, which Anna liked to recite:

Oft her lovely image arises,
Sweetly smiling as she used to be;
She nods and softly advises:
My poor boy, do not grieve for me.[1]

Haeckel was momentarily diverted from his preoccupations by letters waiting in the mail that had accumulated during his trip. Darwin had graciously responded to the gift of the radiolarian monograph and to the clipping describing the Stettin lecture, which Haeckel had sent several months earlier.[2] After a few days, on 7 July 1864, Haeckel composed a long reply that would sympathetically bind Darwin to his new colleague.

In his letter, Haeckel related his own conversion to Darwinism and told of its spread among younger researchers in Germany. He also darkly hinted at, but did not reveal, the great tragedy that had "hardened me against the blame as well as the praise of men, so that I am completely untouched by

1. Haeckel to Hermann Allmers (25 June 1864), in *Ernst Haeckel, Eine Schriftenfolge,* ed. Victor Franz, 2 vols. (Jena: Wilhelm Gronau and W. Agricola, 1943–44), 2:42: "Wie oft kommt mir ihr liebes Bild, / Hold lächelnd wie in bessern Tagen, / Und nicht zu mir, als spräch es mild: / Mein armer Junge, lass dein Klagen!"

2. See chapter 1 for a description of Darwin's initial letters to Haeckel.

external influence of any sort, and only have one goal in life, namely, to work for your descent theory, to support it, and perfect it." He mentioned that to this end he had been laboring on a "general natural history."[3]

Darwin immediately answered Haeckel's long letter, expressing interest in the project and offering condolences for the unnamed injury that so obviously aggrieved his new disciple. Darwin also initiated an academic ritual that in the nineteenth century signaled the beginning of closer collegial ties: he asked for Haeckel's photograph and supplied one of his own.[4] Haeckel quickly posted the portrait, and, as he said he could not refrain, one of his deceased wife. In the accompanying letter, he told Darwin of the great tragedy of his life, the death of Anna, "who held the name Darwin in as high a veneration as I myself do."[5] And in subsequent correspondence in October, he indicated to his new friend that his devotion to evolutionary theory had become not only an effort to comprehend nature but to recover in his work the love that he had lost: "Now in my isolation, which since the death of my wife is so lonesome, this engrossing work is a great consolation, and I toil at it with so great an enthusiasm, as if my Anna herself drove me to its completion and had left this task as a memorial."[6] That memorial was to be his *Generelle Morphologie der Organismen*.[7]

Anna would come to dwell in the pages of the *Generelle Morphologie*, in the revelation of the morphological transformations of the individual, in the reproductive cycles that mirrored phylogenetic development, in the metamorphosis of spirit into matter and man into God. She would be there, too, as the source of anguish at the evanescence of the individual and of anger at the failure of naturalists and other researchers to recognize the transforming truth of evolution.

By contending that Anna would live again in the pages of Haeckel's great treatise, I mean that it would have been a very different work had she not died. The emotional valences would not have shifted to the ac-

3. Haeckel to Darwin (7 July 1864), in the Darwin Correspondence, DAR 166.1, Manuscript Room, Cambridge University Library. I have translated the entire letter in the appendix to this chapter.

4. Darwin to Haeckel (19 July 1864), in the Correspondence of Ernst Haeckel, the Haeckel Correspondence, Institut für Geschichte der Medizin, Naturwissenschaft und Technik, Ernst-Haeckel-Haus, Friedrich-Schiller-Universität, Jena.

5. Haeckel to Darwin (10 August 1864), in the Darwin Correspondence, Cambridge.

6. Haeckel to Darwin (26 October 1864), in ibid.

7. Ernst Haeckel, *Generelle Morphologie der Organismen*, 2 vols. (Berlin: Georg Reimer, 1866).

idly negative. The problem of the individual would not have occupied the theoretical place it does in the book or retain its hold on his subsequent investigations. And the chapter on evolutionary monism might well have been muted or eliminated had he not needed to preserve Anna as yet living in the bosom of a transcendent nature. In the Stettin lecture, Haeckel not only struck a conciliatory note with those opposed to evolution; he even suggested that a divine spark might have ignited the transformation of life on this planet.[8] But in the *Generelle Morphologie,* all of that turned to ashes, and out of the ashes a new and powerful nature arose: one that drew all individuals into a creative unity of *Deus sive natura.* I can, of course, provide no hard demonstration of my assertion and its corresponding counterfactual. I can offer only a tessellated pattern of remarks, confessions, allusions, theoretical divergences—evidence that resonates with a distinctive feeling.

Though Haeckel laid the plan for the *Generelle Morphologie* in the summer of 1864, he was still too distraught to commence the actual writing. In August he escaped to Switzerland, hiking through the Alpine valleys and passes, where once he had whiled away the days and evenings with Anna. He would have lingered there, but in mid-September he aggravated an old knee injury, which forced him to return. Back in Jena, he found it "unremittingly difficult to dwell in my deserted and lonely nest. For here all joy and all happiness, which I experienced in almost ideal measure, have turned to the most bitter grief and most heavy sorrow."[9] Haeckel sought refuge in his students and classes, which he suffused with Darwin's theory.

Haeckel undoubtedly would have appreciated a change in venue, and he got that opportunity when, in March 1865, he received word that he would be offered a full professorship in zoology from Würzburg, his alma mater. However, Moritz Seebeck, the civil administrator and conservator of Jena's ancient glories, quickly moved to thwart the attempt by persuading the university senate to create a new professorship of zoology in the philosophy faculty. After that, Haeckel was awarded a doctorate in philosophy,

8. In his Stettin lecture, Haeckel explicated Darwin's theory as holding that "no species, perhaps with the exception of the first, has thus been independently created." See Ernst Haeckel, "Ueber die Entwickelungstheorie Darwins," in *Amtlicher Bericht über die acht und dreissigste Versammlung Deutscher Naturforscher und Ärzte in Stettin* (Stettin: Hessenland's Buchdruckerei, 1864), 20. In the reprint of this lecture, Haeckel altered the concession: "No species, not even with the exception of the first, has thus been independently created." See Ernst Haeckel, "Ueber die Entwickelungstheorie Darwins," in *Gesammelte populäre Vorträge aus dem Gebiete der Entwickelungslehre* (Bonn: Emil Strauss, 1878), 10.

9. Haeckel to Allmers (20 November 1864), in *Ernst Haeckel, Eine Schriftenfolge,* 2:44.

honoris causa, so that he might take up his new position of ordinarius professor of zoology. Not inconsequentially, he also received a hefty advance in salary to six hundred *Reichstaler.* In subsequent years, Haeckel would often be tempted by other universities (Vienna in 1870 and 1872; Strassburg in 1872; Bonn in 1874); and when Gegenbaur moved to Heidelberg in 1873, he tried to get his old friend to his new university—but Haeckel's Munich lecture of 1877 and his subsequent explosion over Virchow's new illiberalism sank that effort (a matter I will discuss in chapter 8). With each succeeding offer, Haeckel's salary at Jena climbed with generous steps.[10] During 1865, though, he seems to have thought little of his future, only of his happy past and miserable present.

Haeckel weathered his despair more easily through travel. In May 1865 he inquired of his new acquaintance Thomas Henry Huxley whether any British expedition might be planned to the Southern Hemisphere, since he sought to escape "the considerable unhappiness" with which Europe oppressed him.[11] And in August he traveled to Helgoland, returning to those islands where a dozen years before Johannes Müller took him in hand and showed him the wonders of the marine invertebrates. Now Haeckel served as a guide for his student and assistant Anton Dohrn (1840–1909), who became himself a significant researcher and the founder of the Naples Zoological Station. Dohrn's very name (approximately "thorn" in German) might have provided a hint of trouble to come, but only later would Haeckel muse on the warning nominally conveyed.[12] On his return from Helgoland, Haeckel

10. Georg Uschmann details the terms of Haeckel's several calls to other universities in his *Geschichte der Zoologie und der zoologischen Anstalten in Jena 1779–1919* (Jena: Gustav Fischer, 1959), 76–82.

11. Haeckel to Huxley (7 May 1865 and 11 November 1865), in "Der Briefwechsel zwischen Thomas Henry Huxley und Ernst Haeckel," ed. Georg Uschmann and Ilse Jahn, *Wissenschaftliche Zeitschrift der Friedrich-Schiller-Universität Jena* (Mathematisch-Naturwissenschaftliche Reihe) 9 (1959–60): 7–33; citation on 9, 10. The acquaintance with Huxley began by Haeckel sending him, in 1862, a copy of his radiolarian monograph.

12. After a desultory university career, punctuated by brief service in the army, from which he was dismissed because of unbridled political expression, Anton Dohrn came to Jena to work on a medical degree in 1862. He presented a brief paper on hermaphroditism at the Stettin conference at which Haeckel gave his plenary lecture on Darwin. After having difficulty in securing a Ph.D. at Berlin, he finally completed it at Breslau in 1865. He then returned to Jena, where he sought Haeckel's tutelage. The younger man acted as Haeckel's assistant when they traveled to Helgoland in 1865. Haeckel's lectures on Darwin's theory in 1865–66 sparked Dohrn's desire to habilitate at Jena. Gegenbaur, however, told the new applicant that he did not have the talent for scientific work. Haeckel yet took him on, and Dohrn wrote his habilitation on the embryology of arthropods with Haeckel. According to new rules, he had to be tested in this area, as well as two others. He chose physiology and philosophy, the latter of which he undertook with Kuno Fischer. Even after Haeckel agreed to sponsor him, Dohrn could write

could not resist a visit with his old friend Allmers in Rechtenfleth. When the new term began at the end of October, he buried himself in his courses. He was especially devoted to a public lecture series on Darwin's theory, which (from 3 November 1865 to 9 March 1866) drew audiences of between 120 and 150 students and faculty, about twice the size of the next largest public lecture series.[13] And in his free moments, he distracted himself with a study of Kant under the tutelage of the great historian of philosophy Kuno Fischer, who had recently "translated Kant into German."[14] Amidst all of this intense intellectual activity, Haeckel began the actual composition of what became his *Generelle Morphologie der Organismen*. By the next October, he finished the manuscript. In one year's time, he wrote a two-volume book that in its printed form runs to over one thousand pages. It began in despair, advanced through anger, and ended in an encomium to transcendent nature. It contains the foundation for all of Haeckel's later thought.

In what follows, I will characterize the main features of Haeckel's great book. I will be especially concerned to show that these features are quite in harmony with Darwin's own theory of organic transformations in na-

---

the most condescending and vituperative letters to his mentor. When Haeckel suggested that Kant would have written differently had he known of Darwin's theory, Dohrn asked: "Did you read it [Kant's third *Critique*] entirely and thoroughly? Or did you take only fragments from a history of Philosophy?" Dohrn was likely referring to Fischer's volumes on Kant. In 1860 Fischer had published volumes 3 and 4 of his *Geschichte der neuern Philosophie* (Mannheim: Bassermann, 1860)—devoted to Kant's life and works—and *Kant's Leben und die Grundlagen seiner Lehre: 3 Vorträge* (Mannheim: Bassermann, 1860). Dohrn later became a strident opponent of both Haeckel and Gegenbaur, and even launched a campaign in collusion with Du Bois-Reymond to block a possible call of Gegenbaur to Berlin. For further accounts of Dohrn's trajectory, see Uschmann, *Geschichte der Zoologie*, 82–87; and Theodor Heuss, *Anton Dohrn: A Life for Science*, ed. Christiane Groeben, trans. Liselotte Dieckmann (1940; repr., Berlin: Springer, 1991). The quotation above is from the letter to Haeckel (June 1867) in Heuss, *Anton Dohrn*, 352–53. Mario Di Gregorio has an extended discussion of Dohrn's relationship with Gegenbaur and Haeckel in *From Here to Eternity: Ernst Haeckel and Scientific Faith* (Göttingen: Vandenhoeck & Ruprecht, 2005), 324–37.

13. A synopsis of Haeckel's lectures, along with the schedule of meetings and attendance figures, are in the Haeckel Papers, Haeckel-Haus, Jena. The course had nineteen meetings of one hour each. The first part discussed the question of animal organization and species from Aristotle through Cuvier, Lamarck, Geoffroy Saint-Hilaire, Goethe, Oken, Aggasiz, and Darwin. The second part of the course consisted in a more fine-grained analysis of Darwin's theory. He repeated the course the next year, and it became the foundation for his *Natürliche Schöfungsgeschichte* of 1868.

14. August Weismann made this facetious remark to Haeckel, who had recommended Kant for Weismann's recreational reading. The reference was likely to the volumes mentioned in the previous note. See Haeckel to Weismann (29 October 1865) and Weismann to Haeckel (4 December 1865), in "Der Briefwechsel zwischen Ernst Haeckel und August Weismann," ed. Georg Uschmann and Bernhard Hassenstein, in *Kleine Festgabe* (Jenaer Reden und Schriften, Friedrich Schiller-Universität), 12–14.

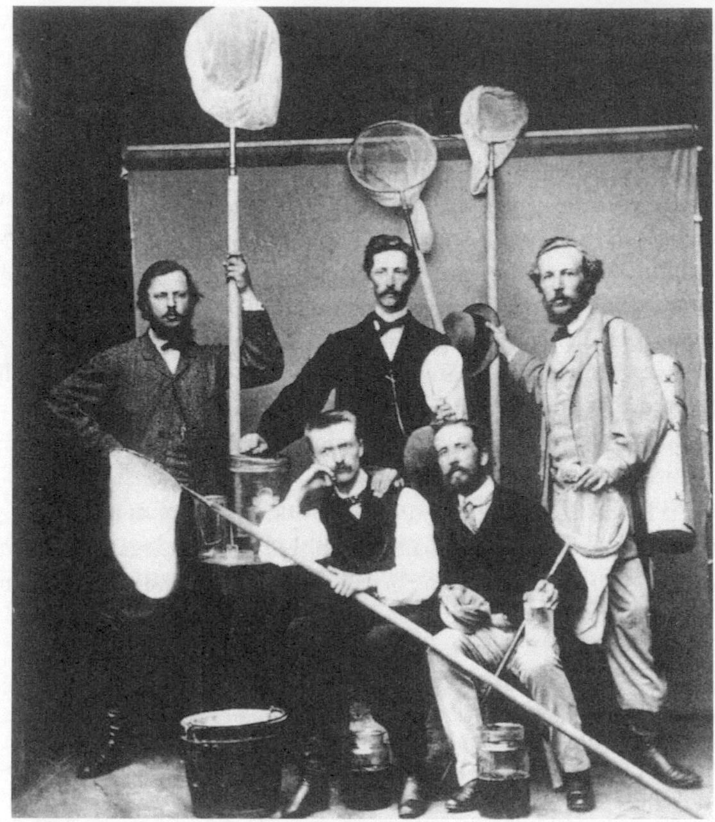

Fig. 5.1. Photo taken on the way to Helgoland (August 1865). *Standing from left:* Anton Dohrn, Richard Greef, Ernst Haeckel; *front from left:* Matthijs Salverda, Pietro Marchi. (Courtesy of the historical archives of Stazione Zoologica Anton Dohrn, Naples.)

ture. Mine is not the usual view of Haeckel's accomplishment; but it is, I believe, the justified view. In rendering this account, I will be mindful of the spirit that animated Haeckel's work and hovered over his life, that of his departed wife.

Haeckel's *Generelle Morphologie der Organismen*

The Reestablishment of Naturphilosophie

Haeckel's *Generelle Morphologie der Organismen* came to birth not simply as a hybrid of Darwinian evolution and personal anguish. The book reveals a considerable lineage of both declared and more cryptic progenitors.[15] Haeckel himself was quite forthcoming about his more obvious intellectual debts and about those individuals whose science provided inspiration for his own. His dedication of the first volume to Gegenbaur certainly indicates a deep affection (and sympathetic concern) but also suggests their protracted conversations about morphology and Darwinian theory. The dedication might also have served as an invitation to his friend for a greater public commitment to the evolutionary framework. He dedicated the second volume to "the three founders of descent theory": Darwin, Goethe, and Lamarck.[16] Lamarck's and Darwin's contributions to descent theory are clear enough. Goethe, according to Haeckel, established the fundamental principles of morphology, especially the proposal that various animal and plant characters could be understood as variations on some basic types. (Haeckel tried to convince Darwin that Goethe had embraced the rudiments of transformation theory and that he was one of the Englishman's predecessors.)[17] Haeckel's commitment to Darwinian theory did not, however, blind him to the important work of some who explicitly rejected transmutation, especially Georges Cuvier, Louis Agassiz, and Karl Ernst von Baer. Even the unanchored speculations of Lorenz Oken yielded usable insights—especially the idea that complex individu-

15. I have tried to make this lineage clear by sketching, in the first appendix to this volume, the history of morphology up to Haeckel's time, emphasizing what I believe to be central ideas that were reborn in his conception of evolution. This history puts in context and provides evidence against the assumption of many current scholars who regard Haeckel as hardly a scientist but rather as a mystical prophet who mesmerized several generations, a charlatan who has to be unmasked. The Haeckel that appears in their works certainly could not look Darwin in the face, so shabby were the notions attributed to him. The history I have sketched, by contrast, indicates that Haeckel not only drew on many of the same sources that formed Darwin's own conceptions, but more importantly, that beneath the distinctively Germanic outer layers, the core of his evolutionary morphology was essentially the same as the Englishman's.

16. Haeckel, *Generelle Morphologie,* 1:vii.

17. Haeckel, I believe, was quite correct. Goethe assumed that within the great classes of organisms substantial species change had occurred. For a discussion of Goethe's evolutionism, see my *Romantic Conception of Life: Science and Philosophy in the Age of Goethe* (Chicago: University of Chicago Press, 2002), chap. 11; see also the first appendix. On 10 August 1864, Haeckel suggested to Darwin that previous German thinkers had, from an a priori perspective, held descent to be the only way to understand the origin of species. He mentioned that "the best [of these German efforts] was by our greatest poet, Goethe, in his essays on morphology and especially in the critique of the 'Principes de Philosophie zoologique' of Geoffroy." (The letter is in the Haeckel Correspondence, Haeckel-Haus, Jena.) Darwin had already accepted Goethe's priority in the historical introduction to the *Origin of Species,* which first appeared in the third edition (1861).

als were decomposable into elemental organic forms that had their antecedents in simpler creatures.[18]

Haeckel's understanding of the nature of science came from the tutelage of his eminent teachers, Schleiden, Virchow, and Müller. Their science was neither the mindless collecting of specimens, of the sort that had been set out in the desiccated descriptions of the mere systematizers, nor was it the befogging *Naturphantasterei* of wild speculators. His teachers, he asserted, took the middle path that combined articulate conception and controlled experience. What Haeckel endorsed was, in the terms of his revered teacher Müller, "thinking experience [*denkenden Erfahrung*]," the kind of theoretically regulated observation that enabled the researcher to extract laws from experience and thus establish a proper *Naturphilosophie*.[19] Darwin himself conducted just this kind of science—a nature philosophy deserving of the name.

The Task of Evolutionary Morphology: The Formation of Organic Laws

Haeckel's definition of morphology followed Goethe's:

> In the widest sense of the term, morphology or the doctrine of forms of the organism is the complete science of the internal and external relations of forms of living natural bodies, of animals and plants.[20]

Haeckel intended, however, to go beyond Goethe's conception: he demanded that morphology become a proper science by specifying the natural laws that governed the formation of external and internal bodily structures. To this end his volume was stuffed with as many lawlike proposals as the municipal code of a small city—well over 140 of them. In the mid-nineteenth century, science demanded the formulation and promulgation of laws. Even Darwin felt compelled to devote an entire chapter of the *Origin of Species* to the "laws of variation"; and he referred to natural selection as one of the laws of nature.[21] The German tendency to legal proclamation was undoubtedly hypertrophied compared to the English; but in this regard, Haeckel

18. Haeckel, *Generelle Morphologie,* 2:161. See the first appendix for further discussion of these individuals.

19. Haeckel, *Generelle Morphologie,* 1:64, 67.

20. Ibid., 1:3.

21. Charles Darwin, *On the Origin of Species* (London: Murray, 1859), chap. 5 ("The Laws of Variation") and 489–90.

was more restrained than commonly thought. He judiciously regarded his "laws," as he remarked, more in the way of "theses" that had to be further developed and tested to produce genuine laws.[22] But what exactly was the status of organic law? Was it mechanistic, teleological, or both? Kant, whose third *Critique* Haeckel knew fairly well, had a rather complicated answer.[23]

According to Kant, the actions of non-organic bodies could, *in principle,* properly and exhaustively be explained by appeal to mechanistic laws alone. Living creatures displayed both mechanical properties—for instance, the refraction of light by the lens of the eye—and nonmechanical properties—for instance, the purposeful situation of the lens in the eye so as to focus light on the retina. Though the goal of the biologist, according to Kant, was to reduce as far as possible the teleological properties of organisms to mechanical properties, that goal, he argued, could not be ultimately achieved. Some features of organisms would forever escape a properly scientific, that is, mechanistic account. In view of such irreducibly telic properties, the biologist would be epistemically required to postulate an idea or plan—*Bauplan*—as the ground of intelligibility. Such postulation logically implied a creative intellect, an *intellectus archetypus,* that brought such plans into existence. Kant nevertheless rejected the theoretical use of the teleological implication to conclude to the actual existence of a supernatural intelligence. The employment of this principle of the *Bauplan* had, he maintained, to be restricted to that of a regulative guide, one that might suggest avenues of approach to and discovery of mechanistic laws that would give further explanatory perchance on organisms. Ultimately, however, Kant presumed that mechanistic principles could never fully account for the teleological structures exhibited by living beings. "A Newton of the grass blade" would never arise.

In the next generation, Romantic thinkers such as Schelling and Goethe perceived the consequence of Kant's analysis: biology failed as a proper science; it must always remain only a set of empirical generalizations tied together heuristically. But Schelling argued—and Goethe was persuaded by his younger colleague—that if teleological assumptions were epistemically necessary to make (biological) experience intelligible, they hardly differed in that regard from mechanistic principles. In the first *Critique,* Kant had justified the categories, whence a priori laws of mechanics were derived, by

22. Haeckel, *Generelle Morphologie,* 1:364–65n1.

23. My depiction of Kant and the Romantics in what follows is based on my *Romantic Conception of Life.*

arguing that they made the very structure of experience comprehensible. Thus both teleological principles and mechanistic principles had equivalent explanatory warrants—or so Schelling maintained.

Haeckel explicitly formulated his own conception of organic law through reflection on the position of his mentor Johannes Müller, who himself was a legatee of the Kantian and Romantic traditions. Haeckel quoted the following passage from Müller's *Handbuch der Physiologie des Menschen* (Handbook of human physiology, 1833–40) to initiate his considerations:

> A mechanistic contrivance is produced by the technician according to the idea that he has in mind, that is, according to the purpose [*Zwecke*] of his action. An idea lies at the foundation of every organism, and according to this idea all its organs become purposively organized. The idea is external to the machine but internal to the organism, and therein it shapes the organism with necessity and without intention [*ohne Absicht*]. The purposively effective cause of the corporeal body, thus, has no [free] choice, and the realization [*Verwirklichung*] of its particular plan is its necessity. Moreover, to operate purposively and to operate necessarily in this effective cause are one and the same.[24]

Müller's conception, as Haeckel recognized, retained the hue of the old *Naturphilosophie* insofar as he attempted to keep telic purposiveness and mechanistic necessity in balance.[25] Haeckel believed, however, that the Müllerian formulation still allowed—if one followed its tracks back to Kant and the Romantics—the implication to remain: the ideas governing organisms must ultimately be those of a creative intelligence. So Haeckel offered this corrective gloss on Müller: the idea that operates "with necessity and without intention" can only be "the force [*Kraft*] that is inseparably bound to the material substrate of the organism"—that is to say, the physical causal force. Therefore, in the final analysis, the teleological cause must disappear into the mechanical: "so the mechanistic conception of the organism is recognized alone as the right one."[26] And with mechanism came reduction.

According to Haeckel, the higher regularities—those characteristic, say, of trait structure and inheritance—expressed more fundamental,

24. Haeckel, *Generelle Morphologie,* 1:94.

25. I have discussed the varieties of ways in which *Naturphilosophie* dealt with the concepts of purpose and mechanism in my *Romantic Conception of Life,* chaps. 5–9.

26. Haeckel, *Generelle Morphologie,* 1:95.

mechanistic relationships at the atomic level. Both living and nonliving matter obeyed the same chemical and physical laws. The active content of the cell (variously called "plasma," "protoplasm," and "cytoplasm") did consist of albumin, a carbon-based compound distinctive of living matter; but beyond such chemical and atomic properties, no special powers, no *Lebenskräften* (vital forces) were required to explain physiological processes or morphological forms.[27] In Haeckel's view, the plasma that composed the body of the cell had nutritive functions and stored, as it were, the effects of adaptations that the organism underwent, while the nucleus held the hereditary material and was chiefly responsible for reproduction. (Haeckel was the first biologist to identify the nucleus as the repository of the hereditary substance).[28] The distinctive properties of life thus arose naturally from the elective affinities of its chemical elements. Even consciousness and thought, Haeckel urged, must ultimately be due to chemical bonds constituting the matter of the organism.[29]

In subsequent decades, as Haeckel squared off against colleagues who objected not to the thesis of evolutionary transformation but to the Darwinian device of natural selection, the status of organic law would play around the edges of their disputes, remaining often in the shadows but controlling the direction of the struggles. His former teacher Kölliker, for instance, would reject what he claimed were Darwin's teleological principles of development; he proposed instead general mechanistic principles, though ones that

27. Ibid., 1:115–20, 275, 364–65.

28. See ibid., 1:287–89: "Accordingly, insofar as we are able to regard the plasma chiefly as the nutritive component of the cell and, on the other hand, the nucleus as the reproductive component, . . . we are justified in regarding the nucleus as the principal organ of inheritance and the plasma as the principal organ of adaptation [*Anpassung*]. In the case of the cytode [the non-nucleated moneron], where nucleus and plasma are not differentiated, we will have to regard the entire plasma as the common organ having both functions." Haeckel's identification of the nucleus as the carrier of hereditary material was based on reasonable supposition, but without definite experimental evidence. Oscar Hertwig, Haeckel's student, is largely responsible for the experimental identification of the joining of the nuclei of egg and sperm during fertilization in the sea urchin. See Oscar Hertwig, "Beiträge zur Kenntniss der Bildung, Befruchtung und Theilung des thierischen Eies," *Morphologisches Jahrbuch* 1 (1876): 347–434. Hertwig recalled that it was in Haeckel's laboratory that he and his brother Richard became interested in the functions of the cell nucleus and protoplasm. See Oscar Hertwig, "Die Geschichte der Zellenlehre," *Deutsche Rundschau* 20 (1879): 417–29. Paul Weindling discusses Hertwig's contribution to cell theory—and much else—in his splendid *Darwinism and Social Darwinism in Imperial Germany: The Contribution of the Cell Biologist Oscar Hertwig (1849–1922)* (Stuttgart: Gustav Fischer, 1991). William Coleman relates the history of cell theory in his comprehensive essay "Cell Nucleus and Inheritance: An Historical Study," *Proceedings of the American Philosophical Association* 109 (1965): 124–58. Coleman, however, failed to recognize Haeckel's contribution to the establishment of the hereditary function of the cell nucleus.

29. Haeckel, *Generelle Morphologie,* 1:119.

silently gazed heavenward. Embryologists such as Wilhelm His (1831–1904) and Alexander Goette (1840–1922) would insist that mechanistic laws, intrinsic to their particular discipline, could fully explain the structures of developing organisms; researchers required no further help from extrinsic evolutionary principles. And Virchow, once Haeckel's revered master and later a politically powerful opponent, would refuse to recognize any causal laws that could not be observed operating in the moment; and natural selection could never be so empirically caught out.[30]

The Metaphysics of Life

Haeckel's reiterated insistence that life exhibited no unique powers appears to rest on a completely materialistic metaphysics. His analyses of the nature of organic laws would seem to endorse this philosophical stance. Certainly many of his critics, especially those of a neo-Kantian bent or of a religious inclination, dismissed his version of evolution as sheer mechanistic materialism.[31] But to contend that the same laws govern both the organic and inorganic could, nonetheless, be interpreted as a vitalization of matter as much as a materialization of life.

The tragic death of Haeckel's wife refracted his line of militant, antitheological remarks, so that its metaphysical source has been displaced from its true position. But the counter-inclinations—the mitigation of death, the discovery of the spirit of the beloved reincarnated in a golden medusa—these forces led Haeckel to the ultimate conviction that the living and nonliving could not be distinguished, that one was simply a phase of the other. Such a conception does not denigrate the wonders of life but ennobles the properties of matter. When he focused directly on the metaphysical question, which he did at the end of his two volumes, he endorsed not sterile materialism but the kind of monism that was rooted firmly in Romantic Jena at the beginning of the nineteenth century and that branched out into many intellectual areas by the end of the century. Not only Haeckel but philosophers and scientists of quite different stripes—such individuals as

30. See chapter 8 for discussions of the objections to Haeckel brought by the aforementioned.

31. Eduard von Hartmann—author of the famous *Philosophie des Unbewussten* (Philosophy of the unconscious)—maintained, in friendly correspondence with Haeckel, that materialism was the fundamental error of Darwinism. In response, Haeckel suggested that their common commitment to monism brought them closer together than Hartmann believed. See Hartmann to Haeckel (30 October 1874) and Haeckel to Hartmann (4 November 1874), in "Metaphysik und Naturphilosophie: Briefwechsel zwischen Eduard von Hartmann und Ernst Haeckel," ed. Bertha Kern-von Hartmann, *Kant Studien* 48 (1956–57): 3–24; letters cited on 4–7.

William James and John Dewey, Bertrand Russell and Ernst Mach—would advance the doctrine of neutral monism. That doctrine held that mind and matter were properties of a more fundamental substrate that was not to be identified with either of its salient traits. Haeckel adopted this metaphysical position earlier on, in the *Generelle Morphologie*; and it would become the foundation for his "monistic religion."

Haeckel credited two sources for these monistic views: his friend, the linguist August Schleicher, and his spiritual guide, Goethe. Haeckel had convinced Schleicher to read Bronn's translation of Darwin's *Origin*; he recommended the book because of his friend's avid interest in gardening. Schleicher the gardener did review the volume for an agricultural journal, but it was the linguist who had become transfixed.[32] He perceived that his own discipline might serve as a perfect complement to the new biology. Linguistics, he thought, could satisfy Bronn's request that concrete evidence lift Darwin's theory from the realm of the merely possible.[33] In a small tract addressed to Haeckel—*Die Darwinsche Theorie und die Sprachwissenschaft* (Darwinian theory and the science of language, 1863)—Schleicher proposed that languages provided the missing evidence to render the idea of historical transmutation a reality.[34] Languages, he maintained, were natural, historical phenomena; and modern languages, it was perfectly obvious, had descended from earlier languages—linguistic fossils existed to demonstrate this descent. Hence we had ample evidence in language of the kind of evolutionary transitions that Darwin's theory could only project but not prove. I will return to this aspect of Schleicher's analysis below, since it had a profound impact on Haeckel's theories of human evolution (and Darwin's as well). But Schleicher's metaphysics, which underlay his argument, provided, if not inspiration, at least confirmation of the monism that became the foundation for Haeckel's evolutionary conceptions.

32. August Schleicher, "Die Darwin'sche Theorie und die Thier- und Pflanzenzucht," *Zeitschrift für deutsche Landwirthe* 15 (1864): 1–11.

33. See chapter 3 for a discussion of Bronn's objections to the *Origin of Species*.

34. August Schleicher, *Die Darwinsche Theorie und die Sprachwissenschaft* (Weimar: Hermann Böhlau, 1863). I have detailed Schleicher's evolutionary linguistics and its sources in "The Linguistic Creation of Man: Charles Darwin, August Schleicher, Ernst Haeckel, and the Missing Link in Nineteenth-Century Evolutionary Theory," in *Experimenting in Tongues: Studies in Science and Language*, ed. Matthias Doerres (Stanford, CA: Stanford University Press, 2002). See also Liba Taub, "Evolutionary Ideas and Empirical Methods: The Analogy between Language and Species in Works by Lyell and Schleicher," *British Journal for the History of Science* 26 (1993): 171–93; and Stephen Alter, *Darwinism and the Linguistic Image* (Baltimore: Johns Hopkins University Press, 1999), especially 73–79.

In *Darwinsche Theorie,* Schleicher expressed his monistic position this way:

> Thought in the contemporary period runs unmistakably in the direction of monism. The dualism, which one conceives as the opposition of mind and nature, content and form, being and appearance, or however one wishes to indicate it—this dualism is for the natural scientific perspective of our day a completely unacceptable position. For the natural scientific perspective there is no matter without mind [*Geist*] (that is, without that necessary power determining matter), nor any mind without matter. Rather there is neither mind nor matter in the usual sense. There is only one thing that is both simultaneously. To accuse this opinion, which rests on observation, of materialism is as perverse as charging it with spiritualism.[35]

For Schleicher, the doctrine of monism provided a metaphysical ground for his theory that the organism of language simply represented the material side of mind—which meant, therefore, that the evolution of one carried the evolution of the other. This organic naturalism had its roots in the German Romantic movement, which likely attracted Haeckel to its possibilities.[36]

Haeckel referred to his friend's monistic doctrine several times in the *Generelle Morphologie* and quoted the above passage, noting that Schleicher's work would have profound consequences for understanding the evolution of the human mind.[37] But Haeckel's deepest and most lasting debt in this, as well as many other areas of his thought, was to Goethe. Each of the thirty chapters of *Generelle Morphologie* begins with a quotation from one of Goethe's scientific or poetic works.[38] The last chapter is introduced with an epigram from *Faust.* The scene occurs in a garden, just before Faust seduces Gretchen. She is in love with him but worries about whether he believes in God. Faust explains his attitude, and Haeckel quotes the passage:

> Who dares name him? And who declare: I believe in him?
> Who might feel so debased as to say: I don't believe in him?

35. Schleicher, *Darwinsche Theorie,* 8. Haeckel quotes this passage in *Generelle Morphologie,* 1:105.

36. See Richards, "The Linguistic Creation of Man," for a sketch of the Romantic sources of Schleicher's thought.

37. Haeckel cites Schleicher in the *Generelle Morphologie,* 1:105–7, 2:448–49.

38. Haeckel, however, never indicated from which of Goethe's works he was quoting. He undoubtedly presumed that his German readers would recognize the passages.

The All-comprehensive, the All-preserver,
Doesn't he comprehend and hold you, me, himself?
Aren't the heavens vaulted all-round?
Doesn't the earth lie steady here below?
And do not the eternal stars, twinkling so companionably, rise above?[39]

This passage concludes on a note to which Haeckel would certainly have resonated. Faust exhorts his love:

Fill your heart [with the wonders of nature], so great are they,
And when you are completely blessed in the feeling,
Call it what you will,
Call it happiness, call it heart or love, call it God!
I have no name for it.
The feeling is all;
The name is noise and smoke
That clouds over the heavenly radiance.[40]

The verse suggests that the poet's heart responded to nature with a feeling that might well be called religious. In a letter to his friend Friedrich von Müller, a Weimar administrator, Goethe expressed his deep-seated, Spinozistic conviction that mind and body, spirit and nature were indissolubly linked. Haeckel cited these remarks as the foundation for his own monistic views:

> Since no matter exists without mind [*Geist*], mind never exists nor can it be effective without matter. And just as matter, like mind, can advance [*steigern*], so mind cannot be denied its ability to attract and repulse. That is to say, an individual is able to think insofar as he can sufficiently analyze in order to synthesize, and can sufficiently synthesize in order again to analyze.[41]

Nature writ large, then, has both its material side and its mental side. But the mental side, Haeckel was quick to indicate, should not be anthropomor-

39. Johann Wolfgang von Goethe, *Faust, Eine Tragödie*, in *Sämtliche Werke nach Epochen seines Schaffens* (Müncher Ausgabe), ed. Karl Richter et al., 21 vols. (Munich: Carl Hanser, 1985–98), 6.1:635 (lines 3433–46).

40. Ibid., 6.1:635–36 (lines 3452–59).

41. Goethe to Müller (24 May 1828), in *Goethe-Briefe*, ed. Philipp Stein, 8 vols. (Berlin: Wertbuchhandel, 1924), 8:251. Haeckel quoted these lines in his *Generelle Morphologie*, 2:449.

phized, lest it turn God into a degraded "gas-bag of a vertebrate [*gasförmige Wirbelthier*]." Rather, according to the elevated doctrine of monism, "God is almighty, the only original Creator, the fundamental cause of all things." Though for Haeckel, this meant: "*God is the comprehensive causal law.*" He is, in this doctrine, "the summation of all power, and consequently of all matter."[42] God is thus one with nature. "Monism," as Haeckel portrayed it, was "the purest kind of monotheism."[43]

For those who believed in a personal God, as Haeckel himself once did, monistic metaphysics could only be viewed as transparently shrouded atheism. But for the new scientific men of the second half of the nineteenth century—Darwin, Huxley, Spencer, Baldwin, Mach, Helmholtz—this doctrine perfectly mirrored their own deep convictions, even if its polemical cast was pure Haeckel. The metaphysics that underlay Haeckel's biology was well-suited to the science of the late nineteenth century and, I think, to that of our own time. It also allowed Haeckel to believe that the force of a once-living soul might be brought back into the beating heart of nature, since the conservation laws indicated that neither force nor matter could be destroyed. Anna would not die forever.

The Biological Individual

The *Generelle Morphologie* deals with many areas of the new evolutionary biology, but none more central, at least to Haeckel's conception of it, than the problem of the biological individual. He was introduced to this problem in his consideration of plants, especially under the guidance of Alexander Braun, with whom he studied briefly in 1852. Though some would find individuality in plants only in the whole species, Braun concluded that the *Sprossung* (bud)—which might give rise to stem, leaf, or flower—met the intuitive criteria for being an individual, namely, the criteria of unitary separability and indissoluble synthesis. In this view, the plant itself formed a "family unity [*Familienverein*] of individual buds."[44]

The problem of the individual—or "techtology," as Haeckel would call

42. Haeckel, *Generelle Morphologie*, 2:451.

43. Ibid., 2:448.

44. Alexander Braun, "Das Individuum der Pflanze in seinem Verhältniss zur Species," *Abhandlungen der königlichen Akademie der Wissenschaften zu Berlin, aus dem Jahre 1853* (1854): 19–122; citation from 29. Ruth Rinard offers several suggestions about the contributions Braun might have made to Haeckel's conception of the relativity of individuality. See her "The Problem of the Organic Individual: Ernst Haeckel and the Development of the Biogenetic Law," *Journal of the History of Biology* 14 (1981): 249–75.

the study—again arose for him when he became a student of Rudolf Virchow and then of Johannes Müller. Virchow had argued that in both plants and animals the ultimate unit of life was the cell. In this conception, the larger organism consisted of a confederation of cells. Just as in the state, individuals had specific functions but operated as a harmonious whole, so in the plant or animal, individual cells with specific tasks formed a kind of social organization, a cell state.[45]

Haeckel also found this idea of the individual as a whole of confederated parts in Johannes Müller. In his *Handbuch der Physiologie der Menschen,* Müller likewise proposed that plants and animals formed a unity of integrated individuals. The various parts of plants, if stuck in the ground, themselves could reproduce the whole. Each part had the quality of the leaf, which Müller, in Goethean fashion, regarded as the true individual.[46] Thus the plant itself constituted not a unique individual but an association of individuals. In the same manner as plants, many kinds of animals, especially marine organisms, had the ability to reproduce themselves from parts; the hydra, for instance, was composed of individual organic parts, such that the whole animal, much like a plant, might be cut in half or quarters and subsequently develop into two or more individuals.[47] Certain

45. Rudolf Virchow, *Die Cellularpathologie in ihrer Begründung auf physiologische und pathologische Gewebelehre* (Berlin: Hirschwald, 1858), 12: "Each animal appears as a sum of vital unities, of which each bears the full character of life. . . . It follows from this that the synthetic unity of a larger body always arises from a kind of social arrangement, an arrangement of a social kind, where a mass of particular existences depend on one another." The most comprehensive study of the cell-state metaphor is by Andrew Reynolds, in his "The Theory of the Cell State and the Question of Cell Autonomy in Nineteenth and Early Twentieth Century Biology," *Science in Context* 20 (2007): 71–95.

46. Johannes Müller, *Handbuch der Physiologie des Menschen für Vorlesungen,* 2 vols. (Coblenz: J. Hölscher, 1833–40), 2:592: "The leaf of the plant must itself be regarded as the individual, containing as it does the entire essence of the plant of a particular kind in respect of its nature and potential and being able to develop its branches. Out of the leaves most of the parts of the plants develop, and the doctrine of metamorphosis indicates that all the parts of the flower are only transformed leaves." Müller's conception of the essence of the plant was identical to Goethe's. (Note: Müller's first volume was published in two parts, in 1833 and 1834; the second in three parts, in 1837, 1838, and 1840.)

47. Ibid., 2:593: "This system of the hydra contains individuals that can move independently and can be separated, so that each can achieve a minimally separate form and no longer constitute a multipla." Müller referred to the famous experiments of Abraham Trembley, who showed that freshwater hydras display many of the characteristics of plants, particularly in their ability to regenerate after being cut in halves, quarters, etc. Trembley, however, was cautious about generalizing his findings. See Abraham Trembley, *Mémoires pour server à l'histoire d'un genre de polypes d'eau douce, à bras en forme de cornes,* 2 vols. (Paris: Durand, 1744). This is translated with excellent reproductions and introduction in *Hydra and the Birth of Experimental Biology—1744,* trans. and ed. Sylvia Lenhoff and Howard Lenhoff (Pacific Grove, CA: Boxwood Press, 1986).

simple worms, though displaying a greater unity of action, could likewise be cut into several pieces with each becoming independent creatures. Even more complex animals—insects, crustaceans, and so on—could regenerate limbs and other organs. This kind of evidence led Müller to suggest that the higher animals must also be understood as *Multipla* (assemblies) of more elemental units, namely cells that had powers of organic specialization during growth and the ability, as sex cells, to reproduce the whole.[48]

Virchow and Müller provided their analyses of biological individuality in light of the newly formulated doctrine of the cell as the fundamental unit of life. Cells had been observed microscopically before the 1830s, but the idea that all structures of plants and animals could be decomposed into cells was confirmed only at the end of that decade by the work of Matthias Jakob Schleiden and Theodor Schwann (1810–1882). In an essay published in Müller's *Archiv* in 1838, "Beiträge zur Phytogenesis" (Contributions to phytogenesis), Schleiden, with the use of a powerful microscope, traced out the development of the cell and propounded the theory that the plant was a community of cells—a *Polypstock,* as he called it.[49] Schwann, who was in personal contact with Schleiden, extended his colleague's theory to animals: every animal structure, he concluded, was formed from cells.[50] Schwann believed the genesis of cells was similar to crystallization, with cells freely forming in intercellular fluid. Despite Schwann's orthodox Catholic conservatism, he was quite content to explain the formation of the fundamental units of life by a mechanical process that appeared to be something like spontaneous generation.[51]

Cell theory established the ground limit for the concept of biological individuality, but certainly did not exhaust, at least for Haeckel, the meaning of the concept. His experience with marine organisms made the problem yet more pressing. Medusae (jellyfish), for instance, have unusual life cycles that do not allow one easily to determine how to apply the concept of individuality. The Discomedusae, for example, go through a process of alternating generation (see fig. 5.2): the free-swimming adult

48. Müller, *Handbuch,* 2:597.

49. Matthias Jakob Schleiden, "Beiträge zur Phytogenesis," *Archiv für Anatomie, Physiologie und wissenschaftliche Medicin* (1838): 137–76.

50. Theodor Schwann, *Mikroskopische Untersuchungen über die Übereinstimmung in der Struktur und dem Wachstum der Tiere und Pflanzen* (Berlin: Sander'schen Buchhandlung, 1839).

51. The fate of cell theory in the latter half of the nineteenth century is given a comprehensive account in Coleman, "Cell, Nucleus, and Inheritance."

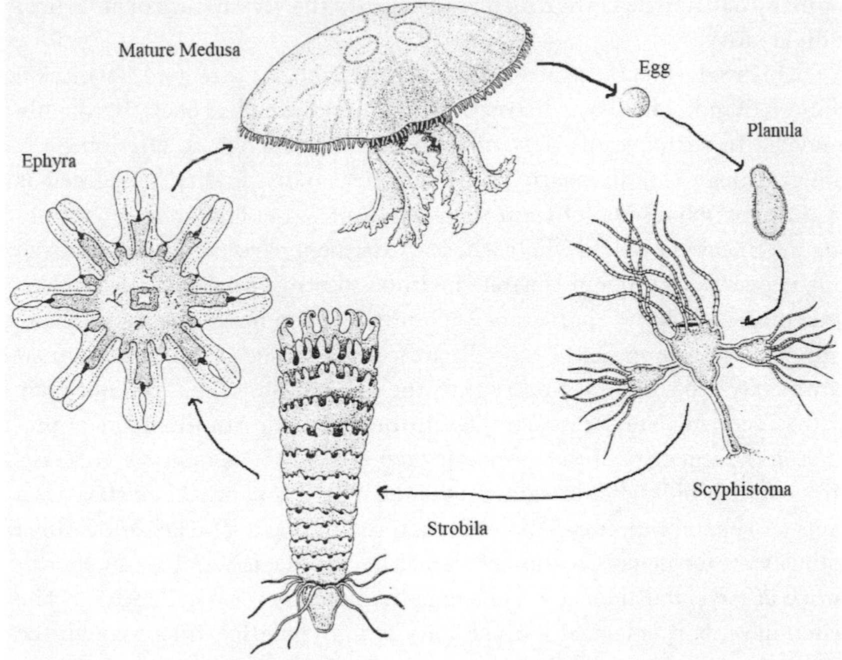

Fig. 5.2. Life cycle of a medusa.

jellyfish reproduces sexually; and from the fertilized egg comes a larva that turns into a hydra-like animal (a scyphistoma) that plants itself on the seafloor; this creature asexually generates buds (and often a stacked structure called a strobila), which release organisms that grow into adult, sexually reproducing medusae—and the cycle begins anew. Some types of free-swimming medusae—siphonophores, for instance—are morphologically colonial animals, with their organs formed by individuals specialized for particular tasks—digestion, motility, reproduction, and so on (see plate 6). Haeckel knew of the basic character of siphonophores from investigations by Kölliker and Gegenbaur; and he himself would undertake a detailed experimental and prizewinning study of the organism in 1867 (see chapter 6).[52] Plants and colonial animals made problematic for Haeckel and

52. Albert von Kölliker, *Die Schwimmpolypen; oder, Siphonophoren von Messina* (Leipzig: Wilhelm Engelmann, 1853); Carl Gegenbaur, "Neue Beiträge zur näheren Kenntniss der Siphonophoren," *Nova Acta Leopoldina* 28 (1859): 333–424. Haeckel's own study is *Zur Entwickelungsgeschichte der Siphonophoren* (Utrecht: C. Van der Post Jr., 1869). The study was conducted during his trip to the Canary Islands in 1866.

for many naturalists of the nineteenth century the very nature of biological individuality.

In his analysis of the concept, Haeckel formulated a threefold distinction of kinds of individuality, a distinction that he believed had been thoroughly muddled in the previous literature: that of morphological, physiological, and genealogical individuality. Morphological individuality he defined as "a unifying expression of form that constitutes a complete and continuously connected whole, a whole whose constituent parts cannot be removed and, in general, cannot be separated without destroying the nature or character of the whole."[53] For instance, a human person has a certain unifying form, which is composed of subordinate forms: organic structures, bilateral symmetry, and serial symmetry (e.g., the form of the backbone). But none of these constituent forms can be eliminated while simultaneously preserving the integrity of the morphological type of the person. By contrast, physiological individuality constitutes the real, living organism that has a unity of general function—that is, it can maintain itself. Haeckel defined it thusly: "a unified expression of forms that is able, for a longer or shorter period of time, to lead its own existence completely independently."[54] The criterion of morphological individuality is indivisibility; thus, though the morphology of the calcareous sponge has component forms (e.g., the form of the spiculae, mouth, pores, etc.; see fig. 6.7), you cannot cut out any of the subordinate forms and retain the morphology of the whole. The criterion of physiological individuality, on the other hand, is self-maintenance; that is, a real sponge might be cut in two, resulting in a pair of creatures that could lead independent lives.[55] Both of these concepts of individuality are obviously closely related: one considers form in the abstract, the other its living embodiment. For both notions, Haeckel distinguished six levels of individuality:

1. The plastid, or elementary form (e.g., the form of the cell)
2. The organ, either the homogeneous form (e.g., structure of skin or bone) or the heterogeneous, the organ system (e.g., structure of heart or stomach)
3. The antimere or oppositional form (e.g., bilateral symmetry)
4. The metamere or segmental form (e.g., backbone)

53. Haeckel, *Generelle Morphologie,* 1:265.
54. Ibid., 1:266.
55. Ibid., 1:268.

5. The person, or *bion* (e.g., the structure of the bud in plants or of a particular vertebrate animal)[56]
6. The colony (e.g., structures of plants or colonial animals)[57]

These levels of individuality can be thought of in the abstract, as structural forms, or as represented by individual organisms. Thus, in ascending order according to the categories, morphological individuals would find their correlate, respectively, in such real animals as one-celled protists, algae, simple plants, worms, higher animals, and most plants and many medusae.

One kind of relationship between the morphological individual and the physiological individual would have crucial significance for Haeckel's fundamental evolutionary project. He observed that physiological individuals, in their ontogenetic development, displayed morphological forms characteristic of the chain of ancient ancestors, a chain that reached back to the beginnings of life. So, for example, the human being starts out as an egg—a form-individual of the first order. At fertilization and with the assembly of the cell mass, it is comparable to an organ form (of homogeneous character); with the primitive streak, it becomes a bilateral individual, or a form-individual of the third order; with the development of the vertebrae, it reaches the metameral stage; and it finally ends as a person displaying the form of the fifth order.[58] Haeckel epitomized this kind of developmental series in his biogenetic law, that ontogeny recapitulates phylogeny. I will discuss the law at greater length below.

In addition to morphological and physiological individuals, Haeckel distinguished genealogical individuals, of which there were three orders. The first order consisted of the reproductive cycle of a person (i.e., a *bion*) from conception to maturity. This type of individual comprises a unity of different morphological stages during the life of a single physiological individual (e.g., a human physiological individual who displays over time the morphology of a single cell, then of an organ structure, right up through the forms of juvenile and adult). The collection of similar reproductive cycles during an extended temporal period constitutes the species, which,

56. Haeckel's friend the linguist August Schleicher mildly criticized Haeckel's choice of the neologism of "Bion" as not derivable from the Greek as Haeckel had thought. See Schleicher to Haeckel (8 December 1866), in the Haeckel Correspondence Haeckel-Haus, Jena. Generally, however, Schleicher approved heartily of the book.

57. Haeckel, *Generelle Morphologie,* 1:266.

58. Ibid., 1:267.

in Haeckel's scheme, is an individual of the second order.[59] Finally, there is the genealogical individual of the third order—the stem (*Stamm*) or, as Haeckel christened it, the "phylum." The stem, or phylum, consists of the series of genealogically related species that sprang from an original parent in the deep evolutionary past.[60] In short: "Every phylum is a plurality of blood-related species and each species is a plurality of the same or rather highly similar reproductive cycles."[61]

Haeckel spent several hundred pages developing these and other more refined distinctions of individuality. The effort bespeaks a kind of mania for puzzle-solving. It required immersion in a large and complex literature, the scope of which is indicated by the considerable depth of the footnotes stacked at the bottom of his pages. I believe, though, the complex task had another function: it helped him bury his melancholy in the details of professional study. He concluded from these examinations that one could define no conception of the "absolute individual."[62] Nonetheless, through the layers of distinctions he deployed, he did seem to detect the traces of one individual of absolute value—Anna, taken up into a myriad of forms. In his later work, as I will indicate in succeeding chapters, these transformations would be expressed in a variety of haunting ways: his enduring love now embodied in a beautiful medusa, now animating the archetypal image of the eternal feminine, now reincarnated in a real person.

The various modes of individuality that Haeckel distinguished in the *Generelle Morphologie* also served a theoretical function in his flowering system. They became the basis for a quite general principle that he had already advanced in his Stettin lecture, namely, the threefold parallel holding among paleontological connections, systematic relationships, and embryological developments. Paleontological connections would be represented in stem-trees indexed for temporal depth, with some branches reaching

59. Ibid., 2:305. The idea that there exist higher-order individuals—e.g., a species as individual—has been argued for by some contemporary philosophers of biology. See, for example, David L. Hull, "Are Species Really Individuals?" *Systematic Zoology* 25 (1974): 174–91; and "A Matter of Individuality," *Philosophy of Science* 45 (1978): 335–60.

60. Haeckel, *Generelle Morphologie,* 2:30: "The organic species or kind is nothing other than a sum of similar generational cycles and is thus constituted from a collection of generational cycles, just as the particular generational cycles are constituted from a collection of morphological stages that a bion . . . goes through during the time of its individual existence. . . . The stem or phylum is the collection of all organic species that have arisen from one and the same spontaneously generated monadic form."

61. Ibid., 2:305.

62. Ibid., 1:250.

the present period and others languishing in the mire of the past (see figs. 5.3 and 5.10). The nodes of such trees would represent speciation events and the common branches the genus forms carrying daughter species that split at more distant nodes. The branches displaying still extant creatures would indicate systematic relatedness. Following a temporally more primitive organic form up through its ascending branches would recapitulate the morphological development during ontogeny of a species-form occupying a higher branch.[63] Such stem-trees unite the several senses of individuality that Haeckel conceived; and they graphically depict the threefold parallel that remains today one of the strongest evidentiary foundations for evolutionary theory (see below).

Haeckel's Darwinism

Haeckel meant his *Generelle Morphologie* to be an exposition and defense of Darwinian evolutionary theory. He was quite well aware of Lamarck's descent theory, Bronn's proto-evolutionary conception, and the incipient transformational views of Goethe. But none of these antecedent proposals had produced any radical alterations in his early biological views. It was only on reading Darwin's *Origin of Species* in the context of his radiolarian work that he became a convert to evolutionary theory—and a true believer in all of the essential features of Darwinism. His concurrence with the Darwinian perspective extended to: (1) descent of species from more primitive forms; (2) natural selection as the principal device for species alteration; (3) divergence of species dependent on ecological and biogeographical relationships; (4) hereditary adaptations produced by selection; (5) progressive advance; (6) recapitulation of phylogeny by ontogeny; and (7) application of selection theory to human beings and human society. Most historians writing during the last thirty years have argued that either Haeckel ignored the most characteristically Darwinian of these areas of concern or that he so grossly distorted them as to produce a monstrous version of the Englishman's scheme. In the following sections, I will sketch Darwin's formulation of those aforementioned subjects and indicate the ways in which Haeckel adapted them to his own uses. While Haeckel certainly had a distinctive way of expressing his ideas, I believe they nonetheless fell essentially within the narrower confines of Darwin's own conception.

63. Ibid., 2:31.

Descent of Species

From a theoretical point of view, Darwin's theory (as well as that of Lamarck) gave concrete determination to the idea that different species were "naturally related." Since Linnaeus, systematists had sought criteria by which to organize plants and animals into a natural system—that is, the system of God's own plan. In his *Systema naturae* (12 eds., 1735–68), Linnaeus had decided that in lieu of the wanted criteria to create the natural system, he would artificially arrange plants according to the characters of their sexual organs. He organized his species into twenty-four classes by reason of the number of stamens and their placement; classes were further divided into orders in view of the number of pistils and other features. His delineation of the plant and animal kingdoms into a hierarchy of species, genera, orders, and classes had produced an intelligible system, but one that relied on a vague and shifting standard of resemblance. Georges-Louis Leclerc, Comte de Buffon (1707–1788), emphasized the artificial character of such systems in his own multivolume *Histoire naturelle* (1749–89) by ordering the various animal species according to their degree of usefulness to human beings.[64] Nonetheless, most naturalists felt that the usual groupings of species into a systematic hierarchy indicated that species were indeed, in some fashion, "related." From Linnaeus's time to the mid-nineteenth century (and even today), biologists have searched for a *natural* way of arranging species into the various higher ordered taxa.

Darwin believed he had discovered the natural way of grouping organisms, namely, by descent from a common progenitor. While this still left the practical problem of actually allocating species to the proper taxa, it yet provided the necessary criterion, at least in principle. When Haeckel began to arrange his large number of radiolarian types, the forms fell neatly together by reason of similarity. But now he knew what similarity indicated—common descent. Moreover, he argued that he could justify the proximity of his groupings by reason of the transitional forms linking them. In the *Generelle Morphologie,* he simply proclaimed: "The natural system of or-

64. In the "Premier discourse: De la manière d'étudier et de traiter l'histoire naturelle," Buffon argued that only the individual existed and that species were created by the minds of men. He would later admit the reality of species, but not that of higher taxa. See Georges-Louis Leclerc, Comte de Buffon, "Initial Discourse," in *From Natural History to the History of Nature,* ed. and trans. John Lyon and Phillip R. Sloan (Notre Dame, IN: University of Notre Dame Press, 1981), 89–130.

ganisms for us is their natural 'stem-tree' (*Stammbaum*), their table of genealogical relationships."[65]

Haeckel believed a stem-tree had to have its roots planted firmly in the ground. In specifying what this might mean, he advanced a proposition of a kind that Darwin only contemplated but did not publicly express: namely, that the most primitive organisms, which Haeckel named "monera," had to have arisen spontaneously from chemical processes of a distinctive kind.[66] He recognized three possible hypotheses concerning the origin of life. First, it could be that only one type of moneron sprang spontaneously to life and that this utterly simple creature—a glob of protoplasm without a nucleus—later differentiated into the basic animal, protist, and plant monera. In this scenario the stem-tree would have a single root and trunk but three main branches that carried animal, protist, and plant kingdoms with their respective phyla. Alternatively, it might be that in the ancient seas, three chemically different sorts of moneron came to life, one that led to the plants, and the other two to protists and animals. In this instance, there would be three original stem-trees, with the major divisions of each forming their branches. Finally it could be that several types of chemical monera sprang to life, each giving rise to one of the several phyla of plants, of protists, or of animals.[67] In this case, each phylum within one of the three kingdoms, represented by a basic *Bauplan,* would form a single stem-tree. Under this last hypothesis, the naturalist would plant a grove of stem-trees to represent all of life. The animal kingdom, for example, would be represented by five stem-trees, one each for the Vertebrata, the Mollusca, the Articulata, the Echinodermata, and the Coelenterata. Of these several hypotheses concerning the primitive origin of species, Darwin initially thought that the last was the most likely; and in the *Origin* he asserted his belief that "animals have descended from at most only four or five progenitors, and plants from an equal or lesser number." He did allow, however, that analogy suggested that "all plants and animals have descended from some one prototype."[68] Haeckel stressed that it was impossible to be conclusive about these speculative hypotheses, but he, like Darwin, thought the last of the three hypotheses the most likely. He

65. Haeckel, *Generelle Morphologie,* 1:196.

66. See, for example, Charles Darwin, "Notebook C" (102), in *Charles Darwin's Notebooks, 1836–1844,* ed. Paul Barrett et al. (Ithaca, NY: Cornell University Press, 1987), 269: "The intimate relation of Life with laws of Chemical combination, & the universality of latter render—spontaneous generation not improbable."

67. Haeckel, *Generelle Morphologie,* 1:199–200.

68. Darwin, *Origin of Species,* 484.

seems initially to have steered clear of the first possibility (i.e., all life from one moneron) because Darwin had intimated in the *Origin* that the original creature had received a kind of artificial respiration, when life was divinely breathed into it. Haeckel considered this a bit of subterfuge on Darwin's part (though he himself had suggested something similar in his Stettin lecture on Darwin's theory).[69] But even by the second volume of the *Generelle Morphologie,* he had begun to reconsider the evidence. He came to the view that the types of plants and animals were similar enough to have arisen from one kind of moneron; though the protists, given their distinctive structures, may still have been generated independently. But he was finally willing to concede that all animal, plant, and protist groups might have germinated from one original moneron and, therefore, all of life could be represented by a single "monophyletic stem-tree of organisms." Yet Haeckel hedged his bets in the *Generelle Morphologie* by his initial graphic representation of his theory. He constructed a single stem-tree that displayed each of these major possibilities: it had its root in a common moneron, but lines drawn further up the trunk (*x-y* and *m-n* in fig. 5.3) were meant to suggest origins at the level of the three main branches or at the level of individual types of plants, protists, and animals. The remaining plates of the volume show several phyla of animals or plants individually rooted.

In those remaining plates, Haeckel constructs two kinds of stem-tree: a genealogical stem-tree "paleontologically grounded" (i.e., showing temporal depth) and a systematic stem-tree, displaying morphological relationships. (The latter kind of tree is often situated as an insert in a plate of the former kind.) The first kind is the tree of physiological individuals, the second that of morphological individuals. (See, for example, the stem-tree of the vertebrates, fig. 5.10).

Haeckel confessed in his monograph that he had not resolved the problem of the number of original monera progenitors: "We have busied ourselves long and hard over these primordial questions without coming to any satisfactory result."[70] And whether the chemical generation of life in the seas happened once, or several times, or whether the process continued still in certain parts of the globe—all of this likewise remained uncertain. Haeckel yet hoped that experimental efforts, comparable to Friedrich Wöhler's synthesis of urea, might answer some of these questions.[71]

69. See above, chapter 4.
70. Haeckel, *Generelle Morphologie,* 2:405.
71. Ibid., 1:187–90.

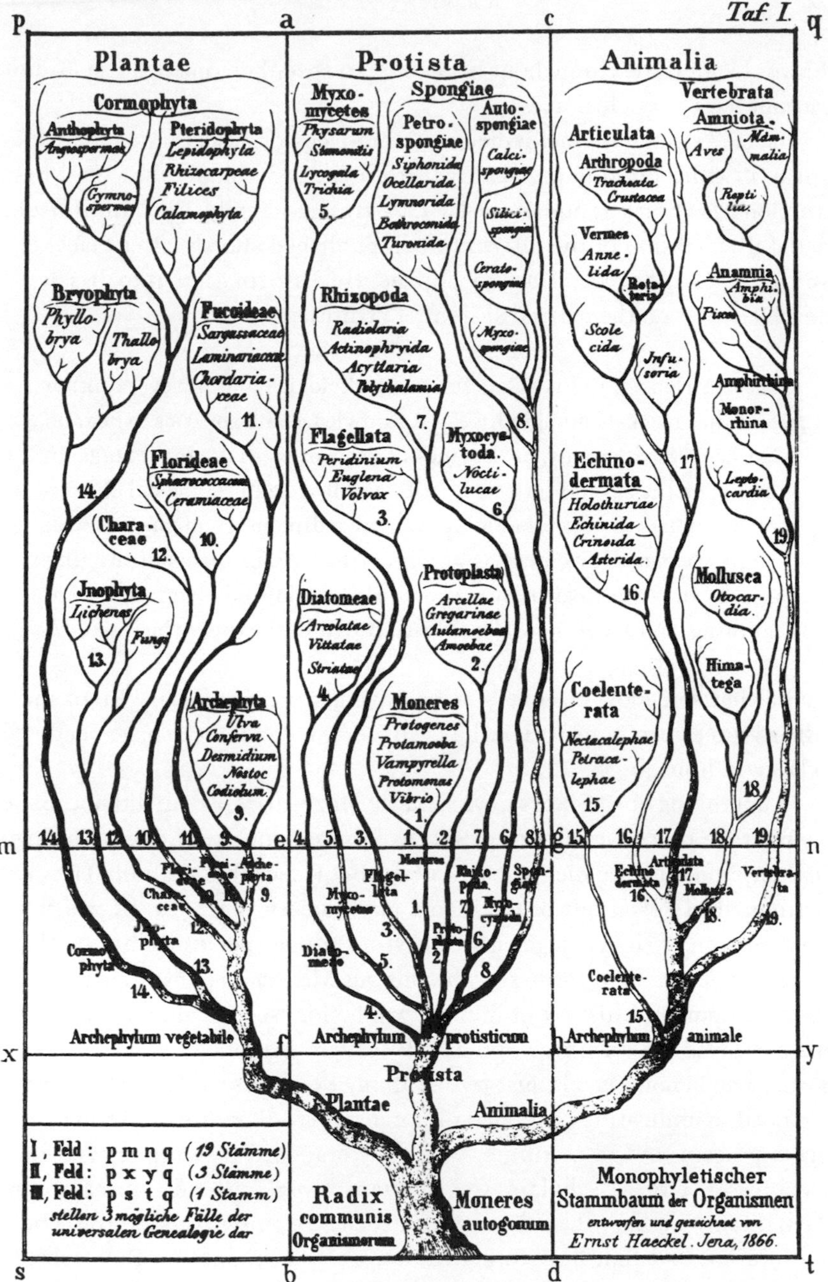

Fig. 5.3. Stem-tree of plants, protists, and animals; with *pstq* representing the hypothesis of one moneron as the source of all life; *pxyq* representing the hypothesis of three monera as the origin, respectively of plants, protists, and animals; and *pmnq* representing the hypothesis that some nineteen different monera gave rise to various phyla of plants, protists, and animals. (From Haeckel, *Generelle Morphologie der Organismen,* 1866.)

He would continue throughout his career to speculate on such issues, but without happy conclusion.

Haeckel's general employment of stem-trees of descent and their justificatory conception found approval from the individual who could best sanction his efforts as authentically Darwinian, namely, Darwin himself. In the fifth edition of the *Origin* (1869), published shortly after Haeckel's *Generelle Morphologie,* Darwin slipped this affirmation into his fourteenth chapter on the natural system of affinities:

> Professor Häckel, in his "Generelle Morphologie" and in several other works, has recently brought his great knowledge and abilities to bear on what he calls phylogeny, or the lines of descent of all organic beings. In drawing up the several series he trusts chiefly to embryological characters, but draws aid from homologous and rudimentary organs, as well as from the successive periods at which the various forms of life first appeared in our geological formations. He has thus boldly made a great beginning, and shows how classification will in the future be treated.[72]

Cavils that somehow Haeckel's trees have distorted the Darwinian message, as some current historians have suggested, seem not to have occurred to Darwin himself.

In approving of Haeckel's phylogenetic proposals, Darwin did not make the mistake that several current scholars have, namely, of confusing two distinct modes of graphic representation that Haeckel undertook. Haeckel distinguished depictions of the whole phylogenetic series of species from the depiction of the phylogenetic ancestors of one species. In his *Anthropogenie; oder, Entwickelungsgeschichte des Menschen* (Anthropogeny; or, the developmental history of man, 1874), he focused on human evolution and portrayed the stem-tree of man's lineal progenitors (see fig. 5.4). Peter Bowler and Benoit Dayrat, just two of many examples, isolate this graphic portrayal as indicating "the idea of a central trunk running through the whole process toward mankind, with side branches drawn in trivial proportions." For these scholars, the diagram reveals "the essentially non-Darwinian view that there is a central theme running through the whole of evolution and that human beings are its end product."[73] But Haeckel

72. Charles Darwin, *The Origin of Species by Charles Darwin: A Variorum Text,* ed. Morse Peckham (Philadelphia: University of Pennsylvania Press, 1959), 676.

73. The quotations are from Peter Bowler, *The Non-Darwinian Revolution* (Baltimore: Johns Hopkins University Press, 1988), 88. Dayrat also charges that Haeckel's trees were not authentically Darwinian, since they depicted lineal development of morphological forms. See

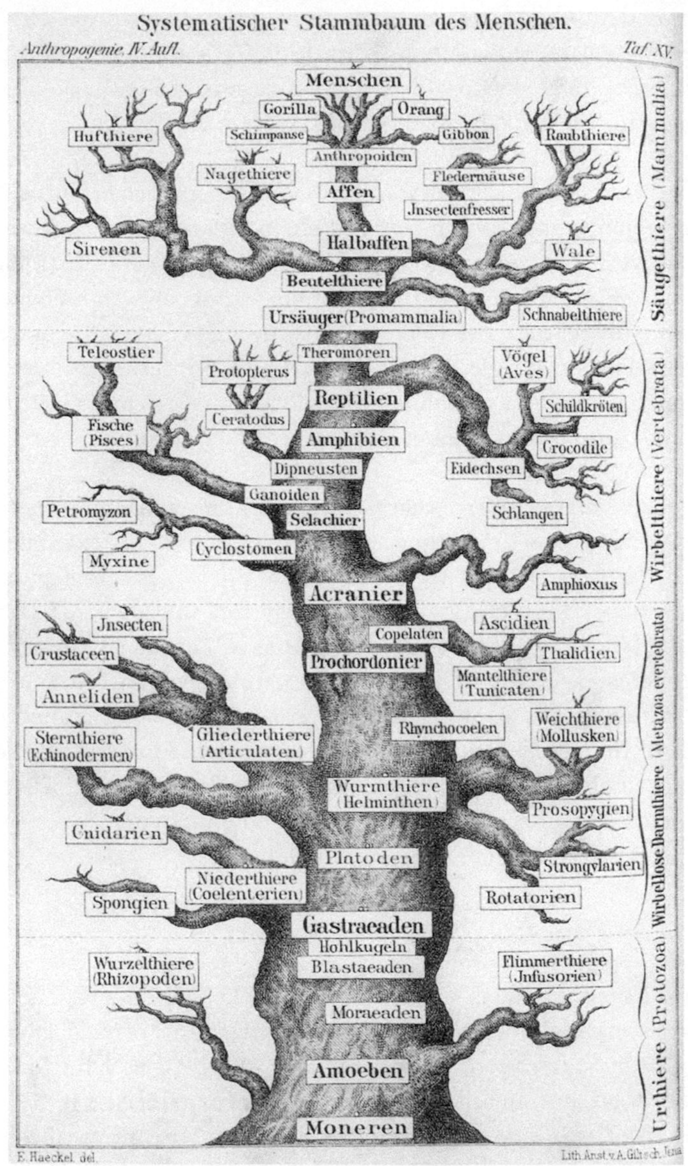

Fig. 5.4. Stem-tree of lineal progenitors of man. (From Haeckel, *Anthropogenie,* 1874.)

differentiated this mode of representation quite clearly from the mode employed in his *Generelle Morphologie* (e.g., fig. 5.10). In the *Natürliche Schöpfungsgeschichte* (Natural history of creation, 1868), the popular presentation of his big book, he made the distinction clear:

> *Ontogeny* or the individual developmental history of each organism . . . forms a simple, *unbranched* or lineal chain of forms; and the case is similar with that *part of phylogeny* which concerns the paleontological developmental history of the direct progenitors of each individual organism. By contrast, the *whole of phylogeny* forms a *branched* or tree-like developmental series, a true stem-tree. In this latter, we see the natural system of each organic stem, or phylum, and can investigate the paleontological development of all branches of the stem.[74]

Thus Haeckel separated a tree that might be constructed simply by tracing back from one species to its direct chain of ancestors—for example, from man to the *Urmensch,* to its narrow-nosed ape predecessor, to its predecessor, and so on—from the fully branched stem-tree that depicted all of the collateral relationships of the range of organisms. The latter, as Haeckel remarked, indicates "a true stem-tree," "the natural system." In figure 5.10, one can barely detect *Homo* crouching on the upper right branch of the stem-tree of the vertebrates (inset), hardly represented as the teleological fulfillment of the evolutionary process. Bowler, Dayrat, and others have simply forgotten the second, more fundamental employment of stem-trees that Haeckel discriminated.

Theory of Natural Selection

Historians have sometimes supposed that Haeckel did not give due accord to Darwin's principle of natural selection, preferring instead Lamarck's notion of direct adaptation.[75] Were this true, it would be hard to fathom why Haeckel became an ardent follower of Darwin, since he regarded the Englishman's distinctive contribution to the development hypothesis to be the very device of natural selection, which had, in Haeckel's estimation,

---

Benoit Dayrat, "The Roots of Phylogeny: How Did Haeckel Build His Trees?" *Systematic Biology* 52 (2003): 515–27.

74. Ernst Haeckel, *Natürliche Schöpfungsgeschichte* (Berlin: Georg Reimer, 1868), 257.

75. See, for example, Peter Bowler, *The Eclipse of Darwinism* (Baltimore: Johns Hopkins University Press, 1983), 68; and Stephen Jay Gould, *Ontogeny and Phylogeny* (Cambridge, MA: Harvard University Press, 1977), 80.

provided an irrefutable mechanical basis for morphological transformation of species.[76] As the following passage indicates, Haeckel's understanding of natural selection conformed to Darwin's own:

> Natural selection consists in this, that the intentionless but effective struggle for existence favors the reproduction of particular individuals that display useful traits that furnish those organisms an advantage. . . . [Such organisms] bestow their individual traits on their descendents and so are preserved, while the others of that species, which did not succeed in reproducing, perished without their individual properties being passed on and so preserved in their offspring.[77]

Haeckel, like Darwin, had a lively sense that the effective environment against which animals would be selected was not an amalgam of the distal circumstances of climate, soil, and inorganic conditions but the proximal organic thicket in which one organism was entwined with others. In agreement with this Darwinian perspective, Haeckel maintained that the environment had always to be considered when determining an animal's fitness. One could not simply view the creature in the abstract.

> In order to comprehend truly the tremendous importance that the struggle for existence possesses for the formation of the entire organic nature and to evaluate correctly its immeasurable significance, one must not conceptually extract the particular life-forms and merely observe them, as is the case with most biologists of the present day, who are wont to do this. Rather one must compare these life-forms in their collective entirety, in their general and continuous mutual interrelationships.[78]

Haeckel mildly criticized even Darwin for sometimes extending the notion of struggle to an isolated organism, a criticism that his English mentor accepted.[79] The results of a struggle for existence within the complex

76. Haeckel, *Generelle Morphologie*, 2:227.

77. Ibid., 2:228, 226.

78. Ibid., 2:240.

79. When Darwin suggested that a plant at the edge of a desert might be said to "struggle for life against the drought" (*Origin of Species*, 62), Haeckel objected that struggle for existence ought to be confined to the competition of organisms for scarce resources (*Generelle Morphologie*, 2:239). In marginal notes to his copy of Haeckel's book, Darwin wrote: "good criticism on my term of struggle for existence—says ought to be confined to struggle between organisms for same end." See Darwin's copy of the *Generelle Morphologie* held in the Manuscript Room of Cambridge University Library.

network of other species would be divergence of organic types, that is, the gradual formation of new species out of old.[80]

Ecology and Biogeography

Haeckel understood the fundamental message of Darwin: that creatures lived entangled in a large network of connections to the inorganic and organic environments. In assessing the advantages traits might bestow on their carriers, the naturalist had to consider all of these relationships. This Darwinian insight led Haeckel to postulate two subdisciplines of evolutionary science, which have perdured and now flourish: ecology (*Oecologie*) and biogeography (which he dubbed *Chorologie*). He defined ecology as "the science of the mutual relationships of organisms to one another, while chorology is the science of the geographic and topographic spread of organisms."[81] Haeckel practiced these subdisciplines even in his marine biology, where he would specify for particular organisms the depths in the seas at which they might be found, the temperatures of the waters where they dwelt, the resources for their survival, and the other organisms immediately surrounding them.

Heredity and Adaptation

Haeckel distinguished two fundamental but distinct modes of heredity: conservative heredity, through which traits were simply passed along the generations without change; and progressive heredity, through which characters newly acquired by parents were transmitted to offspring.[82] Seized by a passion for classification, he discriminated five laws of conservative heredity, including the law of "interrupted heredity" (in which the similarity between progenitors and descendants occurred every other generation, as happened in the tunicate *Salpae*); the law of sexual heredity (in which the secondary sexual characteristics of the female were transmitted to her female offspring, the male characters to the male offspring), and so on. While these were useful distinctions, and similar to ones that Darwin would recognize in *The Variation of Plants and Animals under Domestication* (1868),[83] it was a bit grand to have elevated them to the rank of

80. Haeckel, *Generelle Morphologie*, 2:251.

81. Ibid., 2:287–88.

82. Ibid., 2:177–78.

83. Darwin noticed, for instance, that a female cow having a particular kind of udder or milk would give birth to a male, which in turn would produce a female with characteristics

law. Haeckel advanced the conception of progressive heredity against those who believed in species immutability. Like Darwin, he conceived of this kind of heredity as having two forms: indirect and direct.

Indirect progressive heredity—or as he would call it "indirect accommodation" (*indirecte Anpassung*)—took place when the parent underwent some alteration at the level of sex cells but without expressing that change in some manifest trait, though the change would be passed to and expressed by the offspring. Direct accommodation (*directe Anpassung*) occurred when the parent manifested the alteration and then through heredity passed the change to offspring—this would be the inheritance of an acquired characteristic.[84] In his discussion, it is quite clear that *Anpassung*, which might more usually be translated "adaptation," meant only a change in trait or function. He regarded, for instance, a monstrous birth as an instance of *Anpassung*. Hence *Anpassung*—"accommodation"—carried no necessary implication of preadaptation to environmental circumstances.

Haeckel's explanation of indirect accommodation resembled Darwin's own account of accidental variation. Haeckel maintained that every animal or plant of a given species would be subject to different kinds and amounts of nutrition at the cellular level, including the level of the sex cells.[85] This meant that accommodational differences among organisms were ultimately traceable to differential nutrition, which would occur because of the variable environments in which organisms existed.[86] Both Darwin and Haeckel thus maintained that a small or imperceptible change in the sex organs of an animal—through some environmental impact—might have a significant effect on egg or sperm, which would alter traits in the new embryo. Such indirect accommodations would not come preadapted for a particular function, but would furnish the raw material—accidental variation—on which natural selection might operate.

Some direct accommodations (i.e., when the parent initially expressed the changed trait) would be adaptations in the narrower sense. This kind of inheritance of acquired characters also formed part of Darwin's repertoire

---

of the original female. He also remarked on cases in which a trait was passed along either the male or female line. See Charles Darwin, *Variation of Plants and Animals under Domestication*, 2nd ed., 2 vols. (1868; repr., New York: D. Appleton, 1899), 1:446, 2:60.

84. Haeckel, *Generelle Morphologie*, 2:196.

85. Ibid., 2:199: "Here we observe a fact of considerable note and importance, namely, that a quite small change in nutrition, which in most organs and functions of the parental organism produces no noticeable or only a quite insignificant change, has a relatively tremendous effect on the parental sex organ . . . so that this effect, after successful conception (fructification), introduces a quite obvious change of form and function in the offspring."

86. Ibid., 2:208.

from the beginning of his theorizing to the end. Both Haeckel and Darwin would allow that a habit acquired by the parent to adapt it to some circumstance could be passed to offspring and thus produce a preadaptation. They both would also assume that such inherited habits could likewise be subject to selection.[87]

Both Darwin and Haeckel attempted to dig below the apparent phenomena of heredity to the possible underlying causal structures. Darwin proposed that seedlike units were given off by the cells of various body parts and that these "gemmules," as he called them, might be altered by the continuous exercise of those parts. The gemmules, flowing through the bloodstream, would then collect in the sexual organs and serve as the carriers of either acquired or congenital characters to the next generation. Haeckel, likewise, hypothesized that hereditary determinants were to be found at the molecular level. He suggested that particulate units—*Plastidulen*—composing the protoplasm within the cells would vibrate at different frequencies, which motions might be altered through continuous gross body changes; the sex cells, through those molecular vibratory responses, would thus pass on acquired or congenital characters to offspring. Haeckel's theory of the plastides would later be ridiculed by Virchow—and smirked at by contemporary scholars—but his drive to find a deeper account of the crucial features of heredity moves in the same direction as that charted by Weismann's "ids," Mendel's "factors," and Darwin's "gemmules."[88]

Evolutionary Progress

It is commonly believed that Haeckel conceived evolution as necessarily progressive and that this became an essential feature of the "non-

87. When Darwin considered the sources of variability upon which natural selection might operate, he specified the indirect effects of the environment on the sexual organs of parents, as well as the adaptation of organisms through direct effects and through use and disuse. He believed, however, virtually all complex adaptations to be the result of selection operating on traits introduced through impact on the sexual organs of parents. See Darwin, *Origin of Species*, 131–39.

88. See Charles Darwin, *Variation of Plants and Animals under Domestication*, 2nd ed., 2 vols. (1868; repr., New York: D. Appleton, 1899), 2: chap. 27; August Weismann, *Das Keimplasma: Eine Theorie der Vererbung* (Jena: Gustav Fischer, 1892); Gregor Mendel, "Versuche über Pflanzen-Hybriden," *Verhandlungen des naturforschenden Vereines in Brünn* 4 (1866): 3–47; and Ernst Haeckel, *Die Perigenesis der Plastidule oder die Wellenzeugung der Lebensteilchen* (Berlin: Georg Reimer, 1876). See also the quite insightful essay of Robert Brain on Haeckel's and similar theories, "Protoplasmania: The Vibratory Organism and 'Man's Glassy Essence' in the Later Nineteenth Century," in *Zeichen der Kraft: Wissensformationen 1800–1900*, ed. Thomas Brandstaetter and Christof Windgätter (Berlin: Kadmus, forthcoming).

Darwinian revolution," as Bowler has called it.[89] Haeckel certainly thought that natural selection, since it operated on individuals who had an advantage over others of their kind, would produce "a slow but constant improvement, a progress in organization."[90] However, he denied, as did Darwin, that organisms harbored an *intrinsic* tendency toward improvement. With such a postulation, he warned, "we step onto the slippery slope of teleology, from which we, without hope of salvation, will slide into the abyss of dualistic contradiction and distance ourselves completely from all possible mechanistic natural explanation."[91] Haeckel believed that progressive development, while not necessary, generally occurred through the continuous operations of natural selection. Some organisms might retrogress, but the process would tend toward progression.[92] Haeckel was indeed a progressionist, but then so was Darwin.

Like Haeckel, Darwin thought of natural selection as an engine of progressive development, as he averred in the *Origin:* "As natural selection works solely by and for the good of each being, all corporal and mental endowments will tend to progress towards perfection." This meant for Darwin that "more recent forms must, on my theory, be higher than the more ancient; for each new species is formed by having had some advantage in the struggle for life over other and preceding forms." The operations of natural selection would thus have, as Darwin calculated, the general effect of improving all species and would lead to "the most exalted object which we are capable of conceiving, namely, the production of the higher animals."[93] After Darwin read Haeckel's *Generelle Morphologie,* as best he could, he made a note to himself in the margins of Haeckel's book concerning his future work on human evolution: "In man Chapt I might add as proof of

89. Bowler, *Non-Darwinian Revolution,* 85: "The most powerfully non-Darwinian (and eventually anti-Darwinian) view of evolution arose from the belief that many aspects of the history of life are governed not by haphazard geographical factors but by trends driven on toward a predetermined goal whatever the environmental changes to which the organisms are subjected. Haeckel's evolutionary morphology was essentially progressionist and was even based to some extent on a revival of the old linear image of development ascending a single hierarchy of stages toward its inevitable goal." See also Stephen Jay Gould, "Eternal Metaphors of Palaeontology," in *Patterns of Evolution as Illustrated in the Fossil Record,* ed. A. Hallam (New York: Elsevier, 1977), 13: "An explicit denial of innate progression is the most characteristic feature separating Darwin's theory of natural selection from other nineteenth-century evolutionary theories. Natural selection speaks only of adaptation to local environments, not of directed trends or inherent improvement."

90. Haeckel, *Generelle Morphologie,* 2:169.

91. Ibid., 2:264.

92. Ibid., 2:263.

93. Darwin, *Origin of Species,* 489, 337, 490.

theory—'the progressive perfection or development of organic beings.'"[94] There can be little doubt, I think, that Haeckel and Darwin were in accord concerning the progressive features of evolution by natural selection.[95] To read Darwin otherwise is to make him into a neo-Darwinian, which, needless to say, he was not.

The Biogenetic Law: Ontogeny Recapitulates Phylogeny

A chief feature of Haeckel's evolutionary doctrine that supposedly distinguishes his views from those of Darwin is the principle of recapitulation. Haeckel put the principle thusly:

> The organic individual . . . repeats during the quick and short course of its individual development the most important of those changes in form that its ancestors had gone through during the slow and long course of their paleontological development according to the laws of inheritance and adaptation.[96]

Or as he more succinctly phrased it: "Ontogeny is nothing other than a short recapitulation of phylogeny."[97] The principle of recapitulation—which he later dubbed the "biogenetic law"—became the cardinal hinge connecting Haeckel's multiple evolutionary studies.[98] Many historians, though, suppose recapitulation theory utterly foreign to Darwin's own conception of evolution. Bowler, for instance, simply asserts that "recapitulation theory thus illustrates the non-Darwinian character of Haeckel's evolutionionism."[99]

In one of its forms, the principle of recapitulation antedates its specifically evolutionary use. At the end of the eighteenth century, Carl Friedrich

94. Darwin jotted this remark in the margins of his copy of Haeckel's *Generelle Morphologie,* 2:270. His copy is held in the Manuscript Room of Cambridge University Library.

95. I have more extensively discussed the progressive features of Darwin's theory in my *Meaning of Evolution: The Morphological Construction and Ideological Reconstruction of Darwin's Theory* (Chicago: University of Chicago Press, 1992), chap. 5.

96. Haeckel, *Generelle Morphologie,* 2:300.

97. Ibid., 2:7.

98. Haeckel introduced the term "biogenetic law" (*biogenetisches Grundgesetz*) in his *Natürliche Schöpfungsgeschichte,* 2nd ed. (Berlin: Georg Reimer, 1870), 361–62. Likely he had already formulated the term for the manuscript of his *Die Kalkschwämme,* 3 vols. (Berlin: Georg Reimer, 1872), 1:471.

99. Bowler, *Non-Darwinian Revolution,* 84. E. S. Russell, Stephen Jay Gould, Dov Ospovat, and Ernst Mayr, among others, believe that Darwin rejected the theory of recapitulation, which was the linchpin of Haeckel's evolutionary views. See my discussion of the issue in *The Meaning of Evolution,* 111–64.

Kielmeyer (1765–1844), Johann Heinrich Autenrieth (1772–1835), and Lorenz Oken (1779–1851) had suggested that the embryos of more advanced creatures repeated in their individual morphological development the forms of lower organisms. According to Oken, "The [mammalian] fetus, through the course of the several forms of its existence, is the whole animal," that is, it sequentially passes through stages comparable to the polyp, plant, insect, snail, fish, and amphibian, finally reaching the mammalian form.[100] At the beginning of the nineteenth century, the principle became adapted to evolutionary employment. Friedrich Tiedemann (1781–1861), Johann Friedrich Meckel (1781–1833), and Gottfried Reinhold Treviranus (1776–1837) compared the types of animal forms resident in different fossil layers with the patterns exhibited by the developing embryo. Tiedemann, who studied at the Paris Museum and came into contact with Lamarck's ideas, made the comparison explicit:

> It is clear from the previous propositions that from the oldest strata of the earth to the most recent, there appears a graduated series of fossil remains, from the most simply organized animals, the polyps, to the most complex, the mammals. It is evident too that the entire animal kingdom has its developmental periods [*Entwickelungsperioden*], similar to the periods which are expressed in individual organisms.[101]

Karl Ernst von Baer (1792–1876) understood that for the embryologist, the principle of recapitulation had a seductive power. But that temptation—based, as he thought, on a false analogy—had to be resisted; and he did so with mordant irony:

> One gradually learned to think of the different animal forms as evolving [*entwickelt sich*] out of one another—and then shortly to forget that this metamorphosis was only a mode of conception. Fortified by the fact that in the oldest layers of the earth no remains from vertebrates were to be found, naturalists believed they could prove that such unfolding of different animal forms was historically grounded. They then related with complete seriousness and in detail how such forms arose from one another. Nothing was easier. A fish that swam upon the land wished to go for a stroll, but could not use its fins. The fins shrunk in

100. Lorenz Oken, *Die Zeugung* (Bamberg: Goebhardt, 1805), 146–47. I have discussed the views of Kielmeyer, Autenrieth, and Oken in my *Meaning of Evolution*, 18–20, 39–42.

101. Friedrich Tiedemann, *Zoologie, zu seinen Vorlesungen entworfen*, 3 vols. (Landshut: Weber, 1808–14), 1:64–65.

Fig. 5.5. Karl Ernst von Baer (1792–1876), at about age eighty. (Photo courtesy of the Smithsonian Institution.)

> breadth from want of exercise and grew in length. This went on through generations for a couple of centuries. So it is no wonder that out of fins feet have finally emerged.[102]

Von Baer rejected the principle of recapitulation for two reasons.[103] First, he simply denied the Lamarckian notion that one species might give rise to another. But second, he argued that the analogy between individual and species evolution was false. The vertebrate embryo did not pass through the morphological stages of *adult* forms of more primitive animals. Rather, according to von Baer, the vertebrate embryo was a vertebrate from the beginning: it began as a generalized vertebrate; then, if it were, say, a human being, it became a generalized mammal; then a generalized primate; then a

102. Karl Ernst von Baer, *Über Entwickelungsgeschichte der Thiere: Beobachtung und Reflexion*, 2 vols. (Königsberg: Bornträger, 1828–37), 1:200.

103. See the first appendix for a sketch of von Baer's morphological theories.

specifically human individual. The embryo thus initially displayed a quite general morphological form but developed through stages of greater specification. In England, Richard Owen (1804–1892) followed von Baer in rejecting the principle of recapitulation. In his Hunterian lectures of 1837, Owen marked out the dangerous connection that the unwary might suppose to exist between embryological development and species development: "The doctrine of transmutation of forms during the Embryonal phases," he cautioned, "is closely allied to that still more objectionable one, the transmutation of Species." [104]

Virtually at the same time as Owen railed against species evolution and its supporting principle of recapitulation, one of his colleagues adopted both. On the first page of his initial transmutation notebook, "Notebook B," Charles Darwin proposed that new adaptations sequentially acquired by a species over a long period would be preserved in the embryo: "An originality is given (& power of adaptation) is given by true generation, through means of every step of progressive increase of organization being imitated in the womb, which has been passed through to form that species.—(Man is derived from Monad)." [105] Through his notebooks and essays, Darwin cultivated this idea. He understood the objection of von Baer and Owen that the morphological stages of, for instance, a vertebrate embryo passed from a more general type to a more specific. He simply believed that the general type, which von Baer and Owen had elevated to the status of an ideal archetype, represented a once-existing adult progenitor of the vertebrate line. On the back flyleaf of his copy of Owen's book *On the Nature of Limbs* (1849), Darwin jotted: "I look at Owen's Archetypes as more than idea, as a real representation as far as the most consummate skill & loftiest generalization can represent the parent form of the Vertebrata." [106] In the *Origin of Species,* Darwin reiterated his early thesis that those embryos sheltered in the womb or in the egg would not generally be subject to natural selection and so would not likely acquire new traits; only when an embryo was born and faced a variegated environment would it undergo altering adaptations. The modifications would, as it were, be tacked on to the end of the developmental process. This meant that the embryo, as it gradually developed in the womb, might graphically illustrate the early history of its progenitors;

104. Richard Owen, *The Hunterian Lectures in Comparative Anatomy, May and June 1837,* ed. Phillip Sloan (Chicago: University of Chicago Press, 1992), 192. See also the first appendix.

105. Charles Darwin, "Notebook B," MS p. 1, in *Charles Darwin's Notebooks,* 170.

106. Darwin's pencil annotation occurs on the back flyleaf of his copy of Richard Owen, *On the Nature of Limbs* (London: Van Voorst, 1849). The book is held in the Manuscript Room of Cambridge University Library.

it would go through the various morphological stages attained by them. As Darwin expressed it in the *Origin:*

> The adult differs from its embryo, owing to variations supervening at a not early age, and being inherited at a corresponding age. This process, whilst it leaves the embryo almost unaltered, continually adds, in the course of successive generations, more and more difference to the adult. Thus the embryo comes to be left as a sort of picture, preserved by nature, of the ancient and less modified condition of each animal.[107]

During the composition of the *Origin,* Darwin did experimental research, which he detailed in the penultimate chapter, on neonates of several domestic species—dogs, horses, and pigeons. This man of delicate constitution grew queasy over the gassing of so many birds for his detailed observations. Nonetheless, he persevered in his effort to show that the embryos of several varieties of a given species resembled each other more than they did their own distinctive parents. He thus empirically demonstrated the ways in which early modifications in the history of the species might be retained in embryological development. Yet Darwin remained tentative about the principle of recapitulation, since he had only indirect evidence for its validity. But his hesitations evanesced in 1864, when he received a small book from its author, Fritz Müller (1821–1897), a German naturalist working in South America.

Müller, who had abandoned Prussia because of political and religious dissenting beliefs, settled in Destêrro, an island city off the southern coast of Brazil.[108] He accepted the challenge that Bronn had issued in the epilogue to

107. Darwin, *Origin of Species,* 338.

108. Müller had received a Ph.D. in zoology in Berlin (1844) and then completed study for a medical degree at Greifswald (1849). Though his father was a Lutheran minister, Müller left the church, rejecting what he thought unpalatable superstition; and during the revolution of 1848, he became a leader of the democratic *Verein.* An advocate of free love—or, at least, love not requiring ecclesiastical sanction—he had a daughter in 1849 with Karoline Töllner, whom he later married; they would eventually have a family of ten children. Since he refused to take a religious oath required for the state medical exam, he did not receive his medical degree. After marrying Töllner (to simplify emigration), he, his wife and child, along with his brother and his brother's wife, sailed from Hamburg to the southern coast of South America in May 1852. The trip took two months, during which an epidemic of measles contributed to the deaths of a dozen children on the voyage. Fritz Müller and his growing family eventually settled in the city of Destêrro (now Florianpólis) on the island of Santa Catarina. There he became a teacher in the Jesuit school on the island. He continued work in botany and invertebrate zoology—the island shores offering abundant marine life. He would carry on an extensive correspondence with both Darwin and Haeckel. For the most comprehensive account of Müller's life, see *Fritz Müller, Werke, Briefe und Leben,* ed. Alfred Möller, 3 vols. in 4 (Jena: Gustav Fischer, 1920).

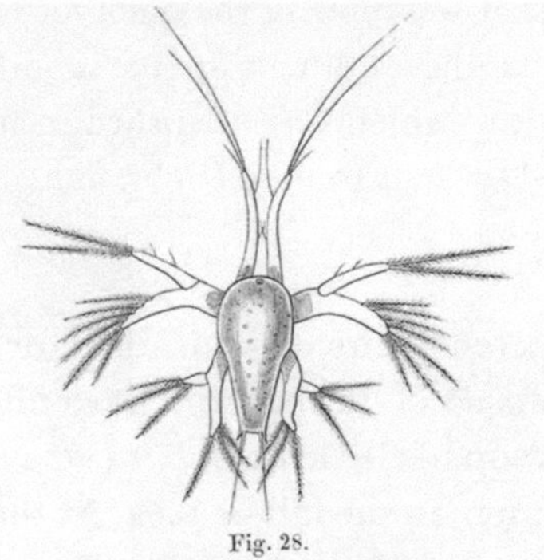

Fig. 28. Nauplius einer Garneele. 45 mal vergr.

Fig. 5.6. Nauplius or larval form of crustaceans. (From Fritz Müller, *Für Darwin*, 1864.)

his translation of the *Origin:* Müller thought Darwin's theory could be tested by a careful examination of one group of animals. He initiated a systematic study of Crustacea, particularly various species of shrimp and crabs. He showed that despite the great differences in adult morphology, the juveniles of these species displayed the same nauplius form (see fig. 5.6). Darwin himself took this kind of evidence as strong support for the recapitulation thesis. In the fourth edition of the *Origin* (1866), he deployed Müller's findings:

> In the enormous class of the *Crustacea*, forms wonderfully distinct from each other, as the suctorial parasites, cirripedes, entomostracan, and even the malacostraca, appear in their larval state under a similar nauplius form; and as these larvae feed and live in the open sea, and are not adapted for any peculiar habits of life, and from other reasons assigned by Fritz Müller, it is probable that an independent adult animal, resembling the nauplius, formerly existed at a remote period and has subsequently produced, through long-continued modification along several divergent lines of descent, the several above named great Crustacean groups.[109]

From Müller's evidence, along with his own experimental studies on the embryos and young of different species, Darwin felt justified in generalizing the principle of recapitulation:

David West has translated parts of Möller's biography and added new material. See his *Fritz Müller: A Naturalist in Brazil* (Blackburg, VA: Pocahontas Press, 2003).

109. Darwin, *Origin of Species: Variorum Text*, 702.

> It is probable from what we know of the embryos of mammals, birds, fishes, and reptiles, that these animals are the modified descendants of some ancient progenitor, which was furnished in its adult state with brachiae, a swim-bladder, four fin-like limbs, and a long tail, all fitted for an aquatic life.[110]

Darwin's formulation here is quite explicit: the embryo goes through the "adult" morphological stages of its ancestral progenitor.

Müller had sent Darwin his little book *Für Darwin,* which detailed his argument for recapitulation, in the fall of 1864. At the same time, Haeckel wrote Darwin to recommend Müller's book.[111] Darwin could not immediately appreciate the significance of the work, since, as he confessed to Haeckel, Müller's German was simply too difficult for him.[112] However, he hired a translator to give him a workable English version, and then in 1868 arranged for a proper English edition by an individual with technical training in zoology.[113] Even with the help of the preliminary translation, Darwin did not likely comprehend the full complexity of Müller's argument, which both supported his version of recapitulation and rejected that of Louis Agassiz (1807–1873).

Agassiz had held that in the course of ages each new species appeared according to a divine plan—a plan that charted the advance of creatures from their primitive condition upward to man. This plan, according to Agassiz, could also be detected in embryonic development, since the ontogeny of a contemporary animal retraced in successive morphological stages the history of divine creation.[114] Müller, however, urged that the evidence told against such a theological interpretation, though not against recapitulation

110. Ibid.

111. Fritz Müller, *Für Darwin* (Leipzig: Wilhelm Engelmann, 1864). See also Haeckel to Darwin (26 October 1864), in the Haeckel Correspondence Haeckel-Haus, Jena; and in *The Correspondence of Charles Darwin,* ed. Frederick Burkhardt et al., 15 vols. to date (Cambridge: Cambridge University Press, 1985–2001), 12:381. Haeckel served as a lifeline to Müller; he sent him numerous publications (including the very large *Challenger* volumes) and kept up a warm and scientifically detailed correspondence with him until Müller's death on 21 May 1897.

112. Darwin to Haeckel (21 November 1864), in the Haeckel Correspondence Haeckel-Haus, Jena; and in *Correspondence of Charles Darwin,* 12:411–12.

113. William Swettland Dallas wrote Darwin to say that he would translate Müller's work. See W. S. Dallas to Darwin (22 February 1868), DAR 162, in the Darwin Papers, Manuscript Room, Cambridge University. Müller's book appeared as *Facts and Arguments for Darwin,* trans. W. S. Dallas (London: John Murray, 1869). Müller provided several additions for the English version.

114. I have discussed Agassiz's version of recapitulation in my *Meaning of Evolution,* 115–21.

itself. He distinguished two kinds of evolutionary patterns to be found in the class of Crustacea: one in which successive species added modifications at the end of the developmental process, so that embryonic development would rather accurately picture the historical development of the group; and one in which modifications were wrought, not on the mature organism but on its larvae, so that embryonic stages would deviate from the ancient pattern.[115] Darwinian theory could explain this latter case as one in which free-living larvae had been subject to competitive struggle, so introducing alterations not characteristic of ancient progenitors. Thus in Müller's view, Darwin's conception could account for both patterns of development in Crustacea. But if one were of Agassiz's inclination, the latter "falsification" of historical development could only be interpreted as the plan of a devious Creator—hardly a palatable conclusion for a believer. Yet the strongest evidence for Darwin's theory, Müller believed, came in the ease with which a stem-tree could be constructed of the various kinds of Crustacea. Their embryonic development displayed natural affinities and relationships of the sort that the Englishman had forecast.[116]

In his Stettin lecture, Haeckel had advanced, as the best proof for Darwin's theory, the threefold parallel of systematics, paleontology, and embryology. But there seems little doubt that Fritz Müller's analysis of recapitulation deepened Haeckel's own understanding of the process, as he himself happily admitted.[117] He came to emphasize two features of embryonic development that made it deviate from an exact duplication of ancient morphological patterns. First, he maintained, citing Müller, that recapitulation would be a *shortened* and *simplified* repetition of previous patterns;

115. Müller, *Für Darwin,* 77: "The historical record containing the developmental history [i.e., retained in embryonic development] would generally be *erased,* since the development detours a straight path from egg to mature animal and is often *falsified* through the struggle for existence, which the free living larvae have to suffer." By the fifth edition of the *Origin* (1869), Darwin had Dallas's professional translation of Müller. And in that fifth edition, he mentions that when the larvae lead an independent existence, selective forces will often have altered the historical record displayed in the embryo. See Darwin, *Origin of Species: Variorum Text,* 705.

116. Müller, *Für Darwin,* 73–74. In his dismantling of Agassiz's general approach to morphology, Müller could not constrain himself from the double-barbed observation of a symmetrical credulity to be found among religious dogmatists and zoological dogmatists. After quoting Agassiz's pious methodological principle that the importance of organs in the animal economy must indicate importance for systematic relationships among organisms, Müller observed: "Just as in Christian lands, each person will mouth a catechismic morality that he does not feel obliged to follow or expect anyone else to follow, so zoology also has its dogmas that each generally acknowledges but disregards in practice" (71).

117. Haeckel had nothing but praise for the "masterful and thorough treatment" of the most difficult questions of morphology undertaken by Müller. See Haeckel, *Generelle Morphologie,* 2:185 and note.

hence the embryo would not be a perfect representation of phylogenetic transformations. Moreover, insofar as the larvae of a particular group lived independently and thus became subject to different environmental forces, the future embryonic forms would give a "false" or incomplete picture of the historical development of the phylum.[118] Haeckel would later elaborate the processes of what he would call "palingenesis" and "cenogenesis," that is, respectively, a rather faithful preservation of the phylogenetic sequence in the embryo and a distorted representation due to adaptations occurring during embryonic or larval development.[119]

Throughout his career, the biogenetic law would govern the vital pulse of Haeckel's many evolutionary studies. It would also provide an inviting target for the attacks of his enemies.

Human Evolution

During the 1865–66 university term, when Haeckel worked feverishly on the composition of his great monograph, he was also lecturing on zoological topics to his university seminar and simultaneously mounting a public lecture series on Darwin's theory. The latter attracted not only students from the various faculties, but other of his colleagues and individuals from the town. He delighted to his friend Allmers that "while the preachers from the chancellery were ripping Darwin apart, they declaimed only to empty benches."[120] His benches, by contrast, were full. During the winter semester, Haeckel also found time to deliver two lectures on human evolution, in October and November, to a more select group.

Though Darwin had refrained from discussing human evolution in the *Origin of Species,* his reviewers, both critics and supporters, immediately perceived the implications of his theory for man. Owen, for example, sneered that "the considerations involved in the attempt to disclose the origin of the worm are inadequate to the requirements of the higher problem of the origin of man."[121] Huxley, happy to do the devil's work, specified the extreme similarity of human anatomy to that of the higher apes in his

118. Ibid., 2:300.

119. Haeckel introduces the terminology of "palingenesis" and "cenogenesis" in "Die Gastrula und die Eifurchung der Thiere," *Jenaische Zeitschrift für Naturwissenschaft* 9 (1875): 409. He again credits Fritz Müller with initially making the distinction.

120. Haeckel to Allmers (16 December 1865), in *Ernst Haeckel, Sein Leben, Denken und Wirken. Eine Schriftenfolge für seine zahlreichen Freunde und Anhänger,* ed. Victor Franz, 2 vols. (Jena: Wilhelm Gronau und W. Agricola, 1943–44): 2:46.

121. [Richard Owen], "Darwin on the *Origin of Species,*" *Edinburgh Review* 111 (1860): 521. The review was anonymously published.

Man's Place in Nature (1863). Another of Darwin's friends, Charles Lyell, in *Antiquity of Man* (1863), yet hesitated over the problem of language. No ape, he thought, could cross the Rubicon of language.[122] Darwin groaned his great disappointment over Lyell's demur. But Haeckel's friend Schleicher, the great comparative linguist, believed he had indeed found the way across that linguistic divide; and it was right through the branching streams of language. (I will discuss Schleicher's specific impact on Haeckel and Darwin in chapter 7.)

Already in his Stettin lecture in 1863, Haeckel had made general application of Darwin's theory to human beings. He more extensively developed the conception of human evolution in the private lectures in the fall of 1865 and in the two brief chapters on man in the *Generelle Morphologie*, which he completed by the summer of the next year. The extension of the *Origin*'s theory to human beings was warranted, Haeckel maintained, by the validity of the general conception: "The proposition that man himself has developed from the lower vertebrates, and most proximately out of the true apes, is a special deductive conclusion that derives with absolute necessity from the general inductive law of descent theory."[123] Darwin had demonstrated, despite the superstitions of the "priests," that "the same, simple mechanically operative causes—pure physical-chemical natural processes—show themselves to be sufficient to accomplish the highest and most difficult of all tasks."[124] Haeckel acknowledged that fossil evidence of human evolution was wanting, but he thought the conceptual force of the general theory, the anatomical similarities between human beings and the higher apes that Huxley had made manifest, and the comparable embryological recapitulation undergone by apes and humans—that all of these had established sufficient warrant for accepting human evolution.[125] From this point forward, however, Haeckel stretched the boundaries of probable conclusion. He suggested that the *Urmenschen*—those missing links—must have arisen from the stem of the catarrhine or narrow-nosed apes, a stem that included in its higher branches the chimpanzees, orangutans, and gorillas. These protomen devoid of speech must have borne similarities, he urged, to contemporary peoples occupying the lowest rungs on the human scale (e.g., Pap-

122. Charles Lyell, *The Geological Evidences of the Antiquity of Man* (London: Murray, 1863), 469.

123. Haeckel, *Generelle Morphologie*, 2:427.

124. Ernst Haeckel, "Ueber die Entstehung des Menschengeschlechts" (October 1865), in *Gesammelte populäre Vorträge aus dem Gebiete der Entwickelungslehre* (Bonn: Emil Strauss, 1878), 36–37.

125. Ibid., 53–57.

uans, Hottentots, Australians), whose mental powers, he surmised, were exceeded by many higher animals (e.g., dogs, elephants, and horses).[126]

Haeckel drew his conclusions about the so-called lower races prior to any real contact with them. Such judgments, though, were common enough among mid-nineteenth-century Europeans. Few German scientists—save Friedrich Tiedemann—made an effort to test empirically what seemed an obvious conclusion.[127] Even Darwin judged the Fuegians and other "natural men" as lacking the mental capacities of the more advanced races. Haeckel would never shed his belief in the hierarchy of races—though he certainly changed his mind about the location of the races in the hierarchy (as I will discuss in chapter 7).

Haeckel, like Darwin, believed that animal mentality differed from the human only in degree, not kind. But the sort of mind human beings possessed stemmed, he argued, from sexual selection, which he regarded as a type of natural selection. Prior to the discussion of human sexual selection that would occupy Darwin's *Descent of Man* (1871), Haeckel maintained that within the advanced races, females would select men of higher mental caliber, thus continually increasing brainpower in the species. In like fashion, active male choice would enhance female beauty. Undoubtedly Haeckel perceived this to be the case in his and Anna's selection of one another, or so he subtly suggested when he remarked: "Thus through generation the mutual advantages of both complementary sexes are raised to a higher grade of ennoblement. That higher grade determines, through the harmonic mutuality of both the ennobled sexes in marriage, the greatest happiness of human life."[128] Throughout the writing of his book, Anna never ceased to flood the cells of his memory.

Stem-Trees

A striking feature of Haeckel's *Generelle Morphologie* is the series of eight plates depicting stem-trees of actual species, living and extinct. He was not the first to use a tree diagram to represent the relations of biological species—Heinrich Georg Bronn had done so unobtrusively in the late 1850s (see fig. 5.7). Darwin, too, had illustrated possible species relationships with a line diagram, the only illustration in the *Origin;* his depiction, though, hardly

126. Ernst Haeckel, "Ueber den Stammbaum des Menschengeschlechts" (November 1865), in *Gesammelte populäre Vorträge,* 84; Haeckel, *Generelle Morphologie,* 2:430.

127. For a brief discussion of Tiedemann's observations on racial hierarchy, see the first appendix.

128. Haeckel, *Generelle Morphologie,* 2:247.

Es ergibt sich aus den vorangehenden Untersuchungen, dass nicht nur die Wirbel-losen Thiere, die Fische, die Reptilien, die warmblüthigen Vögel und Säugthiere und zuletzt der Mensch allmählich erst die einen nach den andern auftreten, — sondern auch in den einzelnen Unterreichen der Strahlenthiere, der Weichthiere, der Kerbthiere, der Fische die höheren Äste des Sytemes erst nach den tieferen erscheinen, — jedoch in der Weise, dass der höhere Zweig eines tieferen Astes sich oft später als der tiefere Zweig eines höheren Astes entwickelt. Will man dieses Verhalten durch ein Bild darstellen, so wird dasselbe einem solchen Baum-förmigen Bilde des Systemes entsprechen.

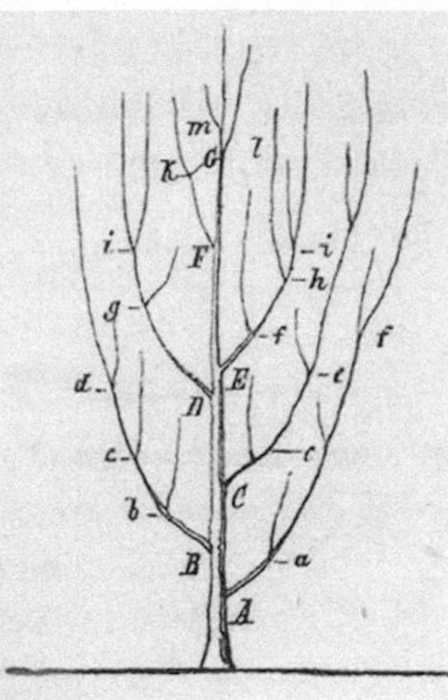

Fig. 5.7. A tree representing on the branches *A* to *G* the larger groups of animals, such as the invertebrates, fish, reptiles, birds, mammals, and man; and the lowercase letters representing various species of those groups. The organisms lower on the tree indicate those found at deeper strata in paleontological deposits. (From Heinrich Georg Bronn, *Untersuchungen über die Entwickelungs-Gesetze der organischen Welt,* 1858.)

has the appearance of a tree. Both Bronn and Darwin had only suggested how species might be represented; they pictured relationships of no actual species. In his *Darwinsche Theorie und die Sprachwissenschaft* (1863), Schleicher, Haeckel's close friend, did represent real species, of a sort, in his stem diagram of the descent of the Indo-German languages (see fig. 5.8).

In the early 1850s, Schleicher had hit on the notion of representing language relationships by tree diagrams; and in his later *Deutsche Sprache* (1860), he rendered these descent relationships in considerable graphic detail, employing certain principles of his own devising. So, for instance, he would represent morphological distance in languages by the acuteness of the angle separating branches of the tree. The principle is illustrated in fig. 5.9: the diagram shows the greater divergence of daughter language *b* from both the mother language *A* and the more lineally descendent daughter language *a*. Accordingly, the time of historical development of languages would be indicated by spatial distance from the root of the stem-tree or from a node.[129]

In comparable fashion, Haeckel also graphically indicated both temporal distance and morphological difference in the evolutionary development of organisms—the former along the vertical axis and the latter by angle of

129. I have discussed Schleicher's theory of language descent, as well as his diagrams, in "The Linguistic Creation of Man.".

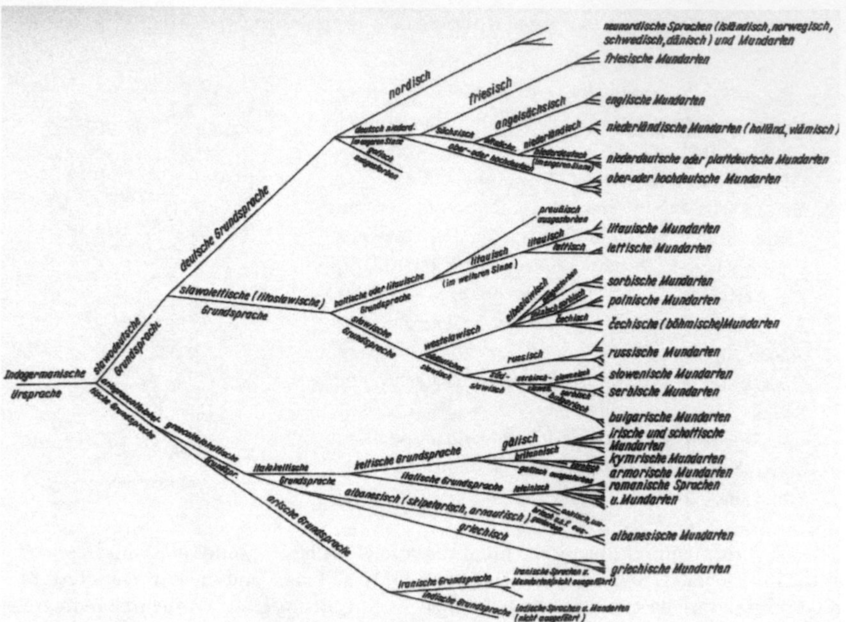

Fig. 5.8. Tree of Indo-German languages. (From August Schleicher, *Die Darwinsche Theorie und die Sprachwissenschaft,* 1863.)

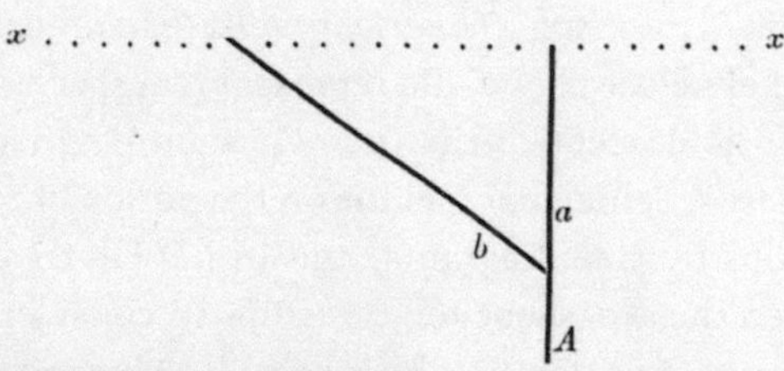

Die Grundſprache A theilt ſich in die Sprachen a und b in der beſchriebenen Weiſe nämlich ſo, daß der Theil des Sprachgebietes b ſtärkeren Veränderungen unterliegt als der mit a bezeichnete. Bis zum Durchſchnitt xx hat alſo b ſich viel weiter von A entfernt als a, und dieß macht eben unſer Schema dadurch anſchaulich, daß es bx ſtärker von der geraden Richtung abweichen läßt als ax, das mehr als eine directe Fortſetzung von A erſcheint.

Fig. 5.9. Schleicher's scheme for illustrating the morphology and descent relations of language: the mother tongue *A,* gives rise to two daughter languages *b* and *a,* with the *b* becoming gradually more distant from *a,* which is a more direct development of the mother tongue. (From August Schleicher, *Die Deutsche Sprache,* 1860.)

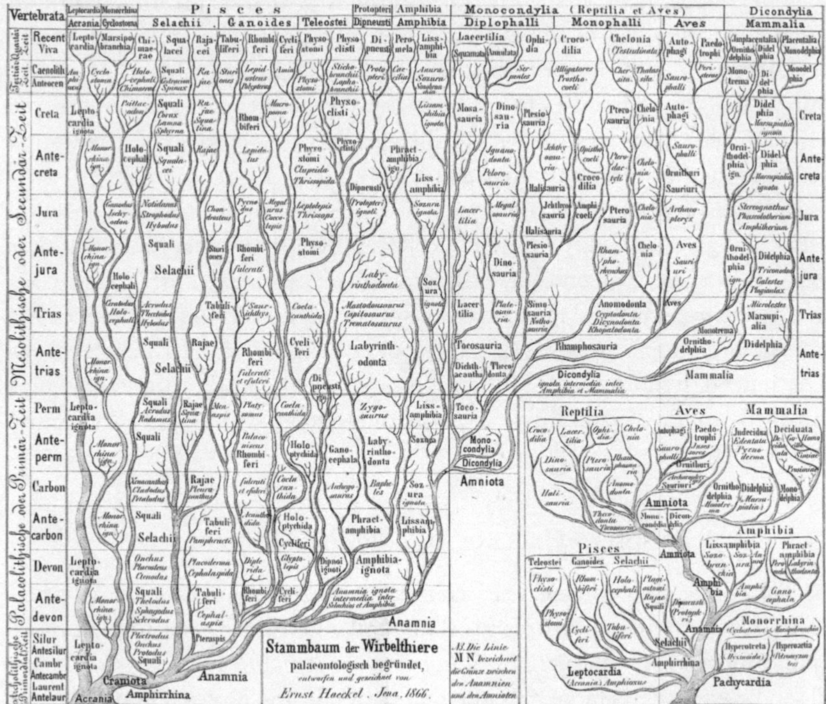

Fig. 5.10. Stem-tree of the vertebrates, paleontologically grounded, with insert of the systematic relations of extant groups. (From Haeckel, *Generelle Morphologie der Organismen*, 1866.)

branching (see fig. 5.10). In addition, he depicted species extinction along the several branches that stopped short of the top level, that is, the contemporary period.

Haeckel's illustrations trump those of Darwin and Schleicher in that they actually look like trees. The illustrations make a powerful rhetorical impact: what is described in endless pages of technical description is summed up in an intuitively clear representation. The measure of the impact can simply be taken by noting that even today species descent is often economically rendered in tree diagrams. In other of his books—especially the two great popular successes, his *Natürliche Schöpfungsgeschichte* (Natural history of creation, 1868) and *Anthropogenie* (1874)—Haeckel used his diagrams for a different purpose, namely, to show the degree of evolutionary progress among the races of mankind. This latter use would reveal more than Haeckel might have realized, a matter I will discuss in chapter 7. Some historians have argued that Haeckel's diagrams demonstrate his teleological perspective, according to which he presumed human beings to be the goal of the evolutionary process.[130] The stem-tree of the vertebrates (fig. 5.10, lower right insert) does not suggest this at all. As is plain to see, *Homo* huddles in the far right corner of the inserted stem-tree of the morphological systematics of the vertebrates—hardly the position of the messiah of evolutionary history.

Haeckel's trees graphically represent the evolutionary history of life on earth, though in an admittedly hypothetical fashion—despite labels that assert them to be "paleontologically grounded." As complaints quickly accumulated, especially about the trees sprouting in various editions of his *Natürliche Schöpfungsgeschichte,* Haeckel acknowledged that the trees were, indeed, hypothetical: "I have expressly declared that my stem-trees claim only to have made a first attempt and to have stimulated further and better efforts."[131]

Reaction to Haeckel's *Generelle Morphologie*

The public reaction to Haeckel's *Generelle Morphologie* grew slowly and did not reach large proportions until he rendered its main features in more popular form by his *Natürliche Schöpfungsgeschichte* two years later. He did send copies of his large monograph to friends and professionals. Gegenbaur, while approving of the biological doctrines of the treatise, admonished his colleague because of the polemical bomblets he sprinkled throughout.[132] August Weismann responded to the gift of the book with surprise because of its great synthetic achievement and with thanks because of its "many new, fruitful ideas."[133] He reported that others of his acquaintance had inveighed against the book (*schimpfen darüber*) but that he was in fundamental accord with all of its theses. He especially liked Haeckel's construction of morphological structures and his proposal that echinoderms

130. See, for example, Bowler, *Non-Darwinian Revolution,* 88; see also the discussion of figure 5.5 above.

131. Ernst Haeckel, *Natürliche Schöpfungsgeschichte,* 2nd ed. (Berlin: Georg Reimer, 1870), xxiv. Gould gives no quarter to Haeckel's graphic invention, pointing out the defects (from a late twentieth-century view) of the inadequate representation of, for instance, the great variability among fishes or the too large of a space given to carnivores and mammals as opposed to rodents. See Stephen Jay Gould, *Wonderful Life: The Burgess Shale and the Nature of History* (New York: Norton, 1989), 263–67.

132. Haeckel mentions Gegenbaur's complaint in a letter to Thomas Henry Huxley (12 May 1867). See Haeckel, "Briefwechsel zwischen Huxley und Haeckel," 12.

133. Weismann to Haeckel (21 May 1867), in "Briefwechsel zwischen Haeckel und Weismann," 19.

derived from worms. Weismann would later reject his friend's claims about heredity, particularly the principle of the inheritance of acquired characteristics; but at the beginning of their acquaintance, there seemed no fundamental disagreements.[134] Ludwig Büchner (1824–1899), whose *Kraft und Stoff* (Force and matter) excited liberal materialists from the time of its publication (1855) to the end of the century, read Haeckel's volumes with "joy and satisfaction." He was especially gratified by "the sharpness and ruthlessness [*Rücksichtslosigkeit*] with which you have confronted the old school and the bloodless empiricists in your capacity as a professional." He did point out to Haeckel that in *Kraft und Stoff*, he himself had "laid the foundation for descent theory five years before Darwin."[135]

Haeckel's English friends also responded positively. Darwin received an advance notice of the book and in anticipation immediately expressed his delight: "Your abstract seems to me wonderfully clear & good; one little fact shews me how clearly you understand my views, namely your bringing prominently forward, which no one else has ever done, the fact & the cause of Divergence of Character."[136] A few months later, after he spent time with Haeckel's publication, he again wrote of his appreciation, though with a cautionary note about the book's strident character:

> What has struck me most is the singular clearness with which all the lesser principles & the general philosophy of the subject have been thought out by you & methodically arranged. . . . I hope that you will not think me impertinent if I make one criticism: some of your remarks on various authors seem to me too severe; but I cannot judge well on this head from being so poor a German scholar. I have however heard complaints from several excellent authorities & admirers of your work on the severity of your criticisms. This seems to me very unfortunate for I have long observed that much severity leads the reader to take the side of the attacked person. I can call to mind distinct instances in which

134. Haeckel would, at the end of the century, become acutely disturbed over Weismann's "ultra-Darwinism." Haeckel wrote Herbert Spencer (10 November 1895): "Our battle against Weismann's false theories (which have now condensed in a pure teleological theory of evolution . . .) seem to have little effect. The need or want of somewhat mystical and irrational dogmatic views seems to be necessary for most naturalists as for most other people." The letter is in the Haeckel Correspondence, Haeckel-Haus, Jena.

135. Büchner to Haeckel (12 August 1867), in *Carl Vogt, Jacob Moleschott, Ludwig Büchner, Ernst Haeckel: Briefwechsel*, ed. Christoph Kockerbeck (Marburg: Basilisken-Presse, 1999), 134.

136. Darwin to Haeckel (18 August 1866), in the Haeckel Correspondence, Haeckel-Haus, Jena.

> severity produced directly the opposite effect to what was intended. I feel sure that our good friend Huxley, though he has much influence, wd have had far more if he had been more moderate & less frequent in his attacks.[137]

While Darwin himself engaged in many disputes, he did not and would not have, unlike Haeckel, referred to his opponents as grubbing after mundane facts, as slumbering in a scientific half-sleep (*wissenschaftlichen Halbschlafe*), or as stumbling through a conceptually impoverished dream life (*gedankenarmes Traumleben*).[138]

Haeckel did stand abashed at Darwin's admonishment. He explained the tone of the work as the result of his mental anguish, both over his beloved wife's death and because of the stupidity of Darwin's opponents. He wrote:

> Certainly you are right to scold me over the excessive hardness of my critical attack and the bitterness of my polemic. My best friend here, Prof. Gegenbaur, has greatly chided me for it. I can only say, by way of excuse, that during the previous summer and winter, as I worked on the book in a gloomy loneliness, I suffered from an extraordinarily bitter attitude and nervous state. Also I was too angry over the unashamed and stupid attacks by your opponents, that I could not let them go unpunished. Because of this controversy, I have suffered in my own person and I have been no less strongly attacked from many sides.[139]

Huxley, unlike Darwin, delighted in the kind of invective Haeckel dished up. Upon receipt of the book, he wrote Haeckel: "I am much inclined to think that it is a good thing for a man, once at any rate in his life, to perform a public war-dance against all sorts of humbug and imposture." But even Huxley thought too many scalps had been taken in Haeckel's *Generelle Morphologie.*[140]

Huxley conferred about a translation of Haeckel's book with Darwin,

137. Darwin to Haeckel (12 April 1867), in ibid.

138. These are typical of the acid remarks made throughout the book. See Haeckel, *Generelle Morphologie,* 1:103. Or consider Haeckel's despoiling of a critic, who has "neither a concept of a 'concept' nor an idea of an 'idea'" (2:330n1).

139. Haeckel to Darwin (12 May 1867), in the Haeckel Correspondence, Haeckel-Haus, Jena.

140. Huxley to Haeckel (20 May 1867), in "Briefwechsel zwischen Huxley und Haeckel," 13.

who generously offered a financial contribution toward the effort. For an English audience, however, the bulk would have to be trimmed and the war paint removed. Haeckel speedily acquiesced to these restrictions and outlined a plan to cut the work by over half and to eliminate all harsh asides.[141] The Ray Society agreed to sponsor the publication, and Haeckel began the task of abridging his two fat volumes and excising the offensive language.[142] The translation, however, never appeared, since a more likely candidate became available.[143] Because the *Generelle Morphologie,* even in Germany, failed to reach beyond a small number of dedicated friends and committed enemies, Haeckel quickly revised a series of popular lectures he gave during the 1867–68 term and from them constructed a more compact and accessible book, which yet advanced essentially the same arguments as his technical monograph. His lectures were published as *Natürliche Schöpfungsgeschichte* in 1868; and after some delay, the first English translation appeared, in 1876, as *The History of Creation.*[144] While Haeckel's friends had initially worried about war whoops reverberating through the pages of his big book, the slimmer English version hardly dampened those bellicose tones; rather it seemed to echo even more prominently his polemical intent. The ferocity of the popular venture caused new enemies to prepare retaliatory measures.

Conclusion

Haeckel's morphological and evolutionary conceptions have become so deeply entrenched in the common, neo-Darwinian tradition that his origi-

141. Haeckel outlined his plan to Darwin in a letter of 25 November 1868 (in the Darwin Correspondence, DAR 166.1, Cambridge). Haeckel said he would undertake "1) to reduce the volume by half; 2) to eliminate all personal polemics; 3) to eliminate all heretical (heterodox) material, all bad jokes and humor, and generally everything that is not strictly to the point."

142. Discussion of the translation of Haeckel's book occupied many letters between Haeckel and Huxley from May 1867 to February 1869. See "Briefwechsel zwischen Huxley und Haeckel," 13–22.

143. Haeckel's abridgement of the *Generelle Morphologie* did finally appear in 1906, published as *Prinzipien der generellen Morphologie der Organismen. Wörtlicher Abdruck eines Teiles der 1866 erschienen Generellen Morphologie* (Berlin: Georg Reimer, 1906).

144. Ernst Haeckel, *The History of Creation,* trans. revised by E. Ray Lankester (London, H. S. King & Co., 1876). Lankester arranged for the publication of the translation—done by "an anonymous but dutiful woman." She was L. Dora Schmitz, a translator of the correspondence between Goethe and Schiller, as well as numerous contemporary German authors. Lankester suggested to Haeckel that the word "natural" be dropped in the English title, "in order not to frighten the pious English public." See Lankester to Haeckel (31 December 1872), in the Haeckel Correspondence, Haeckel-Haus, Jena.

nality may easily be overlooked, especially when he has been depicted by several historians as a "mere popularizer." That he was a popularizer is, of course, true. But he was hardly a *mere* popularizer. No one having just a modicum of awareness of his many research monographs could deny him the additional sobriquet of genius of enormous creative power. Darwin himself recognized Haeckel's great originality, especially concerning the basic conception of phylogeny (see Darwin's remarks above). The *Generelle Morphologie,* though wrought in despair, cascades forth with novel ideas that laid down deep channels right through to the contemporary period. His identification of the protists as a separate kingdom has become canonical; ecology, which he named and practiced, is now a standard part of the discipline of biology; the problem of biological individuality has continued to surface with ever higher profile; he was the first to argue in a systematic and straightforward way that metazoans arose from single-celled creatures through aggregation and then the division of labor;[145] though not the first to employ the biogenetic law (it was part of Darwin's original repertoire), he did more than anyone to make it a staple of nineteenth-century evolutionary thought; his name remains attached to myriads of marine species he discovered and described; and, finally, his tree diagrams became ubiquitous in the biological literature and spawned the many other graphic techniques employed today. And these are just the products of his first few professional years.

If one steps back to view the accomplishments of both Darwin and Haeckel, one can recognize differences of emphasis. Haeckel's morphological and phylogenetic considerations—exemplified in his many large treatises on the systematics of invertebrates—occupied a good portion of his scientific output. One should not forget, however, that Darwin himself initially won scientific recognition as a biologist principally by his four large volumes on the systematics of barnacles, volumes that were not explicitly evolutionary but that employed the recapitulation hypothesis to indicate "relatedness" of disparate groups of those animals. And, of course, the penultimate chapter of the *Origin* is devoted to systematics and mor-

145. Haeckel, *Generelle Morphologie,* 1:289–90: "Many like individuals are able united to develop more power than a lone individual. But out of this quantitative improvement through aggregation the still more important qualitative improvement through differentiation arises. Initially the improvement is slight, but soon significant differences are introduced among the original homogeneous plastids [i.e., one-celled organisms], leading finally to a full-blown division of labor. Since the individual cytodes or cells give up, more or less, their individual independence and enter into the service of the higher unity of the plastid colony, they develop distinctive properties in certain directions."

phology—the latter forming, in Darwin's terms, the "very soul" of natural history.[146] It is nonetheless true that taken as a whole, as Lynn Nyhart has observed, Haeckel's published works do not tend in the same direction as Darwin's.[147] But then none of Darwin's immediate intellectual heirs—Wallace, Romanes, Huxley, or Weismann—followed his trajectory in any more precise ways than did Haeckel. Haeckel moved beyond his predecessor in several areas of creative discovery, yet he always took his bearings from that lodestar of English genius.

During the rest of his career, Haeckel would continue to innovate, to advance new hypotheses, and to argue for them with voluble conviction and Darwinian vision. He would initiate experimental techniques in embryology to demonstrate evolutionary descent, techniques that later became part of the program of *Entwickelungsmechanik*, which his students Wilhelm Roux and Hans Driesch further advanced. He would give currency to the concept of the "missing link" and even correctly suggest where it might be found. Amazingly, his disciple Eugène Dubois discovered it in Java. He would sketch the biogeographical spread of the human races from a common origin off the coast of Africa, a land he would characterize as our biological Paradise. He would identify the importance of language for human evolution, especially for the development of man's big brain. And, during his long career, he would discover countless new species of marine organisms and provide for them detailed phylogenetic accounts. Some of his efforts initiated a legacy that many think unfortunate, for instance: the transformation of evolutionary doctrine into a comprehensive philosophy, and the deep antagonism between evolutionary theory and religious ideology. He has been blamed for the kind of racial and eugenic notions that gave succor to some Nazi biologists. Haeckel's larger cultural impact has, for good or ill, spread through the twentieth and twenty-first centuries. Of course, many of his proposals, even after vigorous tending, failed to produce fruit. Yet it is striking, if little appreciated, what a great garden of still-flourishing ideas were first planted by Ernst Haeckel.

The initial design of that garden came to bitter expression in the aftermath of the death of Anna. Haeckel felt the urgency of its realization not only as balm to a terrible loss but also to lay out a new, reliable path through the tangle of superstition, both religious and scientific, that insinuated false hope and choked off the voice of reason. The path he envisioned

146. Darwin, *Origin of Species*, 434.

147. Lynn Nyhart, *Biology Takes Form: Animal Morphology and the German Universities, 1800–1900* (Chicago: University of Chicago Press, 1995), 150.

would terminate in a new, modern age, one in which religion would be transformed into Darwinian understanding and the individual would find rest in the bosom of Goethean nature.

Appendix: Haeckel's Letter to Darwin

On 7 July 1864, after returning from his convalescence on the Mediterranean, Haeckel wrote Darwin the following letter, which expresses both his commitment to the new theory and his intention to discover what he termed "analytic proof" of the theory through a study of various marine invertebrates. The letter also indicates that the bond linking Haeckel to Darwin was cemented by his great loss.

> My dear Sir,
> I found your letter, which had been written several months ago, when I returned from a zoological trip to the Mediterranean. Your letter has given me great pleasure. It has also provided me opportunity and personally the decided honor, Sir, to express the extraordinary esteem I have for the discovery of the "Struggle for Life" and "Natural selection." Of all the books that I have read, none has made so powerful and marked an impression on me as your theory of the origin of species [*Ihre Theorie über die Entstehung der Arten*]. In this book I find at once the harmonious solution to all the fundamental problems of which I have labored for an explanation since the time I had learned to know nature in her authentic state. Since then I have studied your theory—I say without exaggeration—daily, and whether I study the life of man, animals or plants, I find in your descent theory the satisfactory answer to all my questions no matter how difficult.
>
> Since you must have a certain interest to learn of the spread of your theory in Germany, allow me to impart this. Most of the older zoologists, and among them many of considerable authority, are among your most enthusiastic opponents. These men have, on the one hand, lost, through a life spent in the old accustomed dogmas, the ability to view impartially what is new as worthy and correct—even truth itself; on the other hand, they lack the *courage* to distinguish their convictions from the truth of the descent theory. Many attempt to improve their earlier false views and so, finally, are not able to comprehend the whole of nature with one overview, since the painstaking study of details and the analytic investigation of particulars does not permit a general perception of nature.

Yet among the younger naturalists, the number of your committed and enthusiastic followers grows from day to day; and I believe that in a few years their number will be as large, perhaps, as the number of your committed followers in England itself—since the Germans on the whole (as far as I can judge) are not so constrained by religious and social prejudices as the English; though in respect of political maturity and in relation to full development, they are rather behind. The power of the clergy and religious dogmas and the influence of social prejudice in the educated classes of Germany are small—as I surmise from the great and lively interest your teaching finds, for the most part, among the educated laity. The academic lectures, which I myself and a few of my younger colleagues conduct on your theory, appeal not only to students of natural science and medicine, but are heard also by philosophers and historians, and, yes, even theologians. For the historians a new world is opened, since in the application of descent theory to human beings (as Huxley and Vogt have so happily attempted), they find a way to connect the history of human beings closely with natural history. Indeed, it is here in Jena that we have particularly favorable ground for the development and spread of such reformational teaching, since in all respects we have here the greatest freedom—while at other universities—for instance Göttingen and Berlin—many restrictions and general rules hinder more free intellectual action. Yet one may hope that the progressive development that one hears has begun in all quarters of Germany will defeat, now and again, the opposing elements and that the results of your theory will be correctly understood and adopted.

Perhaps you will allow me to relate to you a few personal matters concerning your theory, since I have devoted my life to it and direct all my activities to making it known. In my first large work, a monograph on the radiolarians (Berlin, Reimer, 1862), I mentioned your theory along the way (p. 232, in the note), and attempted to construct a genealogical table of the relationships of these animals (p. 234). Then, last year, I seized the opportunity in Stettin for the first time, at the meeting of the German Naturalists, to bring the question into discussion; this resulted in a rather lively debate. Though I was strongly attacked by a very eloquent speaker, Dr. Otto Volger from Frankfurt, I yet won many friends for your theory and Virchow, our greatest scientific physician, also spoke on the topic.

Presently I am busy with a large work on coelenterates, the animals which, because of their complicated sort of development, show very well their *common descent* from one original form. On the coast of Nice this spring, I spent a long time studying medusae. I was astonished at the extraordinary spread of individual variations that occurs in some of these animals. One often finds the variation in formation of the essential parts of individuals of one and the same species to be greater than those between different species of a genus and, indeed, between several genera of one and the same family. With your permission, I will send you next year my work on this subject.

In addition to this special work, I have been busy for several years preparing a general natural history, in which I show how your theory illuminates every area of that history and how that theory produces a harmonious integration of the whole. I hope to finish this book next winter. My public academic lectures, which I give here at the university on zoology, comparative anatomy, paleontology, development and histology—these constantly aid me in this undertaking.

Although I am only 30 years old, a terrible fate, which has destroyed my whole happiness in life, has made me mature and resolute. It has hardened me against the blame as well as the praise of men, so that I am completely untouched by external influence of any sort, and only have one goal in life, namely, to work for your descent theory to support it and perfect it.

Please forgive me, Sir, for having taken up your precious time with this long letter. It was for me a vital necessity to express to you at least once those things that move me daily in my tasks and that suffuse all my work. "When the heart is full, the mouth overflows."

My friends and colleagues here, the comparative linguist August Schleicher and the comparative anatomist Carl Gegenbaur, with whom I so often share my strong conviction of the pure truth of your teaching—they send their best wishes. I hope, Sir, that your health improves and that for a long time you will be ready to fight the good fight for the truth and against human prejudice. I remain with the greatest respect, yours very truly,

Ernst Haeckel.[148]

148. Haeckel to Darwin (7 July 1864), in the Darwin Correspondence, DAR 166.1, Cambridge.

CHAPTER SIX

Travel to England and the Canary Islands: Experimental Justification of Evolution

In the mid-1860s, Prussia and Austria momentarily colluded to wrest control of the ancient German duchies of Schleswig and Holstein from the Danish king, Christian IX (1818–1906). After a brief struggle, the German Confederation reclaimed its heritage, with Schleswig coming under administration of Prussia and Holstein under that of Austria. Otto von Bismarck (1815–1898), prime minister of Prussia, had a covert plan, however, to annex both duchies to the kingdom of Wilhelm I (1797–1888).[1] As tensions between the two great powers escalated, Bismarck warned the smaller states of Hanover, Saxony, and Hesse-Cassel that support for the Hapsburg cause would render them enemies of the Hohenzollern king. When war did break out in mid-June 1866, the other nations expected and desired a protracted conflict, of the sort that might weaken the two major forces of central Europe. But within seven weeks, the Austrians, even with the support of several of the smaller German states, collapsed under the heavy weight of Prussian armaments and the brilliant planning of Helmuth, Count von Moltke (1800–1891), the chief of the army's general staff. The victory provided Bismarck opportunity to subject the smaller north German states to Prussian rule, and thus began the unification of the Germanies under the leadership of the Hohenzollern king and the craft of Bismarck. The liberal dream of a united Germany was soaked in blood and shackled in iron.

Haeckel and his father, as good liberals, had long desired unification, but not through the application of the needle gun. As he was seeing the first sections of his *Generelle Morphologie* through the press, Haeckel wrote

1. Wilhelm became king of Prussia at his brother's death in 1861 and emperor of Germany from 1871 to his death in 1888.

Rudolf Virchow, his former teacher and sometime friend, an urgent letter of protest. Virchow, who had returned to Berlin in 1856 and had become a member of the Prussian diet in 1861, was a founder of the new Deutsche Fortschrittspartei (German Progressive Party). Haeckel felt the party was ineffectual and had done little to stop the provocations that had coerced the Austrians into war. "Day after day," he wrote his former teacher,

> we German patriots here have awaited a relief of our German misery by the actions of the German Progressive Party. We have hoped that the party would issue some clear German mandate to halt. In vain! . . . Declare that you wish all to be German and free but that you won't arrive at this goal by way of Bismarckian blood and iron. . . . If this won't work, at least you'll have done your duty. But again, speak out loud and clear! Otherwise we will lose trust in you. If you remain silent, the political coryphées in Berlin will place you generally among the partisans of Bismarckian external politics.[2]

Haeckel's admonition must have stung Virchow, who had often opposed Bismarck in the parliament during the 1860s, and in one instance so irritated the chancellor that the professor was challenged to a duel, an invitation that he prudently did not accept. After the elections in July 1866, as the war with Austria was quickly winding down, the Progressive Party barely held the majority, and even lost the allegiance of many of its members. From this point forward, Virchow would have no potent voice in German political affairs. But he would still have a commanding voice in German scientific affairs, and he would use that voice to denounce Haeckel and his evolutionary conceptions (discussed in chapter 8).

Visit to England and Meeting with Darwin

In the spring of 1866, Thomas Henry Huxley—one of the English scientists to whom Haeckel had sent a copy of his *Radiolarien* monograph—had begun a small correspondence with this new German colleague and invited him to attend the Nottingham meeting of the British Association for the Advancement of Science.[3] Haeckel reluctantly had to decline the invitation

2. Haeckel to Virchow (24 May 1866), in Virchows Nachlaß (N 2, nr. 791), Archive of the Academy of Sciences, Berlin.

3. Huxley to Haeckel (24 April 1866), in "Der Briefwechsel zwischen Thomas Henry Huxley und Ernst Haeckel," ed. Georg Uschmann and Ilse Jahn, *Wissenschaftliche Zeitschrift*

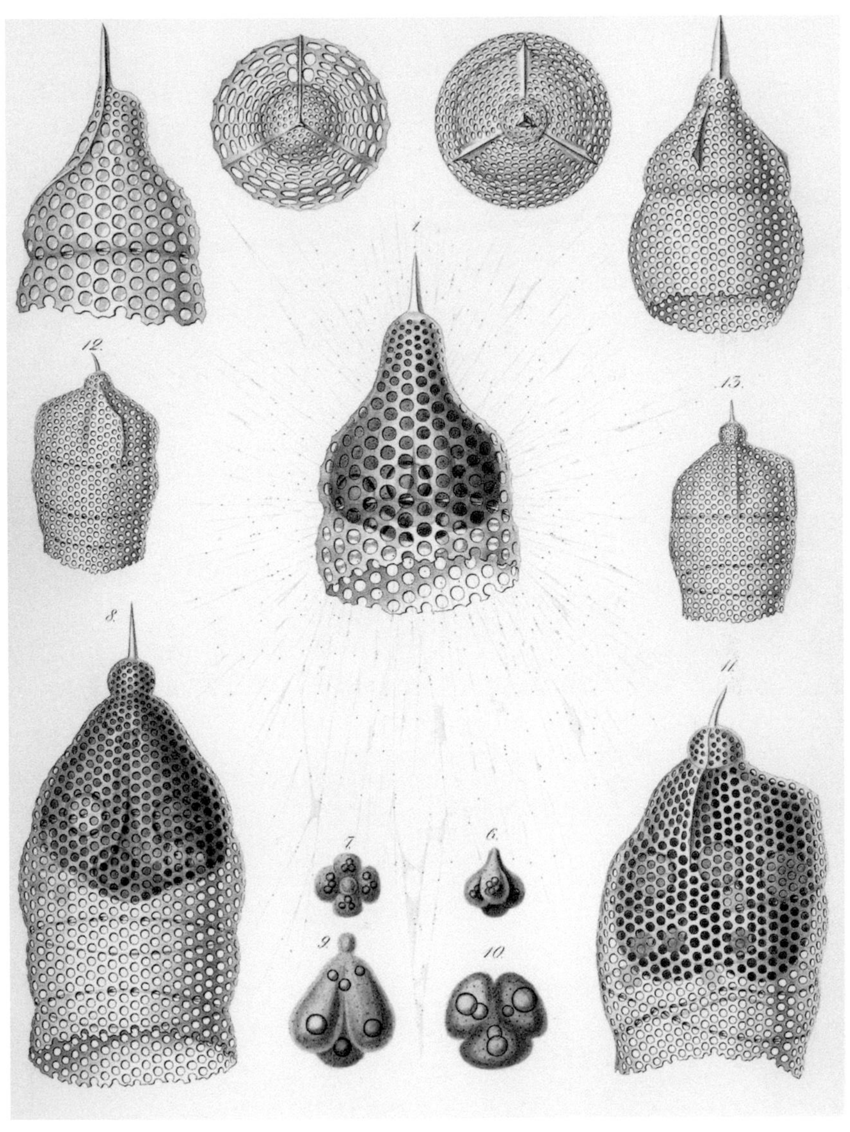

PLATE 1: Radiolaria of the subfamily Eucyrtidium.
(From Haeckel, *Radiolarien*, 1862.)

PLATE 2: Alexander von Humboldt and Aimé Bonpland in their laboratory by the Orinoco. Oil on canvas (c. 1850) by Eduard Ender. (Courtesy of Archiv für Kunst und Geschichte, Berlin.)

Opposite

PLATE 3: Haeckel's portrait (1850) of the *Nationalversammlung der Vögel* (Parliament of birds). (Courtesy of Ernst-Haeckel-Haus, Jena.)

ERNST HÆCKEL
1850
National Versammlung der Vögel
1850.

PLATE 4: Haeckel's watercolor (1859) of his study on Capri.
(Courtesy of Ernst-Haeckel-Haus, Jena.)

PLATE 5: Discomedusa *Desmonema Annasethe*.
(From Haeckel, *System der Medusen*, 1879.)

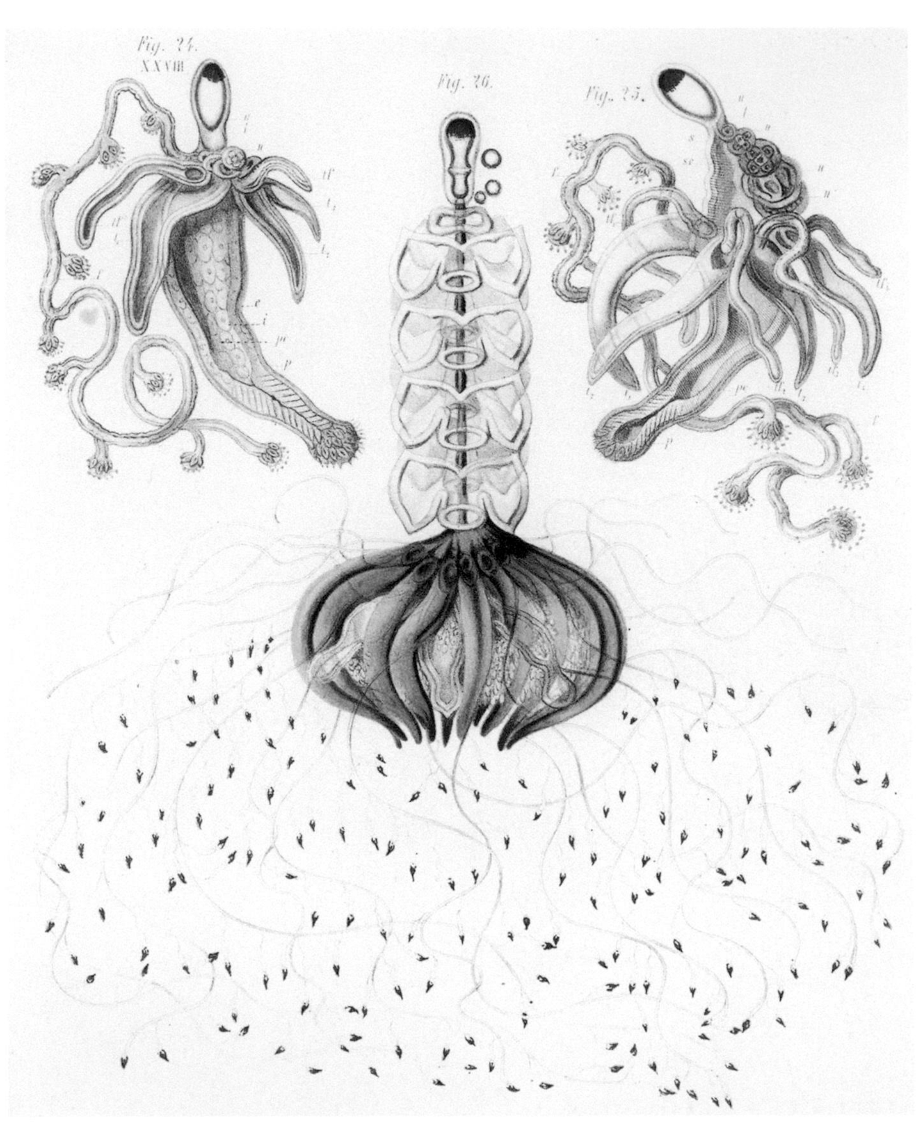

PLATE 6: *Physophora magnifica*, flanked by two larvae at slightly different stages of development. (From Haeckel, *Zur Entwickelungsgeschichte der Siphonophoren*, 1869.)

PLATE 7: Scene painted by Haeckel and lithographed for his *Arabische Korallen*, 1876.

PLATE 8: Highlands of Java; oil landscape by Haeckel.
(From Haeckel, *Wanderbilder*, 1905.)

because of the uncertainties produced by the Austro-Prussian conflict. He harbored some hope, however, of a safer journey to the Mediterranean to work with his friend August Weismann (1834–1914).[4] Surprisingly, though, the unstable situation quickly solidified as the Prussians mounted success after success. Within a few short weeks, Austria capitulated, signing a preliminary peace agreement on 26 July and a final peace accord, the Peace of Prague, on 23 August 1866. This constituted the first stage of German unification, with the kingdom of Prussia annexing the north German states, as well as Thuringia, Saxony, and Darmstadt. By 1871 unification would be completed with the addition of the southern lands of Bavaria, Baden-Württemberg, and Alsace-Lorraine. The triumph of Bismarck in the summer of 1866 meant, for the moment, a more settled political and military climate.

For Haeckel, one wave of tension subsided but another waxed as the palliative of all-consuming work on the *Generelle Morphologie* reached its end. Again he was flooded with memories of happiness crushed by the death of Anna. But with the war over, he could seek refuge in what had become a specific for his depressive moods, exotic travel. Now he planned to follow in the wake of his youthful hero, Alexander von Humboldt, and sail off to the western coast of Africa and the Canary Islands, an area that promised even more biological riches than had the Eldorado of the Italian Mediterranean; it was a land whose beauty, so evocatively described by Humboldt, might bring solace.[5] The trip would also provide opportunity for a pilgrimage to that icon whose devoted acolyte he had become, the sage of Downe.

Haeckel arrived in London on Wednesday, 17 October 1866, and immediately visited with Sir Charles Lyell (1797–1875) and the Huxley family. Henrietta Huxley obviously charmed him sufficiently for him later to remark to Darwin, in his halting English, "I do love Huxley's woman."[6]

---

der Friedrich-Schiller Universität Jena (Mathematisch-Naturwissenschaftliche Reihe) 9 (1959–60): 10.

4. Haeckel broached this collaborative venture to Weismann, whose recent betrothal and degenerating vision precluded the possibility. See Haeckel to Weismann (4 June 1866) and Weismann to Haeckel (12 July 1866), in "Der Briefwechsel zwischen Ernst Haeckel und August Weismann," ed. G. Uschmann and B. Hassenstein, *Jenaer Reden und Schriften* (1965): 16–18.

5. Haeckel wrote his parents, after landing at Tenerife, that he sought the natural beauty so wonderfully described by Humboldt, especially the great volcanic mountain and the ancient dragon tree. See Haeckel to his parents (27 November 1866), in *Ernst Haeckel: Biographie in Briefen mit Erläuterungen*, ed. Georg Uschmann (Gütersloh: Prisma, 1984), 90.

6. In the twilight of their years, Henrietta Huxley reminded Haeckel of this remark. See her letter to Haeckel (9 March 1914), in "Briefwechsel zwischen Huxley und Haeckel," 29.

Early on Sunday, he took the train from London to the village of Downe. Darwin's coach met him at the station, and he was driven to the estate. Haeckel remembered the meeting vividly:

> As the coach pulled up to Darwin's ivy-covered country house, shaded by elms, out of the shadows of the vine-covered entrance came the great scientist himself to meet me. He had a tall, worthy form with the broad shoulders of Atlas, who carries a world of thought. He had a Jupiter-like forehead, high and broadly domed, similar to Goethe's, and with deep furrows from the habit of mental work. His eyes were the friendliest and kindest, beshadowed by the roof of a protruding brow. His sensitive mouth was surrounded by a great silver-white full beard. The welcoming, warm expression of his whole face, the quiet and soft voice, the slow and thoughtful speech, the natural and open flow of ideas in conversation—all of this captured my whole heart during the first hours of our discussion. It was similar to the way his great book on first reading had earlier conquered my understanding by storm. I believed I had before me the kind of noble worldly wisdom of the Greek ancients, that of a Socrates or an Aristotle.[7]

The reciprocal impression made on Darwin—or at least on his family—was also quite positive, though not in the same reverential vein. Darwin's daughter Henrietta wrote her brother George shortly after the meeting to describe the event:

> On Sunday we had a gt visitation. One of Papa's most thoroughgoing disciples, a Jena professor, came to England on his way to Madeira & asked to come down & see Papa. We didn't know whether he cd speak English & our spirits was [*sic*] naturally rather low. He came quite early on Sunday & when first he entered he was so agitated he forgot all the little English he knew & he & Papa shook hands repeatedly, Papa reiteratedly [*sic*] remarking that he was very glad to see him & Haeckel receiving it in dead silence. However afterwards it turned out that he could stumble on very decently—some of his sentences were very fine. Talking of dining in London—"I like a good bit of flesh at a restoration." Of the war in Germany he remarked as an advantage the Russians had

7. This was Haeckel's recollection of the meeting as recorded by his disciple Wilhelm Bölsche, in *Ernst Haeckel: Ein Lebensbild* (Berlin: Georg Bondi, 1909), 179.

Fig. 6.1. Charles Darwin in 1874. Photo by Julia Cameron. (Courtesy of the Smithsonian Institution.)

> from their good education, "Zatven ze officers are deeded ze commons take ze cheap." i.e. when the officers are killed the privates take the chief command. He told us that there are more than 200 medaillons [*sic*] of Papa made by a man from Wm's photo in circulation amongst the students in Jena. Papa has just begun his gt Pangenesis chapter.[8]

Some weeks after his visit, Haeckel's newly published *Generelle Morphologie* reached an appreciative Darwin, who, however, had to struggle with his new friend's prolix German. Despite Darwin's reservations about the length and the polemical character of the book, he consulted with Huxley about an English translation. He offered financial support for a greatly

8. Henrietta Darwin to George Darwin (21 October 1866), in the Darwin Correspondence, DAR 251.326, Manuscript Room, Cambridge University Library.

shortened version, one trimmed of its sour layers of invective.[9] The meeting between the two scientists cemented a relationship that had begun three years earlier and would extend, with many letters, articles, and books passing between them, until Darwin's death sixteen years later.

Travel to the Canary Islands

Haeckel remained in London for two weeks, meeting, through Darwin's good offices, with several other English scientists (e.g., John Lubbock and Joseph Hooker). He also visited the great monuments and sights of the city and was duly impressed by the British Museum, the National Gallery, the zoological and botanical gardens, the Crystal Palace and its attendant park with sculptures of dinosaurs. On 2 November, happy to clear his lungs of the choking London smog, he embarked for Lisbon and Madeira, where he would join up with his assistants for the excursion: Richard Greeff (1829–1892), *Privatdozent* at Bonn, and Hermann Fol (1845–1892) and Nikolai Miklucho (1846–1888), two medical students from Jena. All three would later have distinguished scientific careers, Greeff as professor at Marburg, Fol as embryologist and professor at Geneva, and Miklucho-Maclay (as he became known) as ethnologist of New Guinea and Australia.[10] Though Haeckel regarded the Swiss Fol among his most clever students and the Russian Miklucho among his most beloved, the trip to the Canaries would lightly sprinkle vitriol on these master-pupil relationships.

After sitting out a cholera quarantine in Lisbon for two weeks, the company steamed to Madeira on a small paddle-wheeler, the *Lusitania.* On disembarking, they discovered that their way to the Canary Islands had been blocked by another quarantine. The board of health governing the islands was worried about a cholera epidemic in London and yellow fever on the African coast.[11] Fortuitously the Prussian warship *Niobe* paid

9. See the previous chapter for a discussion of the plans to publish an English translation of Haeckel's book.

10. Miklucho was born Nikolai Nikolaevich Miklouho in St. Petersberg on 17 July 1846. In 1868 he adopted the name Maclay to suggest a connection with a Scots clan. His friends in Germany referred to him as Miklucho, as I will do in the text.

11. Details of Haeckel's sojourn in the Canary Islands are taken from his own travel accounts and letters to his parents and friends as well as from the account of his assistant Richard Greeff. See Ernst Haeckel, "Eine zoologische Excursion nach den canarischen Inseln," *Jenaische Zeitschrift für Medicin und Naturwissenschaft* 3 (1867): 313–28; and "Eine Besteigung der Pik von Teneriffa," *Zeitschrift der Gesellschaft für Erdkunde* 5 (1870): 1–28. See also Haeckel to his parents (27 November 1866 and 27 January 1867), in *Ernst Haeckel: Biographie*

a short visit to Madeira at this time, and a chance meeting with the ship's doctor brought an invitation to come aboard. The captain of the warship turned out to be the grandson of the well-known Jena botanist and friend of Goethe, August Carl Batch (1761–1802). Haeckel's luck was compounded: the ship was en route to Tenerife in the Canaries; and, after hearing of their plans, Batch offered passage to his landsmen. As the ship entered the bay of Santa Cruz on 22 November 1866, the captain saluted the capital with a twenty-one-cannon volley. The Spanish inhabitants received the message that a Prussian warship had arrived.

Santa Cruz and its environs did not display the luxurious vegetation that the company had been led to expect from Portuguese descriptions. They did see the famous dragon tree that Humboldt's portrayal had rendered so exotic, and they were diverted by a curious kind of giant cactus. The bananas were large and delicious, but little else on the volcanic island sparked their interest, save one luxuriant form. These travel-weary biologists quickly became entranced by the Spanish women of the island, whose "famous figures and dark, one might say, volcanic eyes of fire" made the evenings of music in the city plaza a warm delight.[12] After four days of leisure enjoying such pleasures, the company looked to other vistas.

Haeckel inquired about climbing Pico de Tiede, the 12,000-foot volcanic mountain that Humboldt had scaled over half a century before. After securing guides and a mule train, Haeckel and his companions, at 12:30 on the moonlit morning of 26 November, began their approach to the mountain. By 8:30 a.m. they had reached the base of the central cone—what Leopold von Buch had earlier called "the mountain on a mountain"—still some 4,000 feet from the top.[13] Here the mules had to be left, as well as Fol, who had walked too closely behind one of the animals and got kicked in the knee. The further climb, with perilous handholds and fields of ice-covered lava blocks, proved too much for Greeff and Miklucho, who stopped about 1,500 feet from the summit.[14] The company's principal guide, Don Emanuel Reis—who had made over fifty climbs of the peak—balked at the last

in Briefen, 90–100. A series of letters describing the trip to his Jena friends was published as part of a collection: see Ernst Haeckel, *Berg- und Seefahrten: 1857/1883,* ed. Heinrich Schmidt (Leipzig: K. F. Koehler, 1923). Greeff wrote an uninspired and very detailed (mostly about the weather) description of the trip—without even naming his companions. See Richard Greeff, *Reise nach den Canarischen Inseln* (Bonn: Max Cohen & Sohn, 1868).

12. Haeckel, "Besteigung der Pik von Teneriffa."

13. Greeff quotes von Buch in *Reise,* 187.

14. Greeff had earlier suffered a fate similar to Fol's. He got kicked in the chest, fell back, and cracked his elbow so hard that his arm went numb for some time. See ibid., 185–86.

800 feet because of the thick ice. He refused to proceed further. Haeckel, the amazingly fit athlete, and the remaining guide, Hermann Wildpret—a Swiss German who was curator of the local botanical gardens—moved slowly on, hacking out each step with ice axes. Both suffered increasingly from altitude sickness—headaches, congestion, chest pains, and shortness of breath. Three hundred feet from the top, a copious flow of blood gushed from Haeckel's nose, and he passed out. Wildpret revived him and then he himself fainted, requiring Haeckel to perform a like service for his guide. At noon, on shaky feet, they surveyed the beauties of the island from the very peak of the mountain.

Haeckel had to ask himself whether the arduous effort and dangers of such a climb were compensated by the experience of the unparalleled vistas. While he thought those sights were of extraordinary majesty and left an ineradicable memory, he admitted that he could not completely justify the cost in those terms. But there was other coin, which must have entered into the calculation. His own self-image—that of the polymorphic scientist-artist-adventurer, carved from models provided by Humboldt, Goethe, and Darwin—demanded extraordinary accomplishment. And after the death of his wife, he had cast himself beyond the circle of ordinary human consolation and caution. He had become ever more a man of extremes, as further ascents both geological and polemical would prove.

Haeckel and his friends learned that the most abundant marine life could be found off the island of Lanzarote, a forbidding volcanic link in the chain of the Canary archipelago some 150 miles northeast of Tenerife and about 40 miles from the Moroccan coast. They embarked for the island on 4 December; but because of wretched weather and a miserably small boat laden with cows and baskets of foodstuffs, it took them five days to reach the capital Arrecife. From the bay, the city seemed inviting; but walking along its dusty streets, the company discovered a small town in slow collapse. The bareness of the village was complemented by the bareness of the countryside. In the previous century, a large volcanic eruption denuded the land of vegetation; and because of its extremely arid climate and rocky soil, the island could reproduce only a few cactuses here and there. With difficulty, they found an apartment above a store, one of the few on the island with glass windows; and there they set up their headquarters. The house was outfitted with the bare necessities of furniture and a fine collection of parasitical insects—mosquitoes, flies, lice, cockroaches, bedbugs, and so on. They took to estimating the number of fleas killed per day (average 100). The primitive conditions of their living arrangements must have set tempers alight. Haeckel later wrote a friend that during their stay, "Monsieur Fol" had acted so arrogantly that he had to administer the "sternest rebuke" to his student.[15]

The misery of their existence on the island, with its paucity of potable water and its incessant hot winds blowing from the Sahara, was only alleviated by the "great animal soup," as Haeckel called it, found along the coast. And it was a soup unsampled by other naturalists, which gave it even greater savor. They divided their tasks according to the organisms they wished most closely to investigate: Greeff took worms and echinoderms; Fol, ctenophores and mollusks; and Miklucho, sponges and fish. Haeckel concentrated on radiolarians and siphonophores. The siphonophores and sponges would prove the most theoretically interesting.

On 2 March 1867, after four months on the island and unable to take its punishments any longer, the company packed up its over one hundred bottles of specimens and set sail on the English merchant ship *Greatham Hall.* As they left the harbor, they cried out several times, "Auf Nimmerwiedersehen!"—"We'll never see you again!"[16] They disembarked forty hours later at Mogador (now Essaouira), on the southern coast of Morocco. There the two students, Fol and Miklucho, took off on horseback through the country, headed to the fabled city of Marrakesh but disguised as natives to avoid brigands along the way (see fig. 6.2). Haeckel and Greeff, after spending a week in the city visiting the bazaars and the Jewish quarter, sailed up the Moroccan coast, stopping at Mazagan and Tangier. On 17 March they sailed through the Strait of Gibraltar, disembarking at Algeciras in the Spanish state of Cádiz, where after several days they again met up with the students. Enticed by the shoals of the Spanish coast, the company broke out their microscopes and scalpels to dismember a large number of invertebrates. After two weeks of research, they traveled overland to Paris, by train back to Leipzig, and by coach to Jena. The excursion had lasted six months and yielded some of Haeckel's most important discoveries, findings that anticipated the experimental movement of *Entwickelungsmechanik* (developmental mechanics) at the end of the century and provided, or so he thought, the strongest empirical evidence for Darwin's theory.

15. Remarked in a letter to his friend Carl von Siebold, professor of zoology in Munich (14 February 1877), as quoted by Georg Uschmann, in *Geschichte der Zoologie und der zoologischen Anstalten in Jena 1779–1919* (Jena: Gustav Fischer, 1959), 67.

16. Haeckel to his Jena friends (7 March 1867), in *Berg- und Seefahrten: 1857/1883*, 68.

Fig. 6.2. Nikolai Miklucho in 1867, posing in Moroccan dress in Jena. (Courtesy of Ernst-Haeckel-Haus, Jena.)

Research on Siphonophores

From his time in Nice, where he went to recover after the death of his wife, Haeckel had become more interested in the Cnidaria—the stinging aquatic invertebrates—whose various orders included medusa, jellyfish, and the strange colonial hydrozoas that were often mistaken in earlier periods for plants growing from the seafloor. During those three months of sorrow, Haeckel passed the time observing many medusae—including his newly discovered *Mitrocoma Annae,* whose blond tendrils evoked images of his beloved wife.[17] He would later work up extensive systematic accounts of these organisms. Now during his stay on Lanzarote, he pulled up from the sea

17. See chapter 4 for a description of Haeckel's Anna-like medusa.

creatures of the same phylum but decidedly more exotic, the siphonophores. Haeckel described these creatures in a letter to his friends back in Jena:

> The siphonophores surpass all of the animal forms in these waters by their beauty and delicacy, and by their great scientific interest. I myself have chosen them as my special object of investigation. These swimming medusa colonies dwelling in the waters here are quite similar to bouquets of flowers and have an intricate structure indicating a most interesting and rather advanced division of labor. Think of a delicate slim bouquet of flowers, the leaves and colored buds of which are as transparent as glass, a bouquet that winds through the water in a graceful and lively fashion—then you'll have an idea of these wonderful, beautiful, and delicate colonial animals.[18]

The Biology of Siphonophores

Siphonophores are an order of hydrozoa in the phylum of Cnidaria.[19] A typical hydroid animal begins life as a fertilized egg, which then develops into a free-swimming larva. The planula, in modern parlance, settles to the ocean floor and adheres to the substrate. It becomes elongated and may branch (see fig. 6.3), much like a plant, or it may form stalks (now called hydrocauli) stemming from connections (hydrorhiza) that look like runners (see fig. 6.4). In some species the stalks also branch. At the end of each stalk, a polyp forms that has specialized functions: some are feeding polyps, with small tentacles that sting passing prey (see fig. 6.4, polyps marked *a, b,* and *c*); others are reproductive polyps (marked *f*), which generate through asexual budding a free-swimming medusa (marked *g*). The medusae release sperm and eggs into the water, where fertilization takes place to begin the cycle of alternating generation anew. All of the polyps of these hydrozoans display the same basic structure, though adapted to different functions. The structure is similar to that of the freshwater hydra (fig. 6.3), which Abraham Trembley (1700–1784) had shown could be halved or quartered and then would regenerate two or four complete, if smaller, animals.[20] Hydras reproduce asexually by budding, though during some seasons they reproduce

18. Haeckel to his Jena friends (27 January 1867), in *Berg- und Seefahrten: 1857/1883*, 61.

19. The word *cnidaria* is from *knide* (κνίδη)—nettle—a term Aristotle used for medusa, jellyfish, and anemones; he also referred to them as the *acalephe* (ἀκαλήθη), also meaning one that stings. See Aristotle, *Historia animalium*, 548a23, 621a11.

20. See the beautifully produced version of Trembley's memoir on the hydra edited and translated by Sylvia Lenhoff and Howard Lenhoff: *Hydra and the Birth of Experimental Biology*—

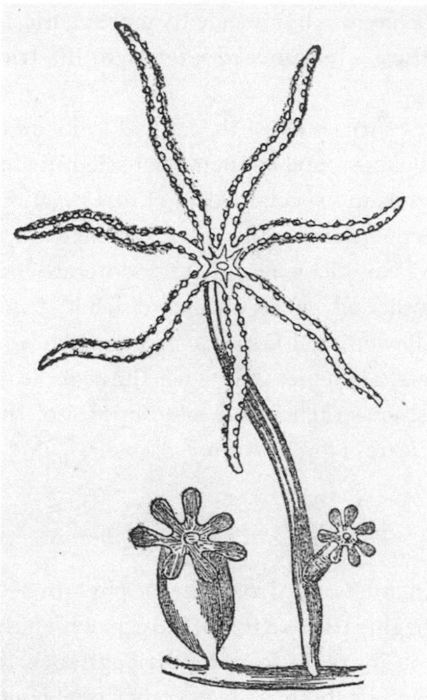

Fig. 6.3. Freshwater hydra; in asexual reproduction, the lower polyps will bud off to produce another hydra. (From Haeckel, *Arbeitsteilung in Natur und Menschenleben,* 1910.)

sexually. Hydras and hydrozoans have only two layers of cells: an outer layer, called an ectoderm, and an inner layer, the endoderm; between them flows a thick fluid that carries nutriments to the various parts.

Siphonophores are quite delicate hydrozoans of extreme polymorphic character (see fig. 6.5). By the mid-nineteenth century, the basic anatomy of these creatures had been established by several naturalists, including Haeckel's colleagues Kölliker and Gegenbaur.[21] Haeckel's own research

---

1744: Abraham Trembley's Mémoires Concerning the Polyps (Pacific Grove, CA: Boxwood Press, 1986).

21. See, for example, Albert von Kölliker, *Die Schwimmpolypen; oder, Siphonophoren von Messina* (Leipzig: Wilhelm Engelmann, 1853); and Carl Gegenbaur, *Beiträge zur näheren Kenntniss der Schwimmpolypen (Siphonophoren)* (Leipzig: Wilhelm Engelmann, 1854). Important as well was Rudolf Leuckart, *Zoologische Untersuchungen, erstes Heft: Siphonophoren* (Giessen: J. Ricker'sche Buchhandlung, 1853); and Thomas Henry Huxley, *The Oceanic Hydrozoa; a Description of the Calycophoridae and Physophoridae observed during the Voyage of H.M.S.*

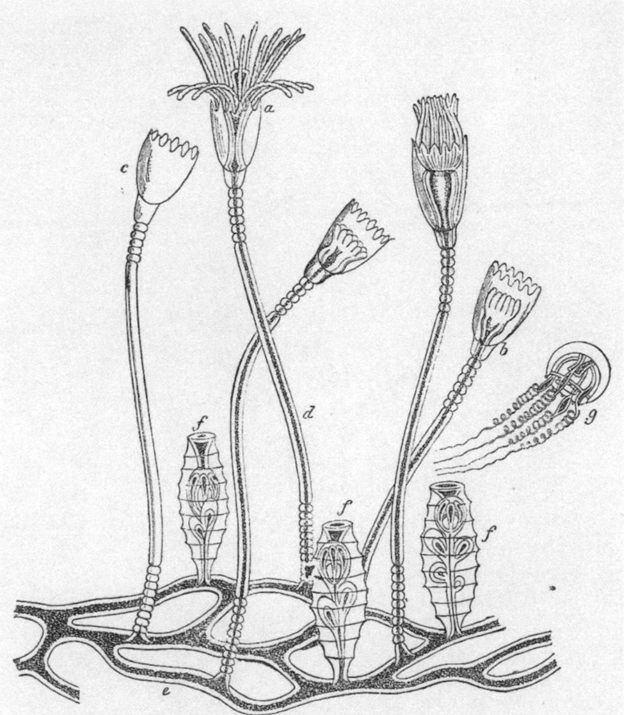

Ein kriechendes Polypenstöckchen (Campanularia Johnstoni). Auf dem kriechenden Wurzelgeflecht (*e*) sitzen zweierlei, durch Arbeitstheilung ganz verschieden entwickelte Hydra-Polypen, langstielige „Nahrungs-Polypen" (*a*—*d*) und kurzstielige „Zeugungs-Polypen" (*f*). Die letzteren bilden Knospen, die sich zu Medusen umbilden und fortschwimmen (*g*). Die ersteren können ihren vorgestreckten Leib (*a*) in eine hornige Kapsel (*c*) zurückziehen (*b*). Ihr Stiel (*d*) ist oben und unten geringelt.

Fig. 6.4. A typical hydrozoan colony: feeding polyps marked *a–d*, which can extend and retract small stinging tentacles; and reproductive polyps marked *f*, which produce free-swimming medusae, marked *g*. (From Haeckel, *Arbeitsteilung in Natur und Menschenleben*, 1910.)

yielded a magnificent monograph (1869) that won a gold medal from the Utrecht Society for Arts and Sciences.[22] Siphonophores were regarded by Haeckel and others as essentially colonial organisms, consisting of pol-

"Rattlesnake" in the years 1846–1850 (London: Ray Society, 1859). Gegenbaur gave Haeckel a copy of Huxley's work for the journey to the Canaries.

22. Ernst Haeckel, *Zur Entwickelungsgeschichte der Siphonophoren* (Utrecht: C. Van der Post Jr., 1869). Haeckel submitted the monograph for the prize competition anonymously.

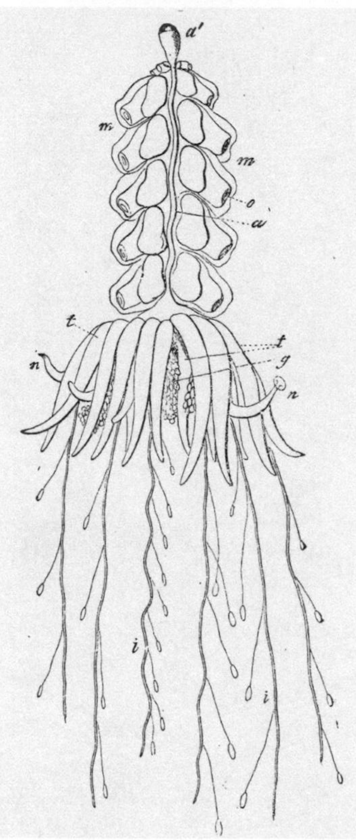

Fig. 6.5. A siphonophore of the genus Physophora: a marks the gas bladder that holds the animal at a certain level in the water, m marks the swim bells that take in and expel water for movement; o marks an opening of the bell; t marks the feelers; g marks egg-producing organs; n marks suctorial tubes for feeding; and i marks stinging tendrils to capture food and for defense. (From Haeckel, Arbeitsteilung in Natur und Menschenleben, 1910.)

yps that had been modified for various functions. At the apex of the stack of polyps is usually a gas bladder (see fig. 6.5, structure marked *a*), which keeps the animal at a given height in the column of water. Below the bladder are often swimming bells (marked *m*), with a mouth opening that allows these organs to move the animal by squeezing out water. Then there are feelers that sense nearby objects (marked *t*), which themselves might overlap an egg or sperm producing organ (marked *g*) and suctorial tubes (marked *n*), through which nutrients are filtered and brought up into the whole organism. In the aptly named *Physophora magnifica* (plate 6), whose illustration served as a centerpiece of Haeckel's monograph, long thin tentacles, with stinging threads (marked *i* on fig. 6.5), hang from the base of the animal; these barbed spaghetti strings serve to capture small prey and to defend against predators.[23]

Haeckel's Experiments on Siphonophores

In the study that won the prize—his *Zur Entwickelungsgeschichte der Siphonophoren* (1869)—Haeckel set out to determine, through observation and experiment, the developmental history of the siphonophores, something known only sketchily up to that point. He performed three kinds of experiment: he cultivated fertilized siphonophore eggs from species of ten different genera, and then followed their larval development from day to day;[24] he altered the conditions (e.g., water temperature, light, movement, etc.) early in development to see what changes were introduced at various later periods;[25] and finally, he divided the cells of the embryos at a very early stage to see if regeneration would occur during development.[26]

Haeckel had but fair luck in cultivating his embryos; only species of three genera got beyond the first day: *Physophora* (a well-known genus), *Athorybia* (a less well-known genus), and *Crystallodes* (a genus newly discovered by Haeckel). None of the larvae reached the adult stage, though that of the species *Physophora magnifica* got to day twenty-eight, a week or so short of full development (see plate 6, two views of larva marked 24 and 25). Haeckel's microscopic investigations showed that in the early stages of development, the larvae of different genera looked very much alike—an observation suggesting, in light of the biogenetic law, that they had descended from the same parent form, a conclusion supported by Haeckel's next set of experiments.

Changing the water temperature, altering the ambient light, strengthening or weakening the salinity, shaking the container holding the embryos—all of these seemingly slight alterations had major effects on the protective bells, the feeding tubes, the air sacs, and other organs of the developing siphonophores. Haeckel wanted to test the variability of the larvae because he believed, as did Darwin, that environmental circumstances could induce mutations in heritable material, which would obviously have

23. Haeckel's *Physophora magnifica* was discovered in the Atlantic Ocean, off the coast of Lanzarote in the Canary Islands. He regarded it as closely related to *Physophora hydrostatica*, a creature found by Gegenbaur in the Mediterranean. Haeckel's *P. magnifica* may only be a variety of *P. hydrostatica*.

24. Haeckel, *Zur Entwickelungsgeschichte der Siphonophoren*, 17–37, 51–72.

25. Ibid., 38–42, 80–92.

26. Ibid., 73–79.

evolutionary consequences.[27] The manipulations not only showed how susceptible the developing embryos were to changed conditions, but the alterations themselves revealed to Haeckel morphologically distant species forms hidden beneath those of the particular species on which he was experimenting. He found that some of the manipulated larvae seemed to display properties of what must have been ancestral organisms, and others appeared to cross over to related species forms or even to different genus forms.[28]

In the final set of ten experiments, Haeckel achieved, as he averred, "an unexpected and, even surprising to me, positive success."[29] To our eyes, opened wider by the subsequent history of developmental biology, the experiments were extraordinary, since they were comparable to what were thought to be unprecedented, groundbreaking experiments conducted by his students Roux and Driesch some twenty years later (discussed below).

In these experiments, Haeckel, with fine needle and microscope, divided two-day-old embryos of his new *Crystallodes* genus, which had already gone through several cleavages, into either two, three, or four groups of cells, and then followed the development of these independent groups. He said he was led to attempt this because the eggs and early embryonic cells of siphonophores exhibited amoeba-like movements, suggesting that the primitive larvae might be capable of regeneration.[30] In six cases, development got to the sixth day; in three of those, the larvae survived to the eighth day; two reached the tenth; and one went to the fifteenth day. In three of the cases where embryos survived to day six, they had been divided in half; in two cases, they were separated into thirds; and in one, quartered. He found that the regenerated larvae, though smaller than usual, essentially developed through the regular stages (see fig. 6.6). They arrived at those stages, however, more slowly than normal larvae; and the smaller they were, the more slowly they developed—though, anomalously, in a few instances, they moved through early stages more rapidly.[31]

Though Haeckel did not explicitly draw the conclusion, his extraordinary experiments demonstrated that all embryonic cells, at least early in development, were totipotent—they had the capacity to develop all the parts of the organism. This would be the marked conclusion reached two

27. Ibid., 38.
28. Ibid., 38–39.
29. Ibid.
30. Ibid., 73.
31. Ibid., 79.

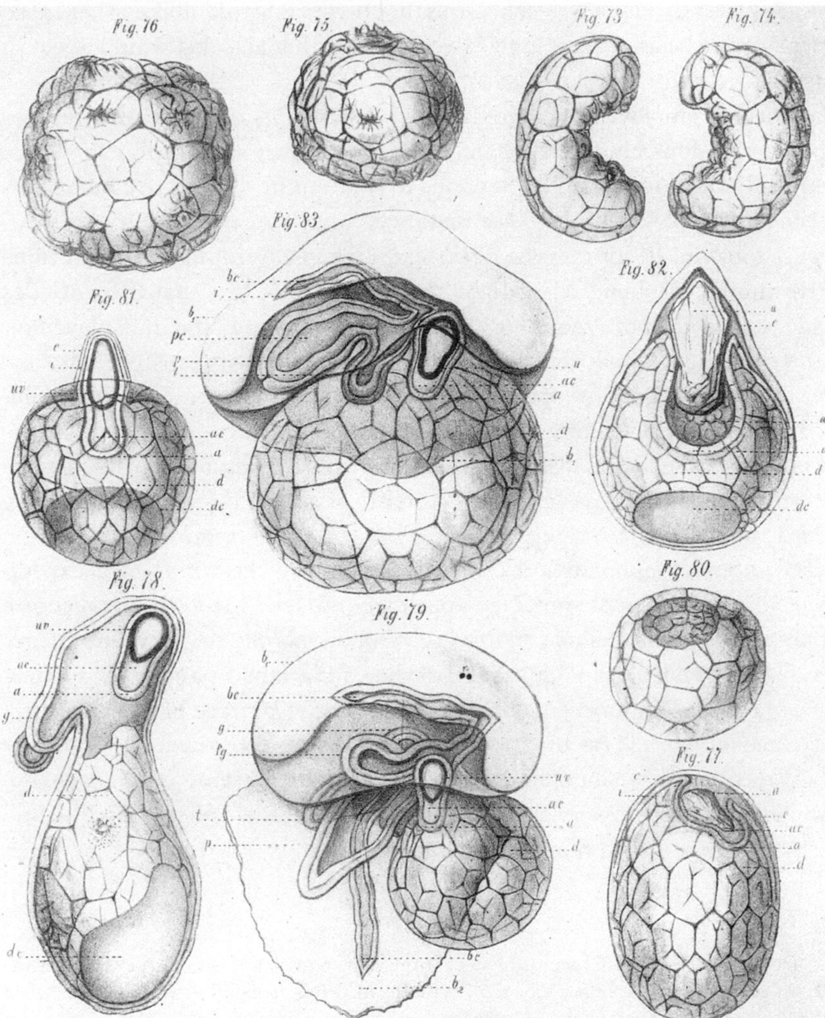

Fig. 6.6. Haeckel's experiments in which he dissected siphonophore larvae and traces their consequent development: figs. 73 and 74, the two halves of a separated *Crystallodes* larva (concavities indicate where the separation occurred); figs. 75 and 76, the same larvae a few hours after the separation (with cells having drawn together in a sphere comparable to normal larvae); figs. 77, 78, 79, *Crystallodes* larva separated into thirds, the eighth day after separation (the smallest, 77, having developed an air sac and two polyps, and the largest, 79, developing normally); figs. 80, 81, 82, 83 the results of a quartered *Crystallodes* larva (with only 83 developing normally). (From Haeckel, *Zur Entwickelungsgeschichte der Siphonophoren*, 1869.)

decades later by Hans Driesch. Driesch, however, would make other inferences on the basis of comparable experiments that Haeckel would reject in high dudgeon.

Haeckel employed his experimental discoveries to draw some reasonable conclusions about the phylogenetic history of siphonophores. These implications ranged from the tenuous to those quite similar to conclusions reached by biologists today. For instance, since some of his *Crytsallodes* larvae took on the form of the *Physophora* when environmental conditions were altered, he supposed, in light of the biogenetic law, that *Crystallodes* had descended from *Physophora.*[32] Because *Physophora* itself displayed organ structures homologous to the medusa-producing organs of hydrozoa, he conjectured that *Physophora* originated from a simple medusa that budded from a hydroid colony.[33] Haeckel recognized that his comparative base of only three genera made his inferences quite chancy, but he thought his research would "at least cast a new light on a field that is as interesting as it is unknown."[34] Some years later, he was asked to provide systematic descriptions of siphonophores dragged up during the *Challenger* expedition.[35] His analysis of some 240 species confirmed his earlier conjecture concerning the medusoid origin of the siphonophores.[36] Today this theory has been extended to encompass all of the Cnidarians—namely, that a free-floating medusoid form constitutes the primitive pattern of the phylum.[37] As Haeckel worked on his monograph, he wrote Darwin to say that the character of the siphonophores, especially their extreme polymorphism, provided "the most excellent demonstration of descent theory."[38] It would be hard to argue with that assessment.

32. Ibid., 99.

33. Ibid., 100. Earlier Gegenbaur had convincingly argued, following Leuckart, that the various organs of the siphonophore were actually modified individuals having a medusoid form. See Carl Gegenbaur, "Neue Beiträge zur näheren Kenntniss Der Siphonophoren," *Nova Acta Leopoldina* 28 (1859): 333–424 (especially 334–37).

34. Haeckel, *Zur Entwickelungsgeschichte der Siphonophoren,* 103.

35. See the appendix to chapter 3 for a description of the *Challenger* expedition.

36. Ernst Haeckel, *Report on the Siphonophorae Collected by H.M.S. Challenger during the Years 1873–76,* vol. 28 of *Report on the Scientific Results of the Voyage of H.M.S. Challenger during the Years 1873–1876* (London: Her Majesty's Stationery Office, 1888), 3–4. Haeckel agreed to write the volume if he could include materials that he had earlier investigated. His monograph remains the most complete systematic description of siphonophores.

37. See, for example, Richard C. Brusca and Gary J. Brusca, *Invertebrates* (Sunderland, MA: Sinauer Associates, 1990), 253.

38. Haeckel to Darwin (February 1868), in the Darwin Correspondence, DAR 166.1, Cambridge.

Entwickelungsmechanik

In his work on siphonophores, Haeckel wove together experimental procedures and evolutionary considerations, particularly the assumption of the biogenetic law. His experiments prefigured the comparable work of two of his students, Wilhelm Roux (1850–1924) and Hans Driesch (1867–1941) some twenty years later. These young embryologists, though, have been described as Haeckel's "apostate students." They performed the kind of experiments that established the movement of *Entwickelungsmechanik* (developmental mechanics), which historians have often portrayed as a dramatic and needed departure from the kind of antiquated biology practiced by Haeckel.[39] To assess this historical thesis, I will spend a moment on Roux, founder of the movement, and Driesch, who advanced it.

Wilhelm Roux

In 1870 Roux, son of a fencing master at Jena, entered medical school there, though almost immediately broke off for two years of military service. Upon resumption of study, he attended lectures by Gegenbaur and Haeckel, and traveled for itinerate research in Berlin with Virchow and

39. See, for example, Stephen Jay Gould, *Ontogeny and Phylogeny* (Cambridge, MA: Harvard University Press, 1977), 194: "By the late 1880s and early 1890s two of Haeckel's apostate students—Wilhelm Roux and Hans Driesch—were advancing experimental methods in embryology and relegating the biogenetic law to a back shelf of outmoded methods. . . . Experimental embryologists rejected all aspects of Haeckel's methodology." William Coleman, whose nuanced judgments always deserve respect, yet prefaced his discussion of Roux's experiments with "The limitations of this approach [Haeckel's use of the biogenetic law] were increasingly evident by 1875. A decade later this low-voiced criticism was converted into a forceful program for a new approach to problems of individual development. No longer might historical explanation, drawing exclusively on descriptive and comparative embryology, suffice for understanding individual development. That understanding would henceforth come from the analysis of causal factors." See William Coleman, *Biology in the Nineteenth Century: Problems of Form, Function, and Transformation* (New York: John Wiley, 1971), 53–55. The judgments of Gould and Coleman have precedent in the opinions of Jane Oppenheimer, the embryologist and historian who wielded considerable power on these questions. In 1955 she observed that "Haeckel himself was never in any sense a professional embryologist. The seduction of embryology by a fanatic who expressed himself even metaphorically in terms of magic represents a darker chapter in its history than any of its earlier or later retreats to mere metaphysics lacking such taint of the mystic." The mystic, she thought, nonetheless intellectually gave birth to Wilhelm Roux, who would "grow far beyond Haeckel's romanticism." See Jane Oppenheimer, "Analysis of Development: Problems, Concepts and Their History" (1955), in her *Essays in the History of Embryology and Biology* (Cambridge, MA: MIT Press, 1967), 154, 160. These judgments are typical.

in Strasbourg with Alexander Goette—both enemies of Haeckel, the former who denied evolutionary theory had any real scientific status and the latter who thought the theory unnecessary to explain embryological development (matters discussed in chapter 8). Back in Jena, Roux completed his dissertation (1878) on the branching of the human vascular system; he worked under the direction of the evolutionary embryologist Wilhelm Preyer (1841–1897).[40] In 1880 Roux habilitated at Breslau with a treatise that sought to show how descent theory could explain the apparently teleological structure of animals.[41] In this monograph, he cited Haeckel liberally to support two propositions: that the environment produced in organisms heritable traits upon which natural selection might work and that animals could more directly accommodate themselves to such environmental conditions through heritable "functional adaptations." Though these ideas were to be found in the *Origin of Species,* he believed Haeckel rightly put a higher value on them than did Darwin.[42] The principle of functional adaptation grounded Roux's considerations in his monograph of 1881, *Der Kampf der Theile im Organismus* (The struggle of parts in the organism). That book extended the principle of natural selection to explain organ development in the individual.[43] Roux acknowledged that the seeds of the idea of competition of parts had been planted in him by the lectures of Haeckel and Preyer.[44] But as indebted to Haeckel as he was, he expressed some reservations about his mentor's biogenetic law. He acknowledged the objection of Wilhelm His, Haeckel's bitter enemy, that the egg of a higher creature must have a different potentiality than that of its phylogenetic ancestor; there must, therefore, be features materially different about it from the beginning. This meant that during ontogenetic development the embryos of advanced creatures could not simply pass through the same forms

40. Wilhelm Roux, "Ueber die Verzweigungen der Blutgefässe des Menschen, eine morphologische Studie," *Jenaische Zeitschrift für Naturwissenschaft* 12 (1878): 205–66.

41. Wilhelm Roux, *Ueber die Leistungsfähigkeit der Principien der Descendenzlehre zur Erklärung der Zweckmässigkeiten des thierischen Organismus* (Breslau: S. Schottänderen, 1880).

42. Ibid., 10.

43. Wilhelm Roux, *Der Kampf der Theile im Organismus, ein Beitrag zur Vervollständigung der mechanischen Zweckmässigkeitslehre* (Leipzig: Wilhelm Engelmann, 1881).

44. Wilhelm Roux, *Der züchtende Kampf der Theile; oder, Die "Theilauslese" im Organismus,* 2nd ed., in *Gesammelte Abhandlungen über Entwickelungsmechanik der Organismen,* 2 vols. (1881; repr., Leipzig: Wilhelm Engelmann, 1895), 1:230n1. He also mentioned that in 1879 he conveyed his plans to publish a work on the selection of parts to Haeckel, who enthusiastically approved (1:227n2).

that lower creatures went through in phylogenetic development. Moreover, the biogenetic law—like Newton's law of gravitation—was, he maintained, empirically false, since all sorts of causes interfered with a perfect repetition of earlier forms.[45] Since Haeckel had already acknowledged these qualifications, he could still praise the work of his former student as "the most important new production of the extensive Darwinian literature."[46]

In the second edition of *Der Kampf der Theile* (1895), Roux announced that some of his "youthful" views had changed. He declared that he no longer regarded the inheritance of acquired characters as certain; and he again objected to what he said was Haeckel's assumption of the homogeneity of the cell's protoplasm.[47] Through the early 1890s, Roux became ever more convinced by Weismann and by his own research that cells of the germ line were distinct from those of the somatic line and thus that inheritance of acquired characters could not occur. Roux's complaint that Haeckel regarded the cell's protoplasm as "homogenous or structureless" is quite odd, since Haeckel certainly thought the protoplasm—that is, the central nucleus—of the fertilized egg contained the hereditary *Anlagen*, or structural dispositions, of the developing organism. What Roux seems to have meant by this objection was that Haeckel did not hold the "mosaic" theory of embryonic development, a theory that Roux's experiments on frog embryos seemed to suggest but that Haeckel's own experiments on siphonophore larvae precluded.

Roux's experiments, usually described as "classic," provided evidence for his mosaic theory of embryological development.[48] In the first set of experiments, using a heated needle, he destroyed one of the two cells, or blastomeres, after the first cell division of the developing frog egg. In other experiments, he destroyed one or more of the four blastomeres after the second cleavage. He discovered that the intact blastomere continued to

45. Roux, *Der Kampf der Theile im Organismus,* 57–58. Roux sent a preliminary study for this work to Haeckel acknowledging his debt. See Roux to Haeckel (21 November 1879), in the Correspondence of Ernst Haeckel, the Haeckel Papers, Institut für Geschichte der Medizin, Naturwissenschaft und Technik, Ernst-Haeckel-Haus, Friedrich-Schiller-Universität, Jena.

46. Ernst Haeckel, *Natürliche Schöpfungsgeschichte,* 2 vols., 8th ed. (Berlin: Wilhelm Engelmann, 1889), 1:277.

47. Roux, *Der züchtende Kampf der Theile,* 140.

48. His experiments were reported in Wilhelm Roux, "Beiträge zur Entwickelungsmechanik des Embryo: Ueber die künstliche Hervorbringung halber Embryonen durch Zerstörung einer der beiden ersten Furchungskugeln, sowie über die Nachentwickelung (Postgeneration) der fehlenden Körperhälfte," *Archiv für pathologische Anatomie und Physiologie und für klinische Medicin* (Virchow's Archive) 94 (1888): 113–53, 246–91.

develop through several more cell divisions, producing a half-embryo (or some fraction of an embryo depending on how many of the four blastomeres were initially destroyed). The longest surviving half-embryo reached a stage in which the neural tube was just forming. Roux concluded from these experiments that the fertilized egg from the beginning had determinants of future development already laid out in fixed spatial positions—a mosaic of causal factors. So with the first cleavage of the egg, determinants of the left and right sides of the embryo would be separated into their respective blastomeres; with the second cleavage determinants of the head and tail halves would be separated into the two top and two bottom blastomeres, and so on. He contended that neither external influences nor location of cells in relation to each other regulated how the embryo would develop: only the fixed mosaic of internal determinants would form cells into organs of the individual.

Roux did not realize that his experiments were confounded by a significant artifact: he left the remnants of the destroyed blastomere attached to the sound one, which inhibited regeneration of missing parts. Had he simply separated the blastomeres, he likely would have gotten a result similar to Haeckel's earlier one, when the separated cells developed into, not half-embryos, but regular embryos of smaller size. Another of Haeckel's former students, Hans Driesch, was able to show the defect in Roux's design.

Hans Driesch

After spending a short time in Freiburg with Weismann, Driesch came to study with Haeckel in 1887. When he arrived, the enthusiastic student looked for the "warrior of the *Natural History of Creation,*" though he only heard a tired professor delivering boring lectures in a monotone. The disappointment vanished, however, in the more intimate setting of Haeckel's seminars, which Driesch remembered as quite exciting. Haeckel comported himself, according to his student, critically: he regarded his "phylogenetic reconstructions as hypotheses and the paleontological material as full of gaps." [49] Driesch received his doctorate under Haeckel in 1889 for work on colonial hydrozoa, a subject that might well have elicited a discussion of or reference to his teacher's earlier experiments on siphonophores. If so, Driesch should have viewed Roux's experiments with some suspi-

49. Hans Driesch, *Lebenserinnerungen: Aufzeichnungen eines Forschers und Denkers in entscheidender Zeit* (Munich: Ernst Reinhardt, 1951), 47, 48.

cion, or at least not with the naive expectation that he initially claimed. In any case, he wished to try Roux's experiment with a different animal, an echinoderm (sea urchin).[50] Instead of killing one of the blastomeres, Driesch put the eggs after first cleavage in a water-filled glass vial and shook it vigorously. The blastomeres of many of the eggs separated; he then followed their subsequent development. Though expecting Roux's half-embryos—or at least that was the rhetoric of his published paper—Driesch got something very different: not half-embryos but complete embryos of half size. By the third day, some of those embryos had developed into larvae, though still of half the normal size. Driesch consequently rejected the general mosaic theory of Roux and maintained that, at least during early cleavage stages, the blastomeres were totipotent: each had the ability to develop all the parts of the organism. Though Driesch did not invoke any unusual forces to explain the ontogeny of development, he indicated in his report of the experiment that "we expect that the mechanistic conception of the entire phenomenal world will certainly become subordinated (non-metaphysical vitalism)."[51] It was perhaps this vague but ominous remark that caused him to mention in a letter to Haeckel that his former teacher "might not agree in all points" of the work.[52] Within a couple of years, he more explicitly repaired to vital forces ("entelechies" in his Aristotelian nomenclature) to explain developmental phenomena that seemed to escape purely mechanical powers.[53]

The Foundations of Entwickelungsmechanik

The experimental procedures of Roux and Driesch, though they came to very different outcomes, fostered comparable efforts among other research biologists at the end of the century—for instance, the American cell biologist E. B. Wilson advanced his own version of what he termed the "Roux-

50. Hans Driesch, "Entwicklungsmechanische Studien: I. Der Werthe der beiden ersten Furchungszellen in der Echinodermenentwicklung. Experimentelle Erzeugung von Theil- und Doppelbildungen. II. Über die Beziehungen des Lichtes zur ersten Etappe der thierischen Formbildung," *Zeitschrift für wissenschaftliche Zoologie* 53 (1891): 160–84.

51. Ibid., 161.

52. Driesch to Haeckel (6 March 1891), in the Haeckel Correspondence, Haeckel-Haus, Jena. Driesch also indicated to Haeckel that he had been thinking about the issues for some time. He said, "I have written what I must write on the basis of my experiments." He hoped that "despite theoretical differences, my personal relationship with you does not change."

53. See, for example, Hans Driesch, *Die Lokalisation morphogenetischer Vorgänge, ein Beweis vitalistischen Geschehens* (Leipzig: Wilhelm Engelmann, 1899), especially 70–72.

Weismann" hypothesis.[54] But Haeckel's own experimental work on the embryology of siphonophores remained largely ignored or unknown; at least it garnered few explicit citations in the literature thereafter. It also has been generally ignored by historians, who frequently depict Haeckel's evolutionary work as having been eclipsed by the new procedures inaugurated by his students. Despite these historical animadversions, Haeckel's experiments, in an obvious way, turned out more successfully than those of either Roux or Driesch. His experimental hand was more sure than that of Roux and his conclusions in support of Darwinian evolution certainly more intellectually satisfying than the vitalism of Driesch. His work, had it been more widely known, would have been a harbinger for today's new area of "evo-devo"—the evolutionary and genetic theory of species and individual development. Perhaps the failure of historians to appreciate Haeckel's groundbreaking effort has occurred, at least in part, because he did not follow up with a train of further studies as did his students, who published extensively under the rubric of *Entwickelungsmechanik*. The bulk of Haeckel's research focused on systematics and evolutionary theory, not on experimental embryology—though the latter nonetheless remained an important feature of his subsequent studies of marine invertebrates. A just historical judgment would not conclude that Haeckel's students initiated new methods that obliterated those of their teacher. Indeed, it is rather hard to believe that Roux and Driesch would not have known of his experimental work; and if they did, then a fair assessment would credit Haeckel with the establishment of this historically important line of inquiry.

A lingering mystery is the absence of any mention of Haeckel's siphonophore experiments by his students. I suspect a number of factors may have led to the lack of reference. First, they undertook their own experimental work several years after their time with Haeckel, and so simply the intervening years may have obscured their memories of what they read under his tutelage. Second, unlike other German professors, Haeckel never insisted that his students slavishly follow out his own research program; he gave them their head. Hence, they simply may not have been terribly attentive to the importance of his early experiments. Third, Roux and Driesch had both moved away from Haeckel doctrinally. Roux had given up the idea of the inheritance of acquired characters, and his mosaic theory

54. Wilson recognized the countervailing considerations of Driesch, but sided with what he termed the "Roux-Weismann hypothesis." See E. B. Wilson, *The Cell in Development and Inheritance* (New York: Macmillan, 1897), 295–323.

would have run exactly counter to Haeckel's conclusions about the totipotency of early embryonic cells. Driesch's experimental result was certainly compatible with Haeckel's; but by the time of his experiments, Driesch had already signed on to vitalism, a view that he knew would separate him from his former teacher. In both cases, memory may well have become suppressed by these marked differences with their onetime mentor. Finally, Haeckel did not follow up on his experiments, so their originality and significance would have faded in the memory of everyone.[55]

Throughout his career, Haeckel drew students of extraordinary ability to his small university in Jena. These were individuals of great industry, intellect, and creativity. They had Haeckel's own aggressive independence as a model. It should hardly be surprising that many of them would absorb his doctrines, techniques—and his attitudes—and then set out in some rugged defiance of their teacher. That this occurred numerous times during his career can hardly be a derogation of his own genius, quite the contrary. Hans Driesch may have strayed further than anyone from the evolutionary principles that defined the intellectual core of Haeckel's universe. As he slipped further into vitalistic metaphysics, Driesch, the erratic genius, greatly disappointed the master; the older man even suggested that his former student might profitably spend time in a sanitarium.[56] A break came that never healed. Yet afterward, Driesch could still render an assessment of Haeckel's impact that tempers the jejune judgments often proffered by contemporary scholars. He wrote of his teacher: "In his forcefulness joined with his child-likeness and openness, Haeckel was, despite everything, a complete human being. And I have never forgotten what I have learned from him, what he himself so often and happily embodied in the words: impavidi progrediamur [let us advance courageously]."[57]

A Polymorphous Sponge: The Analytical Evidence for Darwinian Theory

Haeckel was particularly fond of his impulsive student and research assistant Nikolai Miklucho.[58] The young biologist came from a family of impoverished nobility in St. Petersburg but sought his fortune in the Ger-

55. I have searched for explicit evidence that Roux and Driesch knew of Haeckel's experiments on siphonophores—without success.

56. Driesch, *Lebenserinnerungen*, 71.

57. Ibid.

58. Haeckel mentioned that he was himself moved by Miklucho's close affection. See, for example, Haeckel to his parents (10 May 1867), in *Ernst Haeckel: Biographie in Briefen*, 102.

manies after being expelled from university in Russia. Student peregrinations brought him to the universities of Heidelberg and Leipzig, and, in the winter term of 1865–66, to the medical school at Jena.[59] During his time with Haeckel on Lanzarote, Miklucho worked particularly on sponges and made a discovery—or so it seemed—that charged his teacher's creative imagination. The discovery was of a polymorphic, colonial sponge that seemed to hold the key to understanding the evolution of all the orders of sponge and, indeed, of all the higher animals. The analysis of this sponge led Haeckel to his theory of the gastraea and to his compelling empirical argument for the validity of Darwin's theory.

The Biology of Sponges

Aristotle regarded sponges as plantlike animals. Like animals, they showed sensibility; but like plants, they took their nourishment from the mud in which they were rooted—or so he thought.[60] Through the eighteenth century, researchers variously classified sponges as plants or as plantlike animals (zoophytes). Robert Grant (1793–1874), Darwin's teacher at Edinburgh, inaugurated the modern era of sponge analysis through his exacting observations and experiments. He distinguished three classes of Porifera (as he called the phylum): horny (including the bath sponge), silicious, and calcareous.[61] The classification, largely retained today, depends on the composition of the needle-like spiculae (either of silica or calcium carbonate) and the fibrous texture that gives structure to the bodies of sponges.[62] From

59. The fullest and most diverting account of Miklucho's career can be found in E. M. Webster, *The Moon Man: A Biography of Nikolai Miklouho-Maclay* (Berkeley: University of California Press, 1984).

60. Aristotle, *Historia animalium*, 548a32–549a12. Aristotle recognized three kinds of sponge, one of which was called Achilles because it was used to line helmets and greaves (548b1–3).

61. Grant's name for the phylum stuck. See Robert Grant, "Observations and Experiments on the Structure and Functions of the Sponge," *Edinburgh Philosophical Journal* 13 (1825): 94–107, 333–46; 14 (1826): 113–24, 336–41; "On the Structure and Nature of the Spongilla friabilis," *Edinburgh Philosophical Journal* 14 (1826): 270–84; "Remarks on the Structure of Some Calcareous Sponges," *Edinburgh Philosophical Journal*, n.s. 1 (1826): 166–70; "Observations on the Structure of Some Silicious Sponges," *Edinburgh Philosophical Journal*, n.s. 1 (1826): 341–51; "Observations on the Structure and Functions of the Sponge," *Edinburgh Philosophical Journal*, n.s. 2 (1827): 121–41.

62. In the contemporary period, Grant's classes of sponges are still the main divisions: the Hexactinellida (glass sponges), which have a supporting network of starlike spiculae composed of silica; the Calcarea, which have three pointed spiculae of calcium carbonate; and by far the largest class, the Demospongiae (Grant's horny sponges), which will often contain a few ran-

evidence of ingenious experiment and careful observation by microscope, Grant rejected the belief of some earlier naturalists that larger apertures on the surface of sponges sucked in and expelled water in diastolic and systolic movements. He recognized that small pores on the skin took in currents of water with minute particles of nutriments, passed them through internal channels, and expelled "feculent" matter from the larger apertures.[63] This basic structure would later be modeled by Haeckel on that of a simple calcareous sponge (see fig. 6.7, with pores, marked *p*, on the surface surrounded by spiculae; detritus would be expelled from the larger mouth or osculum at the top).[64] Grant, a disciple of both Erasmus Darwin and Lamarck, suggested, without much elaboration, that the various classes of sponge were connected by descent.[65]

Subsequent naturalists added important details to the growing body of research on sponges. Nathanael Lieberkühn (1822–1887) at Berlin described in detail the external, dermal membrane (ectoderm), which was pitted with pores leading to the small internal canals that Grant had recognized. Lieberkühn, though, discovered that the internal, dermal membrane (the endoderm), which lined the larger cavity (or cavities) of sponges, displayed small openings surrounded by a kind of collar and having a single cilium; the movement of the cilium drew water and food particles through the external pores, into the canals, and then expelled undigested material (see fig. 6.7, no. 8A, B, C). He also recognized that ciliated larvae of the sponge would be expelled in the same fashion.[66] Lieberkühn was able to identify

---

dom spiculae of silica, but more often will display only a fibrous, horny texture. Sometimes a fourth class is recognized, the Sclerospongiae, which have siliceous spiculae in the outer skin but are supported by a massive calcareous matrix on which the animal grows. See Patricia Bergquist, *Sponges* (Berkeley: University of California Press, 1978), 1–15.

63. Grant, "Observations and Experiments" (1825), 101–7.

64. See Ernst Haeckel, *Die Kalkschwämme*, 3 vols. (Berlin: Georg Reimer, 1872), 3: Tafel 1. Haeckel's figure has often been used in modern textbooks. See, for instance, Brusca and Brusca, *Invertebrates*, 188; and Bergquist, *Sponges*, 21.

65. In describing the freshwater *Spongilla friabilis*, Grant observed: "From this greater simplicity of structure, we are forced to consider it as more ancient than the marine sponges, and most probably their original parent; and, as its descendants have greatly improved their organization, during the many changes that have taken place in the composition of the ocean, while the spongilla, living constantly in the same unaltered medium, has retained its primitive simplicity, it is highly probable that the vast abyss, in which the spongilla originated and left its progeny, was fresh, and has gradually become saline, by the materials brought to it by rivers, like the salt lakes of Persia and Siberia." See Grant, "On the Structure and Nature of the Spongilla friabilis," 283–84.

66. Nathanael Lieberkühn, "Neue Beiträge zur Anatomie der Spongien," *Anatomie, Physiologie und wissenschaftliche Medicin* 25 (1859): 353–82, 515–29.

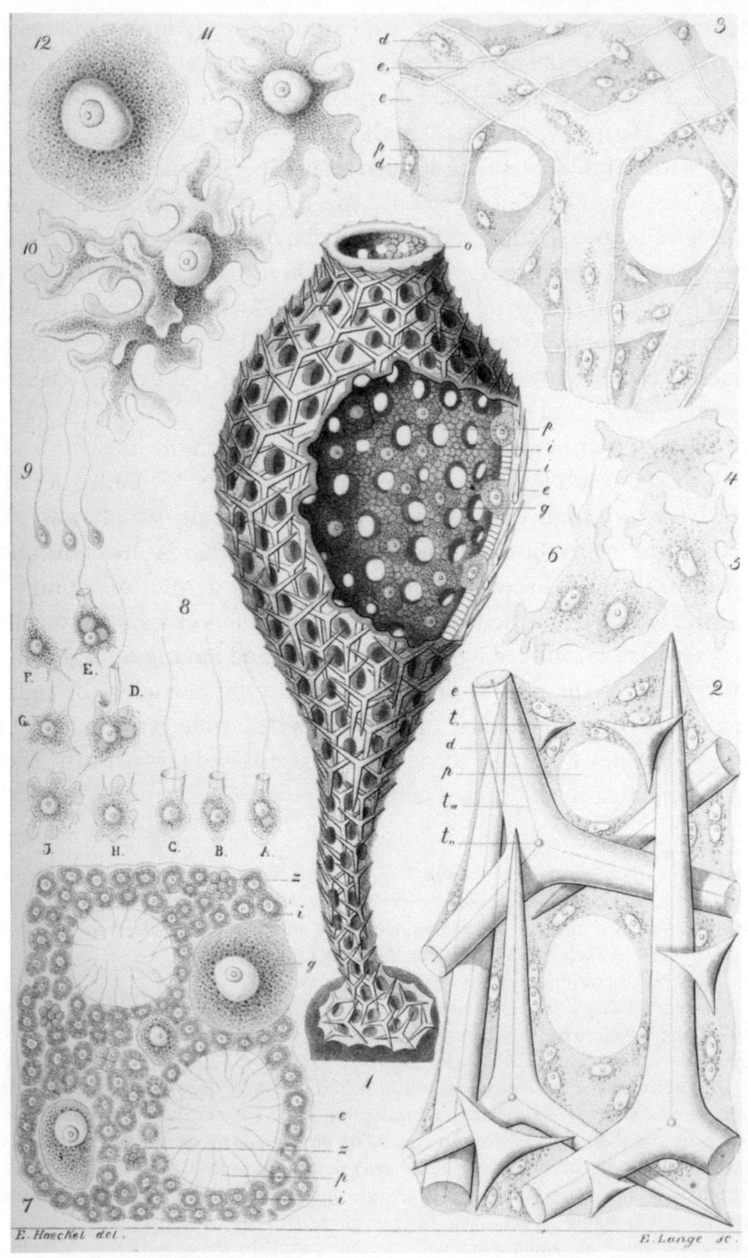

Fig. 6.7. Calcareous sponge person: mouth marked *o*; pores marked *p*; exoderm marked *e*; entoderm marked *i*; egg marked *g*; collared cilia marked *A, B, C*; a section of exoderm with spicula at 2; and spermatozoa at 9. (From Haeckel, *Die Kalkschwämme*, 1872.)

sponge ova but not spermatozoa, which Haeckel himself seems to have been the first to describe accurately.[67] Haeckel's own work on sponges was initiated by Miklucho's discovery of an unusual organism during the trip to the Canary Islands.

Miklucho's Contribution to Haeckel's Project

Miklucho described it as a small colonial sponge, which he named *Guancha blanca.*[68] The animal was a member of the order of calcareous sponges, characterized by a delicate skeleton of interlocking star-shaped spiculae of calcium carbonate—not the sort of sponge with which to share a bath. Miklucho assumed that the colony comprised individual "persons" (in Haeckel's terminology) displaying different morphologies. Two colonies might even exhibit different transitional states (fig. 6.8, colonies marked fig. 1 and fig. 2). Each morphologically distinct person, he argued, represented transformations of what was essentially the same form. He described the basic form as similar to that of a champagne flute (see fig. 6.8, sponges marked fig. 1.A and fig. 3.1) and of about the same size. Haeckel would later call this the *olynthus* form and regard it as having a place comparable to that of the nauplius in the class of crustaceans: it would reveal the descent relationships operative in the families of sponges. This Urform had a mouth opening (marked *m* in 1.A) and what Miklucho called a "digestive

67. Though Lieberkühn believed he had discovered the spermatozoa of the sponge, likely he only identified ciliated epithelial cells. Haeckel quite correctly recognized that the spermatozoa were modified ciliated epithelial cells (now called choanocytes). See Ernst Haeckel, "Ueber die sexuelle Fortpflanzung und das natürliche System der Schwämme," *Jenaische Zeitschrift für Medicin und Naturwissenschaft* 6 (1871): 642–51. Sexually, sponges are very curious animals. Most will produce asexually, should parts break off; some, however, bud. Very many species also reproduce sexually: some will exclusively produce ova, some spermatozoa, but most will produce both (though ova and sperm generation often depend on the season of the year). Usually sperm will be expelled via the mouth of the sponge into the water, and then sucked into the pores of another sponge (in the same way nutriments are brought in). Fertilization takes place in the mesohyl (the space between the two dermal layers), and then the ciliated embryos will be expelled from the mouth of the sponge. Haeckel described these processes, which are essentially the same as those depicted by modern biologists. See Haeckel, *Kalkschwämme,* 1:328–33. For the modern description, see Brusca and Brusca, *Invertebrates,* 198–202.

68. Nikolai Miklucho, "Beiträge zur Kenntniss der Spongien I," *Jenaische Zeitschrift für Medicin und Naturwissenschaft* 4 (1868): 221–40. The title suggests that this was the first of a series of contributions, but no further of his studies on sponges appeared in the journal. Miklucho would publish one further, major paper on sponges: Nikolai Mikloucho-Maclay, "Ueber einige Schwämme des Nördlichen Stillen Oceans und des Eismeres," *Mémoires de l'Académie impériale des sciences de St.-Pétersbourg,* 7th ser., 15 (1870): part 2.

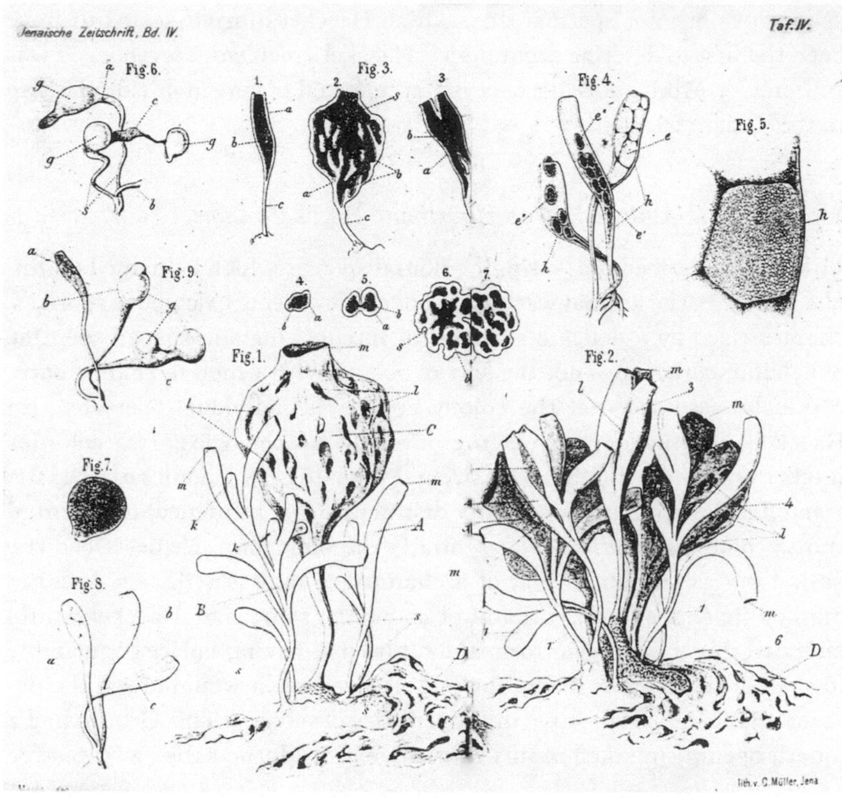

Fig. 6.8. Two colonial groupings (figs. 1 and 2) of the sponge *Guancha blanca*. (From Nikolai Miklucho, "Beiträge zur Kenntniss der Spongien I," 1868.)

cavity," into and out of which water flowed through the mouth. Embryos produced in the inner walls of the sponge would also escape through the mouth (see 4.e) and plant themselves on the seafloor to begin a new sponge colony. The various forms of the individuals making up the colony could be understood, according to Miklucho, as having arisen from the basic form through a twofold process of growth and then a merging and dissolving of sponge walls (as shown in 3.3). A more complicated sponge form (e.g., form 1.C) would, as a consequence, develop through the melding and transformation of basic individuals (e.g., 3.3 and 3.2). Miklucho detected, however, no necessary sequence of changes from one form to another. Moreover, he contended that each of the divergent forms could be found existing independently—a phenomenon, he believed, that had led other researchers to suppose them to be different species, even different genera of sponge.

Miklucho, though a novice researcher, made bold to challenge two deeply held views about sponges.[69] First, he declared that his observations showed that the mouth of his sponge not only expelled water and material but also that it took water and material into the main cavity through the actions of the cilia lining it—an idea that Grant had rejected forty years earlier.[70] Second, he suggested that his sponge (and others as well) were more closely related to the coelenterates (i.e., zoophytes such as hydra and coral) than they were to the rhizopoda (e.g., amoebas and foraminiferans) or other protozoans. The relationship, he thought, was revealed by ingestion and expulsion of material through the one mouth opening, which was similar to the function and structure of the coelenterates. The more standard view had been (and, indeed, has its advocates today) that sponges were really colonies of modified Protista—a conception that Haeckel himself had initially held but would alter in light of his student's work.[71]

Haeckel's first publication on sponges, in 1869, came shortly after that of Miklucho, who was rather surprised and a bit irritated when he initially learned of his mentor's new subject.[72] Yet he likely felt somewhat assuaged at the praise given his own efforts and the confirming support Haeckel offered:

69. Miklucho, "Beiträge zur Kenntniss," 232–34.

70. He qualified his contention by claiming that the phenomenon of ingestion and expulsion through the mouth opening may hold only for his *Guancha blanca.* The modern opinion supposes that the mouth of sponges serves only to expel water and material.

71. Haeckel initially placed sponges, following Lieberkühn, in the phylum of Protista, a phylum that he himself defined (and that remains employed today). See Ernst Haeckel, *Generelle Morphologie der Organismen,* 2 vols. (Berlin: Georg Reimer, 1866), 2:xxix–xxx. Some modern systematists agree with this placement. See, for example, Brusca and Brusca, *Invertebrates,* 205–6. But the issue is complex, and many other modern spongologists place them on the same line as coelenterates and regard both groups as evolving from Protista—and this was essentially Miklucho's view. See, for example, Bergquist, *Sponges,* 240–41.

72. Ernst Haeckel, "Ueber der Organismus der Schwämme und ihre Verwandtschaft mit den Korallen," *Jenaische Zeitschrift für Naturwissenschaft und Medicin* V (1870): 207–35. The issue containing this article came out in September 1869. Anton Dohrn wrote his friend Miklucho to inform him about Haeckel's new interest. Miklucho, who was then traveling in the Middle East, responded on June 9, 1869: "Haeckel's writing has quite surprised me. Psychologically it's interesting—you're right. I had no idea of his plan to write something on sponges. It's too bad that I should not have noted Haeckel's pretty phrases that he laid on Fol in Lanzarote when the latter wished to work on sponges. They would have gone down well here. But it is better for our knowledge, and I'm happy that his results are comparable to mine. He writes me that they [his results] would be to my advantage—he'll perhaps also do work on cartilaginous fish [another of Miklucho's own research areas]! Well, all the better for science. Adios." See *Nikolai Nikolajewitsch Mikloucho-Maclay Briefwechsel mit Anton Dohrn,* ed. Irmgard Müller (Norderstedt: Verlag für Ethnologie, 1980), 43. In subsequent letters, Miklucho asked Dohrn to give greetings to Haeckel, indicating a still cordial relationship.

> The most important finding of this investigation of sponges [Miklucho's], of whose validity I have become convinced through my own observations, was the fact that the sponges stand in a closer relationship to the corals than one had hitherto assumed. . . . Miklucho's investigations show . . . that these digestive systems in both classes [of corals and sponges] are homologous and analogical.[73]

A principal objection to the claim for homology might be that corals draw water and nutriments into the main mouth opening (as well as eliminating detritus through that orifice), while sponges only expel materials from the mouth. But Miklucho had shown, Haeckel agreed, that in some sponges the mouth also took in material. Haeckel built further on this similarity by an observation that had not been hitherto made: the simple endoderm and ectoderm of sponges corresponded closely to the two comparable layers of cells in coelenterates in structure and function, setting them both off from the higher metazoans, which in later developmental stages displayed a middle layer of cells, the mesoderm.[74] Haeckel concluded, not surprisingly, that the homologies between sponges and coelenterates could best be explained by Darwin's theory and the biogenetic law.[75]

Haeckel's Theory of Gastrulation and the Gastraea

In his preliminary paper, Haeckel sketched another important concept, based on extensive and exacting observation, which he would later expand into his theory of the gastraea.[76] He followed the development of ciliated sponge embryos, from their early "mulberry form" during cell cleavage to

73. Haeckel, "Organismus der Schwämme," 210.

74. Jane Oppenheimer provides a comprehensive history of the theory of the germ layers in her "The Non-Specificity of the Germ Layers," in *Essays in the History of Embryology and Biology* (Cambridge, MA: MIT Press, 1967), 256–94.

75. Haeckel, "Organismus der Schwämme," 218–19.

76. Haeckel initially developed his theory in his *Kalkschwämme,* 1:328–60. He subsequently published two further articles on the theory of the gastraea: "Die Gastraea-Theorie, die phylogenetische Classification des Thierreichs und die Homologie der Keimblätter," *Jenaische Zeitschrift für Naturwissenschaft* 8 (1874): 1–55; and "Die Gastrula und die Eifurchung der Thiere," *Jenaische Zeitschrift für Naturwissenschaft* 9 (1875): 402–508. These were collected in his *Biologische Studien: Studien zur Gastraeatheorie* (Jena: Hermann Dufft, 1877). Lynn Nyhart provides an extensive consideration of Haeckel's theory and reactions to it in her *Biology Takes Form: Animal Morphology and the German Universities, 1800–1900* (Chicago: University of Chicago Press, 1995), 181–97.

the process he would later call gastrulation. In his paper, he described the phenomenon succinctly:

> After the egg, through the cleavage process, has formed into a spherical, mulberry-shaped mass of densely packed, similar, naked, spherical cells, the mulberry-shaped embryo, by a stronger growth in one direction, takes on an ellipsoidal or egg-shaped form and covers its surface with cilia. Then within, a small central cavity (the stomach) expands; and at one pole of the long axis, an opening breaks through, the mouth.[77]

Haeckel regarded this embryonic form of sponges as essentially the same as the embryonic form of coelenterates (see fig. 6.9 for an illustration of the formation of gastrula and fig. 6.10 for a comparison of gastrula of several phyla of animals). The young sponge would retain (depending on the species) the cuplike structure of the gastrula. The young coral would also display that structure, but during maturation would develop its distinctive features. These homologies of ontogeny, as interpreted through the biogenetic law, strongly implied that "sponges and corals are blood relations."[78]

Arguments from homology would not be the only support for Darwin's theory provided by sponges. In his preliminary article, Haeckel observed that Miklucho's *Guancha blanca* comprised individual persons "whose structures belong to different species and even different families of sponge, though they derive from one and the same root."[79] In his three-volume monograph on calcareous sponges, *Die Kalkschwämme,* Haeckel would elevate this aspect of Miklucho's polymorphous sponge to a central position

77. Haeckel, "Organismus der Schwämme," 219.

78. Ibid., 220. In his *Kalkschwämme* (1:461n2), Haeckel moderated his claim of a close relationship of sponges and corals. In his monograph, he would represent their connection as occurring only in the form of a common ancestor residing in the distant past. And in his "Die Gastrula und die Eifurchung der Thiere" (422–23), he recognized that in the formation of the gastrula, the blastula becomes invaginated, that is, part of the external derma folds in toward the center to become the endoderm that lines the central cavity. (See fig. 6.9, nos. 118 and 120). Haeckel's student Carl Rabl had initially made this observation in gastropods. He likely had his attention focused by an essay by E. Ray Lankester, who a year or so earlier made the same observations about invagination. See Carl Rabl, "Die Ontogenie der Süsswasser-Pulmonaten," *Jenaische Zeitschrift für Naturwissenschaften* 9 (1875): 195–240 (198–99 refer to the phenomenon of invagination). See also E. Ray Lankester, "On the Primitive Cell-Layers of the Embryo as the Basis of Genealogical Classification of Animals," *Annals and Magazine of Natural History* 11 (1873): 321–38. Haeckel mentions Rabl and Ray Lankester, as well as his own observations.

79. Haeckel, "Organismus der Schwämme," 211.

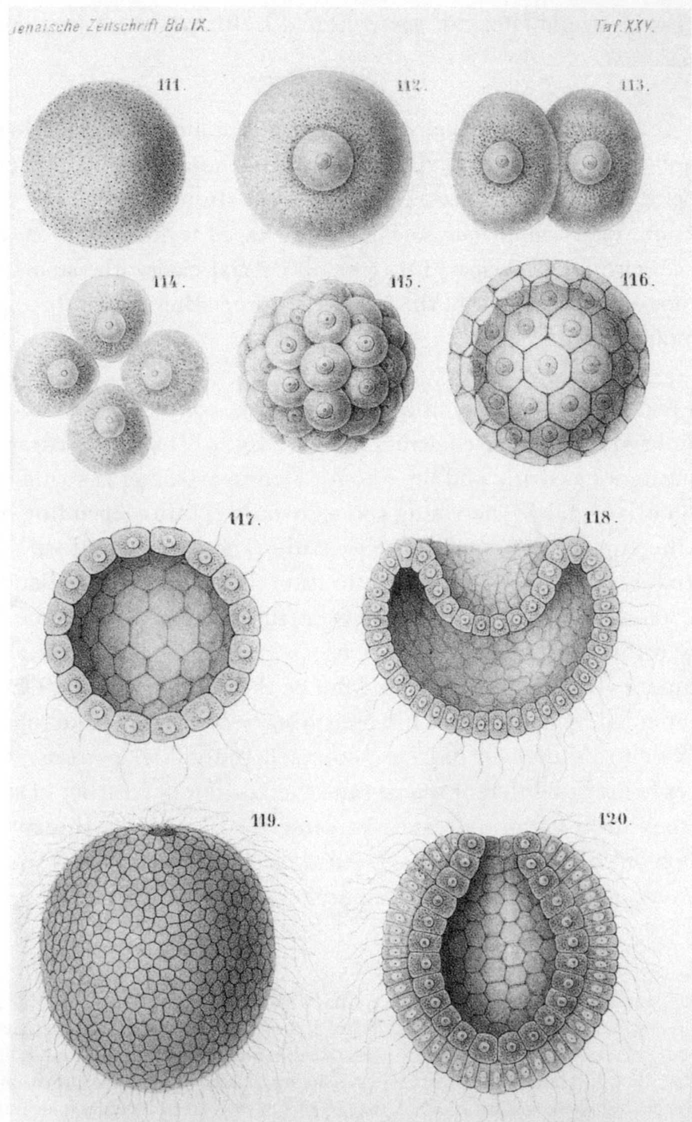

Fig. 6.9. Cell cleavage and formation of the blastula, 111–117; and invagination of the blastula, 118–120. (From Haeckel, "Die Gastrula und die Eifurchung der Thiere," 1875.)

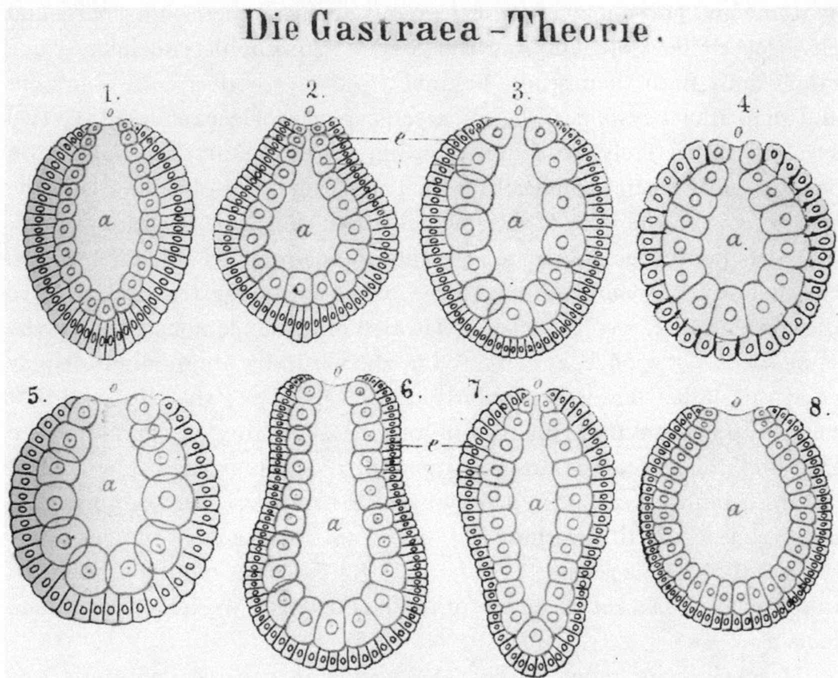

Fig. 6.10. The gastrula of several phyla of animals. (From Haeckel, "Die Gastraea-Theorie, die phylogenetische Classification des Thierreichs und die Homologie der Keimblätter," 1874.)

in the argument for evolution. He believed that this little sponge provided in real time a clear proof of descent, the kind of empirical demonstration of Darwin's theory that the critics had demanded.

The Structure of Haeckel's Monograph on Sponges

Haeckel's *Kalkschwämme* culminated five years of research on calcareous sponges, from the period of his stay in the Canaries to further collecting trips off the coasts of Norway (1869) and Dalmatia (1871), assisted during this latter journey by the brothers Richard (1850–1937) and Oscar Hertwig (1849–1922).[80] Volume 1 of the monograph describes the general

80. Haeckel wrote Darwin about his trip to Norway and Dalmatia. He described an amusing incident that indicated the spread of Darwinian theory. He provided this account: "On the Island of Lerina (in southern Dalmatia), I spent a wonderful month in spring in a Franciscan cloister(!) The prior of the cloister, Father Buoua-Grazia, is an enthusiastic Darwinist! He knows and venerates Goethe's works and Darwin's 'Origin of Species'! Before I stepped onto

anatomy and physiology of the calcareous sponges, their ontogenetic and phylogenetic development, and the general philosophic conclusions one might draw from their study. Volume 2 provides both a natural system and an artificial system of the calcareous sponges. Haeckel based the two systems, respectively, on different principles of organization—either on descent (with special emphasis on the transformations of the walls of the sponges and on the form of the spiculae) or on the vague traditions of systematists (with special focus on the mouth opening). By this duplication, Haeckel meant to show the superiority of the natural system of descent to the conventional, artificial system. He also meant to demonstrate that the same species of sponge, as classified in the natural system, might display great variability. It was particularly the variability of the creatures that seemed to him to argue for transmutation, something he emphasized to Darwin.[81] Volume 3 contains the copper-plate etchings and lithographs of the three families of calcareous sponges: the Ascones, the Leucones, and the Sycones. The illustrations also detail distinctive developmental features of the various genera of the natural system. That natural system fell into place for Haeckel by reason of the character of Miklucho's *Guancha blanca*.

Haeckel rechristened Miklucho's sponge *Ascetta primordialis* (see fig. 6.11). Like his student, he held that all persons of the colony were transformations of a basic form, which he called *olynthus* (fig. 6.11, no. 1). The individual persons of this species existed both colonially (e.g., no. 17) and independently (e.g., nos. 2, 3, 4, 5, 8). Thus in the natural system these individuals would be regarded as polymorphs of the same species; whereas in the artificial system they had been classified as different species, even members of different genera. In the colonial form we could observe, according to Haeckel, the presumed different species "in statu nascendi," a real-time demonstration of Darwin's theory.[82] Moreover,

---

the island, before our boat even had landed, the prior stood on the bank and shouted to me: Isn't it true, Professor, that Darwin was right, that we all derive from one and the same catarrhine apes!! Certainly no naturalist has been so received in a Catholic cloister." Perhaps even Darwin suspected a bit of embellishment in Haeckel's account. But see Haeckel to Darwin (21 December 1871), in the Darwin Correspondence, DAR 166.1, Cambridge. Haeckel records much the same story in the diary of his trip. See *Ernst Haeckel: Biographie in Briefen*, 113.

81. Haeckel to Darwin (6 July 1870), in the Darwin Correspondence, DAR 166.1, Cambridge: "This small animal group [calcareous sponges] is extremely interesting because of their endless changes of form and the unending number of transitional stages between the different forms. One can posit four or twenty or two hundred 'coeval species' of calcareous sponges—or also treat all of the calcareous sponges as one particular species."

82. Haeckel, *Kalkschwämme*, 1:36.

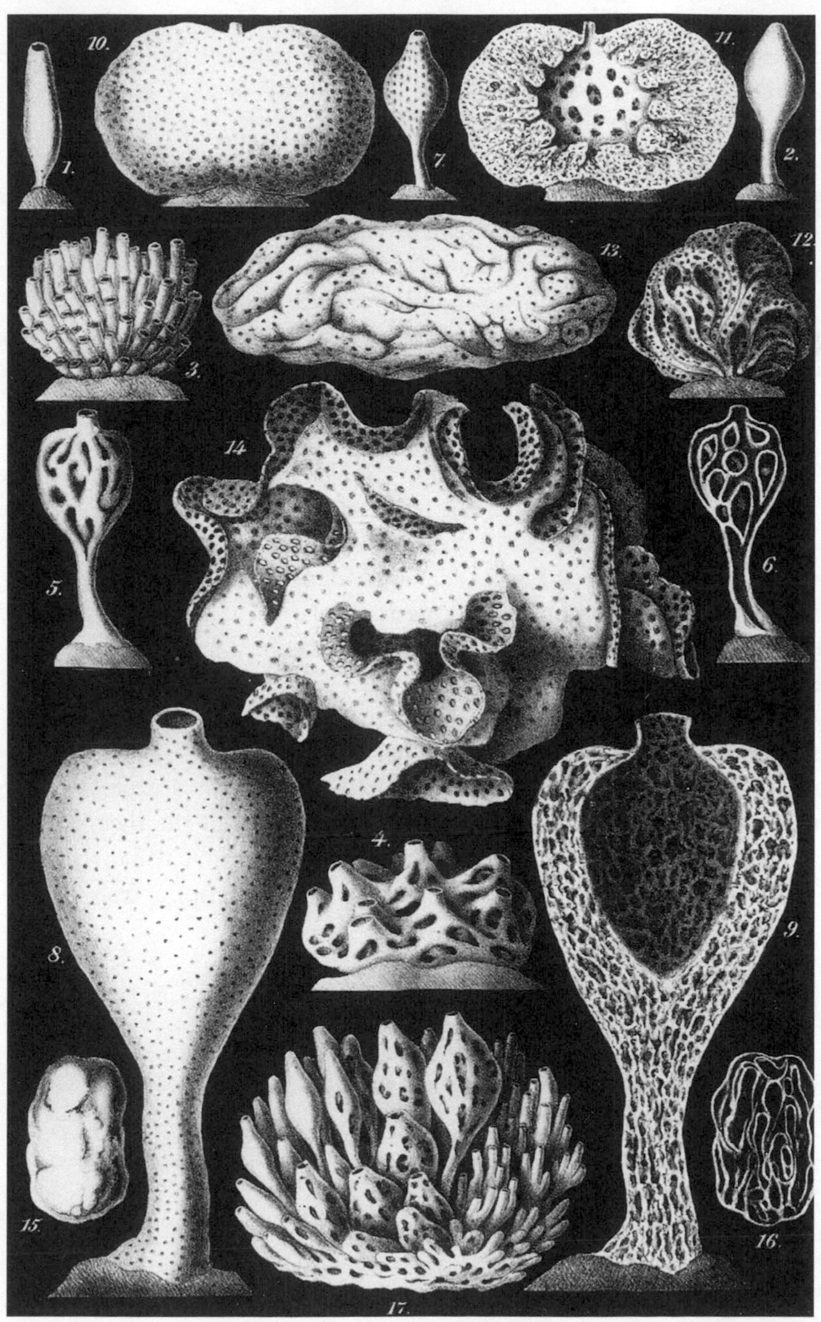

Fig. 6.11. *Ascetta primordialis*, the colonial form, no. 17, and various of its individual persons; the *olynthus* form is no. 1. (From Haeckel, *Die Kalkschwämme*, 1872.)

since the three families of calcareous sponges displayed homologous ontogenies and juvenile stages (i.e., the common *olynthus* form), they could be aligned in reasonable relationships of descent—a view generally held today.[83]

From his analysis of the calcareous sponges, Haeckel drew more far-reaching phylogenetic conclusions: that the siliceous and the horny sponges descended from the calcareous sponges; and that the sponges and coelenterates derived from a common ancestor—as revealed by their similar processes of gastrulation.[84] Moreover, in light of Alexander Kowalevsky's embryological research on *Amphioxus lanceolatus* (1867), which displayed a similar process of gastrulation in the formation of this near vertebrate, the Vertebrata might be joined to the Coelenterata.[85]

Haeckel's now-soaring evolutionary tree required a strong tap root, which he speculatively located: "From this identity of the gastrula among representatives of the various animal stocks—from the sponges to the vertebrates—I conclude that, in accord with the biogenetic law, a common descent of the animal phyla has occurred from one unknown stock form, which in essence was similar to the gastrula: gastraea."[86] He thus assumed this hypothetical organism, the gastraea, plied the ancient seas and gave rise in the course of time to all of the multi-celled animals.

Haeckel's projection of an ancient ancestor to all metazoans drew immediate fire from the enemies of evolution and has since been regarded as one of his typical flights of fancy; but at the time he had expert opinion on his side. When the first installment of his study of the theory of the gastraea appeared in 1874, August Weismann found much that surprised him in the conception and thought that it "had brought our knowledge a

83. For the modern view, see Brusca and Brusca, *Invertebrates,* 186.

84. Modern analysis agrees with this conclusion of Haeckel. See Sally Levs and Dafne Eerkes-Medrano, "Gastrulation in Calcareous Sponges: In Search of Haeckel's Gastraea," *Integrative and Comparative Biology* 45 (2005): 342–51. They write: "While molecular phylogenies agree that the Metazoa is monophyletic, recent comparison of rRNA sequences and of sequences of protein coding genes suggest that in fact calcareous sponges might be more closely related to cnidarians, ctenophores, and other metazoans, than they are to other sponges" (342–43).

85. See Alexander Kowalevsky, "Entwicklungsgeschichte des Amphioxus Lanceolatus," *Memoires de l'Academie Imperiale des Sciences de Saint-Petersbourg* 11 (1867): 1–17. Haeckel would frequently appeal to Kowalevsky's work as supporting his theory of the gastraea. See Haeckel, *Kalkschwämme,* 1:466. For an account of Kowalevsky's accomplishments, see Alexander Mikhailov and Scott Gilbert, "From Development to Evolution: the Re-establishment of the 'Alexander Kowalevsky Medal,'" *International Journal of Developmental Biology* 46 (2002): 693–98.

86. Haeckel, *Kalkschwämme,* 1:467.

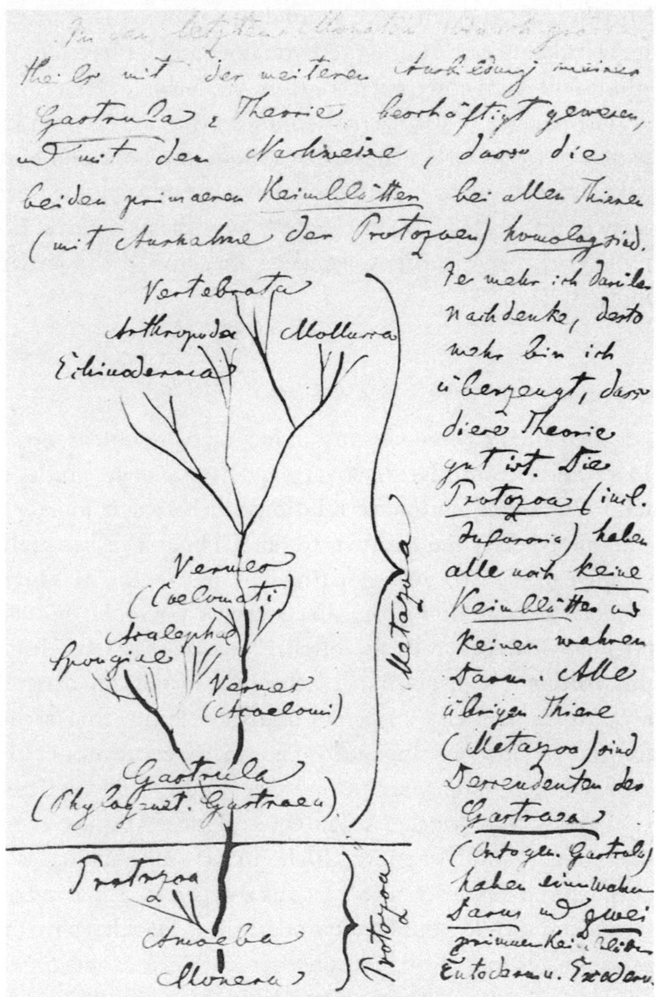

In den letzten Monaten bin ich großtheils mit der weiteren Ausbildung meiner Gastrula-Theorie beschäftigt gewesen, und mit dem Nachweise, dass die beiden primären Keimblätter bei allen Thieren (mit Ausnahme der Protozoen) homolog sind. Je mehr ich darüber nachdenke, desto mehr bin ich überzeugt, dass diese Theorie gut ist. Die Protozoa (incl. Infusoria) haben alle noch keine Keimblätter und keinen wahren Darm. Alle übrigen Thiere (Metazoa) sind Descendenten der Gastraea (Ontogen. Gastrula) haben einen wahren Darm und zwei primäre Keimblätter Entoderm u. Exoderm.

Fig. 6.12. Ernst Haeckel to Charles Darwin (8 October 1873): Sketch of the descent of the metazoans from the protozoans, with the first stage of metazoan development being the gastraea. He writes Darwin: "During the last months, I have been very busy with the further development of my gastraea theory and with the demonstration that the two primary germ layers are homologous in all animals (with the exception of the protozoa). The more I consider this, the more I am convinced that this theory is good." (Courtesy of Manuscript Room, Cambridge University Library.)

good deal further."[87] And with the second installment the next year, he again wrote Haeckel to say: "I must tell you how much I have enjoyed your Gastraea number II. I am convinced. With it, you have taken a mighty step forward."[88] The essence of his theory—namely, that all the metazoans go through a stage of gastrulation that unites them in a common bond of descent—is, of course, now the general orthodoxy in embryology.[89] Moreover, contemporary study suggests that Haeckel's hypothesis was not far off the mark in projecting a spongelike animal as ancestor to the entire multicelled animal kingdom.[90]

The Rejection of a Favorite Student

Haeckel's extraordinary three-volume monograph on calcareous sponges took its inspiration from the discovery by his favorite student, Nikolai Miklucho. But the affectionate relationship between master and pupil slowly dissolved after their return to Jena. Though in his preliminary paper on sponges, Haeckel had nothing but praise for Miklucho—and gave his student full credit for the discovery of *Guancha blanca* in his large monograph—the general tone of the book was, nonetheless, acidic. He thought Miklucho's discussion of the bud formation of the sponge "completely erroneous." His student's analysis of the fine structure of the sponge was "in part very inexact." The spiculae "never occur" where Miklucho placed them. His description of the embryos is "deficient." His later discussion (1870) of another calcareous sponge "is quite inexact and without any value."[91] And when Miklucho in his subsequent essay maintained that the mouth opening of calcareous sponges had evolved out of the small pores—a claim that would have undermined Haeckel's theory of the gastraea—the master meted out the condign discipline: these "nonsensical assertions are utterly false and when Miklucho believes that they have fundamental significance for the understanding of the organization of the sponges . . . I regret I must maintain the opposite, namely, that they have

87. Weismann to Haeckel (27 January 1874), in "Briefwechsel zwischen Haeckel und Weismann," 35. See note 76 for reference to Haeckel's gastraea studies.

88. Weismann to Haeckel (27 December 1875), in ibid., 37.

89. For a lucid discussion of the reaction of Haeckel's contemporaries to his theory of the gastraea, see Nyhart, *Biology Takes Form*, 181–204.

90. See Levs and Eerkes-Medrano, "Gastrulation in Calcareous Sponges," 343: recent analysis suggests that "rather than sponges being a dead-end phylum, a sponge-like animal was indeed ancestral to all metazoans." They deny, however, that the transient cavity formed by invagination is "the future gut of the sponge" (349).

91. Haeckel, *Kalkschwämme*, 1:26–27.

no significance whatsoever and only demonstrate that the author has completely lost his earlier good direction in sponge morphology."[92] These were very harsh judgments aimed at the fragile ego of a young student-scientist. Exactly what pushed Haeckel to issue these incendiary condemnations remains uncertain—letters of neither reveal a precipitating cause. One can imagine that Miklucho's financial irresponsibility may have seemed consistent with his biological irresponsibility in dissenting from Haeckel's fundamental position on the primitive nature of the *olynthus* form. In any case, the breach would become absolute.

In the fall of 1868, after finishing his paper on *Guancha blanca*, Miklucho left Jena and traveled to Sicily, leaving a wake of pressing debts roiling behind. At Messina he stayed with another student of Haeckel, Anton Dohrn, who had grown dilatory in the pursuit of a formal academic career and rested in that southern city on the shifting sands of ill-defined plans for the future. In late winter the mercurial Miklucho left his friend and set off for Suez, where he intended to conduct more research along the Red Sea before the new canal could be completed. He managed to discover several new species of sponge before the hardships and dangers of that part of the world drove him back to Russia, borrowing funds for his passage along the way. In the fall of 1869, he returned to Jena, where he completed his second major paper on sponges—the one Haeckel thought "worthless." The next April he slipped out of town, again owing a considerable amount of money, and set out for London, where he met with Huxley and Wallace. He enlisted them in a scheme to help him finance travel to the South Pacific for further research. By this time he had all but given up his pursuit of a medical degree. He briefly returned to Jena but quit the city in the spring because of the "stupid and absurd rumors" that whirled about him.[93] Miklucho would eventually make his reputation in the South Pacific, particularly in New Guinea and Australia, not as a marine biologist but as an anthropologist of exceeding merit.

Haeckel's Sponge Work

Haeckel's magnificent volumes on calcareous sponges provided the first systematic study of the class, and today more of these species have his

92. Ibid., 1:262–63.

93. Webster quotes this opaque phrase from one of Miklucho's letters home. See her *Moon Man*, 24. No letters of Miklucho to Haeckel in the archives at Haeckel-Haus indicate the cause of the break between the two.

name attached than that of any other investigator. But perhaps of greater significance, the arguments that he rested on these small animals became exceedingly powerful supports for Darwin's theory, showing how both the relationships among sponges could be naturally explained and how this lowly creature might hold the key to the evolution of all the metazoans. The success of Haeckel's study brought him the commission to analyze the *Challenger* expedition's catch of Keratosa (i.e., horny) sponges.[94] Yet Haeckel's great accomplishment appears to have been tethered to a will-o'-the-wisp. The driving inspiration for the whole enterprise, Miklucho's *Guancha blanca* (i.e., Haeckel's *Ascetta primordialis*), was based on an illusion. That supposed colonial sponge, which Miklucho and Haeckel had observed in several other locations, was undoubtedly a number of different species that had simply grown over one another. Robert Grant, in the 1820s, warned of this phenomenon; and contemporary spongologists testify about how difficult it sometimes is to distinguish a true sponge colony from a conglomerate. *Guancha blanca* was not the mother of all sponges.

Virtually all past science, by our contemporary lights, is riddled with fundamental errors: the planets do not travel in circular orbits, as Copernicus believed; Galileo's particular kind of inertia does not exist; space and time are not absolute nor are they part of God's sensorium, as Newton maintained; and functionally acquired traits cannot be inherited as Darwin and most nineteenth-century biologists assumed. Yet from these bold hypotheses and attendant conceptions science has advanced such that we can now make these corrective judgments with confidence. Haeckel's own work on siphonophores and sponges, within a smaller ambit perhaps, moved the area of hydrozoan biology into its modern phase and energized the larger sphere of evolutionary considerations.

Despite Haeckel's accomplishments, even those who have shown some sympathy for his work tend to regard him as not quite the scientist that his colleague Gegenbaur was.[95] More often, though, he is usually perceived as

94. Though the horny sponges form the largest class of sponges, there was doubt about the existence of any deep-sea varieties. Haeckel had the task of examining the very deteriorated and altered condition of the *Challenger* specimens. His anatomical work was made more difficult because of the symbiotic relation these deep-sea sponges had with various hydroid animals, which intersected the whole bodies of the sponges. See Ernst Haeckel, *Report on the Deep-Sea Keratosa collected by H.M.S. Challenger during the Years 1873–1876*, vol. 32 of *Report on the Scientific Results of the Voyage of H.M.S. Challenger during the Years 1873–1876* (London: Her Majesty's Stationery Office, 1889).

95. Nyhart provides an admirable and nuanced comparison of the styles and emphases of Haeckel and his great friend Gegenbaur. She yet comes to a conclusion about their respec-

a mere coryphée, poorly dancing the choreography of the English master. Yet Haeckel's knowledge of systematics was far greater than either Gegenbaur's or Darwin's, and his experimental genius stood with the very best of his times. His industry, his daring, his imagination, and his inventive hypotheses should have made him, in the eyes of historians, Darwin's rival. Yet his own success as a popularizer, ironically, did as much to cast his extraordinary science into the shadows as did the negligent attitude of subsequent scholars, as we will see in the next chapter.

Conclusion: A Naturalist Voyaging

Haeckel's voyage to the Canary Islands was his second major research expedition, the first being his habilitation work in Italy and Sicily. During the course of his professional life, he undertook some twenty or so research trips, several of which involved considerable danger and hardship. His eight-month journey in 1900 to the jungles of Java and Sumatra—the last of his major travels—occurred when he was in his mid-sixties, an excursion that tested his temper and physical endurance. From the Renaissance through the present, naturalists have hazarded great dangers—physical distress, sickness, injury, and even death—in pursuit of scientific discovery. More than once Haeckel's own adventures at sea and in the mountains put his life in jeopardy. Why would anyone do this—not once, but countless times? Haeckel had few rivals in the dangers and difficulties faced during alien travel. Yet his many trips seem almost superfluous for the sheer purpose of acquiring new materials and for advancing a career: organisms could have been obtained through the work of other naturalists and one's own assistants; and after 1870 Haeckel had solidified a reputation as a premier researcher. The acquisition of new materials would always be a justification; but with Haeckel, there was usually more at stake. Here I would like simply to enumerate some of the other reasons this man—raised in the comfortable lap of a very civilized society and enjoying most of its privileges—might undertake research trips that would continue to test his physical and psychological powers. Many of these same motives likely have impelled other naturalists to comparable efforts, but with Haeckel

tive contributions to evolutionary theory that suggests Haeckel was merely the popularizer, while Gegenbaur the real scientist: "If Haeckel was evolutionary morphology's most voluble spokesman, Gegenbaur was its greatest practitioner." I believe this greatly underestimates the extraordinary morphological work Haeckel did in the area of invertebrate biology. See Nyhart, *Biology Takes Form*, 150–67; quotation from 153.

they had an unremitting urgency and a poignant resonance. The list can serve as a guide for subsequent chapters of this book, wherein we will follow him through other exotic venues.

1. Foreign travel opens the way to make discoveries that seem unlikely in your own domain. This remained, in Haeckel's case, the permanent default reason for his many excursions. Often his justification for research trips—against the wishes and constant complaints of his second wife—would be for the recovering of new, not-hitherto-described marine organisms. Yet these discoveries usually only led to ever-greater accumulations of descriptions to fill in his systematic surveys of particular groups of organisms. Few trips would have a payoff comparable to that of his journey to the Canary Islands. The following motives, nonetheless, continued to impel him to gather his equipment and to book passage on ships sailing to faraway places.
2. Travel might serve as a means of sealing the importance of any discoveries made. The model of great voyages of the past suggests that any findings or new ideas derived from a journey would have their significance elevated by the degree of difficulties suffered during the excursion. The assumption is easy: that the importance of results achieved would be commensurate with the dangers chanced. Alexander von Humboldt's reputation as the doyen of German science was made by his near-death experience climbing Chimborazo, highest mountain in the northern Andes, during his South American journey—and he did not even succeed in getting to the top.
3. Travel provides escape from the cares and sorrows of one's own home ground. After Haeckel's second marriage, he often choked on the miasma of the spreading gloom in his house. Through the 1880s and 1890s, his home life would sink ever lower under the hypobaric pressures of the neurasthenic complaints of his wife and his daughter Emma. His long trips allowed escape into sunnier climes, where domestic cares resolved into a dew, melted away by the joys of adventure, new vistas, and all-consuming work.
4. Travel also provides means for a romantic commune with nature. Not every biologist, of course, responds to the aesthetic displays of untamed nature in the manner of a Humboldt or a Haeckel. On every trip Haeckel would bring his sketchbooks and canvases; and during his last travels, these implements of the aesthetic life be-

came even more important than his microscopes, dissecting blades, and spirits of wine.

5. Travel with other naturalists supplies an exclusive, "manly" company. The rugged life with fellows (usually his students, who might be only five to ten years younger than he) served a vital need for Haeckel. His climb of Pico de Tiede engaged the same competitively virile side of his personality already evinced in his earlier participation in gymnastic tournaments. His journeys toward the end of his life often had more of a solitary character. He nonetheless, even in his later years, felt the challenge of every mountain peak in sight.
6. Travel has always opened up possibilities precluded at home, especially erotic connections on foreign soil—the ideal of the sailor's life. Haeckel and his companions were often away for months at a time. There are many hints in his letters and passing remarks that sexual delights—of a sort made difficult in one's own city or country—were not far from his mind, a terrain that harbored not only images of brightly colored hydrozoa but also of voluptuous "figures and dark, one might say, volcanic eyes of fire." In the late 1890s, when he was pursuing a young woman—who radiated the aura of his first wife—he confessed that "many beautiful women flung themselves at him."[96] He did protest in those later years that he resisted such temptations, but the circumstances of his defense make the claim tenuous. Boys, after all, dream of native girls bringing breadfruit.
7. Naturalists have been inspired by predecessors to undertake comparable voyages. Haeckel was certainly aware that Darwin and Humboldt had made their intellectual fortunes by exotic travel. And Goethe, who did not venture so far from home, nonetheless also supplied him an exemplar. These of his predecessors not only inspired Haeckel, but they stood as standards by which to mark his own climb toward that illusive goal of immortal fame. Haeckel felt he had to travel as far as they did; he had to overcome the physical obstacles and dangers that they did; and he had to achieve the intellectual distinction that they did—or even accomplish a bit more than they did. Haeckel was ambitious, and his desire for recognition certainly did not burn less brightly than theirs.

96. See chapter 10.

The motives I have discriminated undoubtedly led many other naturalists to undertake hazardous travel. But with Haeckel, they drove him incessantly over a lifetime. They impelled him headlong into the twentieth century and elevated him to a position of remarkable achievement and significance. They also made him a target for the anxieties, fears, and jealousies of a great number of his peers, as I will portray in the next two chapters.

CHAPTER SEVEN

The Popular Presentation of Evolution

While in the Canary Islands, Haeckel had left loneliness and depression behind, escaping into his research, the camaraderie of his assistants, and the exotic if arduous environment. But when he returned to Jena in the spring of 1867, he found those familiar companions waiting in his desolate house. He sought refuge in the acquaintances he and his wife had cultivated during their brief happiness. Among them was Agnes Huschke (1842–1915), the youngest daughter of the former Jena anatomist and Gegenbaur's predecessor, Emil Huschke (1797–1858). Haeckel mentioned to his friend Allmers that the young Agnes (age twenty-four) reminded him of his "true, unforgettable Anna."[1] In desperate hope and daring haste, he asked Agnes to marry him. They announced their engagement in June and planned an August wedding. His colleague August Weismann responded to the news with hearty congratulations and said that the fates had aligned his and his friend's stars; for he, too, was getting ready to celebrate his own marriage.[2] Allmers, though, had a deeper presentiment about Haeckel's hurried plunge into the union. He wrote his friend that the announcement of the impeding marriage

> so deeply moved me that tears ran down my cheeks as I read your letter, and I don't know why the tears. Was it joy? Was it sorrow? I am forced to think again and again about your beloved Anna and the happy hours

1. Haeckel to Hermann Allmers (22 June 1867), in *Ernst Haeckel: Sein Leben, Denken und Wirken. Eine Schriftenfolge für seine zahlreichen Freunde und Anhänger*, ed. Victor Franz, 2 vols. (Jena: Wilhelm Gronau und W. Agricola, 1943–44), 2:48.

2. Weismann to Haeckel (4 August 1867), in"Der Briefwechsel zwischen Ernst Haeckel und August Weismann," ed. G. Uschmann and B. Hassenstein *Jenaer Reden und Schriften* (1965): 21.

that I once spent with you both, so immensely happy were you. Again, I recalled the lovely excursion to Kunitz and over the border to Lobeda; and I remembered the moonlit evening in the garden where we took our meal, and then back home in a cozy café where we had our heartfelt conversation. The image of Anna, ever more lovely and vivid, passes through my soul.[3]

Those images must have passed through Haeckel's soul as well; a few days after the engagement was announced, as he confessed many years later, he contemplated suicide—neither for the first time nor would it be for the last.[4] He realized that Agnes simply could not replace his Anna. Even in his wooing of Agnes, he hinted to her what role she would play in his life, as these lines, from a long poem he sent her in July, suggest:

What in Anna delighted me,
You have warmly bestowed anew,
A heart of love so full and free,
A feeling so sincere and true.
You best and loveliest of girls,
How can I give you proper thanks?
My heart again its love unfurls,
And mind and soul will join its ranks.[5]

Haeckel's marriage to Agnes would reflect a reverse image of the brief reality and haunting promise of his life with Anna. He described Agnes to Darwin as "simple and natural, a very sensible and cheerful girl, who will, I hope, replace in many ways the loss of my deceased but extraordinary and unforgettable wife."[6] But no woman, certainly not one so unprepossess-

3. Allmers to Haeckel (15 July 1867), in *Haeckel und Allmers: Die Geschichte einer Freundschaft in Briefen der Freunde*, ed. Rudolph Koop (Bremen: Arthur Geist, 1941), 117. Kunitz and Lobeda were two villages close by Jena, now incorporated into the city boundaries.

4. In a letter to Frida von Uslar-Gleichen (11 January 1900), Haeckel mentioned that he had contemplated suicide on Johannistag, June 24, 1867, nine days after his engagement with Agnes. See *Das ungelöste Welträtsel: Frida von Uslar-Gleichen und Ernst Haeckel, Briefe und Tagebücher 1898–1900*, 3 vols., ed. Norbert Elsner (Göttingen: Wallstein, 2000), 1:390.

5. Haeckel to Agnes Huschke (21 July 1867), in *Ernst und Agnes Haeckel: Ein Briefwechsel*, ed. Konrad Huschke (Jena: Urania-Verlag, 1950), 28: "Was mich beglückt' an Anna, / Das gibst Du mir aufs neu: / Ein Herz voll warmer Liebe / Ein' Sinn voll Wahrheit, innig treu! / Du bestes, liebste Mädchen, / Wie dank ich Dir dafür? / Mein ganzes Sein und Wesen, / Mein Herz schenk ich aufs neue Dir!"

6. Haeckel to Darwin (28 June 1867), in the Darwin Correspondence, DAR 166.1, Manuscript Room, Cambridge University Library.

Fig. 7.1. Agnes Haeckel, née Huschke (1842–1915).
(Courtesy of Ernst-Haeckel-Haus, Jena.)

ing, could substitute for that quiet bride of memory, forever young, who grew in beauty and accomplishment over the years. Agnes did not share her husband's scientific enthusiasms and became ever more irritated by his frequent lecture trips and longer research expeditions. She hated the polemics in which he became engaged, especially when his more popular works, with their attacks on organized religion, provoked not only social coolness among her friends but a steady stream of hate mail from anonymous cranks—even threats of death.[7] They had three children: Walter (1868–1939), Elisabeth (1871–1948), and Emma (1873–1946). Walter became a painter of some reputation. Elisabeth married well and established a happy home with several children; her daughter Else would become her grandfather's assistant in his elder years. However, the youngest, Emma, suffered from mental illness and, as a result, created even more tensions in the marriage, tensions that increased toward the end of the century as both Agnes and her daughter fell deeper into the gloomy depths of that Victo-

7. See the "Einführung" by Konrad Huschke to his edition of *Ernst und Agnes Haeckel: Ein Briefwechsel,* 13.

Fig. 7.2. Ernst Haeckel in 1868.
(Courtesy of Ernst-Haeckel-Haus, Jena.)

rian disease of neurasthenia, which confined them to their home. They did their best to keep Haeckel also under house arrest; and when he traveled or engaged in the ordinary professional work of research, writing, and teaching, they made their resentment palpable and stifling.

The formal union began happily enough if quietly. On 20 August 1867, Haeckel and Agnes exchanged vows in a small, picturesque village church just outside of Jena. Immediately after the ceremony, they climbed into a coach and began a five-week honeymoon excursion to southern Germany, Switzerland, and the Tyrol. The journey required Haeckel to abandon plans for a trip to England and another visit with Darwin.[8] During the honeymoon, Haeckel did something typically rash, another example of a man in extremis. While staying at an inn in the southern Tyrol, he decided to scale the 2,700-meter Tristenspitze. Early in the morning of 9 September, he set out with an inexperienced guide to climb the aptly named "gloomy peak." In late morning, as they had passed through a dense cloud cover, a snow-

8. Haeckel to Darwin (28 June 1867), in the Darwin Correspondence, DAR 166.1, Cambridge.

storm broke out. They became lost and exhausted and, as night fell, even more disoriented. So precarious was the situation that Haeckel scribbled out a last will and testament in a small sketchpad, assigning his instruments to Gegenbaur and his books to Allmers. He left instructions for his new wife to cremate his body if it were found and "place the ashes in the grave of my Anna."[9] Miraculously, at 1:00 a.m. the next morning, Haeckel and his guide struggled back to the inn where the couple was staying. Undoubtedly, he never showed Agnes the testament of his despair—and of a love that sought reunion even in the dust of the grave.

When he returned from this ominous honeymoon, Haeckel had to deal with his difficult student, Anton Dohrn. Gegenbaur thought Dohrn had no talent for zoological work, but Haeckel rather liked the recalcitrant fellow and, in 1865, took him on as his assistant. Haeckel was justified, at least in part: Dohrn had over twenty publications on zoology prior to finishing his habilitation. Haeckel feared, though, that his protégé's "fancy," as he wrote Huxley, "flew far beyond his understanding."[10] Moreover, the young researcher could be prickly and rather reckless, willing to burn through the lifeline that held him in the academy. Not long before he had to stand examination for the habilitation, he wrote Haeckel a sharp letter outlining the deficiencies of the older man's philosophical education—he thought his professor simply did not understand Kant.[11]

When Dohrn finished with the two written parts of his habilitation, Haeckel, tolerant and fair, gave them a positive evaluation, if with some qualifications about the speculative nature of the second essay. Dohrn sought an extension of the time to finish the other parts of his examination, in zoology and philosophy; he finally passed them with grades of satisfactory (*befriedigend*).[12] Given the connections severed and those not yet constructed, Dohrn had little chance for a university career; but he was not without scientific acumen and enterprise. He started planning and then later established what became the Naples Zoological Station, a facility he

9. Haeckel wrote his will in a small sketchbook he carried, and so his wife likely never saw it. See *Ernst und Agnes Haeckel: Ein Briefwechsel,* 43.

10. Haeckel to Thomas Henry Huxley (27 January 1868), in "Der Briefwechsel zwischen Thomas Henry Huxley und Ernst Haeckel," ed. Georg Uschmann and Ilse Jahn, *Wissenschaftliche Zeitschrift der Friedrich-Schiller Universität Jena* (Mathematisch-Naturwissenschaftliche Reihe) 9 (1959–60): 15.

11. Dohrn to Haeckel (June 1867), letter quoted in Theodor Heuss, *Anton Dohrn: A Life for Science,* ed. Christiane Groeben, trans. Liselotte Dieckmann (1940; repr., Berlin: Springer, 1991), 351–54. See above, chapter 5, note 12, for a brief description of Dohrn's career.

12. Georg Uschmann, *Geschichte der Zoologie und der zoologischen Anstalten in Jena 1779–1919* (Jena: Gustav Fischer, 1959), 83–84.

would keep afloat through frequent solicitations of funds from members of the biological community and through the renting of research space to various scientific institutions.

In addition to his work with students such as Dohrn and Miklucho, Haeckel prepared in early fall to give a long series of lectures based on his *Generelle Morphologie*. The lectures were an effort to retrieve the fortunes of his big book, which threatened to sink into the intellectual waters without leaving so much as a ripple. With irritation and determination, he set out to reformulate the gist of the book for a general audience. Through the winter semester of 1867–68, he gave some twenty-four lectures to a large population of students, faculty, and townspeople. He spoke from loose notes, but two students made stenographic copies of the lectures. Through the spring and midsummer of 1868, Haeckel worked incessantly on revising and rewriting the stenographs. While he spent time in his study toiling over the manuscript, his wife prepared for the birth of their first child. He finished correcting the page proofs in mid-August; and then, as was often his wont in similar circumstances, he sought refuge from his mental exhaustion in travel.

On 21 August, a month before the birth of his son, Walter, Haeckel and Allmers set out on an excursion to Bavaria and northern Italy. Agnes was both angry and embarrassed by her husband's departure, feelings she vented in a letter to him two days after he left: "Everyone here is completely astonished that this hard-hearted professor would leave his poor little wife [*Frauchen*] so alone *now*; they find it completely incomprehensible."[13] The professor was not so hard-hearted that he neglected to take a picture of his Anna with him, something the current wife happened to notice.[14] The excursion was cut short by Haeckel's suffering a tooth infection in Bolzano, and so he arrived back in Jena early, two weeks before the delivery of his son. Haeckel asked Allmers to serve as godfather at the baptism, but his friend declined the invitation.[15] Allmers simply felt he could not fulfill

13. Agnes Haeckel to Ernst Haeckel (23 August 1868), in *Ernst und Agnes Haeckel: Ein Briefwechsel*, 44.

14. Haeckel told his wife he took down the picture of Anna (and one painted by Allmers, a present given for his first marriage) because the light in his study would fade the colors. He recommended she turn all of the paintings when he was away (something he neglected to do this time!). Presumably the light would not fade the pictures when he was at home. See Ernst Haeckel to Agnes Haeckel (27 August 1868), in ibid., 48. The letters exchanged during Haeckel's excursion turned affectionate on both sides, though with an undercurrent of incipient hostility and anxiety.

15. Allmers to Haeckel (6 October 1868), in *Ernst Haeckel: Sein Leben, Denken und Wirken*, 2:128–31.

the obligation to see the boy raised in the Christian faith, a sentiment that Haeckel must have ultimately appreciated. Allmers, however, did send a long poem for the occasion, which, when Haeckel read it to those at the ceremony, caused tears to stream down his face.[16] This biological birth was followed a few days later by an intellectual delivery: the publication of his *Natürliche Schöpfungsgeschichte,* which reprised his lecture series.

The work proved one of the most successful popular science books of the nineteenth century, only shaded by his own *Welträthsel* at the end of the century. From 1868 through 1920, *Natürliche Schöpfungsgeschichte* (Natural history of creation) went through twelve German editions and was translated into most of the modern languages. At the beginning of the twentieth century, the geneticist and historian of biology Erik Nordenskiöld judged the *Natural History of Creation* "the chief source of the world's knowledge of Darwinism."[17] Even in the English-speaking world, there were as many people who learned of evolutionary theory through Haeckel's book as through Darwin's own. Shorn of its more provocative title—the word "natural" was cut—the book appeared in two English translations as *The History of Creation* and went through numerous reprints up to 1926; the most recent edition appeared as a two-volume paperback in 2007.

In the following sections, I will highlight the main features of Haeckel's book and chart some of the important and dramatic changes—those incorporating new research and arguments—through its several editions. I will also try to isolate the features that seemed to hold the public's attention as no other book of similar intent had. This analysis will lead to a more general discussion of what makes a work an example of "popular science" and of the way Haeckel's volume exemplifies the criteria. In the next chapter, I will consider the rage of the critics and the charges of fraud brought against the book and its author.

Haeckel's *Natural History of Creation*

Aim of the Book

The very title of Haeckel's book—*Natürliche Schöpfungsgeschichte*—may have been inspired by Carl Vogt's (1817–1895) translation of the anonymously published *Vestiges of the Natural History of Creation* (1844), which

16. Allmers to Haeckel (17 November 1868), in ibid., 2:131–34; and Haeckel to Allmers (20 November 1868), in ibid., 2:134–35.

17. Erik Nordenskiöld, *The History of Biology: A Survey* (1920–24), trans. Leonard Eyre, 2nd ed. (New York: Tudor, 1936), 515.

Vogt rendered *Natürliche Geschichte der Schöpfung* (1851 and 1858). Vogt, a marine biologist, began the translation prior to the upheavals of 1848 and offered it to the public from the safe distance of his home in Switzerland. In the preface to his translation, Vogt recommended the consideration of Mr. Vestiges, this "constitutionally minded Englishman." Vestiges had "conceived a constitutionally minded God, who at the beginning, yet like an autocrat, established laws but then of his own accord gave up his autocracy and, without any direct interference, allowed only the law to operate in his land." "A marvelous example for our princes," so Vogt moralized.[18] In the book the anonymous author (later revealed as Robert Chambers) developed a fairly crude theory of species descent, in which one species simply gave birth to another. Vogt salted his footnotes to the translation with numerous corrections, including animadversions on many theological presumptions of Mr. Vestiges.[19]

Haeckel, who knew Vogt personally, sympathized with his friend's political liberalism and certainly with his anti-theological views.[20] Haeckel's own book would cast a dark shadow on aristocratic pretensions (see below); and since it was a "natural" history of creation, the Divinity would be left out of the picture altogether, except as a polemical foil. Like Vestiges, Haeckel did suggest that believers would have more reason to admire the divine inventive power if the Lord operated at a distance, through natural laws, instead of being required to construct every flea and fish of creation; but, unlike Vestiges, he would deflate this suggestion by contending that any effort to introduce the Divinity, even simply to set the world spinning, would be "to take a leap into the inconceivable" (*einen Sprung in das Unbegreifliche thun*).[21]

Haeckel understood that for a general audience (and even for the specialists) the most interesting and fraught question of descent theory would be that of the status of human beings. His *Generelle Morphologie* discussed human evolution only at the very end of two large, technical volumes. The *Natural History of Creation* brought the question to the fore, highlight-

18. Carl Vogt, "Vorrede" to [Robert Chambers], *Natürliche Geschichte der Schöpfung*, trans. Carl Vogt (Braunschweig: Friedrich Vieweg und Sohn, 1851), vi. Vogt added numerous footnotes to "correct" the mistakes in the volume. He also added a wide array of illustrations.

19. See, for example, ibid., the note on 250–51, where Vogt derides any suggestion of the Mosaic story. Vogt and Haeckel would later have a falling out.

20. Haeckel, who had read Vogt's books as a student, visited him in Geneva in 1864. See Haeckel to Vogt (18 October 1864), in Christoph Kockerbeck, ed., *Carl Vogt, Jacob Moleschott, Ludwig Büchner, Ernst Haeckel: Briefwechsel* (Marburg: Basilisken-Presse, 1999), 108–9.

21. Ernst Haeckel, *Natürliche Schöpfungsgeschichte* (Berlin: Georg Reimer, 1868), 261.

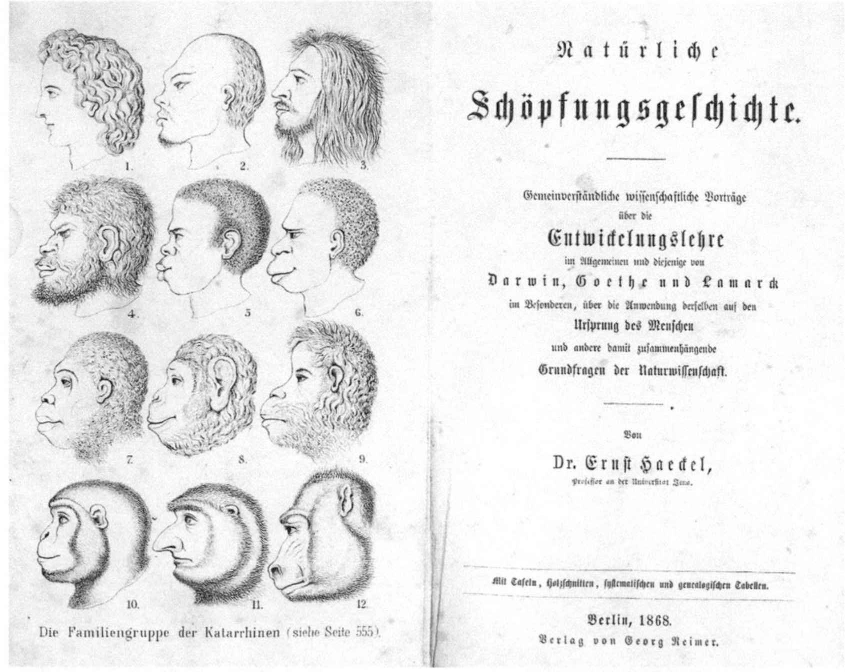

Natürliche
Schöpfungsgeschichte.

Gemeinverständliche wissenschaftliche Vorträge
über die
Entwickelungslehre
im Allgemeinen und diejenige von
Darwin, Goethe und Lamarck
im Besonderen, über die Anwendung derselben auf den
Ursprung des Menschen
und andere damit zusammenhängende
Grundfragen der Naturwissenschaft.

Von
Dr. Ernst Haeckel,
Professor an der Universität Jena.

Mit Tafeln, Holzschnitten, systematischen und genealogischen Tabellen.

Berlin, 1868.
Verlag von Georg Reimer.

Fig. 7.3. Frontispiece and title of the first edition (1868) of Haeckel, *Natürliche Schöpfungsgeschichte.*

ing the subject several years before Darwin himself undertook his study of human evolution in *The Descent of Man.* Haeckel explicitly stated that he intended to provide a "non-miraculous" history of the development of humankind. He shrewdly indicated his aim at the beginning of the volume, returned to the theme at various junctures, and then spent a penultimate chapter discussing human evolution in detail. But even more dramatically, the frontispiece of the book graphically set the races of mankind—or "species" of men, as he regarded them—and their animal forebears in a scale of descent (fig. 7.3). The theory that Haeckel called "the greatest triumph of the human spirit"[22] would be spread over almost six hundred pages; but it was encapsulated on the first page of the volume in a single, bold illustration.

The illustration, however, evoked complaint even from the friends of evolution. Charles Lyell, for instance, thought the Africans had been made too simian. Haeckel somewhat abashedly agreed that "the Australian, Ne-

22. Ibid., 3–6.

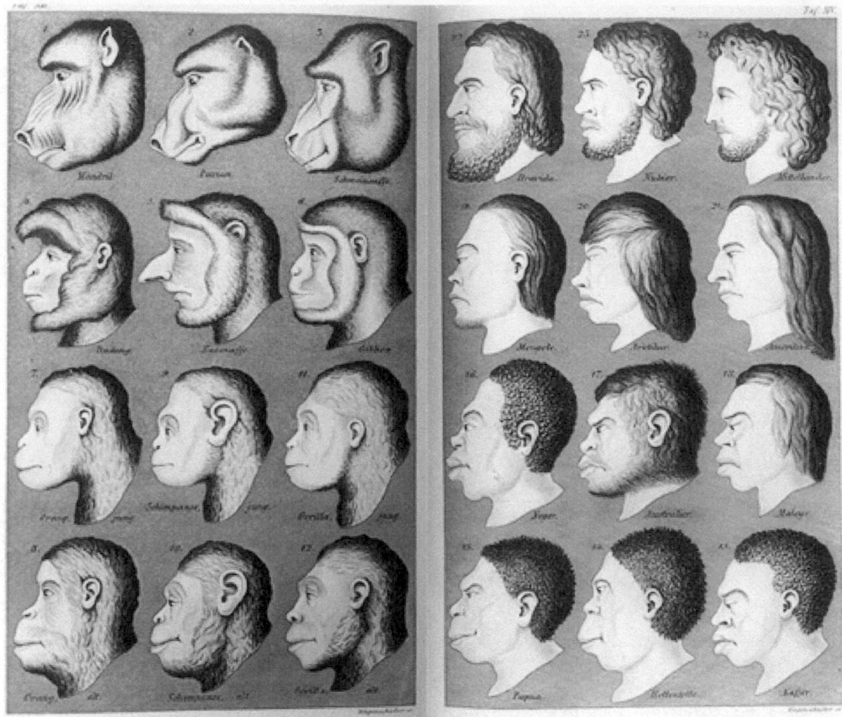

Fig. 7.4. The twelve human species and their descent relations to the narrow-nosed apes. (From Haeckel, *Natürliche Schöpfungsgeschichte,* 2nd ed., 1870.)

gro, and Papuan had been drawn way too pithecoid," and he indicated that he was redrafting the pictures for the second edition of the book (1870).[23] He did modestly reconfigure the human images, while expanding the races of mankind to twelve (fig. 7.4). His new depictions of apes, however, hardly reduced their quite human countenance or the demure look in their eyes. The figure representing European man continued to be modeled on that of a Greek with a near perfectly vertical forehead; though in the second version, the figure sported facial hair, perhaps as a reminder of animal

23. Haeckel to Lyell (27 November 1868), in the Lyell Correspondence, #1798, Manuscript Room, Edinburgh University Library. Thomas Henry Huxley, in an extensive review of the first edition of the *Natürliche Schöpfungsgeschichte,* goes into great detail about Haeckel's geological views and his systematics of plants, animals, and protists; but he not once mentions the German's arguments for human evolution. It may well be that Huxley did not wish to sully the English waters for the imminent appearance of Darwin's *Descent of Man.* See Thomas Henry Huxley, "The Natural History of Creation—by Dr. Ernst Haeckel," *The Academy* 1 (1869): 13–14, 40–43.

origins—or because now Haeckel himself wore a beard. The graphic representation of humanity rising out of ape stock did not last beyond the second edition of the book, but its impact had certainly been felt within the intellectual community and beyond.

Historical Introduction

To ease his audience into the rock-strewn argument, Haeckel employed a device seldom used in science writing up to this time.[24] It is one that scientists today sometimes still employ, though only as a bloodless ritual. Haeckel spent several chapters working out the history of his subject before engaging in a systematic presentation. In those historical chapters, occupying fully 25 percent of the whole, he examined the works of German authors like von Baer, Treviranus, Carus, Büchner, and Kant—indeed, devoting a whole chapter to Kant, the material that provoked Dohrn's tactless critique of his teacher. Haeckel extracted biological considerations from these authors to indicate that the kind of theory Darwin advanced was not foreign to the German mind but had been, in some respects, anticipated. Moreover, he tried to show that both Goethe and Oken had adopted descent theory, though without the large synthetic argument or the causal account that Darwin would furnish. Discussions of Lamarck, Geoffroy, Lyell, Erasmus Darwin, and Spencer further suggested that the modern temper had virtually rendered Darwinian theory inevitable.

Haeckel certainly recognized that great advances had been made in biological science by the likes of Cuvier and Agassiz, even though these two in particular strongly opposed transformation theory. In his discussion of Agassiz, Haeckel derided the naked supernaturalism over which was draped some quite competent biology. He maintained that the anthropomorphic God of Agassiz had to be replaced by the monistic God of Goethe. For Goethe's theology "leads to the most noble and sublime conception of which

24. In the seventeenth and eighteenth centuries, tracts in chemistry and medicine often enough traced the histories of their disciplines, not so much as easy introductions to but as parts of these sciences. With the nineteenth century, however, and the assumption that a new age of scientific observation had arrived, historical prefaces fell out of favor. Charles Lyell's *Principles of Geology* was a major exception. In the initial chapters of his book, Lyell recounted the history of theories of the earth, beginning with the Egyptians. Lyell's technical training was in law, whose dependency on precedents may have suggested the historical mode. And, of course, geology itself in this period had become a distinctively historical science. Haeckel met Lyell in 1866, during his stay in London prior to the trip to the Canary Islands, and the Scotsman's work may have provided a model for the *Natürliche Schöpfungsgeschichte.* See Charles Lyell, *Principles of Geology,* 3 vols. (London: John Murray, 1830–38), 1: chaps. 2–4.

a human being is capable, to the representation of the unity of God and nature."[25] This identity would become for Haeckel the basis for the "monistic religion" he cultivated and would more expressly celebrate in the 1890s.

Haeckel concluded his historical account with two chapters on Darwin's unique contributions to descent theory. These contributions, in Haeckel's estimation, were basically two. First, Darwin provided the most comprehensive synthetic account of descent by any author up to his time, weaving together many strands—artificial selection, embryology, biogeography, systematics, and so on. The second and singular contribution was his causal explanation of descent, his theory of natural selection (*Theorie von der natürlichen Züchtung*). Indeed, Haeckel believed that species transformation—given population pressure and struggle for existence, variability of traits, and inheritance of modifications—had to occur as a matter of mathematical necessity, so that really no further proof was required.[26] What was seen as a virtue in the nineteenth century, the apparent analyticity of the principle of natural selection, would be regarded as a liability in the Popperian intellectual environment of the twentieth—that is, until Popper finally understood what evolutionary theory was all about.[27]

Causes of Species Modification

Like Darwin, Haeckel distinguished two general classes of variable traits that produced changes in species: direct and indirect traits. Direct variations occurred when the parent acquired some adaptive properties and passed them to offspring—the Lamarckian moment. Indirect variations initially arose in the parent through some accidental impingements of the environment; these covert alterations remained unexpressed in the parent but appeared in the offspring and were preserved if they gave advantage to the individual.[28] Since these two sources of variability and adaptation were

25. Haeckel, *Natürliche Schöpfungsgeschichte*, 58.

26. Ibid., 133: ". . . ist mithin eine mathematische Naturnothwendigkeit, welches keines weiteren Beweises bedarf." While working on his book, Haeckel wrote Darwin to mention this same idea: i.e., if one accepts the principles of (1) heredity, (2) modification, and (3) struggle for existence, then natural selection becomes a "necessary truth." He was worried, however, that the premises of the argument, while simple and straightforward, would be yet misunderstood by many in the scientific community. See Haeckel to Darwin (February 1868), in the Darwin Correspondence, DAR 166.1, Cambridge.

27. Michael Ruse provides a lucid account of Popper's various discussions of evolutionary theory; see his "Karl Popper and Evolutionary Biology," in *Is Science Sexist? And Other Problems in the Biomedical Sciences* (Dordrecht: D. Reidel, 1981), 65–84.

28. Haeckel, *Natürliche Schöpfungsgeschichte*, , 158–79. Haeckel characterized indirect variability as due to differences in nutrition, which he generously conceived as "all of the trans-

so closely intertwined, Haeckel did not think one could determine which was the more important. In both instances, though, alterations were ultimately induced in the molecular structure of the heritable protoplasm of egg and sperm. And if the traits caused by the protoplasmic change proved successful, they and, consequently, their carriers would be preserved; that is, they would be naturally selected.

Again like Darwin, Haeckel conceived natural selection as introducing ever more progressive stages in the history of life: from simple monads in the far distant past through the higher metazoans of the present, natural selection had produced a progressive division of labor in organisms. This division of labor could be seen, he thought, even in the various levels of civilization achieved by human beings, from the simple cultures of primitive societies (e.g., that of the Papuans) to the complex cultures of Europe (e.g., that of the Germans). The effects of the division of labor could even be detected in the faces of individuals: "Among the branches of the lower tribes, most of the individuals look so alike in facial features that the European traveler usually cannot distinguish them."[29]

Natural selection, while the engine of progressive development, also showed a negative side. Haeckel, like Darwin, recognized that selection was a powerful reductive force operative on members of all species, including human beings. Were selection not constantly winnowing individuals, the world would lie several layers deep in human biomass; but, in fact, human populations, despite large numbers of children initially produced, actually grew rather slowly. The scythe of selection also meant that as less progressive human groups came into contact with the more progressive ones, the former would suffer. Haeckel thought this would be the case with the American Indians and the Australians, as more European settlers on their lands gradually pushed them into oblivion.[30]

In the second edition of the *Natural History of Creation*, completed

---

formations the organism undergoes by reason of the conditioning of the surrounding external world" (175).

29. Ibid., 228.

30. Ibid., 206. Richard Weikart, in "The Origins of Social Darwinism in Germany, 1859–1895," *Journal of the History of Ideas* 54 (1993): 469–80, tendentiously remarks of Haeckel's observation that he "condoned the extermination of 'primitive' races" (480). A prediction does not imply approval or recommendation. Moreover, Haeckel hardly despised the American Indians, as the passages below will indicate. Darwin himself made comparable predictions, since "when civilized nations come into contact with barbarians the struggle is short." See Charles Darwin, *The Descent of Man and Selection in Relation to Sex*, 2 vols. (London: Murray, 1871), 1:238. In later printings of his book, Darwin charted the demise of the Tasmanians and the decline of the Maoris, Hawaiians, and Australians. These "savage" societies simply could not compete with more civilized nations. See Charles Darwin, *The Descent of Man and Selection*

in the spring of 1870, Haeckel identified two forms of artificial selection that worked against the grain of natural selection, namely, "military selection" and "medical selection."[31] He became acutely aware of the first kind during the Austro-Prussian conflict; and his sensitivities as a new father heightened during the prelude to the Franco-Prussian War. That later conflict proved intensely bloody, if brief.

During the interval between the two editions of Haeckel's book, from 1868 through the early summer of 1870, the antagonisms between Prussia and France grew ever more intense. In late July 1870, the French finally took Bismarck's bait and declared war. It was a colossal miscalculation. They were ill-prepared to stand up against Prussia, and by the end of the year they would face utter defeat, with Paris left in ruins.[32] In early August,

---

in Relation to Sex, 2nd ed., with an introduction by James Moore and Adrian Desmond (1879; repr., London: Penguin, 2004), 211–22.

31. Ernst Haeckel, *Natürliche Schöpfungsgeschichte*, 2nd ed. (Berlin: Georg Reimer, 1870), 152–56.

32. In 1868, after a revolution in Spain, a Hohenzollern was offered the monarchy. The French government, led by Napoleon III, feared that with Germans on both sides of their country, the balance of power would be upset; and so it pressured Spain to withdraw the invitation. The candidate himself decided against taking the throne. The French did not simply accept this diplomatic victory but insisted that King Wilhelm of Prussia declare that no further efforts would be made to reinstate a Prussian prince on the Spanish throne. Wilhelm, then taking the waters at Bad Ems, refused this demand. He sent a telegram to Bismarck, telling the chancellor that he could inform the press, if he thought it necessary, that the impertinent French demand had been rejected. Bismarck released his own version of the Ems telegram, one that suggested the king had not only rejected a humiliating demand but had broken off diplomatic relations with France. The French, in turn, thought their national honor had been sullied. Bismarck readied his chief of staff, Helmuth Karl Bernhard, Graf von Moltke. On 19 July 1870, France declared war and put over a quarter of a million troops in the field. But Prussia had a considerably larger force under arms. On 4 August, the first significant clash occurred at Weissenburg (or as later, Wissembourg) in Alsace, just west of Karlsruhe across the Rhine. Exerting the force of their needle guns, the Germans drove hard against the enemy. Marshal MacMahon, commander of the French forces, had to fall back south to Wörth on the Sauer River. The Prussians, under Prince Friedrich Wilhelm (later Friedrich III of Prussia) routed the French, who numbered some eighty thousand in the field. The war continued to go downhill for the French. On 1 September, the decisive battle occurred at the city of Sedan, near the Belgian border (also the scene of the German breakthrough in 1940). Napoleon III and a hundred thousand of his troops were captured. The Prussians then moved on Paris. They bombarded, blasted, and starved the city; the inhabitants were compelled to eat rats to survive. The siege lasted until 28 January 1871, when Paris capitulated. (But in the spring the shattered city revolted against its own government; and the Paris Commune took power, staving off the old regime for two months.) As the result of the Prussian victory, Baden and Hesse elected to join in confederation with Prussia and the north German states; and after negotiations and some meaningless concessions on the part of Prussia, Bavaria also joined. Alsace and Lorraine were annexed to the confederation as part of the peace settlement with France. With the crowning of Wilhelm as emperor on 1 January 1871, unification of the Germanies became a reality. The liberal dream had turned into a nightmare, which, however, quickly melted away as a new sense of national power and cultural

during the three days of the initial engagements in Alsace, some twenty-five thousand men from both sides fell, either dead or grievously wounded. In September, a few weeks after the fierce battles, Haeckel visited the killing fields near the villages of Weissenburg and Wörth just across the Rhine, where the stench still lingered. He had his prognostication confirmed: military selection had drenched the countryside in precious human blood. Military selection thus occurred, according to the view advanced in the second edition of Haeckel's book, when societies sent their bravest and best to kill one another, while the less brave and weak remained behind to man the bedrooms. This would inevitably lead, in his estimation, to a slow downward slide of moral character and physical ability. The first part of that equation lay before him in Alsace.

Medical selection had a comparable effect. This occurred when physicians used their art, such as it was, to preserve infants who had serious, inherited diseases—syphilis, scrofula, retardation, and the like—the sorts of diseases Haeckel had treated as a young medical student in free clinics in Bavaria and Austria. The Spartans and American Indians, he believed, knew how to correct the momentary lapses of nature. Indeed, the eugenic practices of these natural men might be thought of as nature healing her own. Haeckel regarded the so-called humanitarians who decried this kind of eugenics to be hypocrites, since they very well tolerated the far greater evil of mass death during war. He took some consolation in the conviction that these artificial modes of selection—military and medical—would gradually succumb to the continued action of natural selection on human intelligence. Like Alfred Russel Wallace, Haeckel presumed that the pressures of natural selection had been largely removed from man's body to his mind. He cultivated the hope that "in the long run, the man with the most perfect understanding, not the man with the best revolver, would triumph . . . [and that] he would bequeath to his offspring the properties of brain that had promoted his victory."[33]

Haeckel's eugenic notions—which he would expand at the turn of the century in his book *Die Lebenswunder* (The wonder of life)—certainly burrow beneath the skin of our modern sensibilities. Some historians point to Haeckel's eugenic ideas as clearing the way for the moral hor-

possibility arose in the center of Europe. As Allmers wrote Haeckel in the fall, after lamenting the blood and sacrifice: "How will it [the Fatherland] appear in the glory of victory—from the outside, great and awe-inspiring and mighty, and from the inside, free and happy and united, so that all peoples far and wide should bow before the majesty of the German people." See Allmers to Haeckel (15 November 1870), in *Ernst Haeckel: Sein Leben, Denken und Wirken*, 2:56–57.

33. Haeckel, *Natürliche Schöpfungsgeschichte*, 2nd ed. (1870), 156.

ror of the Nazis.[34] It is worth pausing for a moment, though, to reflect on three features of his position. First, unlike the Nazis, Haeckel regarded the bloodlust of the military as running contrary to natural processes; they relied on the better revolver instead of the better brain. Second, one might have a hard time distinguishing between our contemporary tolerance for therapeutic abortion and his own more primitive solution to the problems of debilitating and degrading chronic disease. Finally, there is no evidence that Haeckel seriously advocated, as a workable policy, the kind of eugenic practice he mentioned. He placed his faith in the corrective hand of natural selection. These considerations may not, however, completely mitigate the acrid taste his remarks leave in the mouth of some modern readers. I will develop in more detail, in the second appendix to this volume, a set of principles that I believe should govern our moral judgments of historical figures, the kind appropriate for the historian to make.

Recapitulation and Its Distortions

From the period of his conversion to Darwinism to the end of his career, Haeckel became ever more convinced that the strongest proof of evolutionary theory lay in the threefold parallel of phylogeny (as represented in paleontological remains), ontogeny, and systematics.[35] The fossil record, though of inestimable value as direct evidence of descent, yet revealed many wide gaps through which substantial doubt might flow. Haeckel believed ontogenetic research helped to narrow our ignorance of earlier times—or at least provide some hints of ancient organisms not yet uncovered in the rocks. Thus the biogenetic law that ontogeny recapitulates phylogeny not only provided evidence for evolutionary theory; it served to unroll the "thread of Ariadne" by which a new path to the past might be followed. But as he insisted in the *Natural History of Creation,* caution was needed in following the trail: ontogeny had its own lacunae. Recapitulation would always be far from perfect because of what he termed the laws of "abbreviated

34. Three books stand out as promoting the argument that Haeckelian biology laid a treacherous path to the Nazis: Daniel Gasman's *The Scientific Origins of National Socialism: Social Darwinism in Ernst Haeckel and the German Monist League* (New York: Science History Publications, 1971); Jürgen Sandmann's *Der Bruch mit der humanitären Tradition: Die Biologisierung der Ethik bei Ernst Haeckel und anderen Darwinisten seiner Zeit* (Stuttgart: Gustav Fischer, 1990); and Richard Weikart's *From Darwin to Hitler: Evolutionary Ethics, Eugenics, and Racism in Germany* (New York: Palgrave Macmillan, 2004). Weikart, while marking out Haeckel as a chief culprit, traces the infection directly back to Darwin himself. I will confront these charges directly in the conclusion to this chapter and in the second appendix.

35. See, for example, Haeckel, *Natürliche Schöpfungsgeschichte,* 1st ed. (1868), 253.

inheritance" and "alternating adaptation."[36] The former stipulates that as adaptations build up over thousands of generations, they elide one another in the smaller space of ontogenetic representation. Thus some morphological structures present in phylogenetic history might simply fall out of ontogenetic development. The law of alternating adaptation indicates that new modifications in phylogenetic history might be introduced earlier or later in ontogeny, thus giving a skewed picture of descent.

In response to critics later on, Haeckel further elaborated the ways in which ontogeny might adequately mirror phylogeny—and also distort it. He undertook this task in the third edition (1877) of his *Anthropogenie; oder, Entwickelungsgeschichte des Menschen* (Anthropogeny; or, the developmental history of man), originally given as a series of lectures in the summer semester 1873. In the third edition of this book, Haeckel distinguished between a "palingenetic" recapitulation in which structures appearing in the embryo (e.g., initially two germ layers) accurately pictured those of ancestor organisms (e.g., the double germ layer of the gastraea), and a "cenogenetic" distortion of the original phylogenetic sequence.[37] These distortions would originate, according to Haeckel, because of special conditions of recent adaptations and thus would not represent the more ancient evolutionary sequence. So, for example, the yolk sac of the embryo, the allantois, the amnion, the chorion, and other features of embryonic existence would not represent structures of ancient adult organisms; they were, rather, adaptations to life in the egg or uterus. Haeckel further distinguished two types of cenogenetic distortions—those of place and those of time. Originally, for instance, male and female sexual organs stemmed from one of the two original germ layers; but in more advanced animals, the sex organs became rooted in the mesoderm—thus a "heterotopic" displacement. Haeckel regarded as an instance of "heterochrony" the early appearance in the human embryo of heart and brain, while in aboriginal creatures (i.e., primitive invertebrates) these would be entirely absent.

In his extensive review of the first edition of the *Natural History of Creation,* Thomas Henry Huxley assessed Haeckel's phylogenetic theories:

> In Professor Haeckel's speculations on Phylogeny, or the genealogy of animal forms, there is much that is profoundly interesting, and his suggestions are always supported by sound knowledge and great ingenuity.

36. Ibid., 166–67.

37. Ernst Haeckel, *Anthropogenie; oder, Entwickelungsgeschichte des Menschen,* 3rd ed. (Leipzig: Wilhelm Engelmann, 1877), 7–13.

> Whether one agrees or disagrees with him, one feels that he has forced the mind into lines of thought in which it is more profitable to go wrong than to stand still.[38]

Through the first half of the twentieth century, a great number of biologists concurred with Huxley. They would further explore the recapitulation hypothesis and advance a multitude of variations on the themes of palingenetic and cenogenetic development. Writing in the second half of the century, Stephen Jay Gould, an implacable foe of Haeckel, composed his own versions of these hypotheses.[39] The modern responses to Haeckel indicate the vitality of his ideas. Like Huxley, even Gould, despite himself, found it "more profitable to go wrong than to stand still."

Illustrations of the Biogenetic Law

The iconic feature of the several editions of Haeckel's *Natural History of Creation* is the set of illustrations of the biogenetic law. The images gave substance to the law in a more striking fashion than any abstract expression could, and those images would initiate the enormous controversies and charges of fraud that followed Haeckel all of his days—indeed, those charges have been renewed in our time both by religious fundamentalists and by orthodox biologists. In the first edition of the *Natural History of Creation,* Haeckel deployed three sets of illustrations to demonstrate one aspect of the biogenetic law, namely, that at early stages of embryonic development organisms displayed proportionately similar morphologies the more closely they were related in their phylogenetic histories. Thus, those organisms that have diverged from each other more recently in their evolutionary past will diverge in morphology later in their embryonic development. In his first set of images, he pictured a dog and human embryo at two stages of development (see fig. 7.5). The younger embryos are virtually identical, while the older have diverged, with the dog having a noticeably longer tail and smaller brain (but as wry a look as the human). He also pictured a turtle and chicken at comparable stages, showing their similarity.

Haeckel believed that ontogeny not only provided compelling evidence of evolutionary descent; it had distinct political and social implications.

38. Huxley, "The Natural History of Creation," 41.

39. See Stephen Jay Gould, *Ontogeny and Phylogeny* (Cambridge, MA: Harvard University Press, 1977), 209–66.

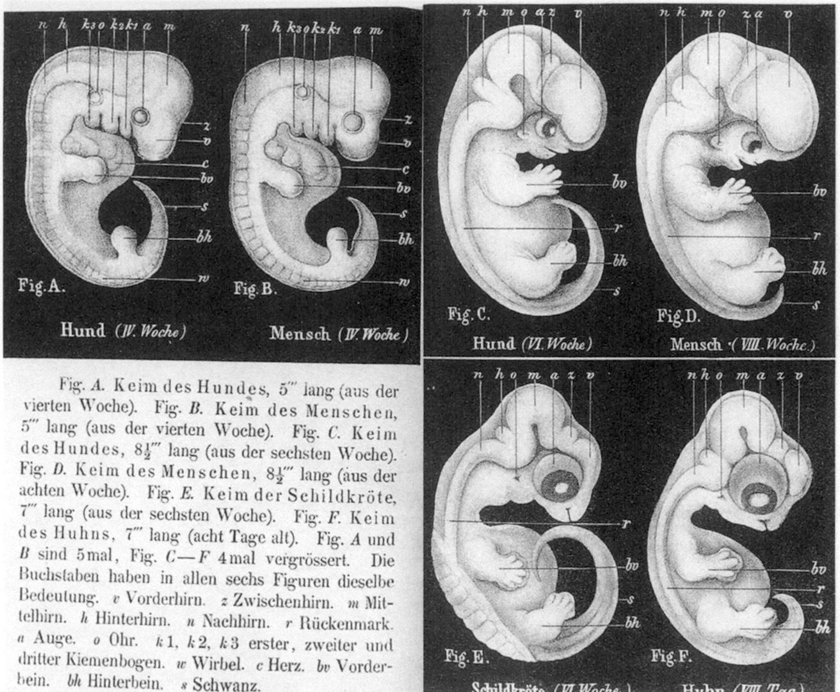

Fig. 7.5. Top row shows a dog and human embryo at two stages of development (at four weeks each and then at six and eight weeks, respectively); bottom row shows a turtle and chicken at a comparable stage (six weeks and eight days, respectively). (From Haeckel, *Natürliche Schöpfungsgeschichte,* 1868.)

He remarked that because of misconceptions about heredity, the aristocratic classes of society often presumed that they were of a different breed from middle- and lower-class folk. "What must these members of the nobility think about that blue blood [*Vollblut*] that rolls through those privileged arteries," he stingingly inquired, "when they learn that all human embryos, noble as well as middle class, during the first two months of development, can hardly be distinguished from the tailed embryos of a dog or other mammals?"[40] For Haeckel, the consequences of embryology leaped beyond the narrow boundaries of biology.

Haeckel's depictions of embryos have a rather schematic character—certainly the images in the early editions of the *Natural History of Creation* lack the intricacy and aesthetic grace of those in his radiolarian and medusa monographs. There are, I believe, two reasons for this. First, the

40. Haeckel, *Natürliche Schöpfungsgeschichte,* 1st ed. (1868), 240.

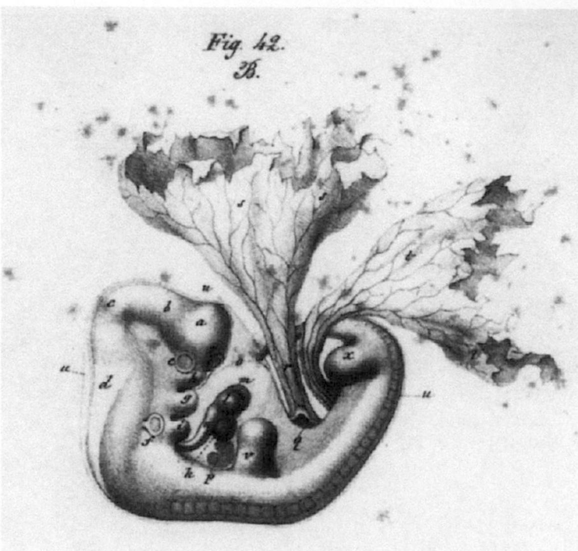

Fig. 7.6. Dog embryo. (From Theodor Bischoff, *Entwicklungsgeschichte des Hunde-Eies*, 1845.)

sketches originally were designed for large wall charts, which Haeckel used in his public lectures. Great intricacy and refinement would have had no practical value. The other reason is that Haeckel likely did not model his illustrations of vertebrates on actual embryos in his possession. Ludwig Rütimeyer (1825–1895) and Wilhelm His (1831–1904), who became dedicated enemies, claimed Haeckel had copied his images from published texts by Theodor Bischoff (1807–1882) and Alexander Ecker (1816–1887).[41] The dog embryo at four weeks does seem to have come from Bischoff's embryology of the dog (see fig. 7.6), and the human embryo at four weeks bears a strong likeness to Ecker's depiction of the human embryo (see fig. 7.7),[42] as well as to a cruder depiction by his former teacher Albert von Kölliker (see

41. Ludwig Rütimeyer, Review of "Ueber die Entstehung und den Stammbaum des Menschengeschlechts" and *Natürliche Schöpfungsgeschichte*, by Ernst Haeckel, *Archiv für Anthropologie* 3 (1868): 301–2; and Wilhelm His, *Unsere Körperform und das physiologische Problem ihrer Entstehung* (Leipzig: F. C. W. Vogel, 1874), 170.

42. Theodor Bischoff, *Entwicklungsgeschichte des Hunde-Eies* (Braunschweig: Friedrich Vieweg und Sohn, 1845), plate XI; and Alexander Ecker, *Icones physiologicae: Erläuterungenstafeln zur Physiologie und Entwickelungsgeschichte* (Leipzig: L. Voss, 1851–59), plates XXX, XXXI. Haeckel explicitly used the illustration by Bischoff (and so labeled it) in his *Anthropogenie; oder, Entwickelungsgeschichte des Menschen* (Leipzig: Wilhelm Engelmann, 1874), 271–72.

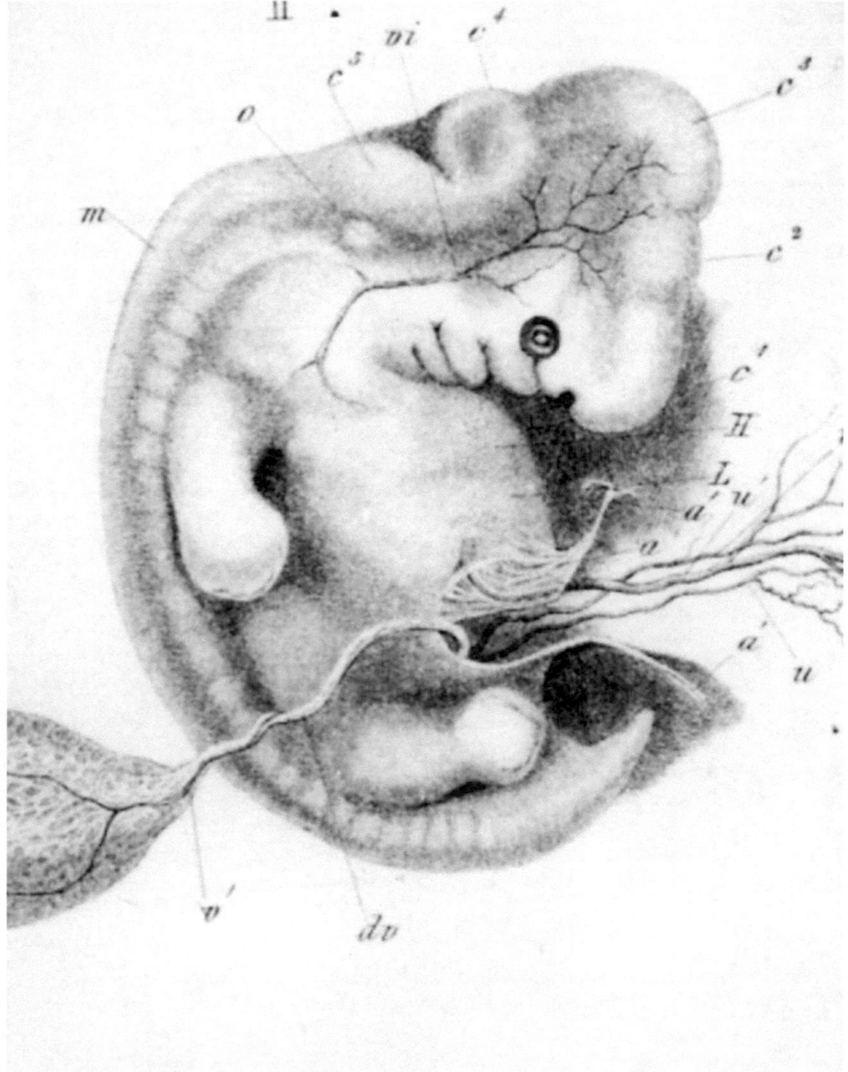

Fig. 7.7. Human embryo. (From Alexander Ecker, *Icones physiologicae,* 1851–59.)

fig. 7.8).[43] The human embryo at eight weeks may also have been modeled on a figure from Ecker—at least the sly smiles on the embryonic faces are similar (see fig. 7.9). In several of his later depictions of embryos in other

43. Albert von Kölliker, *Entwicklungsgeschichte des Menschen und der höheren Thiere* (Leipzig: Wilhelm Engelmann, 1861), 139.

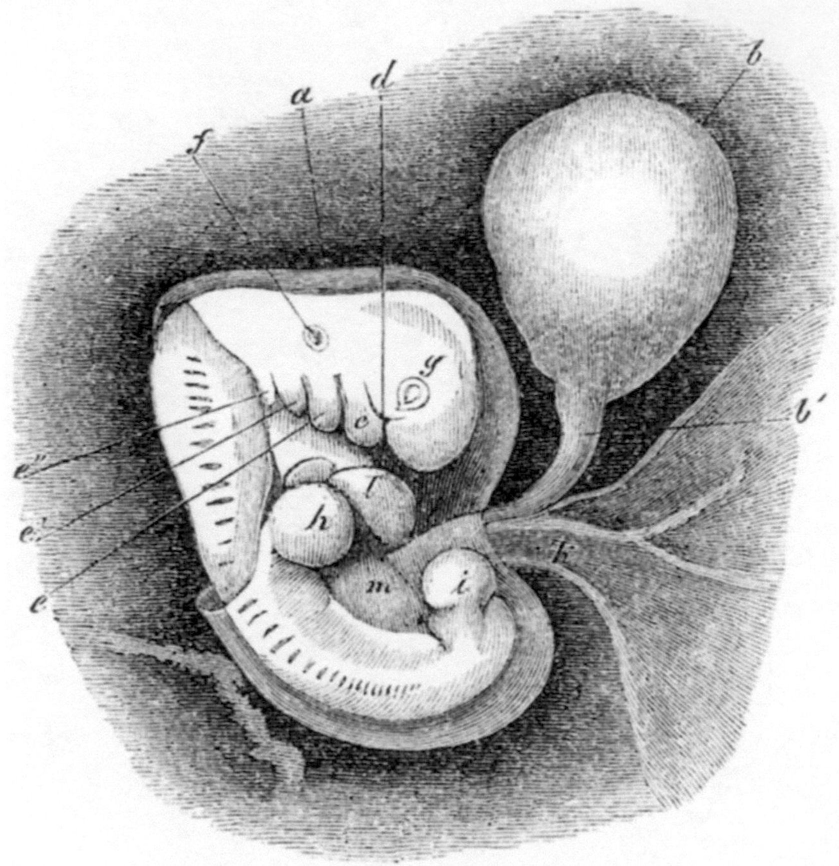

Fig. 7.8. Human embryo. (From Albert von Kölliker, *Entwicklungsgeschichte des Menschen und der höheren Thiere,* 1861.)

works, Haeckel explicitly indicated that he modeled them on published sources.

When Rütimeyer and His pointed out the similarity of Haeckel's embryos to those of other anatomists, they meant it as an indictment—though that was the mildest of their objections. (I will discuss their more serious complaints below and in the next chapter.) It was certainly common enough practice for one textbook writer to use the illustrations of another: Haeckel's mentor and teacher Kölliker, for instance, imported into his volume on human embryology the same illustration of the dog embryo by

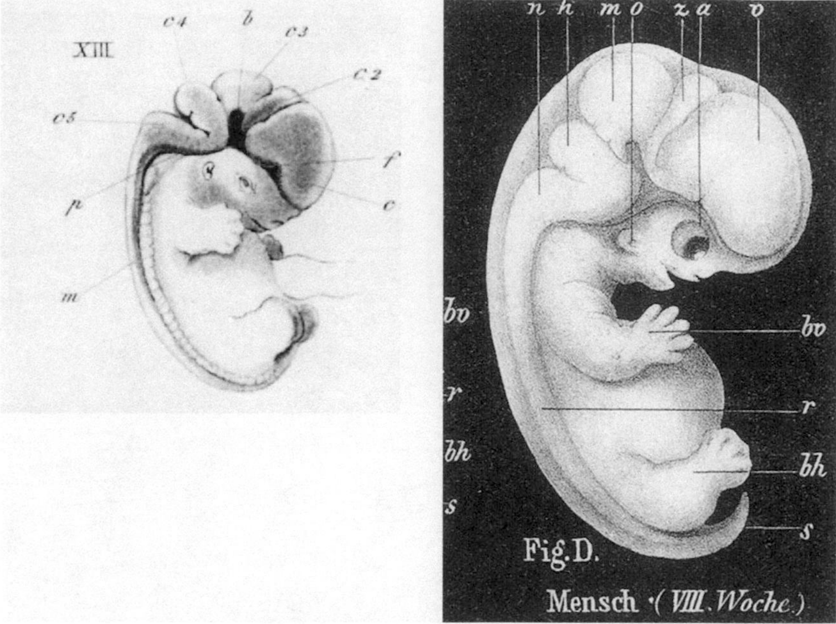

Fig. 7.9. Human embryos from Ecker's *Icones physiologicae* and Haeckel's *Natürliche Schöpfungsgeschichte.*

Bischoff that Haeckel apparently used.[44] Moreover, Kölliker based his image of the human embryo on one by the English anatomist Allen Thomson.[45] Kölliker, though, specifically labeled those illustrations as dependent on the work of other anatomists; Haeckel did not mention any sources for his illustrations in the *Natural History of Creation.* Haeckel's work, however, began as public lectures in which the emphasis was not the originality of depiction of specific embryos but the similarity of their morphologies.

In his *Anthropogenie* (1874), Haeckel greatly expanded his embryological depiction of the biogenetic law, making the row-by-row and column-by-column comparisons more dramatically telling (see fig. 7.10).[46] Yet in the series of editions of the *Natural History of Creation,* he retained the limited sequence of the first edition—though adding images of a turtle and

44. Ibid., 117.

45. Ibid., 130.

46. Nicholas Hopwood remarks that Haeckel's grid depiction of embryos is unprecedented in the literature and that the technique opened a greater space of embryological analysis. See his quite informative and balanced examination of the fraud charges against Haeckel, in his "Pictures of Evolution and Charges of Fraud: Ernst Haeckel's Embryological Illustrations," *Isis* 97 (2006): 260–391.

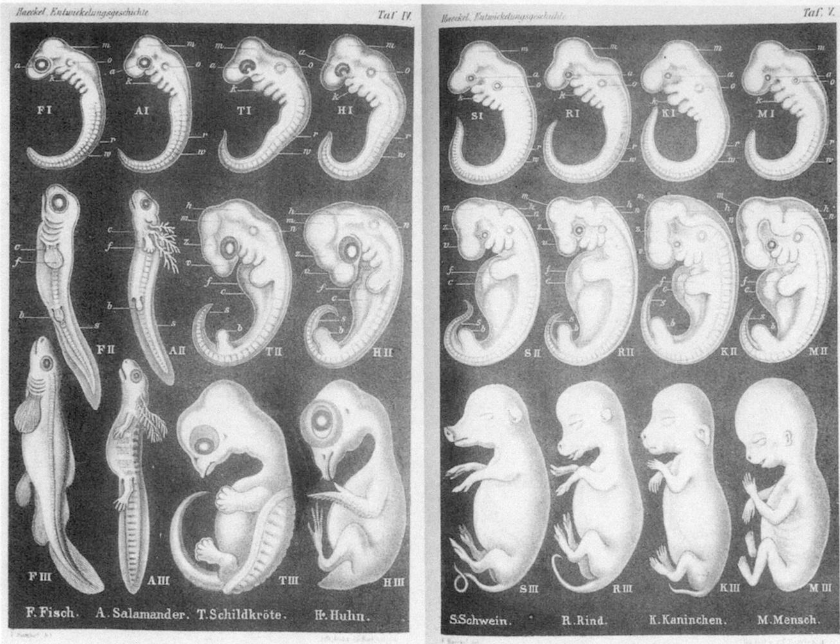

Fig. 7.10. Illustration of the biogenetic law. Vertebrate embryos at three stages of ontogenetic development, showing greater similarity in the earlier stages as evidence of a common ancestor. (From Haeckel, *Anthropogenie,* 1874.)

chick at the earliest stage—right through to the tenth edition (1902). In the *Anthropogenie,* he focused more tightly on human embryology; the whole character of the volume, while designed for an "educated" (*gebildeten*) audience, was much more technical than that of the *Natural History of Creation.* Its pages contain numerous detailed illustrations of isolated embryos at various stages of development. These illustrations, unlike those of the *Natural History of Creation,* almost always carry a tag indicating that they had been drawn after models produced by other authors. However, in an endnote referring to his large comparative illustration (fig. 7.10), he added: "The human embryos (at the third [I], fifth [II], and tenth week [III]) depicted in table five were drawn from very well-preserved preparations in spirits of wine. Most illustrations of human embryos from the first month are drawn from preparations that are in a poor state or defective."[47]

We need to keep in mind that the first edition of Haeckel's *Natural History of Creation* had its incarnation as a series of public lectures in which

47. Haeckel, *Anthropogenie,* 1st ed.(1874), 712.

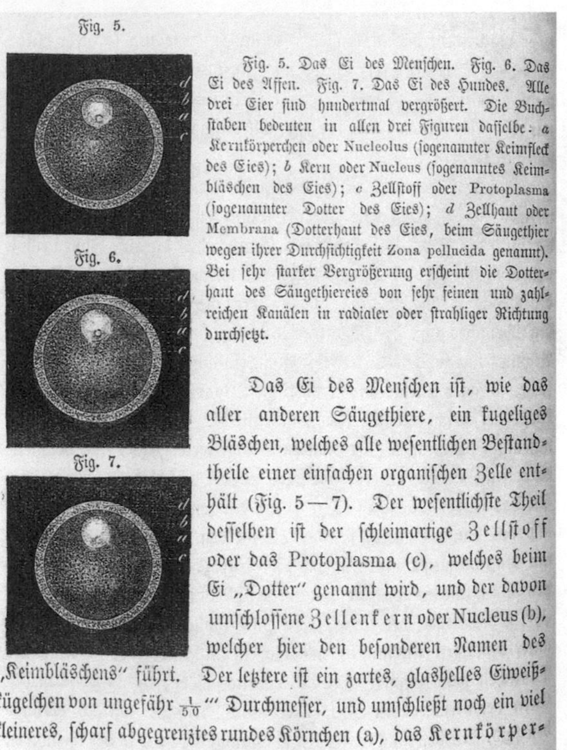

Fig. 5. Das Ei des Menschen. Fig. 6. Das Ei des Affen. Fig. 7. Das Ei des Hundes. Alle drei Eier sind hundertmal vergrößert. Die Buchstaben bedeuten in allen drei Figuren dasselbe: *a* Kernkörperchen oder Nucleolus (sogenannter Keimfleck des Eies); *b* Kern oder Nucleus (sogenanntes Keimbläschen des Eies); *c* Zellstoff oder Protoplasma (sogenannter Dotter des Eies); *d* Zellhaut oder Membrana (Dotterhaut des Eies, beim Säugethier wegen ihrer Durchsichtigkeit Zona pellucida genannt). Bei sehr starker Vergrößerung erscheint die Dotterhaut des Säugethiereies von sehr feinen und zahlreichen Kanälen in radialer oder strahliger Richtung durchsetzt.

Das Ei des Menschen ist, wie das aller anderen Säugethiere, ein kugeliges Bläschen, welches alle wesentlichen Bestandtheile einer einfachen organischen Zelle enthält (Fig. 5—7). Der wesentlichste Theil desselben ist der schleimartige Zellstoff oder das Protoplasma (c), welches beim Ei „Dotter" genannt wird, und der davon umschlossene Zellenkern oder Nucleus (b), welcher hier den besonderen Namen des „Keimbläschens" führt. Der letztere ist ein zartes, glashelles Eiweißkügelchen von ungefähr $\frac{1}{50}$''' Durchmesser, und umschließt noch ein viel kleineres, scharf abgegrenztes rundes Körnchen (a), das Kernkörper-

Fig. 7.11. Eggs of human, ape, and dog. (From Haeckel, *Natürliche Schöpfungsgeschichte*, 1868.)

the niceties of thorough reference could be omitted; and when Haeckel redacted the stenographic notes for publication, he would not have been preoccupied with furnishing detailed citations—especially since the publication was directed at an "educated" but not a professional audience. In other areas of his book, the history sections, for instance, the references were likewise of the most casual sort, again as befitting the genre of literature in which he was working. The other matter to keep in mind is that both Rütimeyer and His were determined opponents of the ingressions of evolutionary explanations into their own areas of embryology—a matter I will discuss in the next chapter. They did not object, therefore, merely to breaches in scientific decorum, even if they shrouded their complaints under that diverting cover.

Though the comparative plate of dog, turtle, chicken, and human being (fig. 7.5) would stick in the imagination (and craw) of both the educated and professional public, that illustration did not prove as damaging to Haeckel's reputation as two others (see figs. 7.11 and 7.12). These depict

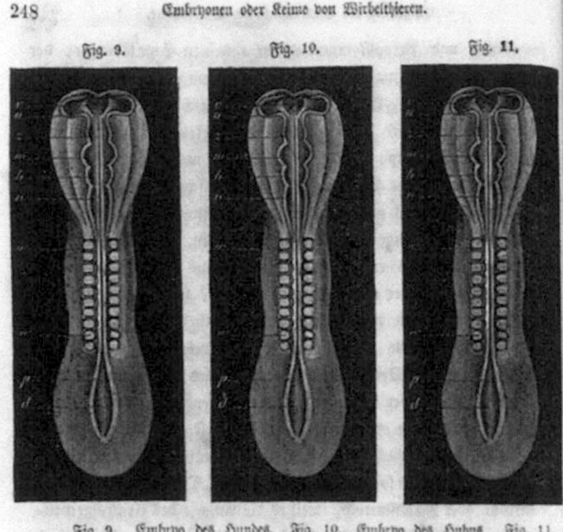

248 Embryonen oder Keime von Wirbelthieren.

Fig. 9. Fig. 10. Fig. 11.

Fig. 9. Embryo des Hundes. Fig. 10. Embryo des Huhns. Fig. 11. Embryo der Schildkröte. Alle drei Embryonen sind genau aus demselben Entwickelungsstadium genommen, in dem soeben die fünf Hirnblasen angelegt sind. Die Buchstaben bedeuten in allen drei Figuren dasselbe: *v* Vorderhirn. *z* Zwischenhirn. *m* Mittelhirn. *h* Hinterhirn. *n* Nachhirn. *p* Rückenmark. *a* Augenblasen. *w* Urwirbel. *d* Rückenstrang oder Chorda.

Die erste Blase, das Vorderhirn (v) ist insofern die wichtigste, als sie vorzugsweise die sogenannten großen Hemisphären, oder die Halbkugeln des großen Gehirns bildet, desjenigen Theiles, welcher der Sitz der höheren Geistesthätigkeiten ist. Je höher diese letzteren sich bei dem Wirbelthier entwickeln, desto mehr wachsen die beiden Seitenhälften des Vorderhirns oder die großen Hemisphären auf Kosten der vier übrigen Blasen und legen sich von vorn und oben her über die anderen herüber. Beim Menschen, wo sie verhältnißmäßig am stärksten entwickelt sind, entsprechend der höheren Geistesentwickelung, bedecken sie später die übrigen Theile von oben her fast ganz.

Bildung und Bedeutung der fünf Hirnblasen der Wirbelthiere. 249

(Vergl. S. 240 c, Fig. C—F.) Die zweite Blase, das Zwischenhirn (z) bildet besonders denjenigen Gehirntheil, welchen man Sehhügel nennt, und steht in der nächsten Beziehung zu den Augen (a), welche als zwei Blasen rechts und links aus dem Vorderhirn hervorwachsen und später am Boden des Zwischenhirns liegen. Die dritte Blase, das Mittelhirn (m) geht größtentheils in der Bildung der sogenannten Vierhügel auf, eines hochgewölbten Gehirntheiles, welcher besonders bei den Reptilien (Fig. E, S. 240 c) und bei den Vögeln (Fig. F) stark ausgebildet ist, während er bei den Säugethieren (C, D) viel mehr zurücktritt. Die vierte Blase, das Hinterhirn (h) bildet die sogenannten kleinen Hemisphären oder die Halbkugeln nebst dem Mitteltheil des kleinen Gehirns (Cerebellum), ein Gehirntheil, über dessen Bedeutung man die widersprechendsten Vermuthungen hegt, der aber vorzugsweise die Coordination der Bewegungen zu regeln scheint. Endlich die fünfte Blase, das Nachhirn (n), bildet sich zu demjenigen sehr wichtigen Theil des Centralnervensystems aus, welchen man das verlängerte Mark (Medulla oblongata) nennt. Es ist das Centralorgan der Athembewegungen und anderer wichtiger Functionen, und seine Verletzung führt sofort den Tod herbei, während man die großen Hemisphären des Vorderhirns (oder die „Seele" im engeren Sinne) stückweise abtragen und zuletzt ganz vernichten kann, ohne daß das Wirbelthier deßhalb stirbt; nur seine höheren Geistesthätigkeiten schwinden dadurch.

Diese fünf Hirnblasen sind ursprünglich bei allen Wirbelthieren, die überhaupt ein Gehirn besitzen, gleichmäßig angelegt, und bilden sich erst allmählich bei den verschiedenen Gruppen so verschiedenartig aus, daß es nachher sehr schwierig ist, in den ganz entwickelten Gehirnen die gleichen Theile wieder zu erkennen. Wenn Sie die jungen Embryonen des Hundes, des Huhns und der Schildkröte in Fig. 9, 10 und 11 vergleichen, werden Sie nicht im Stande sein, einen Unterschied wahrzunehmen. Wenn Sie dagegen die viel weiter entwickelten Embryonen in Fig. C—F mit einander vergleichen, werden Sie schon deutlich die ungleichartige Ausbildung erkennen, und namentlich wahr-

Fig. 7.12. Embryos of a dog, chicken, and turtle at the sandal stage. Text on lower right reads: "If you compare the young embryos of the dog, chicken, and turtle in figs. 9, 10, and 11, you won't be in a position to perceive a difference." (From Haeckel, *Natürliche Schöpfungsgeschichte*, 1868.)

embryos at the earliest stages of development—at the stage of the fertilized egg and at the so-called "sandal" stage (when the embryo appears similar to an hourglass or sole of a sandal). They became for Haeckel a much greater political and professional liability.

Haeckel probably modeled these illustrations of early embryonic development on figures in Kölliker's study, though the examples could have been drawn from a number of sources (e.g., Bischoff, from whose work Kölliker himself drew his sandal embryos).[48] Haeckel indicated that the morphological structures of the eggs of human, ape, and dog were virtually identical, though the sizes varied and, of course, the molecular structure that carried the determinants of inherited traits also differed. He put the issue to his readers: "If you compare the egg of the human (fig. 5) with

48. See Kölliker, *Entwicklungsgeschichte des Menschen*, 23, 78–79.

those of the ape (fig. 6) and the dog (fig. 7), you will not be able to perceive any difference." And concerning the sandal embryos, he reiterated: "If you compare the young embryos of the dog, chicken, and turtle in figs. 9, 10, and 11, you will not be in a position to perceive a difference."[49] In his short review of Haeckel's book, Rütimeyer agreed that one would not be able to perceive any differences in the eggs or in the sandal embryos—because in each instance Haeckel's printer had used the same woodcut three times! Rütimeyer sarcastically judged Haeckel to have played "a game of three-card Monte with the public and with science."[50] His likewise condemned Haeckel, contending that the evolutionist had "forfeited the right to be counted as a citizen of the republic of serious scientists."[51] These charges of fraud would haunt Haeckel throughout the rest of his career.

Haeckel had made a serious mistake, and he obviously knew it. In the second edition of his book, he used each print-block only once, and, referring to the illustration of the single "human" egg, declared: "The egg illustrated in fig. 5 could as well come from a dog, or a horse, or any other mammal as from a human or an ape." And after describing the structures of the sandal embryo (labeled that of "the embryo of a mammal or bird"), he asserted: "It is all the same whether we describe the embryo of a dog, chicken, turtle, or any of the other higher vertebrates."[52] These emendations could not completely eradicate the stain on his scientific reputation.

I will discuss the charges of fraud against Haeckel in the next chapter. But it may be well simply to observe here that he was largely correct: aside from the size of eggs or of embryos, at these very early stages the morphology of higher vertebrates appears essentially the same, at least as resolvable with mid-nineteenth-century equipment. The essential similarity of higher vertebrates at these early stages was never contested by Haeckel's enemies. Of course, his suggestion that the reader could use the illustrations as evidence, as opposed to devices of clarification, remained an error in judgment. It is hard to render a moral evaluation of Haeckel's misstatements, given the circumstances of a popular presentation and the fact that embryos at these early stages cannot be distinguished. In damage to his reputation, however, the error was very grave indeed.

With every succeeding edition of the *Natural History of Creation*, the

49. Haeckel, *Natürliche Schöpfungsgeschichte*, 1st ed. (1868), 241, 249.

50. Rütimeyer, "Review," 302.

51. His, *Unsere Körperform*, 171.

52. Haeckel, *Natürliche Schöpfungsgeschichte*, 2nd ed. (1870), 264, 270.

volumes grew in size, fattened with new discoveries and with larger numbers of illustrations, especially of the biogenetic law. These illustrations were often of creatures of which Haeckel had intimate knowledge, principally marine invertebrates. The second edition (1870), for instance, carried plates depicting different species of echinoderms (*Sternthiere*) at various developmental stages—from egg through larva to adult animal—and the various nauplius forms of different crustaceans. The fourth edition (1873), published just after Haeckel's large work on calcareous sponges, had as its frontispiece—and thus in pride of place—a depiction of the *olynthus* form of the sponge, the various stages of its development out of the gastrula, and finally its supposed colonial form that, he believed, gave rise to different genera and species of sponge (see the previous chapter for a discussion of Haeckel's work on sponges). In the fifth (1874) and all subsequent editions, Haeckel's own portrait, aging gracefully through the editions, served as frontispiece. One might even think of these transformations of his imposing figure as part of the story of the biogenetic law. All of these graphic representations would leave an immediate and strong impression on readers of the *Natural History of Creation*—as witnessed by the reproduction of his illustrations of the biogenetic law in scientific texts right up to the present time. The evocative and suggestive power of Haeckel's graphics made them the chief target of his scientific and religious enemies.

Construction of Stem-Trees of Genealogical Relationships

Prior to the *Origin of Species,* systematic accounts of organisms provided only a pragmatic ordering of species. These accounts were nothing more, according to Haeckel, than a kind of name registry or index of similarity. Darwin furnished the tools to construct something of much greater value, namely, genealogical trees depicting real blood relationships, stem-trees (*Stammbäume*) of descent. Haeckel introduced into the professional biological literature the graphic device of the stem-tree.[53] His *Generelle Morphologie* includes eight plates depicting stem-trees of plant, animal, and protist groups, both extant and extinct. The first edition of his *Natural History of Creation* also carries eight stem-tree plates but with a shift toward more detailed representations of the higher animals; the last plate illustrates the stem-tree of human species in their hierarchy of descent, with Papuans, Hottentots, and Australians (and their respective races) sit-

53. There had been some vague precedents for Haeckel's trees, which I describe at the end of chapter 5.

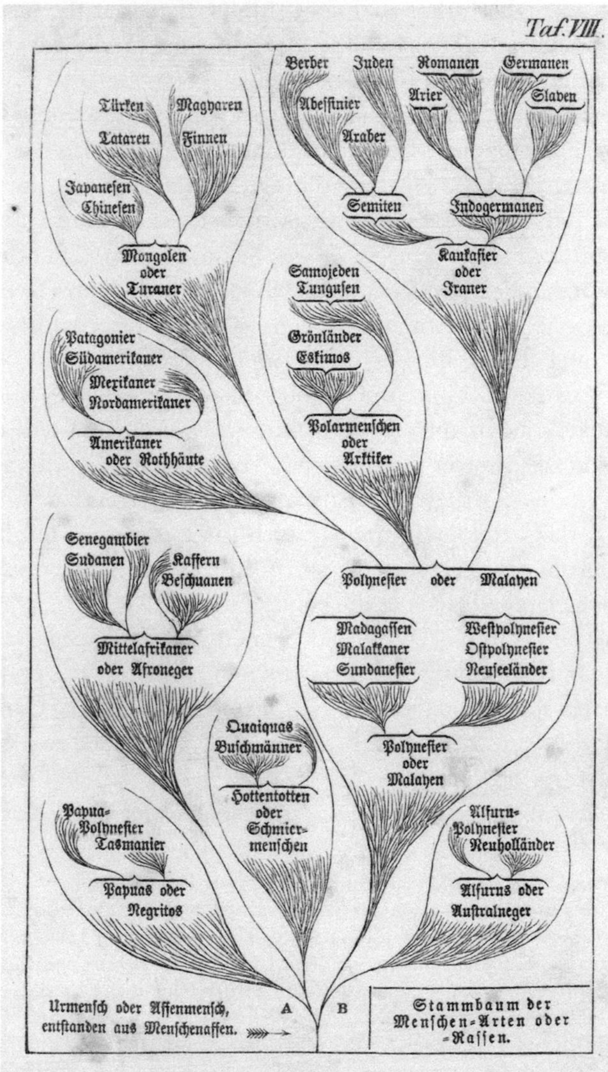

Fig. 7.13. Stem-tree of the human species or races.
(From Haeckel, *Natürliche Schöpfungsgeschichte*, 1868.)

ting on the bottom branches and Caucasians (with their several varieties) on the top (see fig. 7.13). Haeckel meant vertical position in the tree to indicate the level of progressive advance attained by the various species and races. For different reasons, perhaps, neither his nineteenth-century readers nor we would be surprised to see the Germans and Greco-Romans, among the Caucasian races, at the "pinnacle" (*Spitze*) of the human species.[54] But readers, both then and now, might wonder at the placement of the Jews and Berbers. He located them at the same highly developed level as the Germans and within the same species.[55]

Undoubtedly Haeckel's firsthand experience with Jews and Berbers led to his positive estimation of their development. The Berlin of his student years had a significant Jewish population, with a high concentration not far from the university, along Oranienburgerstrasse, where Eduard Knoblauch's magnificent New Synagogue was begun in 1859;[56] and the medical school at Würzburg also had a few Jewish students, whom Haeckel knew. Further, he came into intimate association with Berber and Jewish merchants and intellectuals during his excursions along the coast of Morocco in the spring of 1867.[57] This kind of cultural contact played a considerable role in Haeckel's judgments about evolutionary advance. I will discuss in a later section the criteria that he explicitly employed to situate the various human groups in the evolutionary hierarchy, criteria that "objectively" placed the Berbers and Jews on a par with the Europeans. But here I will simply point out some interesting shifts in the positions of the races as depicted in the later editions of the book.

In the second edition (1870), Haeckel increased the species of mankind from nine (ten, if you count the *Urmensch*) to twelve (or thirteen with the *Urmensch*). He split the Afers into two species, namely, the Cafers (South Africa) and the Negroes (area of the Sudan). He then added two groups: the Dravidians (parts of India and Ceylon—i.e., Sri Lanka) and the Nubians (middle of Africa). Aside from expanding the distinct species of mankind

54. Haeckel, *Natürliche Schöpfungsgeschichte*, 1st ed. (1868), 519.

55. Stem-trees in subsequent editions of the *Natürliche Schöpfungsgeschichte* place the Jews just a bit behind the Indo-Germans. In the text of the first edition, Haeckel does say that it is from "the Indo-German branch that the most highly developed cultural peoples spring." This, he claims, is based on the evidence of comparative linguistics as shown by August Schleicher. See ibid., 520.

56. The New Synagogue, completed in 1866, suffered only relatively minor damage during Kristallnacht (9–10 November 1938) because one police commander, Wilhelm Krützfeld, kept the crowds at bay. The building was heavily damaged during Allied bombing of the city. During the decade after the Berlin Wall came down, the synagogue was restored to its previous splendor.

57. In Mogador, for instance, the population consisted of about one-third Arab Berbers, one-third Jews, and one-third black Africans and individuals of mixed race. Haeckel and his friends stayed in the Jewish quarter, though they also visited the Arab bazaars and mosques. He attended two extremely colorful Orthodox Jewish weddings and made sketches of the participants in the celebrations. He also went to several different synagogues on Shabbat. Through the American Consul, who was Jewish, he was invited to many such events. See Haeckel's "Diary," in *Berg und Seefahrten: 1857/1883* (Leipzig: K. F. Koehler, 1923), 73–78.

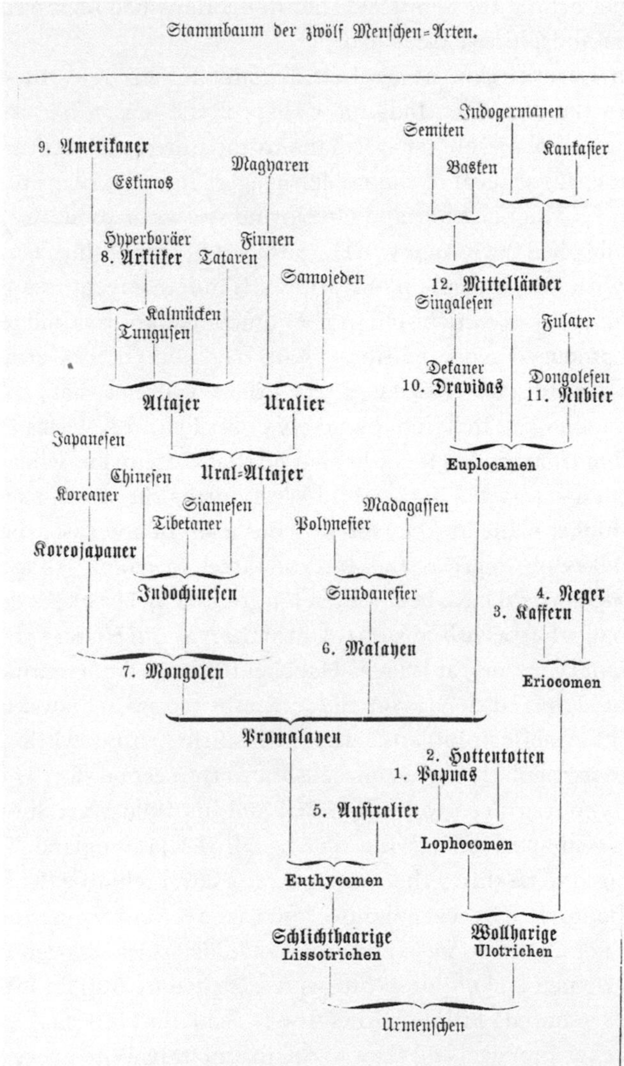

Fig. 7.14. Stem-tree of the human species. (From Haeckel, *Natürliche Schöpfungsgeschichte,* 2nd ed. to 7th ed., 1870–79.)

in the second edition, Haeckel changed significantly his estimate of their evolutionary advance (see fig. 7.14). Among the Midlanders (the Caucasians of the first edition), the Indo-Germans were still in the lead, with the Semites second and only a little behind. But with an unexpected evolutionary surge, the American Indians had advanced to second place among the species (and just behind the Semites). The Mongolians had regressed, at least the Japanese and Chinese races had.

Though Haeckel gave no explicit account for his re-estimate of the character of the American Indians, I suspect the reason had to do with what became an obsession for Germans in the nineteenth century and remains so today: the ideal of the noble savage. From the 1870s to the turn of the century, over one thousand novels and stories of an idealized Indian life were published in Germany.[58] The summit of this enthusiasm came in the 1890s with the publication of Karl May's Indian adventures *Winnetou* and *Old Shatterhand,* which sold in phenomenal quantities and today continue to be produced in new editions. Many a contemporary German father still hands over to sons—and daughters—these volumes that gushed from the imagination of an individual who never set foot in America (and spent a good deal of time in jail). It could also be that Gegenbaur, whose favorite uncle Joseph had traveled in the Wild West during the 1830s, suggested to Haeckel a higher plane for the Native American. In any case, the austere, natural nobility of American Indians and Greek Spartans—that other legendary group—would have been taken for granted by Haeckel's audience.

Up through the seventh edition (1879) of the *Natural History of Creation,* the evolutionary picture, at least as Haeckel illustrated it, remained stable. But with the eighth edition (1889), the scene shifted again dramatically (see fig. 7.15). The American Indians fell significantly behind, while the Mongolians, and particularly the Japanese, shot to the second slot. The Indians may have been sacrificed to Buffalo Bill and his Wild West Show, which played for a year and a half (spring 1886 to fall 1887) in England. The attendance averaged some thirty thousand people a day, including the kings and queens of Denmark, Greece, Belgium, and Saxony. At every performance of the "Drama of Civilization," the savage Indians attacked a cabin holding frightened women and children but were defeated by Buffalo Bill and his troops. In 1889 Buffalo Bill took his show to the Paris Universal Exhibition (which Haeckel attended) and then to Spain and Italy. With every show the progressive values of the higher race triumphed over the lower.

The Japanese likely got a boost in the evolutionary race because during the ten-year interim between the seventh and eighth editions of Haeckel's book, Japan, under the restored Meiji emperor, began a defensive effort

58. This bibliographic survey was conducted by Christian Feest and reported in his "Germany's Indians in a European Perspective," *Germans and Indians: Fantasies, Encounters, Projections,* ed. Colin Calloway, Gerd Gemünden, and Susanne Zantop (Lincoln: University of Nebraska Press, 2002), 37.

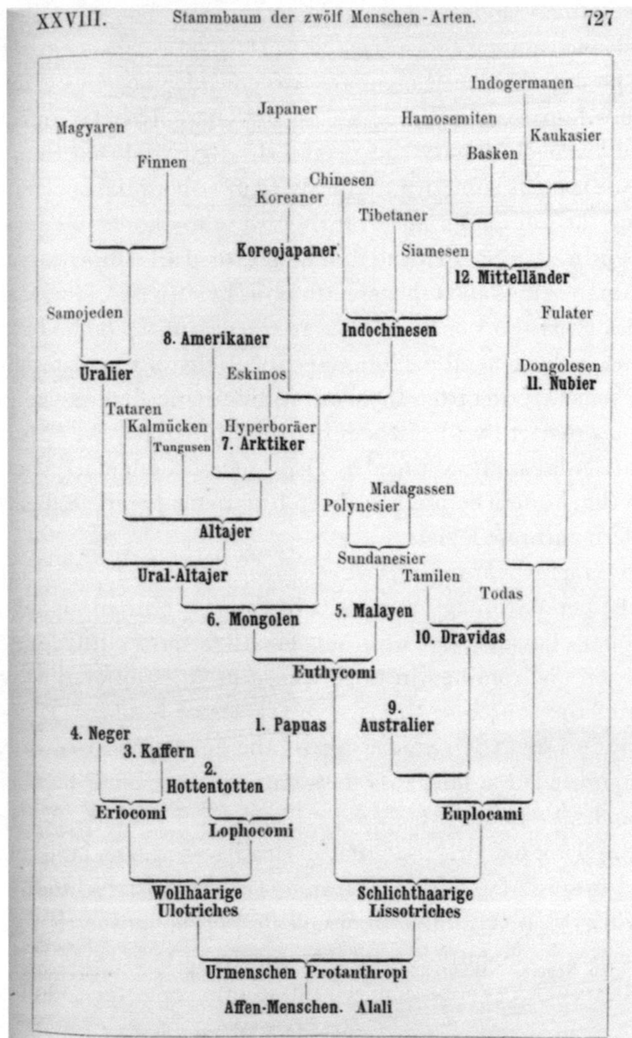

Fig. 7.15. Stem-tree of the human species. (From Haeckel, *Natürliche Schöpfungsgeschichte*, 8th ed. to 12th ed., 1889–1920.)

at modernization. Part of that effort resulted in Japanese translations of Haeckel's *Anthropogenie* in 1882 and his *Natürliche Schöpfungsgeschichte* in 1888.[59] A new constitution for the Japanese nation was constructed explic-

59. Haeckel's *Anthropogenie* was translated as *Seirigaku* (Physiology) by Tai Hasegawa in 1882 and his *Natürliche Schöpfungsgeschichte* by Yamagata in 1888. The latter appears to have preferred the English nature poets to the German, since the translation is prefaced with a poem

itly on the German model and promulgated in 1889. These intellectual and political advances may have suggested to Haeckel that his earlier estimation of the potential of the Mongolians was considerably shy of the reality.

Haeckel's European audience would undoubtedly have found it comforting that the evolutionary race was to the swift and that they were the swift. For us, there is something slightly risible about his accelerating and decelerating the human groups as if they were toy horses in an imagined derby. Yet one must keep in mind that he regarded all of his *Stammbaüme,* those of human beings and other creatures, as hypotheses, emendable when the evidentiary patterns changed.[60] From the beginning he had good scientific reason for the general evolutionary hierarchy of the different human groups, reasons deriving from the work in linguistics by his friend August Schleicher—a matter I will consider below. While today we would not accept the particular analyses urged by Haeckel and Schleicher, their general conviction that human beings have been linguistically created is shared by many modern anthropologists.

Another graphic feature of Haeckel's *Natural History of Creation* that attracted the popular imagination was his map of human dispersal out of an evolutionary Paradise (see fig. 7.16). The illustration, introduced in the second edition and retained in the succeeding ones, offered a hypothetical sketch of the source of the twelve species of humans, their various transformations, and their spread across the earth. The map located the origin of mankind in a land called "Lemuria," so named by the English zoologist and secretary of the London Zoological Society, Philip Sclater (1829–1913). Sclater had proposed that in the distant past a land connection had existed joining Africa, Madagascar, and South Asia—a landmass occupying roughly the area of the Indian Ocean.[61] He supposed that geological processes had subsequently caused this continent to subside into the sea. The hypothesis of a lost continent would explain some very odd features of

---

by Wordsworth. I am grateful to Gerard Clinton Godart for informing me about these translations. Haeckel received a request to have an excerpt of his *Welträthsel* published in Japan. C. Ishikawa to Haeckel (20 January 1904), in the Correspondence of Ernst Haeckel, the Haeckel Papers, Institut für Geschichte der Medizin, Naturwissenschaft und Technik, Ernst-Haeckel-Haus, Friedrich-Schiller-Universität, Jena. The translation was published in 1906.

60. For an explicit statement of his view about the hypothetical nature of his stem-trees, see his *Systematische Phylogenie,* 3 vols. (Berlin: Georg Reimer, 1894–96): "Obviously our stem-history is and remains a hypothetical construction, just like its sister historical geology. For it seeks to achieve a synthetic insight into processes and causes of events long since passed, the direct research into which is impossible for us" (1:vi).

61. P. L. Sclater, "The Mammals of Madagascar," *Quarterly Journal of Science* 1 (1864): 212–19.

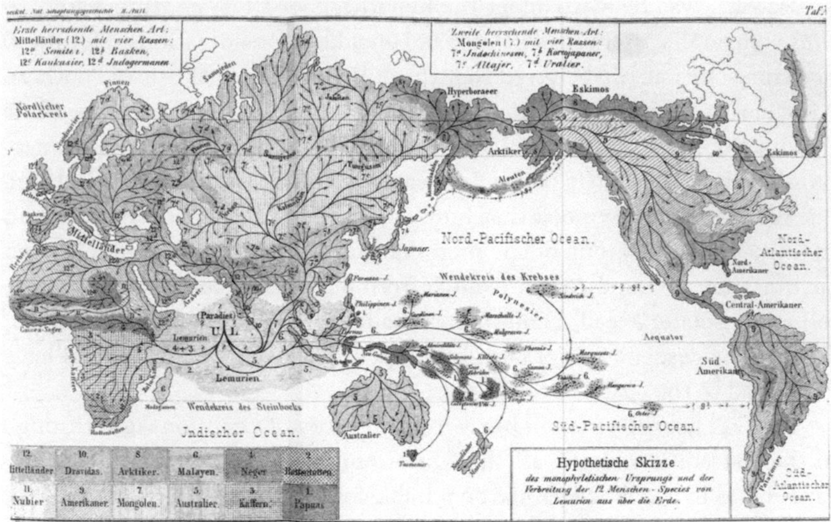

Fig. 7.16. Hypothetical sketch of the origin of the human races in U.L. (unbekantes Land) or Paradise—also Lemuria—and their dispersal over the world. (From Haeckel, *Natürliche Schöpfungsgeschichte,* 2nd ed., 1870.)

the biogeography of these countries. Some thirty species (of eight genera) of lemur exist on Madagascar and some eleven species in Africa, while three different species exist on the Indian subcontinent. One would have expected the fauna on the island of Madagascar to be comparable with that of the African mainland, but certainly unlike that of India. Other oddities of biogeography followed a similar pattern (e.g., the distribution of certain orders of bats, hedgehogs, shrews, and others). These patterns could be explained if "a large continent," Lemuria, had joined the island with Africa and India. Sclater's reasoning was powerful enough to gain the approval of Alfred Russel Wallace, who endorsed the hypothesis in his great two-volume work *The Geographical Distribution of Animals* (1876).[62] Today we recognize that southern Africa, Madagascar, and India were, indeed, joined together; not, however, by a land bridge but via the movement of tectonic plates that formed the massive continent of Gondwanaland, which broke up about 160 million years ago. It is likely that some fauna, like lemurs, spread from Africa to Madagascar and India when these land masses were still in some proximity.

62. See Alfred Russel Wallace, *The Geographical Distribution of Animals,* 2 vols. (New York: Harper and Brothers, 1876), 1:76–77.

Haeckel was ready to adopt the hypothesis of a lost continent for reasons comparable to those of Sclater and for added considerations relevant to human evolution. He believed that the dolichocephalic skulls of southern Africans conformed to the morphology of the great apes in the region, while the brachycephalic skulls of Eastern groups (e.g., Dravidians and Mongolians) fit the morphology of the Asian apes—the orangutan and the gibbon. But he was also convinced that human species had a monogenetic origin in some primitive group of ape-men that forged the missing link between apelike ancestors and African and Eastern humans.[63] Hence, he supposed, following Sclater's lead, a common "Paradise," as he sardonically called it, whence the descendants of both early apes and early ape-men would have decamped to the east and to the west.

Haeckel kept the map in his *Natural History of Creation* right through to the last edition (1920), all the while reminding his readers that he regarded it as hypothetical, since no reliable fossil trail, at least prior to 1890, provided a more secure foundation for his biogeographical speculations. Yet this very graphic portrayal of man's early history captured the imagination of the public, especially as the assumption of a lost continent echoed the old myth of Atlantis. Not all naturalists were quite ready to sign on to the hypothesis, however. Darwin, for instance, had his doubts, which he expressed to Wallace.[64] As the hypothesis waned in plausibility (but gained adherents like Madame Blavatsky),[65] Haeckel, in the eighth edition of his book (1889), deleted that portion of the map depicting the sunken continent. In his subsequent comments on the map, he mentioned the possibility of Lemuria but suggested that mankind might have had two separate roots, one in Africa and one in Southeast Asia.

Haeckel's speculations were taken to heart by the Dutch army doctor Eugène Dubois (1858–1940), who had studied with Haeckel's student Max Fürbringer (1846–1920) and who had become an enthusiastic Darwinian.

63. Haeckel's belief in a single root for the human races in one original group is grounded in a sound biological supposition: interbreeding among the races shows historical kinship.

64. Darwin thought the hypothesis of Lemuria too large for the small evidence upon which it rested. See Darwin to Wallace (17 June 1876), in *More Letters of Charles Darwin*, ed. Francis Darwin, 2 vols. (London: Murray, 1903), 2:15.

65. Madame Blavatsky (Helena Hahn), the founder of theosophy, said she read of Lemuria in a book written in Atlantis, a continent built out of the remains of Lemuria. See her *Secret Doctrine*, 2 vols. (London: Theosophical Society, 1888), 2:323–24, 333. The complex history of the Lemurian hypothesis, as it moved from biogeographical speculation to occult standing, is expertly related by Sumathi Ramaswamy, in *The Lost Land of Lemuria: Fabulous Geographies, Catastrophic Histories* (Berkeley: University of California Press, 2004).

With a grant from his government, Dubois sought the remains of the missing link in one of the locations that Haeckel had predicted—in south Asia. He explored the region around Trinil on the island of Java and, remarkably, found the fossils for which he went searching.[66] He called this ancient apeman *Pithecanthropus erectus* in honor of Haeckel's own designation.[67] The skullcap, femur, and two molars he uncovered (from November 1890 to August 1892) are what we now classify as *Homo erectus* remains—certainly that group of anthropoids having the best claim on the sobriquet "*fehlendes Glied*"—"missing link." Haeckel celebrated and defended Dubois's discovery in the tenth (1898) and subsequent editions of the *Natural History of Creation* and in a lecture delivered at the International Zoological Congress, held at Cambridge University in 1898. Before his Cambridge audience, Haeckel observed:

> The opponents of descent theory and its application to human beings have henceforward been robbed of their favorite objection. They will have to cease referring to the alleged "missing link," since this "missing link between ape and man" lies in the fossil remains of the *Pithecanthropus erectus* right before their eyes. One can thus maintain that this discovery of Dubois holds a greater significance for anthropology than the celebrated discovery of "Röntgen rays" for physics.[68]

Rudolf Virchow, one of the main opponents to whom Haeckel referred, did not retire so easily before the evidence. The great medical anthropologist countered that the skullcap and femur found by Dubois came from

66. See Eugène Dubois, *Pithecanthropus erectus, eine menschenähnliche Übergangsform aus Java* (Batavia: Landesdruckerei, 1894); and "Pithecanthropus Erectus—A Form from the Ancestral Stock of Mankind," *Annual Report, Smithsonian Institution* (1898): 445–59. See also Erika Krauße, "Pithecanthropus erectus Dubois (1891) in Evolutionsbiologie und Kunst," in *Evolutionsbiologie von Darwin bis heute,* ed. Rainer Brömer, Uwe Hoßfeld, and Nicolaas Rupke (Berlin: Verlag für Wissenschaft und Bildung, 2000), 69–88.

67. Eugène Dubois wrote Haeckel (24 December 1895) in thanks for "the inspiration you provide, especially by your Schöpfungsgeschichte, on my entire direction in life. I remember quite well the deep impression which the reading of your book had made on me almost 20 years ago." The letter is in the Haeckel Correspondence, Haeckel-Haus,, Jena.

68. Ernst Haeckel, *Ueber unsere gegenwärtige Kenntnis vom Ursprung des Menschen: Vortrag gehalten auf dem Vierten Internationalen Zoologen-Congress in Cambridge, am 26. August 1898* (Stuttgart: Alfred Kröner, 1905), 47. Haeckel gave his lecture in English but subsequently published a German edition. Dubois thanked Haeckel in a letter (16 December 1898) right after the congress. The letter is held in the Haeckel Correspondence, Haeckel-Haus, Jena.

Fig. 7.17. *Pithecanthropus alalus,* ape-man without speech. (From Haeckel, *Natürliche Schöpfungsgeschichte,* 10th ed. to 12th ed., 1902–20.)

two different creatures, the skullcap from an ape and the femur from a human being.[69] But by the turn of the century, with an accumulation of Neanderthal remains from several locations, Haeckel's old nemesis sounded

69. Benoit Massin discusses Virchow's anti-Darwinist position in his "From Virchow to Fischer: Physical Anthropology and 'Modern Race Theories' in Wilhelmine Germany," *History of Anthropology* 8 (1996): 79–154, especially 114–17. The rejection, while in the context of a quite positive reception, yet stung, as Dubois wrote to Haeckel (24 December 1895): "With great joy I see that Pithecanthropus is already recognized by the majority of researchers for what it really is, a transitional form [*Uebergangsform*] and that those who are against this are

as though he had a fixed conclusion looking for an argument. Not only was the evidence for primitive man quite strong; Haeckel even included a portrait of *Pithecanthropus alalus*—compliments of the Munich artist Gabriel von Max (1840–1915)—in the last three editions of the *Natural History of Creation* (see fig. 7.17). Who could doubt the existence of that contented-looking burgher family?

The Linguistic Creation of Man

Haeckel organized the stem-trees of human descent in the twelve editions of the *Natural History of Creation* according to several implicit and explicit criteria. We have just examined some of the implicit cultural assumptions—for instance, beliefs about the noble savage and so on—that occasionally injected enough force to produce some rearrangements of the stem-trees. The explicit criteria seem more objective and stable, though it takes but little dissecting to expose the cultural sinews structuring their particular contours. Haeckel employed some rather traditional, presumptively objective criteria for classifying human groups: skin color, hair texture, and skull shape. He charted the initial descent of two groups from the *Urmensch:* the woolly-haired, dark-skinned, and dolichocephalic men—the African groups; and the smooth-haired, lighter-skinned, and brachycephalic men—the Mongolians, Malayans, and ultimately the descendant Europeans (whose skulls tended to be mesocephalic). Haeckel recognized that these were ideal types, with considerable variation of skin color, hair texture, and skull shape characterizing individual specimens within the groups.

Haeckel believed that the aesthetics of human morphology allowed a hierarchical placement of the races, but he also acknowledged the uncertain and variable nature of that particular criterion of ordering. He knew the ultimate standard of progressive development of the races of mankind had to be mental quality. But then, how objectively to calibrate high-mindedness? His friend August Schleicher pointed the way.

In the early 1860s, Haeckel had encouraged Schleicher to read the German edition of the *Origin of Species,* thinking it would appeal to his friend's horticultural interests.[70] But it was Schleicher the linguist who res-

very weak in their attacks. I now still remember hearing the words of Virchow that he could not accept my views." The letter is in the Haeckel Correspondence, Haeckel-Haus, Jena.

70. Schleicher was indeed a serious gardener and wrote a review of the *Origin* for an agricultural journal. See August Schleicher, "Die Darwin'sche Theorie und die Thier- und Pflanzenzucht," *Zeitschrift für deutsche Landwirthe* 15 (1864): 1–11. In the review Schleicher sum-

onated more deeply to Darwin's work. He responded to Haeckel in an open letter, which he published as a small tract with the title *Die Darwinsche Theorie und die Sprachwissenschaft* (Darwinian theory and the science of language, 1863).[71]

In the book Schleicher maintained that contemporary languages had gone through a process in which simpler *Ursprachen* had given rise to descendant languages that obeyed natural laws of development. He argued that Darwin's theory was thus perfectly applicable to languages and, indeed, that evolutionary theory itself was confirmed by the facts of language descent. This last point was crucial for Schleicher, since it suggested the singular contribution that the science of language could make to the establishment of Darwin's theory. As we have already seen (chapter 3), Heinrich Bronn, translator of the *Origin,* maintained that Darwin's argument only showed that descent was *possible* but not that it was *actual.* Darwin had, according to Bronn, no direct empirical evidence for his theory, only analogical possibilities.[72] Haeckel judged that Darwin had advanced powerful *synthetic* evidence—embryology, biogeography, systematics, and so on—for descent; but he agreed with Bronn that direct *analytic* evidence was wanted. This he believed he had provided in his work on sponges. Schleicher likewise thought he could furnish direct evidence for Darwin's

---

marized Darwin's argument and added elements that he undoubtedly thought rounded out the theory, including the suggestion that human beings descended from the "higher apes" and differed from them only by reason of language and "high brain development" (6). Schleicher neglected to mention that Darwin himself did not discuss human evolution in the *Origin.* Schleicher sent this review to Darwin, and it is now held in the Manuscript Room of Cambridge University Library. Scorings indicate Darwin read the review.

71. August Schleicher, *Die Darwinsche Theorie und die Sprachwissenschaft* (Weimar: Hermann Böhlau, 1863). I have discussed the origin of Schleicher's theories in "The Linguistic Creation of Man: Charles Darwin, August Schleicher, Ernst Haeckel, and the Missing Link in Nineteenth-Century Evolutionary Theory," in *Experimenting in Tongues: Studies in Science and Language,* ed. Matthias Doerres (Stanford, CA: Stanford University Press, 2002). See also Liba Taub, "Evolutionary Ideas and Empirical Methods: The Analogy between Language and Species in Works by Lyell and Schleicher," *British Journal for the History of Science* 26 (1993): 171–93; and Stephen Alter, *Darwinism and the Linguistic Image* (Baltimore: Johns Hopkins University Press, 1999), especially 73–79.

72. Heinrich Bronn, "Schlusswort des Übersetzers," in Charles Darwin, *Über die Entstehung der Arten im Thier-und Pflanzen-Reich durch natürliche Züchtung; oder, Erhaltung der vervollkommenten Rassen im Kampfe um's Daseyn,* trans. Heinrich Georg Bronn (Stuttgart: Schweizerbart'sche, 1860), 495–520. Bronn brought as a chief objection to Darwin's theory that it was "in its ground-conditions of justification still a thoroughly wanting hypothesis." It remained, according to Bronn "undemonstrated," though also "unrefuted" (502). Bronn did, however, lodge some considerations that militated against the hypothesis, for example, that transitional species were lacking (503–5). Bronn himself was the author of a quasi-evolutionary theory, which he formulated prior to reading Darwin (see the first appendix).

theory. Language descent, he proclaimed, was already an empirically well-established phenomenon. Moreover, the linguist's descent tree might be used as a model for depicting the evolution of plant and animal species; and, as we have seen, Haeckel readily adopted this model (fig. 5.10).

In addition to the graphic model of the stem-tree, Schleicher proposed that linguistics supplied four other supports for Darwinian theory. First, the linguistic system can be used to display a "natural history of the genus *Homo.*" This is because "the developmental history of languages is a main feature of the development of human beings." Second, "languages are natural organisms" (*Naturorganismen)* but have the advantage over other natural organisms since the evidence for earlier forms of language and transitional forms has survived in written records: there are considerably more linguistic fossils than geological fossils. Third, the same processes of competition of languages, the extinction of forms, and the development of more complex languages out of simpler roots—these all suggest mutual confirmation of the basic processes governing such historical entities as species and languages. Finally, since the various language groups descended from more primitive, "cellular" forms, language provides analogous evidence that more advanced species descended from simpler creatures.[73]

Schleicher intended that these four complementary contributions of linguistics to biological theory should buttress an underlying conviction that the pattern of language descent perfectly reflected the pattern of human descent. The monistic point of view, which he assumed in his tract, held that language was simply the material side of mind and thought.[74]

That monistic thesis received fuller elaboration in a small work published two years after *Darwinsche Theorie,* Schleicher's *Über die Bedeutung der Sprache für die Naturgeschichte des Menschen* (On the significance of language for the natural history of mankind, 1865). In this essay he argued that the superficial differences among human beings, which morphologists often exaggerated, proved simply insufficient to classify them. He observed:

> How inconstant are the formation of the skull and other so-called racial differences. Language, by contrast, is always a constant trait. A German can indeed display hair and prognathous jaw to match those of the most distinctive Negro head, but he will never speak a Negro language with native facility. . . . Animals can be ordered according to their mor-

73. Schleicher, *Darwinsche Theorie,* 4–8, 23–24.

74. Schleicher's doctrine of monism is discussed in chapter 5.

> phological character. For man, however, the external form has, to a certain extent, been superseded; as an indicator of his true being, external form is more or less insignificant. To classify human beings we require, I believe, a higher criterion, one which is an exclusive property of man. This we find, as I have mentioned, in language.[75]

Schleicher, who had an amazing capacity for learning languages—some twenty-two at his command—nonetheless believed that even if he had been raised in Africa, his Teutonic brain would not have been able to handle an African language as dexterously as one whose ancestors had spoken the language for generations. The continued use of a language over long periods of time, he assumed, molded the brain to its grammatical and conceptual structures. Hence, the justification for using languages to classify human groups was quite simple: "The formation of language is for us comparable to the evolution of the brain and the organs of speech."[76]

In his first major publication, *Zur vergleichenden Sprachengeschichte* (Toward a comparative history of language, 1848), Schleicher had employed a kind of morphological classification of languages that had been originally formulated by a group of linguists influenced by the Romantic traditions of Jena and Berlin; this classification and his own particular reconstruction of it would frame all of his subsequent work in linguistics. He distinguished three large language families by reason of their forms: isolating languages, agglutinating languages, and flexional languages. Isolating languages (e.g., Chinese and African) have very simple forms, in which grammatical relationships are not expressed in the word; rather, the word consists merely of the one-syllable root (with position or pitch indicating grammatical function). Because of their simple structure, these languages cannot, according to Schleicher, give full expression to the possibilities of thought. Agglutinating languages (e.g., Turkish, Finnish, Magyar) have their relational elements tacked on to the root in a loose fashion (indeed, the relational elements themselves are derived from roots). Flexional languages (e.g., the Indo-Germanic and Semitic families) are the most developed. Roots and relations form an "organic unity," according to Schleicher.[77] So,

75. August Schleicher, *Über die Bedeutung der Sprache für die Naturgeschichte des Menschen* (Weimar: Hermann Böhlau, 1865), 16, 18–19.

76. Ibid., 21.

77. In distinguishing these three forms of language, Schleicher was simply following the lead of Wilhelm von Humboldt, Franz Bopp, and ultimately August Wilhelm Schlegel. Schleicher was certainly familiar with the work of these near-contemporary linguists. In his *Sprachengeschichte,* he cited Humboldt often enough, though not precisely on this distinc-

for example, the Latin word "scriptus" has "scrib" as the root or meaning; "tu" expresses the participial relationship; and "s" indicates the nominative relationship. Schleicher believed that even the most highly developed languages, the flexional group, originated from a simpler stem but continued to develop into varieties with more perfect forms. Isolating and agglutinating languages, on the other hand, simply did not have the potential to move much beyond their more primitive structures.

Since, according to Schleicher, the three basic language types did not evolve from one another, each primitive group speaking those languages became human in distinctively different ways. After the initial establishment of the isolating, agglutinating, and flexional languages, which created the different groups of human beings, they evolved at different rates and in different directions. Only the Indo-Germanic and Semitic languages reached a kind of perfection not realized in the other groups. Here, then, was Haeckel's solution to the evolution of the various human species and their mental capacities.

From the *Urmensch,* who had evolved out of apelike ancestors, came the various species of mankind. Their external morphologies would suggest this simian origin. But even *Pithecanthropus alalus,* the ape-man without speech, would not yet be genuinely human. This proto-man and his clan would require the brain-molding action of language to complete the process of transition from animal to man. At least, this is the hypothesis at which Haeckel arrived in his *Natural History of Creation:*

> We must mention here one of the most important results of the comparative study of languages, which for the *Stammbaum* of the species of men is of the highest significance, namely, that human languages probably had a multiple or polyphyletic origin. Human language as such probably developed only after the species of speechless *Urmenschen* or *Affenmenschen* had split into several species or kinds. With each of these human species, language developed on its own and independently of the others. At least this is the view of Schleicher, one of the foremost authorities

tion. See Wilhelm von Humboldt, *Über die Kawi-Sprache auf der Insel Java,* 3 vols. (Berlin: Königlichen Akademie der Wissenschaften, 1836). The introduction to this famous work on Javanese language made the threefold distinction pivotal (1: cxxxv–cxlviii). August Wilhelm Schlegel, who became professor of linguistics at Bonn, formulated the original distinction in his *Observations sur la langue et la littérature provençales* (Paris: Librarie grecque-latine-allemande, 1818), 14–16. Franz Bopp, whom Humboldt brought to Berlin as professor, canonized the distinction in his *Vergleichende Grammatik des Sanskrit, Zend, Griechischen, Lateinischen, Litthauischen, Gothischen und Deutschen* (Berlin: Königlichen Akademie der Wissenschaften, 1833), 108–13.

> on this subject. . . . If one views the origin of the branches of language as the special and principal act of becoming human, and the species of humankind as distinguished according to their language stem, then one can say that the different species of men arose independently of one another.[78]

The clear implication of this theory is that the languages with the most potential created the human species with the most potential. And, as Haeckel never tired of indicating, that species with the most potential—a potential realized—was constituted by the Semitic and Indo-Germanic groups, with the Berber, Jewish, Greco-Roman, and Germanic varieties in the forefront.[79]

Today linguists generally accept Schleicher's view that languages evolved out of one another, though few would venture to endorse the polygenic theory of language origins and fewer still the notion of a progressive hierarchy of languages. Even Schleicher had some difficulty squaring the hierarchical view with the great reverence he and others had for the classical languages of Greek and Latin, whence the modern and presumptively more advanced languages would have derived.[80] Many contemporary evolutionary theorists would agree that language played a crucial role in the descent of the human brain.[81] No modern evolutionist would, however, have assumed, as Schleicher and Haeckel did, that a kind of Lamarckian use inheritance would have been the instrument of that evolution. But in the nineteenth century, there was another significant biologist who did

78. Haeckel, *Natürliche Schöpfungsgeschichte,* 1st ed. (1868), 511. Haeckel largely depended on the criterion of language for constructing his stem-trees of the varieties of different species of men. So, for instance, the stem-tree of the Indo-German varieties (e.g., Anglo-Saxons, High Germans, Serbs, Poles, Iranians, Albanians, etc.) closely follows Schleicher's own tree of the Indo-German languages. See Haeckel, *Natürliche Schöpfungsgeschichte,* 2nd ed. (1870), 625. Mario Di Gregorio charts some of the other important linguistic authorities who both influenced Haeckel and were influenced by him. See his "Reflections of a Nonpolitical Naturalist: Ernst Haeckel, Wilhelm Bleek, Friedrich Müller and the Meaning of Language," *Journal of the History of Biology* 35 (2002): 79–109.

79. The debate over the monogenic or polygenic origin of man is still played out, if in a slightly different key. See, for instance, Christopher Stringer and Robin McKie, *African Exodus: The Origins of Modern Humanity* (New York: Holt, 1996). See also my review of their book, "Neanderthals Need Not Apply," *New York Times Book Review* (Sunday, 17 August 1997), 10.

80. Schleicher had an ingenious solution to this conundrum, which relied on certain Hegelian assumptions. I discuss his account in "The Linguistic Creation of Man," 38–39.

81. For a wide-ranging and compelling discussion of the language-brain relationship, see Terrence Deacon, *The Symbolic Species: The Co-Evolution of Language and the Brain* (New York: Norton, 1997).

adopt this hereditary principle as well as the Schleicher-Haeckel theory—Charles Darwin.

The Rebounding Impact of the Schleicher-Haeckel Theory on Darwin

During the mid-1860s, Darwin's great friend Alfred Russel Wallace had undergone a conversion to spiritualism—on the basis of experimental evidence, to be sure.[82] In a review article in 1869, Wallace fortified his conviction with some powerful arguments about natural selection's insufficiency to account for man's big brain.[83] Sheer survival, he thought, simply did not require the intellectual capacity demonstrated even by primitive men. Man's mental abilities, he concluded, had to be under the selective influence of higher spiritual powers. Darwin reacted to his friend's article in horror: "But I groan over Man—you write like a metamorphosed (in retrograde direction) naturalist, and you the author of the best paper that ever appeared in the *Anthropological Review*! Eheu! Eheu! Eheu!"[84] Darwin, nonetheless, saw the force of Wallace's argument, and thus the vexing problem it posed—how to explain the complex mind and big brain of human beings when these enlarged traits were not needed for survival. Fortunately during the mid-1860s, another kind of argument came to his attention, through several related sources. This was the Schleicher-Haeckel argument for the linguistic creation of man.

Darwin studied Schleicher's *Darwinsche Theorie,* which he then used and cited in his own considerations of language in *The Descent of Man.*[85] He got two other doses of Schleicher's views more indirectly. The first was through Frederic Farrar, whom Darwin named along with his cousin Hensleigh Wedgwood and Schleicher as contributing to his conception of language. Farrar had made Schleicher's theories known to the British intellectual community through a comprehensive account in the journal *Nature.*[86] The more important source for Schleicher's conceptions, however, was the *Natural History of Creation,* which Haeckel sent Darwin as a gift.

82. See Robert J. Richards, *Darwin and the Emergence of Evolutionary Theories of Mind and Behavior* (Chicago: University of Chicago Press, 1987), 176–84.

83. Wallace first advanced his arguments in a review of new editions of Charles Lyell's works. See Alfred Russel Wallace, Review of *Principles of Geology* and *Elements of Geology,* by Charles Lyell, *Quarterly Review* 126 (1869): 359–94.

84. Darwin to Wallace (26 January 1870), in *Alfred Russel Wallace: Letters and Reminiscences,* 2 vols., ed. James Marchant (London: Cassell, 1916), 1:251.

85. Darwin, *Descent of Man,* 1:56.

86. Frederic Farrar, "Philology and Darwinism," *Nature* 1 (1870): 527–29.

Darwin wrote to a friend after reading Haeckel's volume that it was "one of the most remarkable books of our time."[87] Darwin's notes and underlining in the book are quite extensive. He was particularly interested, as shown by his scorings and marginalia, in Haeckel's account of Schleicher's thesis in *Über die Bedeutung der Sprache für die Naturgeschichte des Menschen.*[88] Here, then, Darwin had a counterargument to Wallace's, one by which he could solidify an evolutionary naturalism: language might modify brain, increasing its size and complexity, with this expansion becoming a permanent, hereditary legacy. As he put it in *The Descent of Man:*

> A great stride in the development of the intellect will have followed, as soon as, through a previous considerable advance, the half-art and half-instinct of language came into use; for the continued use of language will have reacted on the brain, and produced an inherited effect; and this again will have reaction on the improvement of language.[89]

This continued use of a developing language over generations, Darwin thus believed, would produce "heritable modifications" through "the principle of the inherited effects of use."[90] Darwin's adoption of the Schleicher-Haeckel solution to man's enlarged mental capacities is another indication that Darwin was not merely indulging a friendship when he proclaimed in *The Descent of Man:*

87. Darwin to William S. Dallas (9 June 1868), in Darwin Correspondence, DAR 162, Manuscript Room, Cambridge University Library.

88. Darwin's copy of Haeckel's *Natürliche Schöpfungsgeschichte* is held in the Manuscript Room of Cambridge University Library.

89. Darwin, *Descent of Man,* 2:390–91.

90. Ibid, 1:57. In the conclusion to the *Descent of Man,* Darwin referred to an article by Chauncey Wright, which in the last moments of manuscript preparation he had just read. Wright had attacked Wallace's argument that man's big brain could not be given a natural-selection account. See Chauncey Wright, "Limits of Natural Selection," *North American Review* 111 (October 1870): 282–311. Darwin suggested that Wright also endorsed the idea that language operated to produce man's increased intellectual capacity through use inheritance (*Descent of Man,* 2:390–91). Wright's argument is a bit convoluted, but it is clear: he made no such argument as Darwin attributed to him. Quite the contrary. Wright maintained that Wallace had simply misjudged the character of the native's capacities (294–98). Wright rather held that language and so-called higher faculties were merely collateral features of capacities directly useful to the native, and so indirectly acquired through natural selection. "Why may it not be," he asked, "that all that he [the savage] can do with his brains beyond his needs is only incidental to the powers which are directly serviceable?" (295). He further suggested that the difference between the savage and the philosopher "depends on the external inheritances of civilization, rather than on the organic inheritances of the civilized man" (296). Darwin, in his enthusiasm for the Schleicher-Haeckel argument, found its ghost in any text that opposed Wallace's thesis.

> Besides his great work, "Generelle Morphologie" (1866), [Haeckel] has recently (1868, with a second edit. in 1870), published his "Natürliche Schöpfungsgeschichte," in which he fully discusses the genealogy of man. If this work had appeared before my essay had been written, I should probably never have completed it. Almost all the conclusions at which I have arrived I find confirmed by this naturalist, whose knowledge on many points is much fuller than mine.[91]

Sentiments like this, as well as Darwin's use of fundamental Haeckelian ideas, render into very thin beer the claim of many historians that Darwin's and Haeckel's theories were essentially opposed.

The Natural History of Creation *as Popular Science*

Haeckel's *Natural History of Creation* made such an impact on his times that historians have sometimes regarded him as simply a popularizer; they forget the some twenty technical monographs and innumerable specialized articles upon which succeeding editions of his book rested. Even the astute historian Alfred Kelly, who has written on the popularization of Darwinism in Germany, is ready to cast Haeckel with the likes of the novelist Wilhelm Bölsche, who not only wrote general expositions of biology but a popular account of Haeckel's own life and works.[92] Kelly thinks that

> it would be a mistake to search for hidden depths in Haeckel's works; he was often superficial, inconsistent, and just plain muddleheaded. But this does not really matter because popular writers are likely to be judged only by the surface impressions they give. To overanalyze Haeckel would be to misjudge both the man and his popular effect.[93]

The attribution of weak-mindedness to Haeckel, as I have already suggested, stems from several sources: a belief that he had misrepresented Darwin's theory; a lack of appreciation of his technical and empirical accomplish-

91. Darwin, *Descent of Man,* 1:4.

92. Wilhelm Bölsche, *Ernst Haeckel: Ein Lebensbild* (Berlin: Georg Bondi, 1909).

93. Alfred Kelly, *The Descent of Darwin: The Popularization of Darwinism in Germany, 1860–1914* (Chapel Hill: University of North Carolina Press, 1981), 24. For a rather different view of the nature of popular science, especially in Germany around 1900, see Rosemarie Nöthlich, Olaf Breidbach, and Uwe Hoßfeld, "'Was ist die Natur?' Einige Aspekte zur Wissenschaftspopularisierung in Deutschland," in *Klassische Universität und akademische Provinz: Die Universität Jena von der Mitte des 19. bis in die 30er Jahre des 20. Jahrhunderts,* ed. Matthias Steinbach and Stefan Gerber (Jena: Bussert & Stadeler, 2005), 238–50.

ments, particularly in marine biology; and an inadequate measure of his impact on the science of his time—and ours. Moreover, such historical attitudes presume clear distinction between what passes as popular science and what is commonly recognized as *real* science, and, insofar as the distinction can be made, a pejorative assessment of the techniques of popular science—muddleheadedness is permissible in popular science, at least according to Kelly. He declares that the distinction between popular and real science "surprisingly . . . presents relatively little problem."[94] He thinks popular science simply bears its difference from real science like military music from real music. I do not believe the discrimination so easy nor do I accept the presumption, especially in Haeckel's case, that a popular work must necessarily be superficial and misconceived.[95] But for my conviction to be properly evaluated, we must be clear about what constitutes a scientific work as *popular.*

We can discriminate some eight to ten criteria that would characterize works in science as popular. No particular criterion is necessary or sufficient to denominate a given book or article popular; but the more of them that typify a work, the more inclined we should be, I believe, to put it in that genre.

The first criterion is virtually the sole indicator that Kelly mentions, namely, that the author intends the book for a general audience.[96] Not only did Haeckel have that intention, but he delivered the lecture series, whence the book derived, to a general audience. And Haeckel made his intention quite explicit:

> I hold it as the duty of natural scientists [*Naturforscher*] that they should not simply remain circumscribed within the narrow confines of their specialty, improving and making discoveries, that they should not merely cultivate lovingly and carefully the study of their special subjects, but that they should also make their particular study useful for the polity and that they should help to broaden the scientific education of the entire people.[97]

94. Kelly, *The Descent of Darwin,* 6.

95. Rosemarie Nöthlich, Olaf Breidbach, and Uwe Hoßfeld argue, quite persuasively, that popularization in a science runs in complementary tandem with the professionalization of that science. They also suggested the ways in which popularizers can transform the cultural discussion in a period and open up other areas for transformation—so from Darwinism to social Darwinism. See their " 'Was ist die Natur?' "

96. Kelly, *The Descent of Darwin,* 6.

97. Haeckel, *Natürliche Schöpfungsgeschichte,* 1st ed. (1868), 3.

Haeckel's concern to spread the results of modern science beyond the university to a literate audience had behind it the failure of his *Generelle Morphologie.* But he also had the positive motivation of the growing effort by his immediate predecessors at making science accessible to a larger public. As a young boy and adolescent, he had delighted in Alexander von Humboldt's *Ansichten der Natur* (Views of nature, 1808) and Matthias Schleiden's *Die Pflanze und ihr Leben: Populäre Vorträge* (The plant and its life: popular lectures, 1848). More proximately, he was quite familiar with Rudolf Virchow's *Vier Reden über Leben und Kranksein* (Four lectures on life and illness, 1862), Carl Vogt's *Vorlesungen über den Menschen, seine Stellung in der Schöpfung und in der Geschichte der Erde* (Lectures on men, their place in creation and in the history of the earth, 1863), and the first volume of what would be an immensely successful series, Hermann von Helmholtz's *Populäre wissenschaftliche Vorträge* (Popular scientific lectures, 1865). These scientist-popularizers provided models for reaching a general audience and cultivating a public for a new scientific endeavor.[98]

But, of course, there are general audiences and there are general audiences. Haeckel initially had in mind a university audience, composed of faculty and students of different disciplines. This was, after all, the audience for the lectures on which his *Natural History of Creation* was based. Consider a few examples of popular science whose audience could only be regarded as qualifiedly *general.* Voltaire's *Elémens de la philosophie de Neuton*[99] was directed at the company of philosophes, coming as it did from the brain of Madame du Châtelet, who famously remarked to Frederick the Great that "it may be that there are metaphysicians and philosophers whose learning is greater than mine, although I have not met them."[100] The book would have been beyond even rather well-educated members of the upper classes. Alexander von Humboldt's five-volume *Kosmos* was also designed for a general reader—but the reader had to be someone like Charles Darwin, who did devour the tough tomes. The late Nobel Laureate Subramanian Chandrasekhar wrote a book on Newton's physics

98. See Alexander von Humboldt, *Ansichten der Natur, mit wissenschaftliche Erläterungen,* 2nd ed., 2 vols. (Stuttgart: J. G. Cotta'schen Buchhandlung, 1826); M. J. Schleiden, *Die Pflanze und ihr Leben: Populäre Vorträge* (Leipzig: Wilhelm Engelmann, 1848); Rudolf Virchow, *Vier Reden über Leben und Kranksein* (Berlin: Georg Reimer, 1862); Carl Vogt, *Vorlesungen über den Menschen, seine Stellung in der Schöpfung und in der Geschichte der Erde,* 2 vols. (Gieszen: J. Ricker'sche Buchhandlung, 1863); and Hermann von Helmholtz, *Populäre wissenschaftliche Vorträge* (Braunschweig: Friedrich Vieweg und Sohn, 1865).

99. F. Voltaire, *Elémens de la philosophie de Neuton* (Amsterdam : E. Ledet, 1738).

100. Mme. du Châtelet to Frederick the Great of Prussia, quoted in Samuel Edwards, *The Divine Mistress* (New York: McKay, 1970), 1.

designed specifically for "the common reader," as its title declared; but the common reader who might comprehend it could only be found among the mandarins of physics and mathematics.[101] These examples are simply meant to indicate that though intent would certainly be an important criterion for judging some work an example of popular science, it is hardly the sufficient or clear standard that Kelly believes. These examples also suggest that presumptively popular works in science are often meant as well for a more professional audience. And Haeckel's book certainly reached that audience.

In a book of popular science, the author strives for clarity of presentation. The effort to be clear usually involves, of course, a coherent rhetorical and logical introduction of the subject matter. In Haeckel's book the deployment of historical chapters, almost unprecedented in their extent and range, adroitly prepares the reader for the more systematic introduction to the several newly formed branches of evolutionary biology. The historical chapters also have a considerable rhetorical function, namely, to convince readers that this new science has antecedents in the thought of great German predecessors—Kant, Goethe, Oken—and thus is quite congenial to the German mind.

In the interest of clarity, a popular work will minimize the standard jargon of a discipline. Haeckel had fenced off sections of his *Generelle Morphologie* with neologisms that forced many of his nineteenth-century readers into exhausted defeat—for instance, in semi-translation: organology, tectology, promorphology, bions, antimeres, and the like. The task of the modern reader is actually a bit easier than that of his or her nineteenth-century counterpart, since several of Haeckel's new terms did take fruitful root in biology, for example, phylogeny, ontogeny, ecology. While some of these neologisms do appear in the *Natural History of Creation,* Haeckel reduced their variety and quantity—and commensurately readers' comprehension increased.

If a work attempts to reach a more general audience, then it cannot burden its readers with a larger critical apparatus of citations and discussions of the learned literature or suppose too much antecedent knowledge. Haeckel succeeds in avoiding these temptations of the scholarly mind. He does, via a modest set of footnotes, direct readers to other literature—especially his own essays and articles—that supplies a richer load of evidence or a more extended discussion of the issues.

101. Subramanian Chandrasekhar, *Newton's Principia for the Common Reader* (Oxford: Oxford University Press, 1995).

In our time much popular science is written by the light hand of the intelligent journalist, who has read widely in a science, rather than the heavy hand of the professional, whose knowledge may produce a dense thicket. When done well, the nonprofessional's pen can be facile, provocative, and instructive. Haeckel had a talent for rendering his exposition both provocative and instructive. Simply the juxtaposition of the two warring terms of his title—"natural" and "creation"—was a bold move that guaranteed an audience beyond a narrow circle. And the constant baiting of the preachers throughout the volume assured him of a quite attentive and reactive readership. The liveliness of the volume was augmented by the intimate tone of an oral delivery, which flowed smoothly from his pen.

Haeckel also found a medium other than prose with which to craft his arguments. The singular features of his work—made possible by new techniques of copper-plate etching and lithography, as well as by the more traditional technology of the wood-cut—were the startling illustrations accompanying the text. Their number and quality really had no rival among popular scientific works. It is not inevitable, of course, that a book of popular science will be richly illustrated—many are not. But illustrations have singular advantages in works meant for a larger audience. Graphic representations speak a language that transcends the peculiar argot of a given discipline, and they have an immediacy not easily imitated by the written word. Moreover, a picture or graphic compresses into a few strokes many lines of argument, conveying the gist of a particular thesis intuitively and directly. Further, if you can picture a phenomenon, the impression usually lasts. As the verbal descriptions of Haeckel's book fade in memory, the illustrations remain vivid: the frontispiece indicating the descent of races from ape ancestors, the beguiling embryos that silently and slyly illustrate the biogenetic law, the stem-trees that gauge the evolutionary race, and that stoic burgher couple and child, our remote kin. Arguments and claims that withered and died in Haeckel's *Generelle Morphologie* sprang back to life in the illustrations of the *Natural History of Creation*. Most of the critical appraisal of Haeckel's work, along with the charges of fraud and malfeasance, focused on his illustrations.

Unlike more professional literature, popular works in science will often strive for intimacy with the reader. Haeckel's book, born of the lecture, certainly achieves this by frequently addressing the reader directly, for instance: "Each of you will have heard of the name 'Darwin'"; "Since I will later explain to you more fully . . . "; and so on. And this rhetorical strategy will also involve the author, via the first-person personal pronoun, injecting himself frequently into the text—a tactic Haeckel generously employed. From the fifth edition (1874) on, he made his authorial presence even more palpable by the inclusion of his portrait as the frontispiece. The reverse direction was taken by Darwin in the *Origin of Species.* In the first edition, the first-person pronoun is sprinkled through almost every paragraph: "I think we are driven to conclude"; "I am strongly inclined to suspect"; "I may add"; and so forth. By the sixth edition, however, those "I's" have been driven out by Darwin's growing sense of what objective science should look like. In the popular work, the author talks directly and intimately to the general reader; in the professional work, nature talks directly and impersonally to the narrow specialist.

In most professional literature—save for the surveys—the scientist restricts his or her subject to those areas of deep familiarity, especially those topics the individual has explored in firsthand research. The popular work quite often ranges over many subjects, perhaps tangential to the author's research, but certainly extending beyond immediate acquaintance. Haeckel was and, in some areas of marine biology, remains today a principal authority. No one in his own time questioned his understanding or representations of the embryology and developmental stages of radiolaria, medusa, sponges, siphonophores, and other sea-dwelling creatures. But his claims about human embryology and anthropology met stiff resistance from other professionals—much of it, I believe, unmerited. However, the stretch did leave him vulnerable. And when he ingressed, as he increasingly did, into areas of philosophy and theology, the battle was well met.

Works in popular science, especially when authored by scientists who have a claim to expertise in particular areas, quite often reveal a deep metaphysical foundation, a general worldview that the author wishes to convey via the ostensible discussion of scientific topics. Little excavation is required to reveal the hard philosophical-political-social core upon which rest the surface discussions of scientists like Stephen Jay Gould, Richard Dawkins, Ernst Mayr, E. O. Wilson, or Steven Pinker. Haeckel was a prototype for these authors. He grounded his discussions of particular areas of evolutionary biology in a monistic naturalism deriving from Romanticism, on the one hand—especially that of Goethe—and the hyperrationalism of a Spinoza, on the other.

Haeckel's *Natural History of Creation* not only brought the new discipline of evolutionary biology to a wider, literate audience; it brought the author himself to that public stage. The book, like every significant literary accomplishment, embodied the personality of the author: bold, belligerent, inventive, technically adept, and insistently confident. By contrast, Darwin's demeanor in the *Origin of Species* bespoke a man cognizant of the tentative nature of his claims, one who quickly admitted other possibilities, an individual who remained humble before the facts. This public persona won Darwin respect even from arch opponents. Haeckel's aggressive presence in his work ensured that professionals of like disposition would respond, as many did, either in combative opposition or impassioned support. In the next chapter, I will describe the reaction of professionals to Haeckel's various theses and then turn to the religious objectors, whose righteous rage resounds down the generations, echoing in the increasingly tinny voices of contemporary fundamentalists. The opponents in the nineteenth century and today have responded not simply to the particular arguments or positions advanced by Haeckel—seen coolly and dispassionately, his theoretical convictions hardly differed from those of Darwin. Yet the sage of Downe never suffered the kind of obloquy and invective unleashed against his German defender. The hyperbolic reactions were stimulated by the personal qualities that invested Haeckel's arguments. And those qualities had become ever more raw because of the stinging salt of tears shed over a love that still inhabited a fanatic heart.

Conclusion: Evolutionary Theory and Racism

Haeckel's *Natürliche Schöpfungsgeschichte* represented the human species in a hierarchy, from lowest (Papuan and Hottentot) to highest (the Midlanders, including the Indo-German and Semitic races). His stem-tree of human descent and the racial theories that accompany it have been the focus of several recent books—histories arguing that Haeckel had a unique position in the rise of Nazi biology and Fascism during the first part of the twentieth century. In 1971 Daniel Gasman brought the initial bill of particulars: he portrayed Haeckel as having specific responsibility for Nazi racial programs. He argued that Darwin's champion had a distinctive authority at the end of the nineteenth century, throwing in the shadows the myriads of others with similar racial attitudes.[102] But it was not simply a general racism that Haeckel expressed; he was, according to Gasman, a

102. Gasman, *Scientific Origins of National Socialism*, 40: "By bringing biology and anthropology to its support, in works that were widely read and credited, he [Haeckel] succeeded in investing the ideas of racial nationalism with academic respectability and scientific assurance. It was Haeckel, in other words, who was largely responsible for forging the bonds between academic science and racism in Germany in the later decades of the nineteenth century." Gasman's charge of anti-Semitism against Haeckel has been uncritically adopted by many authors. I provide a large list of them in chapter 11.

virulent anti-Semite.[103] In another book, almost thirty years later, Gasman has suggested that this German biologist, dead a decade and a half before Hitler came to power, had virtually begun the work of the Nazis: "For Haeckel, the Jews were the original source of the decadence and morbidity of the modern world and he sought their immediate exclusion from contemporary life and society."[104] Undeterred by a paucity of evidence, Gasman has extended the field of Haeckel's malign influence beyond what this author had originally imagined.[105] Haeckel's monism, Gasman maintains, was chiefly responsible for the rise of Fascism generally throughout Europe in the first part of the twentieth century.[106]

If with somewhat less inflationary rhetoric, Richard Weikart presses similar claims in his tellingly titled book *From Darwin to Hitler.* Weikart's thesis is that "no matter how crooked the road was from Darwin to Hitler, clearly Darwinism and eugenics smoothed the path for Nazi ideology, especially for the Nazi stress on expansion, war, racial struggle, and racial extermination."[107] In Weikart's account, Haeckel simply packed Darwin's evolutionary materialism and racism into his sidecar and delivered their toxic message to Berchtesgaden. In the second appendix to this volume, I will consider the justification for the moral indictments implicitly and explicitly expressed in the claims of Gasman, Weikart, and several others. But here I would like briefly to place Haeckel's racism in the context of nineteenth-century thought.

Haeckel, of course, was hardly alone in calibrating human beings using an intellectual and aesthetic scale. Darwin also aligned the human groups

103. Ibid., 157–59.

104. Daniel Gasman, *Haeckel's Monism and the Birth of Fascist Ideology* (New York: Peter Lang, 1998), 26.

105. Gasman, ibid., 73, believes Haeckel enlisted as a member, in late 1918, of the right-wing Bavarian Thule Society, which became instrumental in the rise of Hitler's Nationalsozialistische Deutsche Arbeiterpartei. The society, as Gasman accurately describes it, was "a political-theosophical-astrological-anti-Semitic secret organization." However, in late 1918, Haeckel was an invalid, not leaving the second floor of his home, and would not likely have joined this Bavarian group. Moreover, he was hardly a devotee of theosophy and astrology—doctrines that he would have regarded as completely antithetic to progressive modernism. Rudolf von Sebottendorff, founder of the Thule Society, lists a one "Ernst Häckel" as a member of his group but distinguishes this colleague from "Ernst Haeckel, Professor in Jena." See Rudolf von Sebottendorff, *Bevor Hitler kam* (Munich: Deukula, 1933), 240. There was in Jena another Ernst Häckel, who was a painter. He wrote a few letters to the Ernst Haeckel of this study, which are contained in the Haeckel Correspondence, Haeckel-Haus, Jena. The designation of the zoologist Haeckel as member of the Thule Society is pure artifact.

106. Gasman, *Haeckel's Monism,* 7: "It was Haeckel's Germanic National Socialism that inspired the rise of Fascist ideas throughout Europe."

107. Weikart, *From Darwin to Hitler,* 6.

on a developmental path, from the "savage" races to the "civilized." In *The Descent of Man,* he urged that it comported better with common usage to speak of one human species with many varieties or races—but the distinction between species and race, he had long since argued, was arbitrary.[108] The significant differences among the human groups, however, were clear. He believed, for instance, that the degenerate human variety inhabiting the Emerald Isle certainly fell below the mark set by the more civilized groups clustered in England and Scotland.[109] As a typical representative of his class and times, Darwin also regarded women as intellectually inferior to men.[110] Sexual selection, he argued, largely accounted for the superiority of the male mind, as well as for the hierarchical distribution of traits through the human groups. There is, then, little question that both Haeckel and Darwin depicted the human races as forming a hierarchy, with some varieties displaying more advanced traits than others. But, of course, neither of these thinkers was original in this respect. Prominent biologists, writing before the advent of Darwinism, proposed schemes of racial classification that reflected prevailing conceptions, namely, assumptions that affirmed the high status of Europeans in the world.

Eighteenth-century naturalists were the first to develop systematic and comprehensive categories to classify human beings and to locate their place in relation to the lower animals. For instance, in the tenth edition (1758) of his *Systema naturae,* Linnaeus (1707–1778) placed the genus *Homo* within the order Primates (which included monkeys, bats, and sloths) and distinguished two species: *Homo sapiens* and *Homo troglodytes* (anthropoid apes). He divided *Homo sapiens* (wise man) into four varieties: American (copper-colored, choleric, upright [*rectus*], regulated by custom), Asiatic (sooty, melancholic, stiff, and governed by opinions), African (black, phlegmatic, languid [*laxus*], and governed by caprice), and European (fair, sanguine, muscular, and governed by laws). Linnaeus conceived such differences as expressive of divine intent.[111] Johann Friedrich Blumenbach (1752–1840), the most influential theorist on this question at the turn of the nineteenth century, argued that human beings constituted one species, with five varieties merging into one another. His *De generis humani varie-*

108. Darwin, *Descent of Man,* 1:235.

109. I discuss Darwin's consideration of the Irish in *Darwin and the Emergence of Evolutionary Theories,* 172–76.

110. Darwin, *Descent of Man,* 1:188–89.

111. Carolus Linnaeus, *Systema naturae per regna tria naturae, secundum classes, ordines, genera, species, cum characteribus, differentiis, synonymis, locis,* 3 vols. (Hale: Curt, 1760–70), 1:20–24.

tate nativa liber (3rd ed., 1795) distinguished the Caucasian (originating in Georgia), the Mongolian (including Greenlanders and Eskimos), the Ethiopian (Africans), the American (Indians of North and South America), and the Malayan (including the islanders of the South Pacific).[112] These groups differed in skin color, facial traits, hair texture, stature, and skull shape. Blumenbach speculated that the large penis of his Ethiopian specimen would support tales of sexual prowess, but he did not venture whether the trait generally characterized the variety.[113] He thought the Caucasian race the most beautiful and inferred that it might constitute the original people, whence the others had altered and declined through the effects of climate.

During the early nineteenth century, the clear delineation of distinct races, each with special characteristics, held firm. Georges Cuvier (1769–1832), who wielded extraordinary influence on such questions, distinguished three "races" of the one human species: the white race, or Caucasians; the yellow race, or Mongolians; and the black race, or Ethiopians. Not surprisingly, he regarded the Caucasians as the most beautiful and the most cultured. The Mongolian race—the Chinese and Japanese—had founded great civilizations, which, however, stagnated in an unprogressive mode. The Ethiopian race had a "reduced skull" (*crâne comprimé*) with facial features that "approach those of monkeys"; and its peoples had "remained barbarians."[114] The singular exception to such common racial judgments during the early nineteenth century was the work of Friedrich Tiedemann (1781–1861). Stimulated by debates in the British Parliament over slavery, he made numerous experimental measurements, both absolute and relative to body size, of the brains and skulls of the several human groups. In his *Das Hirn des Negers* (The brain of the Negro, 1837), he found no significant differences among Caucasians, Mongolians, Malays, American Indians, and Negros.[115] He completed his study with accounts of Negroes who had received an education and had made important contributions to the sciences and literature.[116]

The anthropological treatises of the nineteenth century, both evolutionary and non-evolutionary tracts, carried forward the earlier assump-

112. Johann Friedrich Blumenbach, *De generis humani varietate nativa liber*, 3rd ed. (Göttingen: Vandenhoek et Ruprecht, 1795).

113. Ibid., 240–41.

114. Georges Cuvier, *Le Régne animal*, 2nd ed., 5 vols. (Paris: Deterville Libraire, 1829–30), 1:80.

115. Friedrich Tiedemann, *Das Hirn des Negers mit dem des Europäers und Orang-outangs verglichen* (Heidelberg: Karl Winter, 1837). See the first appendix for a discussion of Tiedemann's study.

116. Ibid., 79–82.

tions that human beings formed distinct groupings and that these stood in hierarchical relationships. James Hunt (1833–1869), founder of the Anthropological Society of London in the early 1860s and no friend of the Darwinians, continued the older polygenist tradition of regarding the races as separate creations. In his 1863 presidential address to the society, Hunt declared the African a distinct species, much closer to the apes than to Europeans.[117] Right through the early part of the twentieth century—and beyond—most naturalists and anthropologists, evolutionists and non-evolutionists, simply took racial hierarchy as empirically given. What the evolutionists attempted to do was to explain the phenomenon of racial differences in a comprehensive theory. Haeckel, likewise, took this as part of his task. He called himself "utterly and completely a child of the nineteenth century."[118] In his racial thought, he was certainly that but hardly uniquely that, despite the suggestions of Gasman and Weikart. These historians have simply unveiled to a startled world that the founders of evolutionary theory lived in the nineteenth century.

One final aspect of Haeckel's racialism ought here to be considered, namely, the tendentious charge of anti-Semitism by Gasman and Weikart.[119] On its face, the indictment seems unlikely, since the most rabid anti-Semites during Haeckel's time were conservative Christians, such as the Berlin court preacher Adolf Stöcker (1835–1909). It is unlikely that Haeckel would be allied with such Christian apologists, and he loathed Stöcker in particular.[120] Moreover, after Haeckel's death, a onetime student turned opponent, Ludwig Plate (1862–1937), declared that while he was "an idealist, freethinking Christian, German populist, and anti-Semite," Haeckel was "a crass materialist and atheist, and one who ridiculed Christianity in every way possible. For this reason he was celebrated at every opportunity by the Social Democrats and Jews as the world-famous light of true science."[121]

117. James Hunt, "On the Negro's Place in Nature," *Memoirs of the Anthropological Society* 1 (1863): 1–64. For a comprehensive discussion of Hunt and the assumptions of racial hierarchy in late nineteenth-century anthropology, see George Stocking, *Victorian Anthropology* (New York: Free Press, 1987), especially 245–54.

118. Ernst Haeckel, *Die Welträthsel, gemeinverständliche Studien über Monistische Philosophie*] (Bonn: Emil Strauss, 1899), vii.

119. Even Weikart thinks Gasman has overemphasized Haeckel's "anti-Semitism." He is, nonetheless, convinced that Haeckel was an anti-Semite. See Weikart, *From Darwin to Hitler*, 216–17.

120. See Ernst Haeckel, *Die Naturanschauung von Darwin, Goethe und Lamarck* (Jena: Gustav Fischer, 1882), 60n18. Stöcker had demanded the employment only of religiously orthodox professors at Jena.

121. Ludwig Plate, as cited by Heinrich Schmidt, *Ernst Haeckel und sein Nachfolger Prof. Dr. Ludwig Plate* (Jena: Volksbuchhandlung, 1921), 19.

And, of course, there is Haeckel's placement of Semites at the pinnacle of his tree of human progress.

Yet, in addition to evidence of the aforementioned kind, we have direct testimony about Haeckel's attitude concerning Jews. In the early 1890s, he discussed the phenomenon of anti-Semitism with the Austrian novelist and journalist Hermann Bahr (1863–1934), who collected almost forty interviews with European notables on the subject, such individuals as August Bebel, Theodor Mommsen, Arthur James Balfour, and Henrik Ibsen. Haeckel's bonhomie and unpretentious style dispelled Bahr's fear that he would be speaking with a "German professor." Haeckel mentioned that he had several students who were quite anti-Semitic, but that he had many good friends among Jews, "admirable and excellent men," and that these acquaintances had rendered him without this prejudice. He believed that a certain uniformity of religion and social custom was demanded by many countries because of the growth of nationalism, with the exception of France and especially Paris, where the youth of that country had cultivated the "ideal" of cosmopolitanism. Jewish immigrants from the East, particularly Russia, did, he observed, fail to adopt the prevailing customs in Germany and thus provoked distrust and dislike; their behavior, he thought, justified protective restrictions on immigration, though not because they were Jews but because they could not be assimilated. He believed that a growing number of native German Jews held the same opinion. He then concluded his discussion with an encomium to the educated (*gebildeten*) Jews who had always been vital to German social and intellectual life:

> I hold these refined and noble Jews to be important elements in German culture. One should not forget that they have always stood bravely for enlightenment and freedom against the forces of reaction, inexhaustible opponents, as often as needed, against the obscurantists [*Dunkelmänner*]. And now in the dangers of these perilous times, when Papism again rears up mightily everywhere, we cannot do without their tried-and-true courage.[122]

122. Hermann Bahr, "Ernst Haeckel," in *Der Antisemitismus: Ein internationals Interview* (Berlin: S. Fischer, 1894), 62–69; quotation from 69. By "*Dunkelmänner*" (dark men) Haeckel meant those opposed to enlightenment, but with sly reference, perhaps, to the Roman Catholic clergy. Gasman refers to Bahr's interview with Haeckel but gives it a spin that makes Haeckel's concerns "racial" and not behavioral. Gasman omits mention of the longer passage I quote in the text. See Gasman, *Scientific Origins of National Socialism*, 157–59.

Any tincture of what might be thought anti-Semitism has to be placed within the scope of Haeckel's more broadly directed animus: namely, against all orthodox religions, including Judaism, but with special disdain, as the above passage indicates, for Catholicism (which I will more thoroughly discuss in the next chapter). His tangential reservations were not racial or biological, certainly not of the sort favored by the Nazis, but behavioral and attitudinal, more in keeping with the distaste of the German mandarins for the lower classes of any sort.[123]

Another small example might suffice to indicate that as a freethinker, Haeckel harbored no egregious anti-Semitic attitudes, and rather expressed views that would be completely anathema to the Nazis. In his later years, from 1912 until his death in 1919, he became especially friendly with Magnus Hirschfeld, a Jewish physician and freethinker who specialized in research on various sexual practices (especially transvestitism and homosexuality) that would be strictly condemned and regarded as executable crimes by the Nazis.[124] Hirschfeld dedicated his book *Naturgesetze der Liebe* (Natural laws of love) to Haeckel after securing the latter's permission.[125] The book urged that homosexuality was an innate condition and a natural form of love. Hirschfeld visited Haeckel in Jena several times between 1912 and 1917, and lectured on "Ernst Haeckel, ein deutscher Geistesheld" (Ernst Haeckel, a German spiritual hero, 1914).[126] Not the kind of company a proto-Nazi should keep.

123. A small anecdote, stemming from Haeckel's later years, reflects his open-minded attitude toward Jews. He had fallen in love with a woman of the minor nobility (an affair discussed in chapter 10), who wrote of her own anti-Semitic prejudice and contrasted her shameful attitude to that of Haeckel's: "You know, the Hamburg lady is a Jew! She certainly doesn't look like one . . . rather she looks like a Frenchwoman rather than a Jewish woman! That's a healthy lesson for me! I have a great antipathy toward Jews—and had I known, I would not have traveled and gone with her. But now, since I am acquainted with her, and regard her as an open, freethinking woman, I should be ashamed if I despised her on account of her nature, which she can do nothing about. Isn't that right, my dear? Gradually I am learning to ascend to your great, open worldview." The passage comes from a diary page (13 May 1901) that Frida von Uslar-Gleichen sent to Haeckel. See *Das ungelöste Welträtsel*, 2:660.

124. Almost as soon as the Nazis came to power, they burned Hirschfeld's Institut für Sexualwissenschaft and his library in May 1933. He had been away lecturing and never returned to Germany.

125. See the letters of Hirschfeld to Haeckel (21 February 1912 and 20 March 1912), wherein Hirschfeld asks permission for the dedication and Haeckel enthusiastically grants it after reading page proofs for the book. The letters are in the Haeckel Correspondence, Haeckel-Haus, Jena.

126. Hirschfeld mentions his lecture in a letter to Haeckel after the outbreak of the Great War (17 December 1914), in ibid.

The racism that dominated anthropological thinking—as well as religious and popular discourse—during the eighteenth and nineteenth centuries undoubtedly contributed to the ideology of the Nazis, but no less to the cant of American exclusionists during the period of heavy immigration to the United States in the early twentieth century. In the case of the Nazis, there is no compelling evidence that evolutionary ideas (as opposed to genetic and eugenic ideas) played a dominant role in forming their attitudes, especially since Hitler and his immediate circle expressed no particularly favorable disposition toward the theory. The retrospective fight against evolutionary theory in general and Haeckel in particular has its inclement source in our contemporary political and religious climate; but the chilly mists of our cultural atmosphere hardly compare to the raging storms invoked by the religious and political rain dances of the late nineteenth century, which are the subjects of the next two chapters.

CHAPTER EIGHT

The Rage of the Critics

The tide of opposition to Haeckel's evolutionary ideas reached high water during the period from 1868, when the first edition of his *Natürliche Schöpfungsgeschichte* appeared, through the mid-1870s, with recurrent antagonisms welling up during the next three decades.[1] The initial surge of hostility threatened to wash away Haeckel's professional career and would have swept under men of less combative personalities or men lacking the protection of hardened emotional scars. Haeckel did have formidable allies, such stalwarts as Darwin and Huxley in England and Gegenbaur and Weismann at home.[2] Had the opponents attacked using only

1. A search of the OCLC (Online Computer Library Center) indicates that about ninety books were published on Haeckel's theories, both pro and con, during his lifetime. Journal articles, of course, were a multiple of the book numbers.

2. Haeckel and Weismann remained close friends throughout their careers, even though Weismann's views would gradually diverge from Haeckel's, especially over the inheritance of acquired characters. Weismann, however, always accepted Haeckel's central thesis of the biogenetic law. During the period when Haeckel came under severe attack, from the late 1860s through the 1870s, the correspondence between the two remained quite cordial. Weismann greatly appreciated Haeckel's work on siphonophores and admired his three-volume monograph on calcareous sponges. Even the speculative gastraea theory drew support. Weismann wrote Haeckel on 27 January 1874 to thank him for the long monograph on the gastraea theory. Weismann observed: "I have thoroughly studied your gastraea theory with real joy. Although I knew the chief features from our conversation, yet much has surprised me; and I do not doubt that you have advanced our contemporary knowledge a good deal with this synthesis. Nor do I doubt that tremendous screams will echo about 'this theory constructed out of the blue,' etc. Since I have heard my honorable colleagues this past autumn in Wiesbaden talk (for instance, [Anton] Schneider [professor in Giessen] and [Richard] Greef [professor in Marburg and Haeckel's companion in the Canary Islands]), I am in this respect fully aware. Since they can't do it, then no one else should. It's easy to refrain from proposing a theory under the pretext that such would sully the necessary grounding in facts—as if it weren't theory that must show the way in

antique weapons of a creaking theology or of a rusting natural philosophy, the threat would have been easily deflected. But several of Haeckel's most severe critics came from formidable scientific quarters. And their objections were not simply based on differences in theory. They charged Haeckel with conscious fraud and with betraying the trust that the public had placed in science. These attacks have been renewed in our own time, both by religious opponents and, as in the nineteenth century, by academics of considerable prowess.

In this chapter I will examine in detail the most serious charges against Haeckel and assess his responses. I believe the evidence will show that the charges have quite unstable foundations and that they have gained force only through certain methodological, epistemological, and historiographical assumptions that were and remain quite unwarranted. Additionally, one must recognize that religious beliefs, harbored by Haeckel's scientific contemporaries, often crawled beneath objections having a different surface character. Haeckel made mistakes in his depictions of embryos and was not always forthcoming in his responses to critics. While to understand is not, ipso facto, to excuse, the charges against him have to be placed within the larger context of opposition to Darwinian theory and the abreaction to Haeckel's aggressive personality. Both of these factors elevated the vehemence of the original objections and continue to keep the heat at levels quite extraordinary for a scientific controversy that took place almost a century and a half ago.

Critical Objections and Charges of Fraud

Ludwig Rütimeyer

The first individual to strike a serious blow against Haeckel was Ludwig Rütimeyer (1825–1895), an anatomist at Basel. Rütimeyer reviewed the first edition of Haeckel's *Natürliche Schöpfungsgeschichte* shortly after it appeared in the early fall of 1868.[3] In his brief review, Rütimeyer began by comparing Haeckel's work to the anonymously published *Telliamed*, which

---

the first place and indicate what facts are to be sought!" See Weismann to Haeckel (27 January 1874), in "Der Briefwechsel zwischen Ernst Haeckel und August Weismann," ed. G. Uschmann and B. Hassenstein, *Jenaer Reden und Schriften* (1965): 35.

3. Ludwig Rütimeyer, Review of "Ueber die Entstehung und den Stammbaum des Menschengeschlechts" and *Natürliche Schöpfungsgeschichte*, by Ernst Haeckel, *Archiv für Anthropologie* 3 (1868): 301–2.

a century before had spun a fantastical tale of the gradual development of the world and of man through quasi-evolutionary forces.[4] Rütimeyer thought Haeckel's book formed a piece with such fictional literature. He feigned that readers could not have hoped for original research in a work such as Haeckel's; but they might have expected, he supposed, that speculative claims would have been grounded on something more solid than images filched from other scientists, such as Bischoff, Ecker, and Agassiz.[5] Fiction parading as science was one thing, but Haeckel had stooped even lower. He had replicated the same woodcut under three different titles, as embryos of a dog, a chicken, and a turtle (see fig. 7.12). This, Rütimeyer sneered, could only be described as "playing a game of three-card monte with the public and with science."[6] What Rütimeyer found so damning about this tactic was that Haeckel had not labeled the three illustrations of the same embryonic structure as "raw schemata" but instead asserted: "If you compare the young embryos of the dog, chicken, and turtle in figs. 9, 10, and 11, you will not be in a position to perceive a difference."[7] Here was a clear violation of the public's trust.

Though Haeckel would receive consoling words from Darwin about the review, he immediately felt the razor slash and recognized his missteps concerning the replication of clichés.[8] In the second and subsequent editions of the *Natürliche Schöpfungsgeschichte*, as I have indicated in the previous chapter, he deployed only one woodcut for the egg of vertebrates and one for the vertebrate embryo at the sandal stage, remarking about the latter: "It is all the same whether we describe the embryo of a dog, chicken, turtle, or any of the other higher vertebrates. For the embryos . . . at the represented stage certainly cannot be distinguished."[9] This rapid

4. See Benoît de Maillet, *Telliamed; ou, Entretiens d'un philosophe indien avec un missionnaire François*, 2 vols. (Amsterdam: L'honore & Fils, 1748). Interestingly, Rütimeyer (or the printer) names the book *Velliamed*.

5. See the previous chapter for a discussion of Haeckel's use of illustrations from other authors.

6. Rütimeyer, review, 302.

7. Ernst Haeckel, *Natürliche Schöpfungsgeschichte* (Berlin: Georg Reimer, 1868), 249; Rütimeyer quotes this passage.

8. Darwin did try to console him a few years later. See Darwin to Haeckel (2 September 1872), in the Correspondence of Ernst Haeckel, in the Haeckel Papers, Institut für Geschichte der Medizin, Naturwissenschaft und Technik, Ernst-Haeckel-Haus, Friedrich-Schiller-Universität, Jena. Darwin wrote: "It grieved me to read a year or two ago a review by Rueti-meyer on you. I am sorry that he is so retrograde, as I felt much respect for him."

9. Ernst Haeckel, *Natürliche Schöpfungsgeschichte*, 2nd ed. (Berlin: Georg Reimer, 1870), 270.

Fig. 8.1. Wilhelm His (1831–1904). (Courtesy of the Smithsonian Institution.)

correction, however, would not remove the stigma of apparent falsification. The embarrassment might have faded had Rütimeyer been the only critic and had the defamation been confined to a journal of modest circulation. But a more famous scientist, Wilhelm His, who had been a colleague of Rütimeyer, renewed and extended the indictment a few years later in a significant monograph. During the second half of the century, His became one of Germany's leading embryologists and one of Haeckel's most formidable opponents.

Wilhelm His

Wilhelm His (1831-1904) was the sixth child of Eduard His and Anna Katharina (née La Roche), and the grandson of Peter Ochs, a well-known Basel statesman and historian. The father, who changed the family name to His, was a fairly successful merchant and member of the appellate court in Basel.[10] In 1849 his son Wilhelm began medical school in his hometown and the next year traveled to Berlin, where he fell under the spell of Johannes Müller. He attended lectures by the great Berlin anatomist and performed dissections under the watchful eye of the master. In the summer of 1851, he had an intimate colloquium on embryological development in the home of Robert Remak (1815–1865), who was just embarking on his groundbreaking analysis of the germ layers of vertebrates.[11] Remak (and Karl Ernst von Baer) inspired His's theory of embryogenesis, which proposed a mechanical process in which a differentiated germ layer expanded, contracted, and folded to produce the morphological structures of the developing organism.[12] His's association with Müller and Remak revealed to him just "how exciting studying could be when the teacher provides one with insight into his own intellectual endeavors." The experience of real science in Berlin and the enjoyment of friendship with a variety of students in theology and law left him, as he attested in a memoir, with little desire to follow "sterile medical lectures" or to pursue clinical practice.[13]

His's disappointment with medicine hardly abated when he move to Würzburg in 1852 to begin clinical study. While engaged in the latter, he was quickly drawn again to the scientific side of medicine, especially the work of the famous professors then in residence: Albert von Kölliker, Rudolf Virchow, and Franz Leydig in anatomy and microscopy, and Johann Scherer in chemistry. Because of clinical requirements, His only got a taste of the exciting theoretical work conducted at Würzburg at the time. He ex-

10. The father changed the family name to His (his mother's maiden name) because of the unhappy associations with the name Ochs ("ox" in German).

11. See Robert Remak, *Untersuchungen über die Entwickelung der Wirbelthiere*, 3 vols. (Berlin: Georg Reimer, 1850–55). Because he was Jewish, Remak had been prohibited from teaching at university. In 1847, through the intercession of Alexander von Humboldt with the Prussian king—and because of the brilliant research he had done—Remak finally got license to lecture at the University of Berlin.

12. His made clear his debt to Remak and von Baer in the monograph in which he developed his theory of mechanical formation of parts from the germ layers: Wilhelm His, *Untersuchungen über die erste Anlage des Wirbelthierleibes* (Leipzig: F. C. W. Vogel, 1868), v–vi. In 1866 His had invented a microtome for finely slicing cross and longitudinal sections of quite young chick embryos, so that he could follow sequentially in space and time the gradual formation of embryonic parts. In partnership with Adolf Ziegler, a medical-supply modeler, His produced a series of wax models of chick development that gave researchers and students three-dimensional standards by which more intuitively to understand development. Nick Hopwood has detailed the work of Ziegler in his beautifully illustrated and meticulously researched *Embryos in Wax: Models from the Ziegler Studio* (Cambridge: Whipple Museum of the History of Science, 2002).

13. Wilhelm His, *Lebenserinnerungen* (Leipzig: Fischer & Wittig, 1903), 24, 28.

perienced merely secondhand the frisson of discovery, principally through the informal gatherings of a physical-medical society and through friendship with students of the famous professors.[14] Had he been able to attend the regular lectures and the laboratories that occupied beginning students, he might have met the young Ernst Haeckel, who also matriculated at Würzburg in 1852.[15]

His continued clinical study in Prague (1853), where he cultivated an interest in ophthalmology, and in Vienna (1854), where he attended the physiology lectures of Ernst Brücke (1819–1892) and was invited to work in Brücke's laboratory. (Brücke would later serve as mentor to Freud, helping to launch the latter's career as a developmental biologist.) His passed his oral exams back in Basel at the end of summer 1854 and completed his medical dissertation the next year. He then spent time in Paris with Claude Bernard (1813–1878), who received the young student graciously. Bernard would later write in his *Introduction à l'étude de la médicine expérimentale* (1865) that scientific ideas must be strictly controlled by facts, which are recorded passively by the observer as if he were a photographer.[16] Bernard's abstemious epistemology would be adopted by His, especially the notion that acceptable theory must be necessitated by observed facts. In later years he would even take Bernard's metaphor literally in his pioneering use of photography for embryological research (which I will discuss below). In 1856 His habilitated at Basel; and the following year his great opportunity arose: after the departure of the professor of anatomy and physiology, he received the call to the chair at his alma mater. Though he felt inadequate to the position—since his formal training only skirted the central areas of the subjects he was to teach—he yet believed he had at last found his life's work. His gradually changed his research orientation at Basel to embryology, the discipline in which he would achieve fame.

In 1869 His became rector at Basel and delivered an address that laid out his essential views about the independence of embryological research and its need to defend against foreign invasion from beyond the discipline, a threat he saw especially coming from some of Darwin's younger follow-

14. Ibid., 31–39.

15. There is no indication by either His or Haeckel that they knew each other during their three overlapping terms at Würzburg, but it is hard to believe that they would not have met within the confines of a small group of roughly two hundred medical students.

16. See, for example, Claude Bernard, *Introduction à l'étude de la médicine expérimentale* (1865; repr., Paris: Éditions Garnier-Flammarion, 1966), 29: "The observer must be a photographer of the phenomena; his observations must represent nature exactly. He must observe without a preconceived idea; the mind of the observer must be passive, that is to say, it must remain silent; it listens to nature and writes down what nature dictates."

ers. These latter had tried to argue that the development of the individual organism retraced the morphological stages of the development of the species.

> Though one might disregard the many possible doubts about the absolute truth of this proposition [i.e., the biogenetic law], one would not be in a position to hold it for the reason it was offered, namely, as a physiological explanation of the developmental history of the observed facts. The task remains for this science [embryology] to deliver, completely independently of the teachings on species formation, the physiological explanatory foundations of individual corporeal formation according to its own methods and experience.[17]

The mechanical principles that His proposed as proper to embryology held that anatomical structures were formed through differential growth of the various segments of the germ layers. (I will say more about these principles below.) No extra-embryological principles were needed to give a fully explanatory account of development.

A burgeoning reputation brought His a call to the chair of anatomy at Leipzig in 1872, and there he established a modern laboratory for anatomical research and produced his important studies of human ontogenetic development. In his inaugural address at Leipzig, he reiterated views about the task of embryology, which he believed was beginning to achieve a new foundation in exact science. Anatomy required precise description and measurement of parts, and the use of new devices to depict development and make relevant comparisons across species.[18] He again remarked that his science was too immature to adjudicate the thesis of embryological recapitulation of the ancestral stem-tree; more research about inheritance and the mechanisms of variation was still required.[19] But one area, in His's estimate, simply fell beyond the boundaries of scientific investigation: "whether a guiding consciousness is responsible for the origin of the world and our own origin."[20] Such questions, he thought, belonged to the realm of nonscientific belief.

In 1874 His publicly joined the attack on Haeckel that had been initially

17. Wilhelm His, *Ueber die Bedeutung der Entwickelungsgeschichte für die Auffassung der Organischen Natur* (Leipzig: F. C. W. Vogel, 1870), 30.

18. Wilhelm His, *Über die Aufgaben und Zielpunkte der wissenschaftlichen Anatomie* (Leipzig: F.C.W. Vogel, 1872), 4–11.

19. Ibid., 16.

20. Ibid., 13.

launched by his colleague Rütimeyer. He did so in *Unsere Körperform und das physiologische Problem ihrer Entstehung* (Our bodily form and the physiological problem of its origin), a book that embedded specific charges against Haeckel within a set of conservative epistemological principles.[21] These principles erected fact-braced walls against incursions into the professional sphere of embryology and into the private domain of consciousness. Indeed, virtually all the indictments of Haeckel through the turn of the century have to be understood as indices of three larger concerns and fears: about the nature and stability of science, about professional boundaries of competence in science and philosophy, and about the transcendent character of human beings. All three areas had been shaken by the seismic jolts whose epicenter was the library of Charles Darwin.

His intended *Unsere Körperform* for both a professional and a lay audience; and to that end, he filled its pages with numerous, finely crafted illustrations of embryos and embryonic structures. At the very beginning of the book, he made his purpose explicit: he would confront the challenge of descent theory with a set of epistemological considerations that would show the priority of a proximate mechanistic account to a remote phylogenetic account.[22]

In the book's first several chapters, His described the structures of the developing chicken embryo, which was to serve as the model for ontogenetic development; and he argued for a topological and mechanical explanation of the chick's emerging parts. He supposed that various areas of the germ layers, whence the structures and organs of the chick arose, displayed differential rates of growth. Thus in the embryo, "one part grows more quickly, another more slowly, one stops its growth sooner, another later, such that each grows in accord with its particular law."[23] As a result of these areas of uneven growth, regions of the germ layers would begin folding over on themselves, forming tubular structures (e.g., the neural tube, the vascular system) and other features of the developing organism.[24] His supposed that this mechanical activity of organ formation was due to different dispositions (*Anlagen*) scattered in distinct locations throughout the germinal layers:

21. Wilhelm His, *Unsere Körperform und das physiologische Problem ihrer Entstehung: Briefe an einen befreundeten Naturforscher* (Leipzig: F. C. W. Vogel, 1875 [1874]). The book bears two dates of publication, one on the cover and another on the title page, indicating a delay of publication.

22. Ibid., v.

23. Ibid., 19.

24. Ibid., 20–21.

> Every point in the embryonal spheres of the germinal layers must correspond to a larger organ or organ part; and every organ that stems from the germinal layer has its preformed disposition in some spatially determinable sphere on the plane of the layer.[25]

His called this the "principle of the organ-forming germinal regions" (*Princip der organbildenden Keimbezirke*). Though he undoubtedly thought the principle inductively grounded in what he understood as the mechanics of embryo development, it had no further warrant in his work. Two decades later, however, Wilhelm Roux would reactivate a similar principle and attempt to give it experimental justification in his version of *Entwickelungsmechanik*.[26]

In His's analysis, the theory of physiological mechanics conceptually derived the whole form of the organism from the proximate development of individual parts, while descent theory derived the development of the particular parts from more remote, antecedent forms.[27] Accordingly, His argued that the more proximate and, thus, the more certain had priority over the more remote and, consequently, the more speculative. He remained, however, oblivious to the analytical vacuity of his proposal of dispositional *Anlagen* to explain organ formation—that is, properties that had no experimental or observational justification other than what they were called on to explain. The implicit justification, if it can be called that, for invoking *Anlagen* was simply the assumption that modern science required any generalizations to be drawn from immediately observed facts by something like a necessary induction. The assumption of spatially fixed *Anlagen*, he supposed, would furnish the needed facts, despite the insubstantial character of these hypothetical dispositions. A bit later Haeckel would try to distinguish another kind of modern science, what he called "historical natural science" (see below). This new kind of science argued from a different set of assumptions, especially regarding the supposed separation of fact from theory and the prescribed derivation of the latter from the former.

Organic structures must be transmitted from generation to generation through inheritance. But the concepts of heredity found among descent theorists were, in His's estimation, utterly defective. Darwin's theory of pangenesis hardly differed from that of the Hippocratics and was subject to the same devastating criticisms that Aristotle directed against the ancient

25. Ibid., 19.
26. See chapter 6 for a discussion of Roux's theory.
27. His, *Unsere Körperform*, 1–2.

doctrine.[28] And when Haeckel reduced the actions of the hereditary material ultimately to attraction and repulsion of molecular elements—the same forces as those operative in the inorganic—His expostulated: "Where in the world, I ask you seriously, is a public to be found in 1874 that can make any sense of Mr. Haeckel's torrent of words?"[29]

His directed a chapter of his book to the question of the biogenetic law, which he regarded as the principal intrusion of descent theory into the realm of professional embryology. To establish a law in the proper sense, Haeckel would have had to produce a series of exceptionless observations, whence anything called a law would have to be demonstrated. Firstly, one would have to discount the many adaptations of the embryo to the maternal womb (e.g., the umbilical cord, the amnion, etc.), traits that simply could not be ascribed to adult ancestors who lived in vastly different circumstances.[30] Secondly, and more importantly, Haeckel simply failed to provide an observational series of paleontological forms and a parallel series of embryonic forms.[31] Only from immediate observations of such parallel series could a law be derived that would authorize phylogenetic conclusions from ontogenetic evidence. Finally, what embryonic forms Haeckel did depict were based, as Rütimeyer showed, on repeated woodcuts and distorted illustrations drawn from the likes of Bischoff and Ecker.

In an attempt to make the indictment of Haeckel even more damning, His provided measurements of embryo *illustrations*. He concluded that either Haeckel or his printer had lengthened the forehead of Bischoff's dog by 3.5 millimeters and reduced the forehead of Ecker's human embryo by 2 millimeters and doubled the length of the tail—all of this to make the illustrated embryos of dog and human more similar.[32] Yet what His neglected to mention—or failed to recognize—was Haeckel's normalization of size in his embryo representations. Haeckel explicitly forewarned his reader that he would enlarge or reduce one embryo to make it comparable in size to another in order to make morphological comparisons easier. He tried to keep the proportions of each image correct, but their absolute sizes (on which His seems to have based his complaint) might differ from the original depictions from which he borrowed. Moreover, he likely used several images of a given species, drawn from different authors, to construct a compound or standardized morphological representation. His simply

28. Ibid., 132–36.
29. Ibid., 141.
30. Ibid., 167.
31. Ibid., 166–68.
32. Ibid., 170.

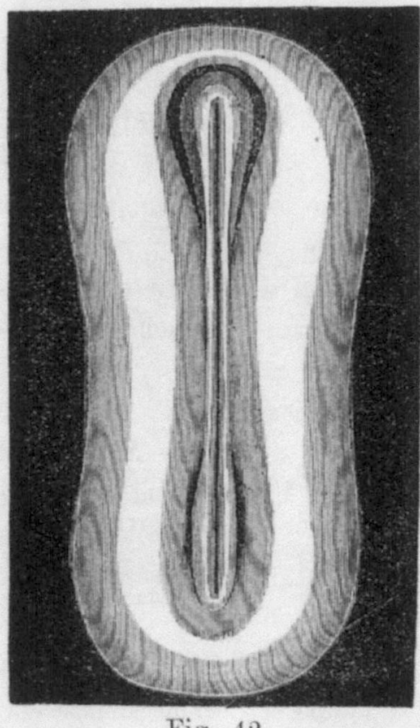

Fig. 8.2. Human sandal embryo. (From Haeckel, *Anthropogenie*, 1874.)

assumed that Haeckel should have transferred the basic illustrations of Bischoff and Ecker to the pages of the *Natürliche Schöpfungsgeschichte* without any comparative standardizing or regularizing of the images.

In his bill of particulars, His pointed to two illustrations from Haeckel's *Anthropogenie* that he contended were not merely modified versions of other illustrations but completely invented (*erfunden*). One was of a human embryo (about two weeks old) at the sandal stage showing the primitive streak where the notochord would form (see fig. 8.2). His declared: "Fig. 42, the primitive conceptus of man [*Urkeim des Menschen*] in the form of a shoe sole, magnified 40 times, is invented. No observer has seen this stage."[33] He suspected that Haeckel simply altered Bischoff's illustration of a rabbit embryo with a primitive streak and passed it off as human. Haeckel, perhaps intimidated by His's confident rejection of this stage of development in human beings, omitted the figure from the third (1877) and fourth (1891) editions of the *Anthropogenie*.

33. Ibid. The figure was repeated by Haeckel as figure 60 and compared with a comparable representation of a chicken and rabbit at the sandal stage.

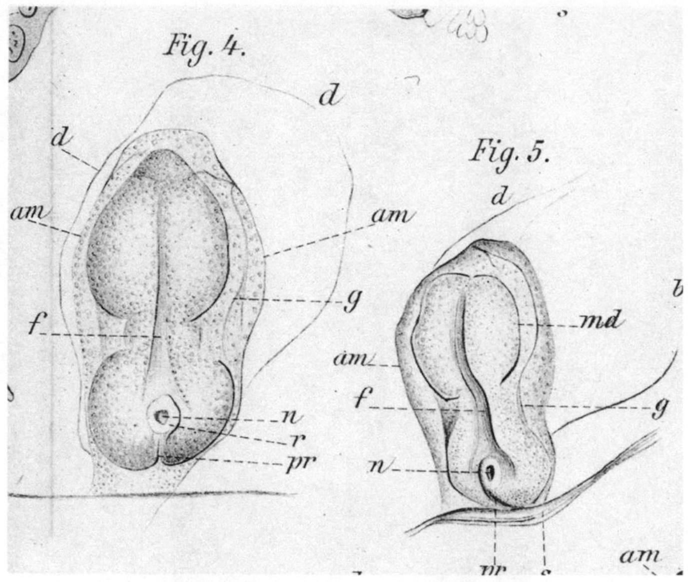

Fig. 8.3. Human sandal embryo with primitive streak (marked *f*). (From Ferdinand Graf von Spee, "Beobachtungen an einer menschlichen Keimscheibe," 1889.)

However, in the late 1890s, another embryologist of growing reputation, Ferdinand Graf von Spee (1855–1937), identified the medulary stage in human beings and provided a figure of the structure (fig. 8.3, *f* marks the *Medullarfurche*, or primitive streak).[34] Undoubtedly feeling vindicated, Haeckel adopted (with attribution) von Spee's illustration in the fifth (1903) and sixth (1910) editions of the *Anthropogenie* (fig. 8.4).

The other illustration that His questioned pictured two human embryos at early stages, each showing a freely projecting allantois sac (see fig. 8.5; the allantois is the small pouch in fig. 82 extending to the right of the embryo, just above the larger yolk sac, and in fig. 83, below the body of the embryo). His asserted: "Further invented are the two figures of human embryos, p. 272, in which an allantois is not only pictured but expressly described as a 'considerable little sac' [*ansehnliches Bläschen*] (but in man it is known that the sac form is never visible)."[35] In this latter instance, His's own supposition was disputed by the Göttingen anatomist

34. Ferdinand Graf von Spee, "Beobachtungen an einer menschlichen Keimscheibe mit offener Medullarrinne und Canalis neurentericus," *Archiv für Anatomie und Entwickelungsgeschichte* (1889): 159–76 and plate 11. Haeckel probably did not initially observe the primitive streak in the early human embryo and only projected it on the basis of the biogenetic law.

35. His, *Unsere Körperform*, 70.

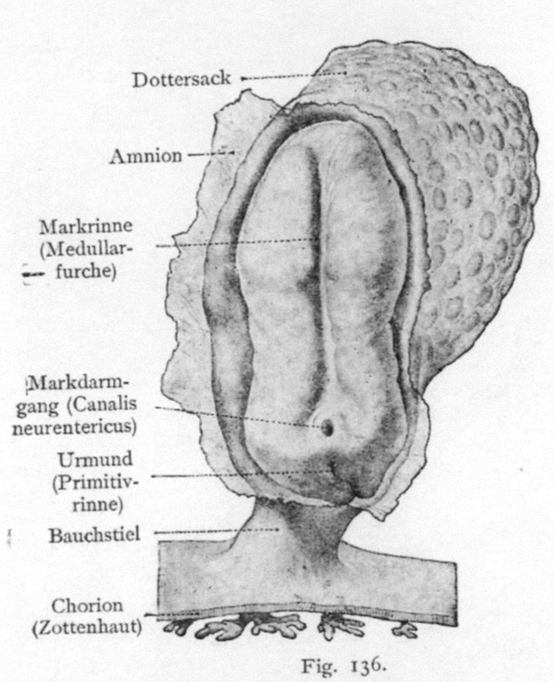

Fig. 8.4. Human sandal embryo with primitive streak (*Medularfurche*). (From Haeckel, *Anthropogenie*, 5th ed., 1903.)

Wilhelm Krause (1833–1910), who countered that he had a human embryo that showed the allantois as a quite visible sac. For several years thereafter, embryologists debated the presence of a pendant allantois, and Haeckel, feeling justified for what seems to have been more a prediction than an observation, retained the illustration.[36]

Both of the charges by His do seem to have caught Haeckel projecting features onto the human embryo on the basis of general vertebrate embryology. The projections were ultimately vindicated, since the human embryo does show a primitive streak at about the stage Haeckel proposed and the allantois is a distinguishable saclike structure in the very early human

36. Nick Hopwood follows out in illuminating detail the battle that Krause initiated. See his "Producing Development: The Anatomy of Human Embryos and the Norms of Wilhelm His," *Bulletin of the History of Medicine* 74 (2000): 29–79. Oscar Hertwig—the Hertwig brother who fell away from his teacher Haeckel—entered into the discussion and in his textbook of human development supported the conclusion of His. See Oscar Hertwig, *Lehrbuch der Entwicklungsgeschichte des Menschen und der Wirbelthiere*, 2nd ed. (Jena: Gustav Fischer, 1888), 190.

272 Mutterkuchen oder Placenta. XII.

hat (Fig. 78, 5, *al*), bald in ein sehr wichtiges Organ, das die Ernährung des Embryokörpers durch das Blut der Mutter vermittelt. Dies Organ ist der sogenannte Aderkuchen oder Mutterkuchen (*Placenta*) (Fig. 83, Fig. 84, *pl*). Die Blutgefässe, welche sich in dem Darmfaserblatte auf diesem Sacke ausbreiten, wachsen sehr stark, und entwickeln sich besonders mächtig an der Stelle, wo der Sack die innere Fläche der

Fig. 82.

Fig. 83.

Fig. 8.5. Human embryo with allantois sac extending to the right. (From Haeckel, *Anthropogenie,* 1874.)

embryo.[37] Yet he does seem to have gone beyond what could be clearly demonstrated at the time. He thus felt the rhetorical punches of His's series of expostulations of *erfunden*—invented!

His finished his extensive indictment of Haeckel with a series of de-

37. See Roland Bender, "Der Streit um Ernst Haeckels Embryonenbilder," *Biologie in unserer Zeit* 28 (1998): 157–65.

tailed illustrations of embryos based on his own measurements, those of a human and of an array of animals similar to the ones that Haeckel displayed at various stages of development in the *Anthropogenie*. With these graphic devices, the Leipzig anatomist attempted to show that "an identity in the external form of the animal embryos, which has so often been maintained, does not exist."[38] His did neglect to mention, however, that Haeckel, in the *Anthropogenie*, never claimed the embryos were identical at early stages, only very similar; and in later editions of his book, the displayed embryos grew more dissimilar as the depictions grew more graphically refined.

Alexander Goette

Haeckel's two popular books incited negative responses from many other biologists, though usually without the same vitriol and obloquy as dispensed by Rütimeyer and His. One such researcher was Alexander Goette (1840–1922), a *Privatdozent* (1872) and then *Extraordinarius* professor (1877) at Strassburg. Goette had the experience to render a considered judgment on Haeckel's embryological theses and the stature to garner a hearing in the scientific community. He had published, in 1875, a comprehensive and extraordinarily detailed study of a species of toad, the *Bombinator igneus*—a study that brought him advancement to extraordinary professor. He concluded his study of the *Unke*, or toad, with a chapter devoted to evolutionary theory and Haeckel's biogenetic law.[39]

Like many leading biologists, Goette professed to be stunned upon reading Darwin's *Origin of Species* and could only express his respect for the accomplishment. He still could not, however, endorse the letter of the Englishman's theory, but neither could he turn back to unregenerate creationism. Like His and Kölliker, he acknowledged that species must have altered over time; he conceived of the transformations as guided, in a vaguely unspecified way, by a law-governed process.

Goette rejected the Darwinian-Haeckelian device of natural selection as based on a contradiction: the principle assumed that creatures must initially display a modified or varied form, which was then supposed to be selected; but if the modification had to be prior to the operation of selection, selection could not explain the altered form in the first place.[40] Moreover,

38. His, *Unsere Körperform*, 201.

39. Alexander Goette, *Die Entwickelungsgeschichte der Unke (Bombinator igneus)*, 2 vols. (Leipzig: Leopold Voss, 1875).

40. Ibid., 1:891.

Fig. 8.6. Alexander Goette (1840–1922). (Courtesy of the Smithsonian Institution.)

Haeckel offered no real explanation of inheritance. Since he believed the protoplasmic substance to be unorganized, to be without form—having only molecular properties of attraction and repulsion—some causal law of form (*Formgesetz*) would have to be operative in the developing embryo to reproduce likeness to the parent. Yet Haeckel specified no such mechanical law.[41] This meant that transmission of the parental phylogenetic pattern to the offspring was left without account, and so the biogenetic law failed of any explanatory power.[42] Therefore, according to Goette, the principles of

41. Ibid., 1:589–90.

42. Goette applied the same reasoning to Haeckel's assumption of the spontaneous generation of the first living monad: "Since I have explained that a real life is unthinkable without a law of form (*Formgesetz*), which in no way is a property of the material molecule in itself . . . I obviously cannot endorse Haeckel's hypothesis of the autogenesis of the first organism." See ibid., 1:889.

heredity and adaptation that Darwin and Haeckel's theory proposed simply could not render intelligible transformations either in the individual or the species.

Goette's own theory of embryogenesis made many of the same assumptions as that of His. Like His, Goette argued that the form of the embryo developed from mechanical causes under the guidance of a law, which he termed the "law of form" (*Formgesetz*). Since this law operated repeatedly in the offspring, "small and rather inconspicuous changes [would be] unavoidable"—thus new modifications would be introduced in the offspring and the species would gradually alter.[43] Goette concluded his analysis of Haeckel's biogenetic law by sounding what to an evolutionist's ears could only be a dissonant theological chord: "It is thus clear that if one does not allow in place of that connection [between phylogeny and ontogeny] supernatural [*übernatürliche*] adaptational and hereditary processes, an individual developmental course according to a type cannot begin."[44] Perhaps Goette only meant that law as such must stand above natural events; and perhaps, like His, he was merely erecting a professional wall against outside enemies. Haeckel would nonetheless hear in that remark the sound of Gabriel's horn.

Albert von Kölliker

The critic that disappointed Haeckel rather than enraged him was Albert von Kölliker (1817–1905), his onetime teacher and friend at Würzburg (fig. 8.7). Kölliker had introduced the young medical student to the field in which he would subsequently exercise his research abilities, marine biology; and the professor even enticed the tyro into repeating a course in developmental biology, so "extraordinarily clear, morphologically insightful, and beautifully illustrated were Kölliker's lectures."[45] After Haeckel left Würzburg, the relationship between the two grew even warmer, as the protégé found ever greater success in research. In 1865 Kölliker tried to lure his former student back to Würzburg as professor of zoology, declaring: "I, for my part, will do everything to make your position here comfortable."[46]

43. Ibid., 1:901.

44. Ibid., 1:903.

45. Haeckel to his parents (17 May 1855), in *Entwicklungsgeschichte einer Jugend: Briefe an die Eltern, 1852–1856*, ed. Heinrich Schmidt (Leipzig: K. F. Koehler, 1921), 137.

46. Kölliker to Haeckel (3 April 1865), in "Über die Beziehungen zwischen Albert Koelliker und Ernst Haeckel," ed. Georg Uschmann, *Wissenschaftliche Zeitschrift der Friedrich-Schiller-Universität* (Mathematisch-Naturwissenschaftliche Reihe) 25 (1976): 127.

Fig. 8.7. Albert von Kölliker (1817–1905). (Courtesy of the Smithsonian Institution.)

The invitation came a year after Kölliker had analyzed the Darwinian hypothesis and found it wanting.

In that 1864 article, "Ueber die Darwin'sche Schöpfungstheorie" (On the Darwinian theory of creation), Kölliker evaluated several objections that could be brought against the theory and concurred with some of them. He found special fault with what he regarded as Darwin's *teleological* mode of arguing.[47] According to Kölliker, Darwin assumed that all traits of an organism were "the best" for that creature, and he further supposed that the unities of type organized species into a "wonderful harmony."[48] Darwin,

47. Albert von Kölliker, "Ueber die Darwin'sche Schöpfungstheorie," *Zeitschrift für wissenschaftliche Zoologie* 14 (1864): 174–86. See also Remigius Stölzle, *A. Von Köllikers Stellung zur Descendenzlehre: Ein Beitrag zur Geschichte moderner Naturphilosophie* (Munster: Aschendorffschen Buchhandlung, 1901).

48. Kölliker, "Ueber die Darwin'sche Schöpfungstheorie," 175. The offending passage in Darwin's *Origin* to which Kölliker referred reads: "Hence every detail of structure in every living creature (making some little allowance for the direct action of physical conditions) may be

in Kölliker's view, presumed that only natural selection could explain these teleological relationships. Yet, Kölliker observed, we find a general harmony in the non-organic world—chemical affinities, for instance—that cannot be explained by natural selection. Moreover, there was no reason to suspect that all traits of organisms should be useful; we needed only hold that creatures would have enough useful properties to survive and to function. General law seemed to Kölliker a sufficient explanation of harmony in the world. He remained mute about the nature of general law, except to suggest it derived from higher powers.

As an alternative to Darwin's conception, Kölliker advanced a "theory of heterogeneous generation."[49] According to this view, from a few primitive types might spring organisms of a different and more advanced sort—similar to the way in which medusae at one stage in their generational cycle would produce simple polyps from which would bud more complex medusoid forms. Kölliker assumed that "at the foundation of the origin of the whole organized world lay a great developmental plan that drove the simplest forms to ever more variable unfolding."[50] Obviously Kölliker had yet to abandon a theological foundation for his conception of species change.[51]

Given his assessment of Darwinian theory generally, it is not surprising that Kölliker would find Haeckel's biogenetic law unpersuasive. In the second edition (1879) of his *Entwicklungsgeschichte des Menschen und der höheren Tiere* (Developmental history of man and the higher animals),[52] he directed a mildly expressed but firmly stated set of objections to his former

viewed, either as having been of special use to some ancestral form, or as being now of special use to the descendants of this form—either directly, or indirectly through the complex laws of growth." See Charles Darwin, *On the Origin of Species* (London: Murray, 1859), 200. Kölliker used the first English edition. What Kölliker failed to appreciate is that Darwin did not claim that all *extant* traits of an organism were useful or the best. Huxley thought Kölliker's argument a remarkable misreading of the *Origin*. And he was no more satisfied with Kölliker's own hypothesis of alternating generation as a model for the production of new species out of old. See Thomas Henry Huxley, "Criticisms on 'The Origin of Species,'" *Natural History Review* (1864): 566–80.

49. Kölliker, "Ueber die Darwin'sche Schöpfungstheorie," 181–86.

50. Ibid., 184.

51. In her lucid and otherwise persuasive account of Kölliker's developmental theory, Lynn Nyhart suggests that he avoided altogether any reference to a theological support for natural laws. See her *Biology Takes Form: Animal Morphology and the German Universities, 1800–1900* (Chicago: University of Chicago Press, 1995), 127. It could be, however, that Kölliker's reference to "a great developmental plan" had less theological import than I am attributing to it.

52. Albert von Kölliker, *Entwicklungsgeschichte des Menschen und der höheren Tiere*, 2nd ed. (Leipzig: Wilhelm Engelmann, 1879).

student's principal conception. While he agreed that the ontogeny of organisms generally displayed features of the evolutionary history of the phylum, he yet cautioned that many properties specific to the organism's life in the womb (e.g., amnion, allantois, etc.) would mislead anyone attempting a phylogenetic reconstruction. Moreover, even such closely related species as rabbit and guinea pig had quite different ontogenetic histories, something not explained by phylogenetic causation.[53] The biogenetic law simply did not and could not explain exactly how hereditary transmission of earlier forms operated; nor could it make clear what would be elided or altered in the cenogenetic compression of embryological development.[54]

For His, Goette, and Kölliker, Haeckel's biogenetic law displayed two related defects: it focused on external and remote determination of individual development, instead of respecting more proximate causes; and it lacked the epistemological stability of the sort provided by mechanical laws in the physical sciences. Behind these explicit objections, however, lay both the professional defense of a discipline and subtle theological concerns. The new professional embryologists argued that developmental biology had its own set of intrinsic principles and should not brook ingressions from outside the discipline, such as those represented by evolutionists and comparative morphologists. Right through the early twentieth century, embryologists would stand watch over the integrity of their domain. In the case of the three embryologists who contributed significantly to the early foundation of the discipline, each revealed a lingering and surprising attraction to higher powers. Haeckel would mount biting responses to these objections, and in the wake of his dispute with two other adversaries—Du Bois-Reymond and Virchow—he would formulate a new epistemological conception, that of "historical natural science" (discussed below). Then there was, of course, the objection mounted initially by Rütimeyer and elaborated by His that Haeckel's argument depended on fraudulent evidence.

Haeckel's Responses to His Critics

In the second edition (1870) of his *Natürliche Schöpfungsgeschichte*, Haeckel removed the triplicate reproduction of clichés but did not otherwise respond to Rütimeyer's charge of fraud. Instead, in a diversionary

53. Ibid., 6–7, 392–93.
54. Ibid., 391–92.

move, he took up three more general objections to the first edition of his book: namely, that he had gone beyond what Darwin had proposed, that he had advanced an utterly materialistic theory, and that he had thus settled into an atheistic construction of nature. Haeckel responded to the first charge by admitting that he did move beyond the founder's original doctrine, though insisting that he nonetheless worked within the Darwinian framework and that he offered his phylogenetic reconstructions only as a tentative basis for further development.[55] Those protesting that evolutionary theory was materialistic failed to distinguish two kinds of materialism: ethical materialism and natural-philosophic materialism, or monism. The former regarded sensuous pleasure rather than knowledge of nature to be the goal of life and, as Haeckel provocatively suggested, was more often to be found in the "palaces of ecclesiastical princes."[56] The latter kind of materialism simply recognized what Kant took to be the chief explanatory principle of natural science, namely, mechanism. The monistic philosophy, however, regarded whatever might be called matter as equally entitled to be considered spirit. This meant, in Haeckel's view, that there was a fundamental unity between the inorganic and the organic, between body and mind—and thus no insurmountable barriers had to be traversed at the dawn of life or at the inception of consciousness. Here again Haeckel invoked the Goethean understanding of reality, a metaphysics that he had sketched in the *Generelle Morphologie*. He also reiterated his endorsement of Goethe's Spinozistic naturalism, which he regarded as the only true monotheism: *Deus sive natura*.[57]

Between the second and third editions of the *Natürliche Schöpfungsgeschichte* fell a crucial event, the publication of Darwin's *Descent of Man* (1871). Haeckel considered this to be a propitious vindication of his own application of evolutionary doctrine to human beings. It also offered him an opportunity to respond to Rütimeyer without quite answering the charge of fraud. His nemesis had suggested that he had strayed beyond the confines that Darwin himself had established; but this contention could now be dismissed by pointing to the *Descent*. With that objection neatly dispatched, Haeckel then took aim, not at the main complaint but perhaps one close enough to make it appear as though he had dealt with the original charge of fraud. He pictured Rütimeyer as so disoriented in his furor against the

55. Haeckel, *Natürliche Schöpfungsgeschichte*, 2nd ed. (1870), xxiv–xxv.
56. Ibid., 33.
57. Ibid., xxvi–xxx.

new theory as to "deny the formal identity of the eggs and early embryos of human beings and closely related mammals." But this flew in the face of common anatomical knowledge: "That no individual is in a position to distinguish the human egg from those of closely related mammals—even with the help of the best microscope—has been a long-acknowledged, if not properly respected, fact that can be found in almost every handbook."[58] Haeckel thus replied to Rütimeyer's objections and rather dexterously deflected the central charge of fraud.

Perhaps stimulated by the cogency of his own response to Rütimeyer, Haeckel also took a jab at His, saying that the Basel professor had offered a mechanistic theory of embryogenesis of a kind that "those acquainted with the facts of comparative anatomy and ontogeny could only regard with a smile."[59] Likely it was this insulting slight that impelled His to undertake a full-length study of the mooted questions in his formidable *Unsere Körperform*.

With the publication of His's book, Haeckel recognized that it would not suffice to issue offhand responses to his critics. Within a few months, he composed a small monograph to deal with mounting objections to evolutionary theory and his particular interpretation of it. In October 1875 he published *Ziele und Wege der heutigen Entwickelungsgeschichte* (Aims and methods of contemporary developmental history), whose title echoed His's 1872 inaugural address at Leipzig. Pointedly, the monograph was dedicated to "the old master of developmental history, Carl Ernst Baer."

After indicating what his monistic theory entailed, especially concerning the biogenetic law, Haeckel began his critical considerations on a conciliatory note, admitting that his principal opponents, His and Goette, had made great contributions to the "purely empirical and especially the technical part" of embryology.[60] Haeckel then rehearsed the essential positions of each, emphasizing their doubts about the general validity of Darwinian theory, especially what they regarded as the uncertain subtheses about inheritance and adaptation. Haeckel placed these professed uncertainties against Goette's certain avowal that adaptation and heredity were "supernatural processes."[61] He located the nub of his dispute with His at the insufficiency of the latter's developmental mechanics:

58. Ernst Haeckel, *Natürliche Schöpfungsgeschichte*, 3rd ed. (Berlin: Georg Reimer, 1872), xxxiv.

59. Ibid., xxxvi.

60. Ernst Haeckel, *Ziele und Wege der heutigen Entwickelungsgeschichte* (Jena: Hermann Dufft, 1875), 11.

61. Ibid., 16–17. Haeckel cites Goette's remark in the *Unke*, 901. See the discussion above.

> Here I depart fundamentally from the explanatory path of His. I turn to phylogeny to clarify the historical origin of the different forms of growth and seek their completely sufficient explanatory foundation in the mutual causality of inheritance and adaptation. His holds this "roundabout way" to be utterly superfluous and seeks to clarify ontogeny directly from itself.[62]

To this summation, Haeckel could not help but add: "Thereby he adopts the well-known technique of Münchhausen, who drew himself out of the swamp by his own pig-tail."[63]

The essence of Haeckel's criticism of His, as well as of Goette and Kölliker, is not that the mechanical processes of differential growth and folding do not occur—though how these actually work, "we get no word from His." Rather, what we want to know is why the processes function in the way they supposedly do: "For each of these simple ontogenetic folding processes is a highly developed historical result that has been determined causally through thousands of phylogenetic modifications by processes of inheritance and adaptation."[64] Ontogenetic development, Haeckel insisted, is a historical phenomenon and requires a historical explanation. Such explanation does not abandon mechanical causes of the kind His and Goette propose but seeks to understand them through the adaptational and hereditary history of the species.

In this extensive response to his critics, Haeckel could not avoid the charges, recently reiterated by His, of falsification of his embryo illustrations. He said he would offer a fuller explanation at another time but would respond at this point with the observation that "I believe that for didactic purposes simple schematic figures (especially for a larger public) are far more useful and instructive than illustrations that are as faithful to nature as possible and most carefully executed." In producing such didactic models, one must, Haeckel asserted, include only the most important features and leave out the unimportant. In this sense, "all schematic illustrations are as such invented [*erfunden*]." Haeckel, nonetheless, confessed that "in the use of schematic figures, I have here and there gone too far and regret very much that many of them (partly through my own fault, partly through the fault of the woodcutter) have rather badly missed the mark."[65] A year later, in 1876, Haeckel returned again to reply to the charge of falsification, this time in the

62. Haeckel, *Ziele und Wege*, 21.
63. Ibid.
64. Ibid., 24.
65. Ibid., 37, 38.

foreword to the third edition of his *Anthropogenie*. The response, though, was basically in the same terms that he had offered in *Ziele und Wege*—that is, the usefulness of schematic drawings for didactic purposes.[66]

Though Haeckel's critics continued their attacks on various of his ideas, they did not join personal antagonism with professional rebuke in the same measures as doled out by Rütimeyer and His[67]—until, that is, 1891, when the Kiel zoologist Victor Hensen reignited the smoldering matter of the three woodcuts. Haeckel himself had provoked Hensen by disparaging the organization and methods of a marine research expedition led by the latter in the summer of 1889. Haeckel thought that the enormous expenditure of funds from the government and from the Humboldt-Stiftung—in amounts unavailable to other German biologists[68]—was a complete boondoggle, especially when the results were compared to the success of the British *Challenger* expedition. Hensen had botched everything: from season of study to inadequate techniques of dredging to making improbable counts of individual infusoria, diatoms, and crustaceans. Haeckel maintained that all of these errors rendered "completely worthless" the statistical analyses undertaken by Hensen and his colleagues.[69] Hensen rejoined with a broadside

66. Ernst Haeckel, *Anthropogenie; oder. Entwickelungsgeschichte des Menschen*, 3rd ed. (Leipzig: Wilhelm Engelmann, 1877), xxv.

67. Personal attacks continued, though without the bruising effect of his principal antagonists. The Würzburg anatomist Carl Semper, for instance, had issued two critical pamphlets whose roundhouse swings only produced a small breeze: *Der Haeckelismus in der Zoologie* (Hamburg: W. Mauke Söhne, 1876) and *Offener Brief an Herrn Prof. Haeckel in Jena* (Hamburg: W. Mauke Söhne, 1877). The latter berated Haeckel for making "Darwinism into a religion" (8).

68. Hensen received over 100,000 marks from the Kaiser and the Humboldt-Stiftung for his expedition. Earlier Haeckel himself had been turned down for research money from the same sources. For his trip to Ceylon, his friends had made the request, on his behalf, of 12,000 marks for a five-month excursion to study medusae and coral. He learned that Helmholtz, among others, had voted for the grant, but that Virchow and Du Bois-Reymond had objected, protesting that Haeckel, "though a prominent researcher of the Darwinian error-doctrine [*Darwin'schen Irrlehren*] and scientific materialism, had injured science more often than aided it and that [his] monographs (radiolarian, sponges, medusae, etc.) were without real value." This, at least, is the account of the debate in the Academy of Science that Haeckel had heard about and reported to his friend Huxley. See Haeckel to Huxley (21 June 1881), in "Der Briefwechsel zwischen Thomas Henry Huxley und Ernst Haeckel," ed. Georg Uschmann and Ilse Jahn, *Wissenschaftliche Zeitschrift der Friedrich-Schiller Universität Jena* (Mathematisch-Naturwissenschaftliche Reihe) 9 (1959–60): 7–33; citation from 28. The opposition by Virchow and Du Bois-Reymond to Haeckel's funding came just a few years after the great row between Virchow and his former student, a matter discussed below.

69. Ernst Haeckel, *Plankton-Studien: Vergleichende Untersuchungen über die Bedeutung und Zusammensetzung der Pelagischen Fauna und Flora* (Jena: Gustav Fischer, 1890), 9–10, 20, 36, 56–57, 58–59 (especially), 81–82, 88–89 (especially), 96–97. Haeckel took special exception to Hensen's conclusion that his statistical studies would demonstrate the clear distinctions among species (101). Haeckel on the contrary believed that his morphological studies showed

that sprayed the Jena Darwinian with moralizing invective from one end of his reputation to the other, without, however, squarely hitting any of the methodological objections. Hensen not only rehearsed the charges made two decades earlier; he flung enough other poisoned-tipped criticisms so as to make response a matter of honor.

Hensen began the indictment by offering his own bona fides. He had, he modestly admitted, both sufficient expertise to confront Haeckel—based on fishing up plankton from the high seas for some nine years—and courage to stand up to one of the foremost researchers in Germany. Despite the challenge, he thought duty required him to make the effort. So he began the discussion by reminding his readers of matters that had nothing to do with the dispute about methods of plankton study, namely, Haeckel's "deception" with the three woodcuts of eggs and sandal embryos and thus "the great shameful sin he had committed against scientific truth," which His and Rütimeyer had uncovered years ago. Moreover, His had demonstrated that Haeckel had not only copied human and dog embryos from Ecker and Bischoff but had altered them "so that they, contrary to truth, would be as similar as possible."[70] Hensen also attempted to sink the biogenetic law and Haeckel's gastraea theory. He maintained that apparent embryological similarity of vertebrates, for instance, was the result of limited possibilities for the arrangement of parts and of their small size—though His's more exact observations showed considerable differences among them.[71] This was also the case for the assumed similar stage of gastrulation in a range of animals. Hensen dismissively remarked that Haeckel himself found so many exceptions to the general rule of gastrulation that he ran out of Greek words to describe them.[72]

Haeckel took the opportunity of his about-to-be-published fourth edition (1891) of the *Anthropogenie* to answer the resurrected charges. In the "Concluding Apologia," he described His's objections to the illustrations of embryos, which the Leipzig anatomist depicted as "part utterly untrue and part completely invented."[73] Haeckel said his friends urged him to take legal action against Hensen, but he knew an ordinary court would

---

that Darwin was correct, that the concept of species was arbitrary, and that presumed species faded into one another.

70. Victor Hensen, *Die Plankton-Expedition und Haeckel's Darwinismus, ueber einige Aufgaben und Ziele der beschreibenden Naturwissenschaften* (Kiel: Lipsius & Tisher, 1891), 10–11.

71. Ibid., 55–59.

72. Ibid., 59–63.

73. Ernst Haeckel, *Anthropogenie; oder, Entwickelungsgeschichte des Menschen*, 4th ed., 2 parts (Leipzig: Wilhelm Engelmann, 1891), 2:858.

not find for him, even if a court of his professional peers would. He mentioned that he had long ago responded to these charges in *Ziele und Wege*, but that Hensen's libel required him once again to give account, which he proceeded to do in five points.[74]

First, he acknowledged that most of his illustrations in the *Anthropogenie* were adopted from well-known biologists—a practice with considerable precedent (and, in fact, he had labeled his borrowings as such).[75] Second, most of the figures were not "exact and perfectly faithful-to-nature illustrations," but "diagrams and schematic figures, that is, illustrations, which show only the essential features of the object and dispense with the inessential"—of the sort His himself employed. Third, all such illustrations found in handbooks and used for instruction are as such "invented" (*erfunden*), that is, the researcher "alters the real form of his object to correspond to the conception that he has formed of the essence of the thing and dispenses with all unnecessary and distorting accidents." Fourth, His should recognize that all thinking morphologists—von Baer, Müller, Gegenbaur, Huxley—produce comparably contrived illustrations. "They all represent the object to be illustrated in their diagrams, not as they actually *see* it but as they *think* it!"[76] Fifth, "accordingly, only the photograph is, in the view of His (and many other 'exact' pedants), completely free of blemish and virtuously pure." With these five points Haeckel asked his readers, figuratively, of course, to render a verdict. These points do allow a verdict of sorts, especially concerning crucial epistemological issues, to which I'll return below.

In his "Apologia," Haeckel did come to the sensationally damning charge of the repeated woodcuts. He confessed that

> professional colleagues know that it was a highly rash kind of madness that I committed, in good faith, with the quite rushed presentation of a few illustrations in the first edition of my *Natural History of Creation*. I illustrated with three *identical* figures three *very similar* objects, so similar that, as is known, no embryologist is in a position to distinguish them. . . . Already in the second edition I corrected this formal lapse, which gave me the external appearance of making a scientifically false representation.[77]

74. Ibid., 2:859–60.

75. The illustrations in the subsequent editions of the *Natürliche Schöpfungsgeschichte* remained, however, without attribution.

76. Ibid., 2:860: "Sie alle stellen den abzubildenden Gegenstand in ihren Diagrammen nicht so dar, wie sie ihn wirklich *sehen*, sondern wie sie ihn sich *denken*!"

77. Ibid., 2:161–62.

This was the fullest confession that Haeckel made of the damaging misstep that he took some quarter of a century earlier. He lamented that Hensen renewed the charge, which was picked up in the newspapers. The *Leipziger Zeitung* sprang the trapdoor once again: "Haeckel has only one fault—he is no researcher or scholar [*kein Forscher und kein Gelehrter*]. As a worker in science he long ago lost his credit, and as a Darwinist his position is completely in dispute."[78]

Haeckel made a mistake, but not one that should have damned him professionally for life. While some segments of the scientific community continued to disparage him, the tide of accolades through the turn of the century reached a flood and swamped the lingering opposition. He accumulated honorary degrees from Edinburgh (1884), Cambridge (1898), Uppsala (1907), and Geneva (1909); and he received awards from the Natural and Medical Society in Amsterdam (1890), the Linnaean Society of London (1894), the Royal Society of England (1899), and the Academy of Science of Turin (1900).[79] In finding the just balance in weighing the scientific accomplishments of the man, these celebrations by peers need to be kept in mind. The weight of his scientific work marked him as a man of tremendous industry, and the ranging power of his ideas revealed a scientific genius of remarkable capacities. Yet should we regard him as a flawed genius, one who looked not to the future but to the past, one who attempted to construct his science using implements that would inevitably lead, if not to conscious deception, then to something like deception in practice? If he were quite seriously intent on precision and fact-based science, should he not have employed the available technology of photography as His had, and thus have avoided the liabilities of subjective illustration?

The Epistemology of Photograph and Fact: Renewed Charges of Fraud

In an influential article, Lorraine Daston and Peter Galison have argued that the second half of the nineteenth century became dominated by efforts of scientists to introduce "mechanical objectivity" into the study of a great variety of subjects, but especially living organisms. The kind of objectiv-

78. Ibid., 2:162, citing the *Leipziger Zeitung* (no. 124).

79. When word leaked out in 1908 that the Nobel Prize in literature would be awarded to someone at Jena, congratulations began pouring in to Haeckel. Much to Haeckel's disappointment and chagrin, the prize actually went to his colleague in philosophy Rudolf Eucken. See Uwe Hoßfeld, Rosemarie Nöthlich, and Lennart Olsson, "Haeckel's Literary Hopes Dashed by Materialism," *Nature* 424 (2003): 875.

ity that the photograph represented would remove subjective temptations to particular interpretations, inclinations to see the world in a preferred way, and would introduce precise, dispassionate, and honest depictions of objects of concern. "As the photograph promised to replace the meddling, weary artist, so the self-recording instruments promised to replace the meddling, weary observer." According to the authors, while eighteenth-century scientists, such as Goethe, attempted to represent the archetypal pattern of an organism, "late-nineteenth-century anatomists and paleontologists believed that only particulars were real, and that to stray from particulars was to invite distortions in the interests of dubious theories or systems."[80] Photography and other mechanical devices enforced scientific humility in the face of particular facts.

In respect of Daston and Galison's historical argument, Haeckel seems a throwback to an earlier time. And Daston, in a more recent article, does represent him as epistemologically atavistic, cast more in the mold of Goethe than of contemporaries like His.[81] Haeckel, admittedly, contrived to represent in graphic form the essence of embryological stages, what the scientist "thought" was necessary, and to dispense with accidental and particularizing features of organisms. Examined through the lens of Goethean morphology, he might appear to have stumbled at the vertiginous edge of subjective prejudice, while His, microtome in hand, might be seen striding along the ascending road to modern science.

This understanding of Haeckel and his times has supported renewals of the charge of fraud in the present day, especially when the mechanical objectivity of photographs is brought into play. The embryologist Michael Richardson and his colleagues compared Haeckel's illustrations in the *Anthropogenie* (see fig. 7.10) with their own exact photographs of embryos of equivalent species and deemed Haeckel to have misrepresented his specimens (see fig. 8.8). *Science* magazine, describing Richardson's analysis and interviewing him, labeled its report "Haeckel's Embryos: Fraud Rediscovered." Richardson said of Haeckel's illustrations: "It looks like it's turning out to be one of the most famous fakes in biology."[82] The popular press repeated the indictment, carrying such headlines as the London *Times*

80. Lorraine Daston and Peter Galison, "The Image of Objectivity," *Representations* 40 (1992): 83, 91.

81. Lorraine Daston, "Objectivity versus Truth," in *Wissenschaft als kulturelle Praxis, 1750–1900*, ed. Hans Bödeker, Peter Reill, and Jürgen Schlumbohm (Göttingen: Vandenhoeck & Ruprecht, 1999), 28–29.

82. Michael Richardson as quoted by Elizabeth Pennisi, "Haeckel's Embryos: Fraud Rediscovered," *Science* 277 (1997): 1435.

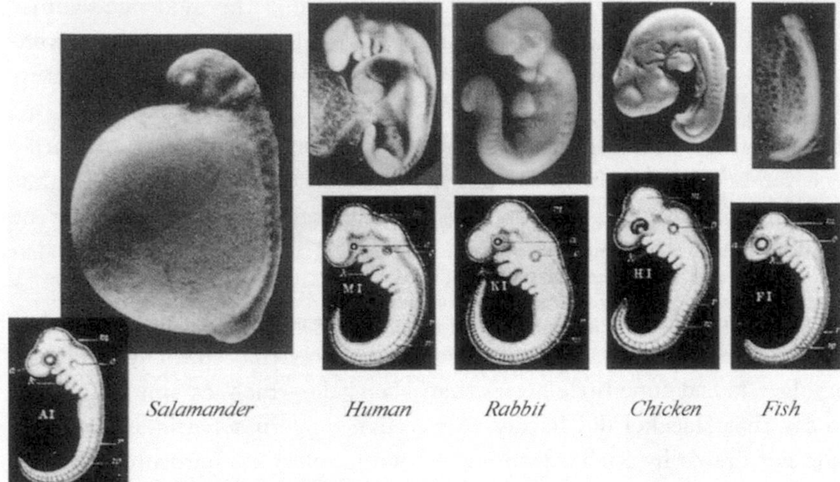

Fig. 8.8. The photographic comparisons by juxtaposition as they appeared in a *Science* magazine report (1997) of the work of Michael Richardson (photos courtesy of Michael Richardson). *Top row:* photos of embryos by Richardson comparable in species and developmental stage to those depicted by Haeckel; *bottom row:* in his *Anthropogenie* (1874); see Fig. 7.10.

lead: "An Embryonic Liar."[83] Though Richardson more recently has greatly moderated his initial assessment,[84] creationists have cited his work as evidence not simply for the supposed fraud of the nineteenth-century scientist but for the continuing "fraud" of evolutionary theory more generally.[85]

The issues concerning Haeckel's supposedly unregenerate practice would remain quite muddled if one simply accepted without qualification either Richardson's photographs or Daston and Galison's generalizations.

83. Nigel Hawkes, "An Embryonic Liar," *(London) Times*, 11 August 1997, 14.

84. See the letter to *Science* 280 (15 May 1998): 983, by Michael Richardson, James Hanken, Lynne Selwood, Glenda Wright, Robert Richards, Claude Pieau, and Albert Raynaud. Richardson more recently has extensively reviewed the question in light of modern theory. See Michael Richardson and Gerhard Keuck, "Haeckel's ABC of Evolution and Development," *Biological Review* 77 (2002): 495–528. Richardson and Keuck write: "The Biogenetic Law is supported by several recent studies—if applied to single characters only. . . . Haeckel's much-criticized embryo drawings are important as phylogenetic hypotheses, teaching aids, and evidence for evolution. While some criticisms of the drawings are legitimate, others are more tendentious" (495).

85. See, for example, Jonathan Wells, *Icons of Evolution* (Washington, DC: Regnery, 2000), 81–109. Michael Behe, one of the architects of the intelligent design movement, cited Richardson's paper in a *New York Times* op-ed as indicating the sort of faked evidence for evolution that schoolchildren need to be informed about. See his "Teach Evolution and Ask Hard Questions," *New York Times*, 13 August 1999, A21.

Richardson, along with colleagues, first presented the evidence against Haeckel in a paper arguing that the early stages of vertebrate embryos were not as highly conserved as most embryologists of both past and present have supposed. Their argument thus concluded that embryos of evolutionarily related species were, at early stages, not as morphologically similar as commonly represented.[86] Richardson and his collaborators claimed that not only did Haeckel's illustrations of early embryo stages misrepresent the true situation but that His's did as well. His, they maintained, also exaggerated the similarities of embryos and ignored their differences—a telltale assertion left mute in the many accounts of Richardson's work.[87] The burden of the Richardson paper was to show that contemporary embryologists had done little better than Haeckel—which, by implication, is to say that Haeckel did hardly worse than modern scientists (a conclusion not drawn by Richardson and his colleagues). Richardson and company made their argument by a comparison of Haeckel's illustrations with photographs taken by highly specialized optical equipment—obviously the kind of instrumentation not available to Haeckel or anyone else in his time. (Features of embryos at these early stages are virtually invisible to the naked eye.) The popular and even the scientific press—joined by a myriad of gleeful religious fundamentalists—simply duplicated Richardson's photographs and posed them against Haeckel's illustrations as if these two sets of representations had the same technical status. Accordingly, the judgment came swiftly: the only source of difference had to be willful fraud on the part of Haeckel—a judgment endorsed by Richardson in the *Science* magazine interview. No mention was made by *Science* or other reputable sources that, according to the logic of Richardson's argument, His and many modern-day embryologists would have to be held complicit in the fraud as well.

There are yet even deeper problems with Richardson's argument for Haeckel's supposed fraud, at least as represented by *Science*. In the photographs of the fish, salamander, and human being that *Science* selected from the original article by Richardson and colleagues, the yolk remains attached to the embryos (see fig. 8.8). The salamander, for instance, stands out like a lopsided beach ball, as pictured against Haeckel's slender embryo

86. Michael K. Richardson et al., "There Is No Highly Conserved Embryonic Stage in the Vertebrates: Implications for Current Theories of Evolution and Development," *Anatomy and Embryology* 196 (1997): 91–106.

87. Ibid., 92–93.

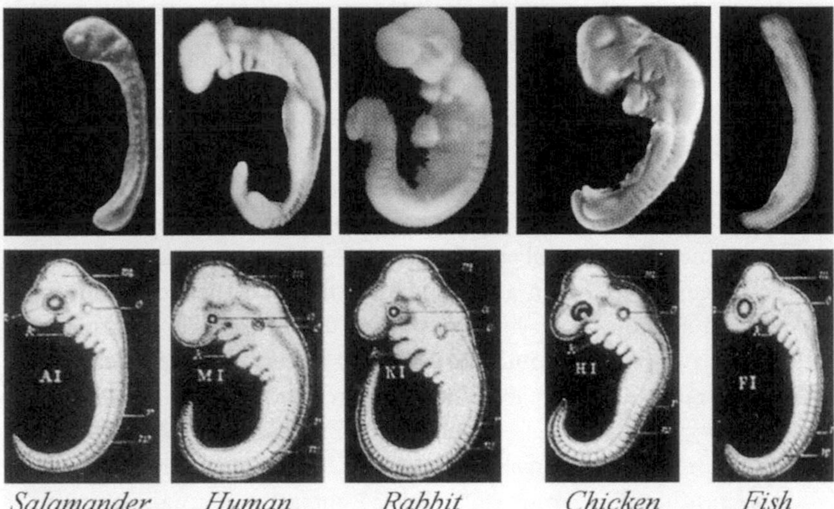

Fig. 8.9. Reengineered photographs of embryos in fig. 8.8 with yolk material removed, comparable scaling, and orientation.

of the same organism. Haeckel's supposed fraud appears startlingly evident in the comparison. But the photographs smother an important historical fact: Haeckel illustrated his embryos without yolk material. Moreover, the photograph of the chicken shows the embryo in a different orientation from the other embryos displayed and from Haeckel's, and the salamander embryo is scaled proportionately larger than the other embryos. These artifacts make the modern images appear to diverge from Haeckel's much more than they actually do.[88]

The presumption of the many discussions of Richardson's work, in the *Science* magazine article as well as in the popular press and websites, is that photographs speak for themselves. Of course, they do not. Richardson's material might, for instance, be made more justly comparable to Haeckel's illustrations by removing the yolk sacs, reducing proportionately the size of the salamander, and uncurling the chicken embryo, all of which can be done with a computer program designed to "enhance" the realism of the photography. With these alterations, the differences between Haeckel's embryos and those of Richardson are not nearly as pronounced (see fig. 8.9).

Daston and Galison argue that researchers in the last part of the nine-

88. Alessandro Pajewski pointed out to me the artifact of attached yolk in Richardson's pictures, and Christopher DiTeresi and Trevor Pearce added further revealing discriminations.

teenth century replaced "subjectively" derived archetypes—essential types—with "objective," mechanical reproductions of particular objects through photography. The question here is not so much of the veracity of Haeckel (and certainly not that of evolutionary theory) but of the "objectivity" of illustrations and of the use of photographs in biological science in the nineteenth century and today. The views of Wilhelm His himself, one of Haeckel's most ardent opponents, should make us hesitate to accept any easy generalizations about the supposed abandonment, in the late nineteenth century, of mental essentialism in the depiction of scientific objects. In a section discussing methods in his *Anatomie menschlicher Embryonen* (Anatomy of human embryos, 1880), His considered the use of drawn illustrations as compared with photographs:

> Illustrations and photographs complement each other without replacing each other. The advantage and disadvantage of every illustration over the photograph lie in the subjective elements that play a role in its formation. In every competent illustration, the essential [*Wesentliche*] is separated from the inessential in consciousness and the connections of the represented image-form are put in the correct light according to the conception of the illustrator. The illustration is thus more or less a translation [*Deutung*] of the object; it arises from the mental labor of the illustrating and embodies this for the observer. The photograph, on the other hand, renders the object in all of its particularities and contingent traits; to a certain extent it guarantees, as the raw material, the absolute truthfulness.[89]

His's ultimate task in the *Anatomie menschlicher Embryonen* was to produce a table of *norms* [*Normen-Tafel*] for the stages of embryonic development through the first two months of pregnancy, norms that could be used by researchers and clinicians for understanding human development and for the comparative use of embryological materials they might have at hand. These standards were represented by His's drawings, which were printed in the three atlases that accompanied his book. The final atlas (1885) provided a normal table of human embryos at twenty-five stages of development (fig. 8.10). The three-volume text itself was also replete with line drawings of his norms and various structures of embryos at different stages.

89. Wilhelm His, *Anatomie menschlicher Embryonen*, 3 vols. with 3 atlases (Leipzig: F. C. W. Vogel, 1880–85), 1:6.

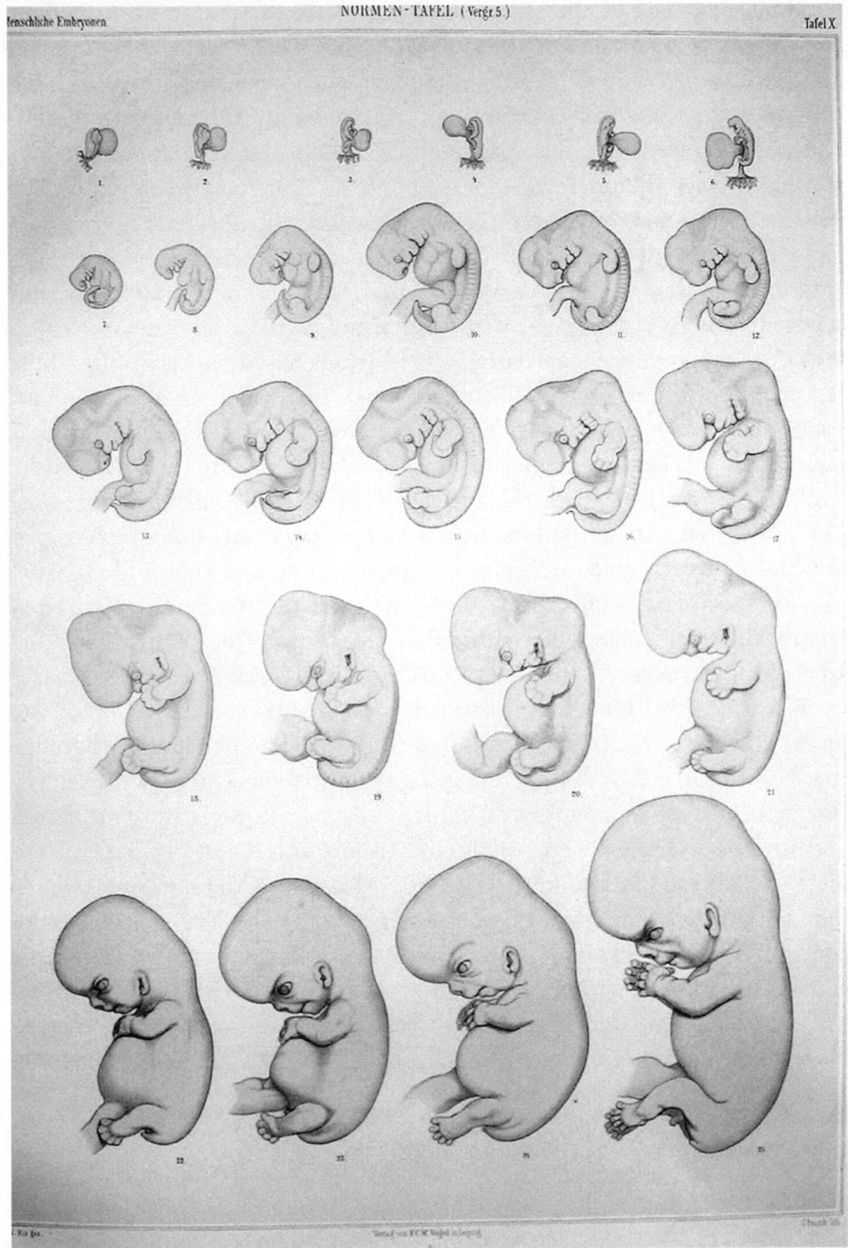

Fig. 8.10. Normal table of embryos during first two months of pregnancy. (From His, *Anatomie menschlicher Embryonen*, Atlas, 1880–85.)

His based each of the represented stages on particular embryos that he had gotten from another researcher or from a hospital, which supplied him with spontaneously aborted material. With laborious effort, as Nick Hopwood has shown in extensive detail, His would take a series of photographs and use them in preparing a drawing of the human embryo at a particular stage.[90] Drawings were made with the help of a camera lucida, which allowed more precisely detailed sketches of embryonic sections he had prepared with a microtome. Yet while each of the stages was guaranteed, as it were, by a particular embryo and photographs of that embryo, the embryos as represented went through His's own mental filter, that is, through his conceptualizing reductions based on his observations of a multitude of embryos at different stages of development. By implicit comparative reconstruction, grounded in extensive experience, the normal representations became refined and further developed. By this "mental labor," the norms lost all of their particularities and contingent features. The illustrations thus filled in structures missing from photographs of individual embryos, made irregular features regular, and clearly delineated important structures only dimly visible or absent in photographs. Through His's archetypal conceptions, the embryo became *normal*, that is, became an essential structure without particularizing features.[91]

When Haeckel produced illustrations of medusae in his atlas for the *System der Medusen* (1879), these, too, were based on particular examples that he himself either discovered or, in a few instances, that were sent to him by other researchers. The structures of the specimens were measured and all of the pertinent data, including time and place of discovery, were meticulously tabulated in his text.[92] Haeckel also used the same techniques in producing what amounted to tables of norms for the several volumes he contributed to the reports of the *Challenger* expedition. Like His, Haeckel standardized the structures he depicted on the basis of his own extensive knowledge of comparable specimens. True, he did not use photography in his work nor did he employ a camera lucida. But then it is hardly obvious

90. Hopwood, "Producing Development."

91. Hopwood makes the puzzling judgment that His's embryo representations in his *Normen-Tafel* were particulars: "His's *Normentafel* shows not ideal types but specific individual embryos" (ibid., 36). His, rather, based his illustrations on particular embryos but made them "normal"—i.e., general—in his representations. Hopwood later states precisely that His "converted them into what, according to his own analysis of the process of drawing, were views he had selected on the basis of his understanding of their structure" (44).

92. See Ernst Haeckel, *Das System der Medusen*, 2 vols. (Jena: Gustav Fischer, 1879).

how either of these instruments would have allowed him to make more precise drawings of the large organisms he had ready to hand, sitting on his dissecting table. He certainly did use a microscope extensively in depictions of radiolaria and of hydrozoan larvae and structures. The microscope was Haeckel's instrument of precision.

The epistemological situations for His and Haeckel were thus exactly the same: on the basis of individual examples, they produced through judgment and experience a standard organism; each reproduced what he *thought*, not what he immediately *perceived*. Even today college textbooks in biology, while stuffed with colorful photographs, yet accompany the important ones with line illustrations that convey the essential information. And throughout such manuals, *typical* organisms appear in highly standardized drawings. Mechanically reproduced photographs have not replaced the mental labor of the illustrator or the production of standard, general models—and this for many other good reasons, as will be suggested in the next chapter.[93]

On questions of epistemology, His and Hensen, as well as Virchow and Du Bois-Reymond, profoundly differed from Haeckel, Goethe, and Darwin. But the differences were not centered on mechanical versus nonmechanical means of representation but on the roles of fact and of theory in science. The former group, who might be dubbed "paleo-positivists," maintained that theory and facts were quite independent of one another and that theory had to be derived by a kind of mathematical induction from a comprehensive and virtually exceptionless set of facts. They regarded anything less as speculation, not science. In decided contrast, Haeckel, along with Goethe and Darwin, held that theory and fact inextricably interpenetrated one another and that theory led to the discovery of new facts. The deep epistemological differences between the two groups became ever more manifest in Haeckel's next major confrontation, which occurred at

93. In their book *Objectivity* (New York: Zone Books, 2007), Daston and Galison have moderated their notions about the replacement of anatomical illustrations by photographs. Nonetheless, in respect to the His-Haeckel controversy, they maintain that "any amendments or idealizations of the drawings or models that slipped through this system of multiple controls His equated with 'conscious bungling.'" They conclude: "When Haeckel used his drawings to extract 'the essential,' or what he believed to be the true idea hidden beneath potentially false or confusing appearances, His indicted him for sinning against objectivity" (194–95). Of course, the very notion of a "norm," as His made abundantly clear, was that of an idealized standard. The standard provided the essential structure, which the particulars—especially in the case of aborted embryos—only partially realized. Both Haeckel and His illustrated what they "thought."

the Munich meeting in 1877 of the Society of German Natural Scientists and Physicians, where Virchow suggested that Haeckel's evolutionary doctrines threatened to foment a communist revolution.

The Munich Confrontation with Virchow: Science vs. Socialism

The fiftieth meeting of the Gesellschaft deutscher Naturforscher und Aerzte—the Society of German Natural Scientists and Physicians—took place in Munich between 17 and 22 September 1877. Almost eighteen hundred participants would register (many with their families) from Germany, from other nations on the Continent (e.g., Claude Bernard and Henri Milne-Edwards), from England (e.g., John Lubbock), and from the United States (e.g., Asa Gray). There would be over 120 lectures given in the separate sections of the natural, biological, and medical sciences. Haeckel was invited to give a plenary lecture on the opening day of the proceedings (18 September); and two other prominent scientists whose topics touched intimately on one another would speak to the full assembly on subsequent days: Carl Nägeli (1817–1891), the eminent Munich botanist and correspondent of both Darwin and Mendel (20 September), and Rudolf Virchow, whose fame reserved the closing day of the meeting for him to speak (22 September). Haeckel had to leave the conference shortly after his own lecture, so he was not present for either Nägeli's or Virchow's talks; but he soon learned of them.

The business of the congress began in the morning of Tuesday, 18 September, with a celebration of the work of Karl Ernst von Baer by Heinrich Wilhelm Gottfried Waldeyer-Hartz (1836–1921), professor and physician at Strassburg.[94] Waldeyer-Hartz reminded the assembly of von Baer's great embryological work *Über Entwickelungsgeschichte der Thiere*, published fifty years prior, and of other of his accomplishments. Waldeyer-Hartz could not refrain from adding that the great anatomist opposed, not transmutation theory per se—he allowed it could occur within the various animal archetypes—but only Darwinian selection theory, which presumed those archetypal barriers could be breached and which regarded the history of life to be merely a material process. The audience was thus primed for the featured speaker of that first morning, Ernst Haeckel.

94. H. W. Waldeyer, "C. E. von Baer und seine Bedeutung für Entwicklungsgeschichte," in *Amtlicher Bericht der 50. Versammlung deutscher Naturforscher und Aerzte in München* (Munich: Akademische Buchdruckerei von F. Straub, 1877), 4–11.

Haeckel's Lecture in Munich

Haeckel prefaced his primary concerns with a brief general account of Darwinian theory. He then laid out the three principal theses he wished to argue: that evolutionary theory was a historical science; that it gave an account of human origins, particularly of man's mind; and that the theory ought to be part of the biology curriculum in the German lower schools and universities.[95]

Haeckel maintained that evolutionary theory along with significant parts of biology and geology were not experimental or exact (which is to say, mathematical) sciences, but historical sciences—similar in that respect to archaeology and linguistics. These historical sciences, as Haeckel conceived them, existed in the conceptual space between the *Naturwissenschaften*—the exact natural sciences—and the *Geisteswissenschaften*—mental sciences such as psychology, political economy, and history.

At this time in Germany, the distinction between these two kinds of systematic knowledge was still being established and argued over. Natural science, in the Kantian mold, was ultimately a set of mathematically fixed propositions rendered in deductive fashion—it dealt with universal theorems of physics. Wilhelm Dilthey urged that the mental sciences had to recognize the nonmaterial, particular character of their subject, ultimately the human mind and its activities; those sciences required not the application of the mechanistic causal principle and mathematics but of understanding (*Verstehen*) and hermeneutical interpretation.[96]

Haeckel did not propose abandoning mechanistic causal methods but wished to have those methods applied to historical phenomena. This would involve the acquisition of inductive evidence for the application of fixed laws of heredity and selection in order to yield phylogenetic knowledge of animal and human morphology and behavior. Especially crucial for determining phylogenetic history would be embryology, comparative anatomy, and paleontology. Historical science, as Haeckel would forcefully argue later, did not regard theory as insubstantial speculation—a wan specter hovering over the absence of real knowledge—but as the guide in the search for evidence and, when the evidence was sufficient, as the established structure of inductively confirmed science. He thought that "an

95. Ernst Haeckel, "Ueber die heutige Entwickelungslehre im Verhältnisse zur Gesamtwissenschaft," in *Amtlicher Bericht der 50. Versammlung,* 14–22.

96. See in particular, Wilhelm Dilthey, *Der Aufbau der geschichtlichen Welt in den Geisteswissenschaften,* vol. 7: *Gesammelte Schriften* (1883; repr., Göttingen: Vandenhoeck & Ruprecht, 1992).

objective science consisting merely of facts without subjective theory is simply inconceivable."[97] Theory and fact would always be mingled in any putative scientific knowledge; the admixture became, perhaps, only a little more obvious in historical science.

Haeckel's second thesis asserted the descent of man from lower creatures. In his presentation, he simply reiterated the principal contention of his earlier lectures and books, namely, that nothing suggested human beings had escaped the grasp of Darwin's theory. Moreover, August Schleicher's studies in linguistics, he observed, indicated a common source for such diverse languages as German, Russian, Latin, Greek, and Indian—forceful evidence for the common descent of the speakers of these languages. And those who might allow that man's body evolved but not his mind would have to explain how the functions of an organ such as the brain could be independent of the phylogenetic development of the organ itself.[98]

Haeckel argued that the foundations of the human psyche ran down to the very bedrock of life, the cell. If the cell was the basic unit of life, as Virchow maintained, then all of our psychic abilities must ultimately stem from the cell, which is to say, they must derive from the molecules of protoplasm, the cell's own ultimate constituents. Haeckel called this reductive view his *Plastidule* theory and attributed, half-jokingly, a *Plastidule* soul to cellular protoplasm. The proposal of a *Plastidule* soul would serve up an irresistible target for ridicule, and Virchow would not, a few days later, try to resist. The doctrine, though, was simply a version of the philosophy of monism—that is, the view that mind and matter were attributes of an underlying stuff. No mind without matter and no matter without mind. It was a philosophy shared by the likes of Spinoza and Goethe, though without the fanciful designation.[99]

In his lecture Haeckel decried the teaching of religious myth in the schools instead of sound science. He urged that evolutionary theory be taught as part of the biology curriculum. This would not entail the elimination of a religious perspective, only the scuttling of doctrinaire church religion (*Kirchenreligion*). Dogma would be replaced by a "rationally based natural religion" whose "highest commandment is love [and] a limitation

97. Ernst Haeckel, *Freie Wissenschaft und freie Lehre: Eine Entgegnung auf Rudolf Virchow's Münchener Rede über "Die Freiheit der Wissenschaft im modernen Staat"* (Stuttgart: E. Schweizerbart'sche, 1878), 52.

98. Haeckel, "Ueber die heutige Entwickelungslehre," 17.

99. Ibid, 17–18. Weismann gave Haeckel's *Plastidule* hypothesis a sympathetic hearing, while recognizing its tentative and incomplete character. See August Weismann, *Studien zur Descendenz Theorie,* 2 vols. (Leipzig: Wilhelm Engelmann, 1875–76), 2:298–99.

of natural egoism for the betterment of human society."[100] Haeckel's monistic philosophy, braced by a natural religion of love, was meant to suggest the views of Goethe and Darwin; but his auditors would be encouraged by Virchow to hear only Feuerbach and Marx.

The Lectures of Nägeli and Du Bois-Reymond

After his presentation on 18 September, Haeckel had to return to Jena; so he was not present for either Nägeli's or Virchow's lectures. Nägeli's lecture, which took place on 20 September, is important for understanding the variety of epistemological views that characterized the biological community in the second half of the nineteenth century in Germany. So it is worthwhile lingering a bit over his presentation and that of his principal antagonist, Emil Du Bois-Reymond.

Nägeli, as well as Haeckel and Virchow, would refer to a lecture that took place five years earlier, in 1872, at the meeting of the German Natural Scientists and Physicians at Leipzig. This was the famous presentation by Emil Du Bois-Reymond, "Über die Grenzen der Naturerkennens" (On the limits of natural knowledge). Du Bois-Reymond maintained that there were certain things that we would never know—though these seemed to be within the purview of natural science: namely, the relationship between matter and force, and that between consciousness and brain. In the case of the first set of relationships, their connections would remain opaque because our concepts of matter and force were formed from experience and such concepts when used to construct notions of atoms and central forces must always be fictions. Of necessity they would involve contradictions and paradoxes. In the case of consciousness and brain, though we might be able to associate every mental event with a brain event, we would still not be able to understand how the matter of the brain could cause a nonmaterial response in the mind. Du Bois-Reymond concluded his lecture of 1872 with the remark that would become a battle cry: *Ignoramus et ignorabimus*—we are ignorant and will be ignorant.[101] In later writings he would considerably expand the sphere of our ignorance.[102]

100. Haeckel, "Ueber die heutige Entwickelungslehre," 19.

101. See Emil Du Bois-Reymond, "Über die Grenzen des Naturerkennens," in *Vorträge über Philosophie und Gesellschaft*, ed. Reymond Siegfried Wollgast (Hamburg: Felix Meiner Verlag, 1974), 54–76.

102. In 1880 Du Bois-Reymond contended there were five additional "world puzzles" that would remain either insoluble because they transcended the range of natural knowledge or because they had failed continuous efforts to resolve them. The seven world puzzles he discrimi-

Fig. 8.11. Emil Du Bois-Reymond (1818–1896). (Courtesy of the Smithsonian Institution.)

At the Munich meeting, Nägeli squared off against Du Bois-Reymond's delimitation of knowledge, with a lecture entitled "Ueber die Schranken der naturwissenschaftlichen Erkenntniss" (On the restrictions of natural-scientific knowledge).[103] In a lengthy, calm analysis, Nägeli suggested that Du Bois-Reymond had first surreptitiously introduced into the natural realm unbridgeable metaphysical disjunctions and that he then asserted those gaps could not be crossed by natural knowledge. But Du Bois-

nated, then, were the following: the relationship between matter and force; the origin of the first movement of matter; the first development of life; the teleological orientation of nature; the origin of sensibility; the origin of consciousness and language; and finally, the freedom of will. He believed that the relationship between force and movement of matter, as well as the origin of sensibility, surpassed human understanding, while the others, at least for now, showed no clear path to solution. See Emil Du Bois-Reymond, "Die sieben Welträtsel," in *Vorträge über Philosophie und Gesellschaft*, 159–86. The challenge of these world puzzles would furnish Haeckel the stimulus, at the end of the century, to produce one of the single most successful popular publications of the modern period: *Die Welträthsel* (The world puzzles, 1899).

103. Carl von Nägeli, "Ueber die Schranken der naturwissenschaftlichen Erkenntniss," in *Amtlicher Bericht der 50. Versammlung*, 25–41. Nägeli was not originally scheduled for a plenary lecture, but the original invitee, Gustav Tschermak von Seysenegg (1836–1927), was not able to make the journey from Vienna.

Fig. 8.12. Carl Nägeli (1817–1891). (Courtesy of the Smithsonian Institution.)

Reymond required of natural knowledge a kind of property it nowhere evinced. In scientific understanding, Nägeli observed, the posited relationship between a cause and its effect (e.g., a negatively charged and a positively charged body moving toward each other) could be determined only on the basis of repeated experience. This kind of experience made nature mechanically intelligible, something Du Bois-Reymond did not question. But, according to Nägeli (and his unspoken guide, David Hume), we never had any further insight into the link between cause and effect, just the experience of universal conjunction. Yet this kind of knowledge, he maintained, did not differ from that acquired by experience of the conjunction of brain alterations being followed by conscious changes in awareness. Causal knowledge in all domains had the same epistemological character, whether it was that of gravitational attraction of planets or that of the relationship of brain and mind. If the former was scientifically tractable, so was the latter.[104] Moreover, a perfect continuity existed, he maintained, between the

104. Nägeli's response to Du Bois-Reymond's assertion of an unbridgeable causal gap is precisely the same as John Searle's to a similar objection concerning the gap between neural

higher mental life of the adult human, the simpler mental life of the child, and the lower mental life of the animal. Further, the reactions of the lowest animals to their environments could not really be distinguished from that of plants to theirs; and the very lowest plant hardly differed from inorganic material. Nägeli thus concluded that either no natural scientific knowledge was possible or that all finite areas of inquiry could be known by human understanding. He ended his talk with an epigram that answered Du Bois-Reymond's: *Wir wissen und wir warden wissen*—we know and we will know.

Virchow's Indictment of Evolutionary Theory as Socialistic and Nonscientific

On 22 September Virchow took the podium to give the closing plenary lecture.[105] He had arrived in Munich after the lectures of Haeckel and Nägeli. Their lectures, however, had been quickly printed, and Virchow had time to adjust his own remarks to those of his predecessors. He orchestrated a rhetorically powerful response—at least for the audience so assembled. The fallout from the incendiary bomb that he tossed at his onetime student shrouded his own defensive redoubt—a Potemkin construction of science.

He began his lecture by reminding the audience of the "dramatic events" that had occurred across the Rhine just a few years back, events that had bearing on the current disputes within German science. His audience was, of course, perfectly cognizant of the great civil upheavals that had shaken France after the Franco-Prussian War of 1870–71. With most of the defeated French army outside of Paris awaiting negotiations with the Prussians for war reparations, the local government in Paris, disdainful of the failure of the national government, organized committees to resist occupation and to inaugurate social reforms. They passed legislation forgiving debts and advancing the prospects of laborers and the middle class of the city, with the promise of separating religion from government and canceling rents. They also secured the large arms caches and cannons of the regular French army. On 28 March 1871, the Paris Commune established itself as an independent government. The body included many socialist, anarchist, and

---

cause and conscious effect. See John Searle, "Reply to 'What Is Consciousness?'" *New York Review of Books* 52, no. 11 (June 23, 2005): 56–57.

105. Rudolf Virchow, "Die Freiheit der Wissenschaft im modernen Staatsleben," in *Amtlicher Bericht der 50. Versammlung*, 65–78.

Fig. 8.13. Rudolf Virchow in his anthropology laboratory. (Courtesy of the National Institutes of Health.)

republican elements. The council of the Commune occupied the Paris city hall, the Hôtel de Ville, and raised the red flag of socialism.

On 2 April the national government at Versailles ordered the bombardment of the city in an effort to take back the capital. The barricades went up, and the National Guard of the city, along with countless citizens, defended their new republic. A month and a half later, on 21 May, the walls of the city were breached; and street fighting began, with perhaps some thirty thousand Communards dying in a futile effort to fend off the regular army. By 28 May the last barricade fell. Some fifty thousand citizens directly involved in the revolution were executed or jailed, and thousands were sent to prisons set up on Pacific islands. The three daughters of Karl Marx and his son-in-law were momentarily swept up in the search for members of the International; only cunning and British passports saved them. The

Fig. 8.14. Hôtel de Ville, with Communards standing in front, spring 1871. (Courtesy of Northwestern University.)

Paris Commune became a symbol under which later anarchistic and communistic efforts at revolution would be conducted.

With this political reminder of the chaotic events in France, Virchow began his attack on descent theory from three directions—its political danger, its epistemological confusion, and its pedagogical malfeasance. First was the dangerous political consequence of unrestrained theorizing. He declared to the shouts of approval from his audience: "I am of the opinion that we are indeed in danger from a too-liberal utilization of freedom that

Fig. 8.15. Barricades and cannons on a Paris street.
(Courtesy of Northwestern University.)

the present circumstances offer, a danger that threatens our future. I would caution against indulging in arbitrary personal speculation of the sort that is abroad in many areas of natural science." [106] He made clear exactly what he thought the political danger might be:

> Just imagine how descent theory is conceived in the head of a socialist. Yes, Gentleman, that may seem laughable to many, but it is seriously meant. And I would hope that descent theory won't bring to us all the tribulation that similar theories have actually created in our neighboring country. Indeed, if you follow out the consequences of this theory,

106. Ibid., 66.

> I hope it will not have escaped your notice that it has an uncommonly worrisome side and that socialism has become sympathetic toward it.[107]

With the still-lingering smell of revolution wafting over the Rhine, these egregious assertions, in the political atmosphere of Bismarck's Germany, imbued Darwinian theory with the stench of a repugnant political doctrine. Huxley would later decry this blatant appeal to political prejudice.

Both Haeckel and Virchow alluded to the government's agenda for new educational legislation for lower schools and universities in the wake of the creation of new gymnasia and the expansion of higher education. By associating evolutionary theory with communism and socialism, Virchow obviously was attempting to explode any suggestion that Darwinian notions should be introduced into the curriculum. He urged his fellow scientists that in regard to this pending legislation, "we indeed must be reasonable about what we can demand and wish to demand."[108]

The political dangers of evolutionary theory were compounded, in Virchow's view, and his warning justified because the theory did not meet the standards of authentically valid science. He asserted that a border had to be sharply drawn between "what we disseminate as real science in the strictest sense of the word . . . and that greater province, which belongs to speculative extension."[109] He vigorously maintained that evolutionary doctrine exceeded the limits of established scientific fact and gamboled in fields of unbridled theory, thus violating the distinction between demonstrated science, which is the realm of certified facts, and speculative ideas. And this is a distinction, he argued, that had to be maintained in teaching science, which should only extend to the realm of the proven.

107. Ibid, 69. In asserting that socialism had become sympathetic to Darwinism, Virchow may have been referring to one or both of the following discussions: Albert Lange's *Die Arbeiterfrage in ihrer Bedeutung für Gegenwart und Zukunft* (Duisburg: W. Falk & Volmer, 1865), with subsequent editions in 1870 and 1875; or August Bebel's *Die Parlamentarische Tätigkeit des Deutschen Reichstages und der Landtage und die Sozialdemokratie von 1874 bis 1876*, in *August Bebel Ausgewählte Reden und Schriften*, ed. Horst Bartel, Rolf Dlubek, and Heinrich Gemkow, 10 vols. in 14 (Berlin: Dietz, 1970–97), 1:343–439. Lange applied Darwinism to the human condition, arguing that the struggle for existence would eventually lead to peace and harmonious relations among individuals. Bebel cited Haeckel's notion of "military selection," in which the bravest would be slaughtered on the fields of battle and those of lesser virtue would be left behind to foster the next, more debilitated generation (359).

108. Virchow, "Die Freiheit der Wissenschaft," 68.

109. Ibid., 66.

> When Mr. Haeckel says it's a question of pedagogy whether one now places descent theory at the foundation of instruction and accepts the *Plastidule* soul as the groundwork of all mental nature, whether one should follow the phylogeny of man down through the lowest kinds of life, even down to the original generation, then I think this is an abandonment of the task [of pedagogy]. When descent theory becomes as certain as Mr. Haeckel assumes, then we must demand—then it is a strict requirement—that it be taught in the schools.[110]

In attempting to show that evolutionary theory as a whole was without evidentiary foundation, Virchow set out to undermine two principal areas: the first was that of the origin of life, which the theory supposed (at least in Haeckel's construction) must occur through spontaneous generation; and the second was the descent of man. Concerning the first, he asserted that "one knows of not one positive fact, which has been established, that such spontaneous generation [*Generatio aequivoca*] has occurred, that an originating production has taken place such that unorganized masses had ever developed themselves—thus, for instance, Messrs. Carbon and Co.—by free choice into organized masses." At the other end of the supposed line of descent, no true "ape-man" has ever been found. Thus here, too, Virchow took his stand: "We cannot teach, we can not maintain as a discovery of science that man has descended from apes or any other animals." [111]

Virchow did allow that evolutionary theory could be a proper subject for research, in an effort to find demonstrable evidence for it, but stipulated that it should never be a subject for teaching, which ought only propound authenticated science. This small bone of appeasement, though, had no real meat. Virchow had already declared, in his Stettin lecture of 1863, that the gap between consciousness and brain could never be closed by science—a view he shared with Du Bois-Reymond—and that positive fact militated against spontaneous generation.[112] And in controversies over fossil men in the coming years, he never recognized any fossils that suggested a protohuman group had once roamed the earth.[113] His view remained constant

110. Ibid., 68.

111. Ibid., 72, 77.

112. See chapter 4, above.

113. See the recapitulation of Virchow's anti-Darwinist position in Benoit Massin, "From Virchow to Fischer: Physical Anthropology and 'Modern Race Theories' in Wilhelmine Germany," *History of Anthropology* 8 (1996): 79–154, especially 114–17.

that "no factual demonstration of the descent of man from the apes has been provided."[114]

Virchow closed with a final admonition: "Gentlemen, we would misuse our power, we would endanger our power, if we, in teaching, did not draw back to the perfectly justified, the perfectly secure, the impregnable zone [of verified science]."[115]

Response to the Controversy with Virchow

It was during his return to Jena when Haeckel first discovered that Virchow had aimed his lecture against his old student, though he may have suspected something like that was afoot and decided to depart early. Haeckel wrote Weismann that "on the return I happened to get number 220 of *Germania* (supplement), in which I saw, to my great amusement, that my 'friend' Virchow had, in Munich, knocked me dead [*todtgemacht hat*] with 'a really heavy blow.' Well, as you see, I'm still alive and feel pretty well."[116] When Haeckel finally got a copy of his former teacher's lecture, he felt considerably less well. To have this eminent researcher, this leader of the scientific establishment, this onetime teacher and friend treat him with such scorn and invective—that did deliver "a really heavy blow." At the beginning of December, Haeckel wrote Allmers, who had attempted to console his friend, a letter revealing both the depression and the defiance that settled upon him:

> My dear Allmers!
> You have shown yourself a true friend with your heartfelt last letter. I can see from it that you belong to the few, the very few friends who completely and perfectly understand me. You can't believe the sad experience I continually have among the narrower circle of my friends, and more so among my professional colleagues. Although the principle concepts for which I have fought during these last eighteen years have spread far and wide—or rather, *because* they have spread chiefly through my energetic and fearless propaganda [*unerschrockene Propaganda*], because I have not shied away from the implications of the natural worldview, *because* I did not treat the old scientific assumptions as gently

114. Rudolf Virchow, *Menschen- und Affenschädel* (Berlin: Luderitz'sche Verlagsbuchhandlung, 1870), 33.

115. Virchow, "Die Freiheit der Wissenschaft," 77.

116. Haeckel to Weismann (16 October 1877), in "Briefwechsel zwischen Haeckel und Weismann," 41. *Germania* was a Catholic Bavarian newspaper. The article appeared in *Germania* (25 September 1877), supplement.

> as had the complaining Philistines—*for these reasons* the number of my enemies has grown from day to day; and increasingly my so-called "friends" have seized every opportunity to hang something on me. . . . But what does it matter? I now already enjoy the great satisfaction of not having lived otherwise and of anticipating the victory of my ideas in the distant future.[117]

After several months of not knowing exactly what to do about Virchow's attack, which had been printed as a small book, Haeckel composed his own small tract in response—his *Freie Wissenschaft und freie Lehre* (The freedom of science and the freedom to teach, 1878).[118] The book defended evolutionary theory, especially the descent of man, the evidence for which came from many quarters. More pointedly, however, Haeckel argued that the procedures of science could not be neatly separated into fixed and certain facts, on the one hand, and speculative theory, on the other. All of our science, he maintained, was shot through with theory, which provided explanation for the facts—and it was explanations that we sought in science, as a brief survey of the great accomplishments of past science, from Copernicus and Newton to modern chemistry showed.[119] Our teaching, he urged, would be sterile and dead if we did not include the theoretical side of science—as if we could dispense with it.

> Where is there and where has there ever been a great teacher who has limited his instruction to the dry deliverances of certain, indubitably fixed facts? Who has not rather found the charm and value of his instruction precisely in the teaching of problems that are connected with every fact, in the teaching of uncertain theories and wavering hypotheses that serve to resolve the problems? And is there something more formative and better for the young, striving mind than the exercise of thought on the problems of research?[120]

In his little book, Haeckel, of course, denied that Darwinian theory gave any succor to socialism. It seemed obvious to him—as it would to

117. Haeckel to Allmers (2 December 1877), in *Ernst Haeckel: Sein Leben, Denken und Wirken. Eine Schriftenfolge für seine zahlreichen Freunde und Anhänger*, ed. Victor Franz, 2 vols. (Jena: Wilhelm Gronau und W. Agricola, 1943–44), 2:74.

118. Haeckel, *Freie Wissenschaft*. Virchow's book was entitled *Die Freiheit der Wissenschaft im modernen Staat: Rede gehalten in der dritten allgemeinen Sitzung der fünfzigsten Versammlung Deutscher Naturforscher und Aerzte zu München am 22. September 1877* (Berlin: Wiegandt, Hempel & Parey, 1877).

119. Haeckel, *Freie Wissenschaft*, 53–56.

120. Ibid., 63.

many in the next century—that evolutionary theory defied a socialist political construction:

> The theory of descent proclaims more clearly than any other scientific theory that the equality of the individual striven for by socialism is an impossibility, that this kind of equality stands in irresolvable contradiction with the empirically given and necessary inequality of individuals. Socialism demands for all citizens the same rights, the same duties, the same benefits, the same pleasures; descent theory, shows precisely the opposite, that the reality of this demand is a sheer impossibility, that in the social organization of men, as well as of animals, neither the rights and duties nor the benefits and pleasures of all citizens will ever be the same nor can they be the same.[121]

Haeckel focused on the evolutionary assumption of variability of organisms even of the same species as a ground for the denial of socialism's demands. He also advanced the fundamental instinct of competition as another barrier to the success of extreme forms of that political doctrine.[122] If rights and duties are founded in the nature of human beings and if that nature is fundamentally biological (and surely it is), then one might forthwith conclude with Haeckel that socialism and communism (at least in their extreme forms) must fail as political and economic philosophies by which to organize human beings. In the last part of the twentieth century, the Western democracies—as well as former Eastern Bloc nations—seem to have drawn this same political conclusion. In the last part of the nineteenth century, however, a judgment of this kind was strongly contested, as were the implications Haeckel drew about the relationship of evolutionary theory to socialism.

The great socialist thinker August Bebel (1840–1913), for instance, rejected Haeckel's interpretation of the political implications of Darwinism and sided with Virchow, though without the latter's negative evaluation of socialism: that is, Bebel argued that evolutionary theory did, indeed, support socialism, which thus demonstrated the conformity of that political doctrine to the scientific conception of human nature and to the assessment of its reasonable aspirations. He, as well as other socialists—such as Karl Kautsky (1854–1938) and Eduard Bernstein (1850–1932)—maintained that Darwinian theory, with its law of struggle, actually led to the improvement of men and their social conditions:

121. Ibid., 72.
122. Ibid., 73–74.

> The Darwinian law of struggle for existence—which holds in essence that the more organized and stronger creatures exclude and eliminate the lower—finds in human beings this result, that men, as thinking and knowing beings, continually change, improve, and perfect with conscious intent their living conditions—that is, their social environment and everything that depends on it—so that finally the same favorable conditions of existence are available for all human beings.[123]

Curiously, this was a goal toward which Haeckel's own application of evolutionary theory to human society also pointed.

As against the likes of both Virchow and Bebel, one must, however, carefully consider Haeckel's ultimate conclusion about the positive political implications of evolutionary theory—namely, that there were none! Or at least, as he contended, the use of science to form policy for practical political decisions had to be hedged with a multitude of philosophical, social, and ethical considerations:

> We would not like to miss this opportunity, moreover, to point out how dangerous it is directly to transfer scientific theory of this sort to the realm of practical politics. The highly developed relationships of our cultural life today demand from our practical politician very circumspect and impartial considerations, a quite fundamental historical education and critical ability, such that he might, only with the greatest care and hesitation, ever dare apply a "natural law" to the practical features of that cultural life.[124]

While Haeckel disavowed drawing social Darwinist conclusions about specific political policy, he nonetheless had firm ideas about the nature of human beings and would continue to develop ethical considerations consistent with his proposed scientific account of man.[125]

To Du Bois-Reymond's cry of "*ignoramus et ignorabimus*," Haeckel ended his *Freie Wissenschaft und freie Lehre* with "*impavidi progrediamur*"—courageously let us go forward. And he went forward with some stalwart allies, at least among the English.

123. August Bebel, *Die Frau und der Sozialismus* (1879), in *August Bebel Ausgewählte Reden und Schriften*, 10.1:56.

124. Haeckel, *Freie Wissenschaft*, 74.

125. Haeckel's own political involvements increased after the turn of the century as he detected the greater influence of religion on state government. Uwe Hoßfeld and Olaf Breidbach sketch out Haeckel's political entanglements after 1900 in their "Ernst Haeckels Politisierung der Biologie," *Thüringen: Blätter zur Landeskunde* 54 (2005).

When Haeckel's book appeared in English translation in 1879, Darwin immediately wrote his friend: "I agree with all of it." He thought Virchow's conduct was "shameful" and added: "I hope he will some day feel the shame."[126] Huxley was even more vigorous in support. He supplied the preface for the English edition of the book and expressed his "sympathy with his [Haeckel's] defense of the freedom of learning and teaching." Huxley also wished to "vent my reprobation of the introduction of the sinister arts of unscrupulous political warfare into scientific controversy, manifested in the attempt to connect the doctrines he advocates with those of a political party which is, at present, the object of hatred and persecution in his native land."[127]

In Germany the religious journals, such as the *Neuen evangelischen Kirchenzeitung* and the Catholic *Germania*, delighted in Virchow's drubbing of the "dogma of the Affenmensch." Other newspapers, like the *Frankfurter Zeitung*, while not endorsing Haeckel's desire to have the doctrine of monera introduced to twelve-year-olds, yet decried Virchow's defense of the myths of religion being inculcated as science.[128] Ludwig Büchner wrote Haeckel to thank him in the name of all "freethinkers" and suggested that Virchow rejected descent theory simply because the great pathologist had not discovered it first.[129] At least one author found some humor in the controversy. Moritz von Reymond penned a long series of rhyming reflections in his neatly titled *Laienbrevier des Häckelismus* (Lay breviary of Haeckelism)—for instance:

I will storm the primary school institution
To teach girls and boys a bit of evolution,
Bringing up a cadre of wee selectionists
In descent theory to be tomorrow's socialists;
And therefore a novel religion will I found
With natural civilization as its ground.[130]

126. Darwin to Haeckel (29 April 1879), in the Haeckel Correspondence, Haeckel-Haus, Jena.

127. T. H. Huxley, preface to *Freedom in Science and Teaching*, by Ernst Haeckel (New York: D. Appleton, 1879), xviii.

128. See "Berliner Hofprediger," *Neue evangelischen Kirchenzeitung*, 20 October 1877, 659; *Germania*, 25 September 1877, supplement; and *Frankfurter Zeitung*, 28 September 1877. Haeckel included these and other accounts in his *Freie Wissenschaft*, 94–106.

129. Büchner to Haeckel (21 October 1878), in *Carl Vogt, Jacob Moleschott, Ludwig Büchner, Ernst Haeckel: Briefwechsel*, ed. Christoph Kockerbeck (Marburg: Basilisken-Presse, 1999), 147.

130. Moritz von Reymond, *Laienbrevier des Häckelismus: Jubiläumsausgabe, 1862–1882–1912* (Munich: Ernst Reinhardt, 1912), 132: "Ich stürme die A-B-C-Schulstuben, / Lehre 'Enst-

Perhaps the most significant response in Germany came obliquely a month after the Munich meeting in the inaugural address of Hermann von Helmholtz as rector of the University of Berlin. In describing the character of the German universities, Helmholtz pointedly celebrated *Lehrfreiheit*—the freedom to teach. He declared: "Currently at the German universities, the extreme consequences of materialistic metaphysics, the boldest speculations grounded in Darwin's evolutionary theory may be lectured on with as little restraint as the most extreme pontifications on papal infallibility." The allusion to the Haeckel-Virchow dispute seems clear, as in the next sentence with its implied admonition: "As in the assemblies of the European parliaments, accusations of suspect motives or denigration of the personal traits of the opponent—neither of which has anything obviously to do with decisions about scientific propositions—are forbidden."[131] Even for Helmholtz, Virchow had stepped over a line.

Conclusion

The Decade of the 1870s

In response to rumors about the gathering of Haeckel's enemies, Huxley offered this consoling curse: "May your shadow never be less, and may all your enemies, unbelieving dogs who resist the Prophet of Evolution, be defiled by the sitting of jackasses upon their grandmothers' graves!"[132] Haeckel must have muttered that curse continually during the decade of the 1870s, as several portions of the intellectual community engaged in unremitting attacks on his scientific accomplishments and personal integrity. The generally hostile atmosphere that arose in the wake of the brutal conflicts of the earlier part of the decade condensed into a raging storm after Munich. The personal attacks were so deeply wounding that in 1874 and 1877, as he later confessed, he had quite seriously contemplated suicide.[133] And Haeckel felt the continuing howl of the storm even after the

wicklung' den Mädels und Buben, / Erziehe mir Zukunftssozialisten / Aus Deszendenzlern und Selektionisten, / Und stifte die neue Religion / Der naturgemässen Zivilisation." Even the introduction to the third, Jubilee edition of this comic send-up is in rhyming couplets.

131. Hermann von Helmholtz, *Über die Akademische Freiheit der deutschen Universitäten* (Berlin: August Hirschwald, 1878), 22.

132. Huxley to Haeckel (28 December 1874), in "Briefwechsel zwischen Huxley und Haeckel," 23.

133. In a letter to Frida von Uslar-Gleichen (11 January 1900), he mentioned that he had contemplated suicide in 1874 and 1877. See *Das ungelöste Welträtsel: Frida von Uslar-Gleichen und Ernst Haeckel, Briefe und Tagebücher 1898–1900*, ed. Norbert Elsner, 3 vols. (Berlin: Wallstein, 2000), 1:390.

worst had passed. He had desired to join Gegenbaur at Heidelberg, to which his friend had been called in 1873. Gegenbaur, for his part, had striven to prepare the way for the appointment. But the philosophy faculty, as Gegenbaur reported to Haeckel in January 1878, balked at the effort. Gegenbaur attributed the rejection both to the dispute with Virchow and, more significantly, to those who were simply jealous of Haeckel's accomplishments: "Dear friend, it is not the opponents alone who work against you, not the opponents of your direction or aim, but also the envious who become enemies and constitute a good half of your opponents. These envious individuals are worse than the others, since they go along a darkened way and practice their evil unrecognized."[134] The call to Heidelberg never came.

Though under constant siege and with the many disappointments and depressions that this brought, Haeckel yet remained extraordinarily productive: in 1872 his three-volume *Kalkschwämme* appeared; in 1874 the first of the six editions of his *Anthropogenie* was published, with two more being reworked during the decade; from 1870 to 1879, the second to seventh editions of the *Natürliche Schöpfungsgeschichte* appeared, each having been revised from the previous edition; in 1877 and 1878, he saw through the presses two volumes of his collected essays, with five of the essays having been composed during the eight years prior; and in 1879 his giant monograph *System der Medusen* came out, with an atlas of extraordinarily beautiful depictions of these animals—it would be followed in 1881 by a second large volume, *Die Tiefsee-Medusen der Challenger-Reise und der Organismus der Medusen* and in 1882 by the first of his four *Challenger* volumes.[135] And during all of this, he maintained his teaching schedule and yet managed numerous research expeditions, of the sort his wife Agnes detested.

In March and April 1873, Haeckel traveled to Egypt and down to the Red Sea to study coral—and three years later his beautiful *Arabische Korallen*, part research monograph, part travel book (see plate 7), emerged from the publishers.[136] During March and April 1875, he took the "golden brothers" Hertwig on an expedition to Sardinia and Corsica, which yielded material for his gastraea theory and, later, for his work on medusae. In the fall of

134. Gegenbaur to Haeckel (14 January 1878), in *Ernst Haeckel: Biographie in Briefen mit Erläuterungen*, ed. Georg Uschmann (Güttersloh: Prisma, 1984), 154.

135. Ernst Haeckel, *Die Tiefsee-Medusen der Challenger-Reise und der Organismus der Medusen*, with *Atlas* (Jena: Gustav Fischer, 1881); and *Report on the Deep-Sea Medusae dredged by H.M.S. Challenger*, vol. 14 of *Report on the Scientific Results of the Voyage of H.M.S. Challenger during the Years 1873–1876* (London: Her Majesty's Stationery Office, 1882).

136. Ernst Haeckel, *Arabische Korallen* (Berlin: Georg Reimer, 1876). For Haeckel, the trip was also one of cultural exploration. He dedicated the book to the khedive of Egypt, Ismail Pascha, who arranged for an Egyptian naval ship to provide transportation for the research party.

the next year, he journeyed to Glasgow, where he received a very warm reception at the meeting of the British Association for the Advancement of Science, and while in England he traveled again to the village of Downe to visit Darwin. During his stay in Britain, he got his first look at the *Challenger* material, and his own multivolume contribution to the *Challenger Reports* would occupy him fully for the next decade.[137] In February 1877 he sailed to Corfu. In February and March 1878, he lectured throughout Germany (Mannheim, Kassel, Cologne, Frankfurt, Leipzig, etc.) and also in Trieste and Vienna. In late summer he spent a week in Paris, working at the Museum Nationale, and then several weeks off the coasts of Normandy and the Isle of Jersey, collecting ever more species of radiolaria and medusae. He wrote his wife from Paris of his success in finding some new medusae at the museum and to remind her, as if she needed it, that 20 August was their eleventh wedding anniversary. She wrote to him shortly thereafter: "I am very happy for you that your ceaseless efforts have finally been crowned with success. But I am tired of your eternal travels and separations."[138] But he was not tired of them. In September 1879 he returned to Scotland and England for consultation about the *Challenger* assignment and a final visit to his friend at Downe. The new decade saw Haeckel preparing for the first of his expeditions to the tropics of southern Asia—India and Ceylon (Sri Lanka)—where he would spend November 1881 to March 1882.

Evaluation of the Charges of the Critics

The amount of research and the number of significant publications that Haeckel turned out in the decade of the 1870s might well have occupied other scientists for a full career and have brought them considerable fame. Haeckel, of course, did achieve enormous success and found recognition for his work among the scientific elite of Europe. The commission to contribute to the *Challenger* volumes, his many invitations to lecture throughout Germany and England, his glittering list of correspondents, his numerous honorary degrees and awards, and the caliber of students who flocked to Jena—these all testify to his prestige and fame. His reputation

137. Haeckel's volumes treated medusae, radiolaria, siphonophores, and calcareous sponges: Haeckel, *Report on the Deep-Sea Medusae*; *Report on Radiolaria*, vol. 18, parts 1 and 2 of *Report on the Scientific Results of the Voyage of H.M.S. Challenger during the years 1873–1876* (London: Her Majesty's Stationery Office, 1887); *Report on the Siphonophorae Collected by H.M.S. Challenger during the Years 1873–76*, vol. 28 of ibid. (1888); and *Report on the Deep-Sea Keratosa Collected by H.M.S. Challenger during the Years 1873–1876*, vol. 32 of ibid. (1889).

138. Agnes Haeckel to Ernst Haeckel (16 September 1878), in *Ernst und Agnes Haeckel: Ein Briefwechsel*, ed. Konrad Huschke (Jena: Urania, 1950), 138.

amongst his colleagues was, as we have seen, mixed, with some in the scientific community charging him with fraud and accusing him of passing off rank speculation as sound science. Rumors of these charges have, since the 1870s and up to the present day, grown in phantasmagoric ways so as to obscure the accomplishments of an undeniable genius.

There are several distinct sources for the toxic cloud that has come to settle over the man: the lingering notoriety his combative personality produced; the sustained efforts of embryologists, right through the twentieth century, to preserve professional boundaries; the failure of recent critics to recognize the disputed epistemological relationship between fact and theory in the nineteenth century; an importation of modern scientific assumptions into that earlier period; the continued religious reaction to evolutionary theory, especially in its resurgent form as "scientific creationism" and "intelligent design" ideology; the adoption of certain of Haeckel's ideas by some Nazis; and, finally, the charges of fraud that have been periodically renewed over the course of the last 150 years. The first of these sources has been a constant theme of the preceding chapters. I have touched on the next three—the professional, epistemological, and methodological—in this chapter; and I will consider the religious reaction to Haeckel in the next. I will reserve the penultimate issue, the use of Haeckel's work by Nazi biologists, for the concluding chapter and the second appendix.

The current chapter has also been devoted to consideration of the charges of fraud made against Haeckel. Here I would like to come to summary judgment about those charges. There were two kinds of indictment: that Haeckel had replicated the three woodcuts of vertebrate eggs and vertebrate embryos; and that he had altered illustrations he had borrowed from other authors, so as to make embryos of animals and humans appear more similar. As to the first of these damning claims, one has to keep in mind that the initial edition of Haeckel's *Natürliche Schöpfungsgeschichte* started life as a lecture series in which large, instructive wall charts were employed. The redaction of notes taken by Haeckel's students yielded the book manuscript, which was completed by Haeckel in just a few months. Schematic wall charts served as the model for the illustrations that appeared in the book. Given the cost of woodcuts and the fact that the morphological structures of vertebrate eggs and early embryos are almost impossible to distinguish, it might seem a reasonable economy simply to replicate the same images and use them as a device to pound home the message in a popular work meant for a nonprofessional audience. The book, nonetheless, quickly became enmeshed in the professional literature. And Haeckel made the mistake of suggesting that the woodcuts constituted *evidence* of similarity. This was a lapse that he regretted. He quickly altered the second edition of the *Natürliche Schöpfungsgeschichte* to eliminate the replication, though not his conclusion that at the earliest stages of embryonic development, related creatures were, indeed, quite similar. Yet there is little doubt that his use of the replicated woodcuts was a moral failure, one borne out of rushed confidence in his conclusions and condescension to his readers. The failure, though, slid along at a rather low level, a level where carelessness, inattention, and precipitous choice occluded sensitive awareness of the perilous way. The lapse did not, I believe, approach standards of gross fraud. Moreover, the conclusions he attempted to draw about embryonic similarity, given the level of then-contemporary knowledge, were sound enough.

In the early editions of *Natürliche Schöpfungsgeschichte* and *Anthropogenie,* Haeckel set out illustrations of embryos borrowed from various authors. Though he had observed vertebrate embryos at firsthand, he did not attempt to render them with the precision he found already exhibited by experts. When the illustrations taken from others were lined up, the morphological structures of, say, Bischoff's dog embryo and Ecker's human embryo appeared strikingly similar. Neither Bischoff nor Ecker intended the kind of argument advanced by Haeckel, yet their juxtaposed illustrations seemed to support the basic claim of the biogenetic law. Haeckel did alter some of these images (particularly standardizing their sizes to make similarities more obvious), and he reduced his images to the essential features of the organisms he represented. The authors from whom he adapted the images had already themselves eliminated most of the particularizing and non-essential features. Thus at one level, Haeckel's biogenetic law had more convincing support by reason of the comparisons of borrowed images than if he himself had rendered the illustrations from firsthand examination of vertebrate embryos.

Later in his career, Haeckel engaged in more extensive dissection observations of various vertebrate species. During his travels to Malaya in 1900–1901, when in his late sixties, he spent two months at the research station of Buitenzorg, just outside of Jakarta (discussed in chapter 10). He had injured his knee, so he had to postpone more extensive travel throughout the islands. While convalescing, he engaged in embryological research, especially on such animals as tuataras (large lizards), aquatic turtles, echidnas, dolphins, and gibbons. In the fifth edition (1903) of the *Anthropogenie,* he added these creatures to his already enlarged and improved representations of the biogenetic law (fig. 8.16). In all, he portrayed twenty different creatures at three stages of ontogenetic development. In these depictions, creatures at some distance from human beings displayed, even at the earliest stages, distinctive differences instead of the indistinct sameness of embryos represented in his earlier texts. In the case of the human embryo, Haeckel represented several stages of ontogenetic development through the use of images adapted directly from His's *Normen-Taffe* (see fig. 8.10). Undoubtedly he thought His could not then object to his illustrations of the human embryo. Despite these finer articulations of distinction, one feature of Haeckel's general principle remained graphically patent: those creatures more closely related to human beings during their evolutionary history manifested quite similar morphologies during early stages of embryonic development, with similarity falling away for those individuals more distantly related.

In the paper by Michael Richardson and his colleagues that laid the foundation for the renewed charge of fraud against Haeckel, the illustrations they employed came from the latter's *Anthropogenie* of 1874, a volume based on a series of lecture he had given at Jena during the previous year. The contention of Richardson and company was that embryo morphology at earlier stages of development was more differentiated than Haeckel, His, or modern researchers had supposed. That argument and the *Science* magazine indictment of Haeckel would have been much more difficult to sustain had Richardson used illustrations from the fifth edition (1903) of the *Anthropogenie*, since those illustrations of early embryos are quite differentiated. Some judicious selection from Haeckel's repertoire of images was required to give the modern charge of fraud an air of plausibility.

In making a judgment of fraud, one must consider *intention*. Did the individual intend to mislead in a way that he or she believed to be contrary to nature? Intention has recently become the concern of Richardson and his colleague Gerhard Keuck; they recognize that discrepancies between Haeckel's illustrations and modern photographs do not allow one to assess intention.[139] Instead they now suggest that fraudulent intention can be revealed if one can detect dramatic alterations of borrowed illustrations. The kind of slight alterations that His complained of simply do not allow the impartial critic to make a confident judgment—especially when both Haeckel and His recognized the need to rectify abnormalities in a specimen, fill in gaps, and produce an "ideal" representation. Richardson and Keuck have focused on the kind of change made by Haeckel that has the requisite dramatic character. They have chosen as a principal exhibit an image from the fifth edition (1903) of the *Anthropogenie*, that of an echidna at three

139. Michael Richardson and Gerhard Keuck, "A Question of Intent: When Is a 'Schematic' Illustration a Fraud?" *Nature* 410 (2001): 144.

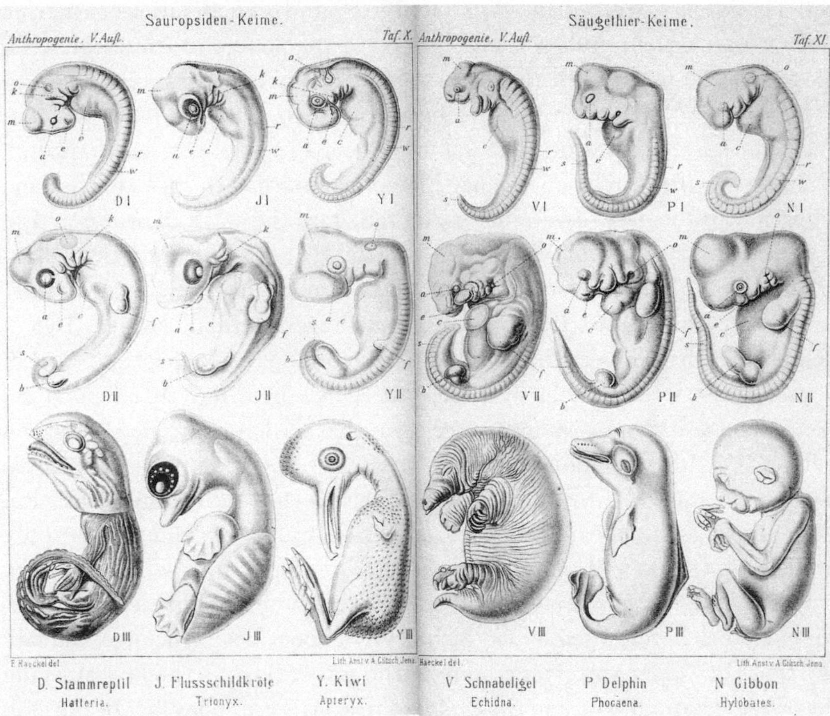

Fig. 8.16. Illustration of the biogenetic law. (From Haeckel, *Anthropogenie,* 5th ed., 1903.)

stages of development (see fig. 8.16).[140] All three depictions originated from Haeckel's onetime student Richard Semon (see fig. 8.17).[141] Haeckel, in his version, eliminated the limb buds in V1 (see fig. 8.16), the earliest stage, so that the echidna embryo would more closely resemble the other vertebrates depicted in the first row. This, Richardson and Keuck judge, could not be an unconscious or unintended act. But does it mean fraud? I do not think so.

In assessing intent, the historian must examine not only a specific action but the broader context of the subject's activities as well. Though Haeckel did embryological work on vertebrates (and the plate in question seems to be the result of his trip to Malaya), nonetheless virtually all of his images of vertebrate embryos were derived from embryologists who were leading authorities on the particular organisms depicted. He him-

140. Ernst Haeckel, *Anthropogenie; oder, Entwickelungsgeschichte des Menschen,* 5th ed., 2 vols. (Leipzig: Wilhelm Engelmann, 1903). Plates are in vol. 1, between 376–77.

141. Richard Semon, *Zoologische Forschungsreisen in Australien und dem Malayischen Archipel,* 4 vols. (Jena: Gustav Fischer, 1893–), 2: plate 10.

self, of course, dissected and illustrated firsthand a great variety of marine invertebrates, their embryos, and larvae. In the sphere of invertebrate biology, no one questioned the verisimilitude of his illustrations (well, almost no one).[142] The commission he received to work on material from the *Challenger* expedition—radiolaria, medusae, sponges, and siphonophores—testifies to the confidence his colleagues had in his expertise and integrity. In the case of vertebrates, he relied on the work of others and so had to make the best of what was available. If there were sufficient illustrations of embryos at some stages of development, he might alter another image to yield an illustration needed to represent a further stage. Two of Semon's illustrations of the echidna had achieved the appropriate level of development for Haeckel's purpose, but the earliest one that Semon depicted was at a more advanced stage than that which he needed.[143] So, as I believe, he altered Semon's image by removing the limb buds, in order to illustrate a stage of development comparable to that of his other examples. After all, it is simply true that for all vertebrates, limb buds appear only after head, trunk, and tail are present. At very early stages, all vertebrate embryos lack these incipient limbs. Haeckel thought this fact about human and other vertebrate embryos had phylogenetic significance, indicating that the "older vertebrates were without feet, just as the lowest living vertebrates (amphioxus and the cyclostomes) still are today."[144] In the case of the echidna, Haeckel did not change any other morphological features of the embryo at the earliest stage; indeed, he left it distinctively different in appearance from the other vertebrate embryos at a comparable stage, though still retaining, of course, fundamental similarity. Hence that elision of limb buds itself would have had negligible force in persuading a reader of the general validity of the biogenetic law. And, after all, there would be little dispute about the fact that the earliest vertebrates (e.g., jawless fish) lacked extremities. Thus there would be no point in fraudulently depicting the echidna embryo at the early stage.

One must also consider Haeckel's personal circumstances. He was not only a former teacher and friend of Semon; he wrote the long introduction to his onetime student's study of Malayan and Australian vertebrates—

142. As mentioned in chapter 3, Gould had an unhinged response to Haeckel's early work on radiolaria. See Stephen Jay Gould, "*Abscheulich!* (Atrocious!): Haeckel's Distortions Did Not Help Darwin," *Natural History* 109, no. 2 (2000): 43. In this article Gould concentrated on the illustrations of the *Natürliche Schöpfungsgeschichte.*

143. Semon's illustration, he indicated, was in natural size. The early embryo in question is comparatively quite large, showing its advanced state.

144. Haeckel, *Anthropogenie,* 5th ed. (1903), 1:371.

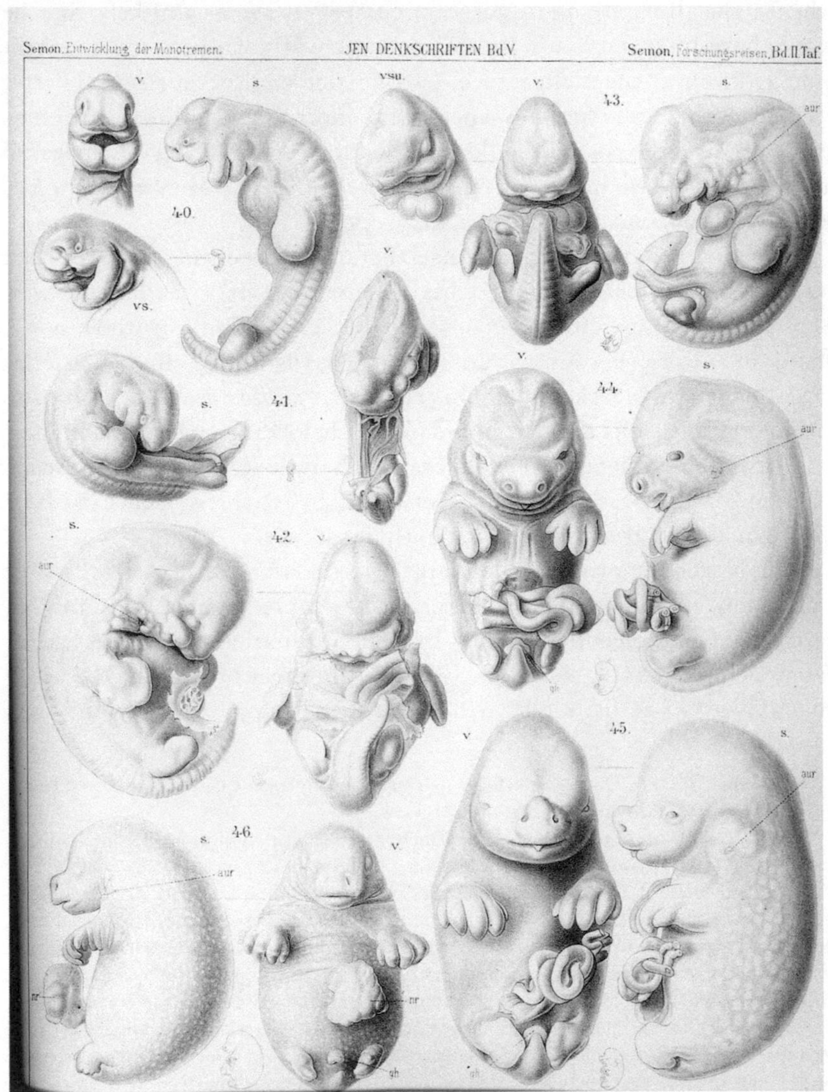

Fig. 8.17. Echidna embryos at various stages of development. (From Semon, *Zoologische Forschungsreisen in Australien und dem Malayischen Archipel,* vol. 2, 1893–.)

whence the illustrations in question came.[145] It seems unlikely that he would have intended to use fraudulently the work of a person who might quickly discover the malfeasance. Further, one can see in Haeckel's later editions of the *Anthropogenie* a persistent effort to render his illustrations as accurately as he possibly could, even to the extent of using the images of human embryos from His, his most bitter enemy. He was ever alert to the scrutiny his opponents would give his work.

I do not believe, therefore, that there is compelling evidence that Haeckel *intentionally* distorted his illustrations in a malfeasant way. That scientists are often *subconsciously* carried along by their hopes and desires to see certain patterns in the data is an old story in science (e.g., Mendel). In Haeckel's case, his *cacoëthes scribendi* often led him to precipitous decisions that left him dangling below the level of the highest standards and vulnerable to enemies. But charges of fraud, as Richardson and Keuck rightly assume, must be sustained by reliable evidence of intention, which I believe to be wanting in this instance.

The biogenetic principle, due largely to the work of Haeckel, became a dominant if controverted hypothesis, one adopted by many leading scientists of the nineteenth century—by such luminaries as Darwin, Gegenbaur, and Weismann—as well as by a great number of twentieth-century embryologists (see fig. 8.18).[146] An interesting example in this latter cat-

145. Ernst Haeckel, "Systematische Einleitung: Zur Phylogenie der Australischen Fana," in Semon, *Zoologische Forschungsreisen*, 1:iii–xxiv.

146. I detail Darwin's adoption of the principle of recapitulation in my *Meaning of Evolution: The Morphological Construction and Ideological Reconstruction of Darwin's Theory* (Chicago: University of Chicago Press, 1992), chap. 5. Gegenbaur first explicitly adopted the principle in his *Grundzüge der vergleichenden Anatomie*, 2nd ed. (Leipzig: Wilhelm Engelmann, 1870), 75: "We perceive in the course of the individual development of the higher organisms an arrangement, in itself transitory, that agrees with the permanent condition of lower organisms. This arrangement indicates to us the conditions that the higher organism had at one time expressed and that have been transformed through the series of generations from which they have been inherited." Weismann noted that the biogenetic law had to be used with great care since "in my view in every case [there is] a strongly altered repetition. . . . But Haeckel also was himself fully conscious of this difficulty." See August Weismann, *Vorträge über Deszendenztheorie*, 3rd ed., 2 vols. (Jena: Gustav Fischer, 1913), 2:157. Weismann went so far as to suggest that a classical education for younger children—stressing Latin and Greek—made good pedagogical sense in light of the biogenetic law. See Weismann to Hans Vaihinger (9 April 1889), in *August Weismann: Ausgewählte Briefe und Dokumente*, ed. Frederick Churchill and Helmut Risler, 2 vols. (Freiburg: Universitätsbibliothek, 1999), 2:134. The principle of recapitulation still held sway in the first half of the twentieth century, enough so that Gavin de Beer felt compelled to write a small book that sought to undermine the idea. See Gavin R. de Beer, *Embryos and Ancestors* (Oxford: Clarendon Press, 1940), especially 4–9. Richardson and Keuck discuss the degree to which contemporary embryologists do or do not accept something like Haeckel's law. See Richardson and Keuck, "Haeckel's ABC of Evolution and Development," 501–8.

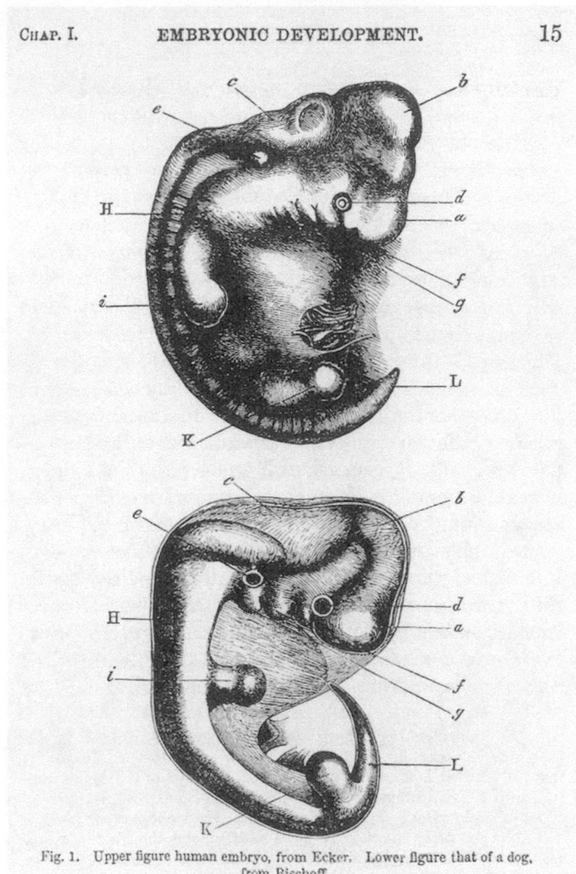

Fig. 8.18. Illustration of embryonic similarity of human (taken from Ecker) and dog (taken from Bischoff). (From Darwin, *The Descent of Man,* 1871.) Darwin remarks: "Häckel has also given analogous drawings in his 'Schöpfungsgeschichte.'"

egory is Scott Gilbert, a historically minded, eminent developmental biologist. Gilbert includes in his *Developmental Biology* (second through fifth editions, 1985–97) a plate taken from George Romanes's *Darwin and After Darwin.*[147] The plate depicts the developmental sequence of eight vertebrate embryos at three stages of development (fig. 8.19).[148] The illustration,

147. George Romanes, *Darwin and After Darwin,* 3 vols. (Chicago: Open Court, 1892–97), 1:152–53.

148. Scott Gilbert, *Developmental Biology,* 2nd ed. (Sunderland, MA: Sinauer Associates, 1985), 153. There is an unintended irony in that Gilbert mentions in a footnote that Haeckel's

1. *The general features of a large group of animals appear earlier in the ıbryo than do the specialized features.* All developing vertebrates (fishes, ptiles, amphibians, birds, and mammals) appear very similar shortly ter gastrulation. It is only later in development that the special features ˙ class, order, and finally species emerge (Figure 1). All vertebrate

. E. von Baer discovered the notochord, the mammalian egg, and the human egg, as ell as making the conceptual advances described here.

FIGURE 1
Illustration of von Baer's law. Early vertebrate embryos exhibit features common to the entire subphylum. As development progresses, embryos become recognizable as members of their class, their order, their family, and finally their species. (From Romanes, 1901.)

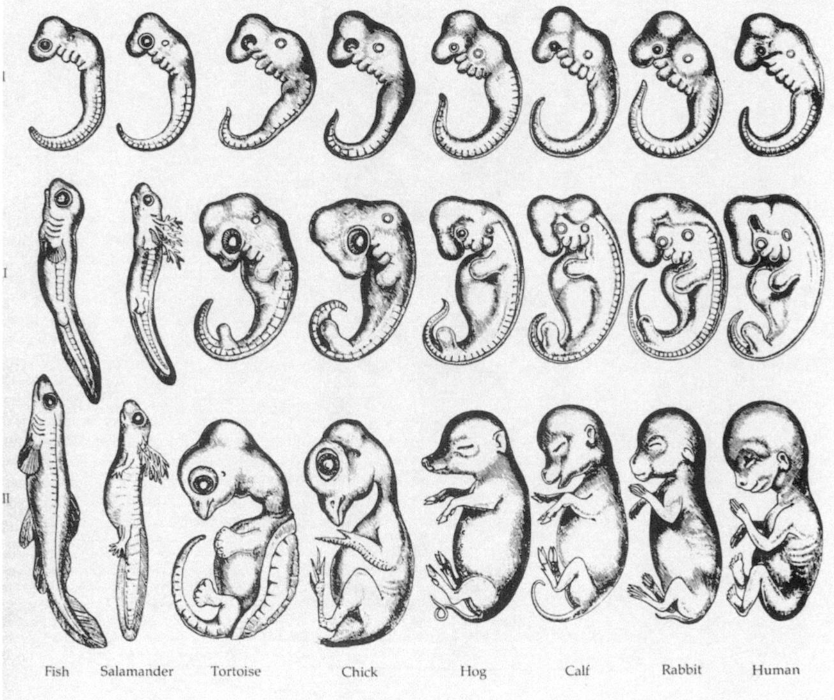

EARLY VERTEBRATE DEVELOPMENT: NEURULATION AND ECTODERM 153

Fig. 8.19. Illustration of von Baer's law that organisms develop from a general to a more specific morphology during ontogeny; from the special features of class, order, and genus to species-specific traits and individual details. The illustration is adapted from George Romanes, who took it from Haeckel. (From Gilbert, *Developmental Biology*, 2nd ed. to 5th ed., 1985–97.)

however, actually came from Haeckel (as Romanes had discreetly indicated). Gilbert uses the Romanes-Haeckel illustration without qualm—and without recognition of its true provenance—to represent what he takes to be von Baer's law that embryological development in an archetype goes from the general morphological structure of a group to increasingly more particular specification, from characteristics of the class and order to those of the species and individual. To the eye of this experienced embryologist, Haeckel's images (in the guise of an illustration from Romanes) conformed to the best biological knowledge of the late twentieth century. However, in an interview done by *Science* magazine, devoted to Richardson's reassessment of Haeckel, Gilbert endorsed Richardson's view that embryos at early stages show more variation than Haeckel had misleadingly represented.[149] Thus, Haeckel's images seemed, by the light of the best scientific knowledge of the late twentieth century, perfectly adequate to indicate embryological similarity—until, that is, they were discovered to be Haeckel's.

Haeckel did not possess the knowledge and facts about development that we have accumulated since his death in 1919. Among his severe critics today, this lack of knowledge seems almost an indictable offense. But all past science stands comparably guilty. When Haeckel's ideas are resituated within his own time, his "errors" do not cast the same long, dark shadows. Nonetheless, the myth of Haeckel's gross malfeasance in promoting his embryological views dies hard, especially when the myth is fed by those whose purpose is religion, not science—a matter to be explored in the next chapter.

---

biogenetic law of *adult* recapitulation had been discredited (154n). In the fifth edition (1997), the plate occurs on 245.

149. Pennisi, "Haeckel's Embryos," 1435. Gilbert dropped the illustration from Haeckel (via Romanes) from the sixth edition of his textbook. He left in its place a cryptic remark: "The acceptance of von Baer's principles and their interpretation over the past hundred years has varied enormously. Recent evidence suggests that one important researcher in the 1800s even fabricated data when his own theory went against these postulates." See Scott Gilbert, *Developmental Biology*, 6th ed. (Sunderland, MA: Sinauer Associates, 2000), 10. Gilbert has, however, included Haeckel's illustration in the website that was designed for the eighth edition (2006) of his book (http://8e.devbio.com/article.php?id=242&search=Richardson). He contrasts the illustrations with Richardson's photographs and remarks: "Yet, the idea that early vertebrate embryos are essentially identical has survived. I think there were two reasons for the survival. First, Haeckel's illustration was reproduced in Romane's (1901) *Darwin and After Darwin*. From here, the illustration entered Anglophone biology, 'sanitized' from Haeckel. Second, the picture can be used (as it has been in several developmental biology books) to illustrate von Baer's principles rather than Haeckel's biogenetic law." Of course, if his illustration had been so used—and Gilbert, indeed, so used them—then even by contemporary lights, Haeckel had fairly represented embryos at an early stage.

Fig. 9.1. Ernst Haeckel on the way to Ceylon, 1881–82.
(Courtesy of Ernst-Haeckel-Haus, Jena.)

CHAPTER NINE

The Religious Response to Evolutionism: Ants, Embryos, and Jesuits

If religion means a commitment to a set of theological propositions regarding the nature of God, the soul, and an afterlife, Haeckel was never a religious enthusiast. The influence of Schleiermacher on his family kept religious observance decorous and commitment vague.[1] The theologian had maintained that true religion lay deep in the heart, where the inner person experienced a feeling of absolute dependence. Dogmatic tenets, he argued, served merely as inadequate symbols of this fundamental experience. Religious feeling, according to Schleiermacher's *Über die Religion* (On religion, 1799), might best be cultivated by seeking after truth, experiencing beauty, and contemplating nature.[2] Haeckel practiced this kind of Schleiermacherian religion all of his life.

Haeckel's association with the Evangelical Church, even as a youth, had been conventional.[3] The death of his first wife severed the loose threads still holding him to formal observance. The power of that death, his obsession with a life that might have been, and the dark feeling of love forever lost—these haunting remains drove him to find a more enduring and rational substitute for orthodox religion in Goethean nature and Darwinian evolution. The passions that had bound him to one individual and her lingering shadow became transformed into acid recriminations against

1. Wilhelm Bölsche, who interviewed Haeckel's aunt Bertha Sethe (sister of his mother), describes the impact of the Schleiermacherian view on the family in his *Ernst Haeckel: Ein Lebensbild* (Berlin: Georg Bondi, 1909), 10–11.

2. I have discussed Schleiermacher's religious ideas in *The Romantic Conception of Life: Science and Philosophy in the Age of Goethe* (Chicago: University of Chicago Press, 2002), 94–105.

3. In Germany the Evangelische Kirche consisted of the Lutheran and Lutheran Reformed communities.

any person or institution promoting what he saw, through Darwinian eyes, as cynical superstition. The antagonism between conservative religion and evolutionary theory, brought to incandescence at the turn of the century and burning still brightly in our own time, can be attributed, in large part, to Haeckel's fierce broadsides launched against orthodoxy in his popular books and lectures. These attacks and reactions to them escalated to a new level during the period from 1880 to his death in 1919.

It is a sociological commonplace that researchers who study the patterns of belief exhibited by different social and ethnic groups generally harbor little religious conviction themselves.[4] Their skepticism results from awareness of the great diversity of systems of belief entertained and fought over by adherents—systems that are in conflict with each other and often with themselves. Anthropologists have recognized that such systems usually have disguised practical and frequently political functions. Haeckel's own casual anthropological observations, made during his many research trips, brought him to detect venomous political creatures scrambling under the cover of agitated doctrine. He became ever more convinced that aggressive measures, grounded on the most advanced science, were required to root out superstition and to protect modern cultural life against such dangers. In the early 1880s, his first journey to tropical lands, where Eastern and Western religions met in steamy jungle settings, brought his sensitivity to religious irrationality to a new pitch of intolerance.

Haeckel's Journey to the Tropics: The Footprint of Religion

Haeckel made it a condition of his contributions to the reports of the Challenger expedition that he be allowed to include his own specimens in his systematic descriptions. One important area that the *Challenger* neglected to dredge was the Indian Ocean. With this omission as a modest incentive, Haeckel decided to travel to the coasts of Ceylon (now Sri Lanka) to bring back fauna that would render his own additions to the *Reports* as comprehensive as possible. But as in several other instances, scientific pretext served to cover the fundamental text, a desire welling up from his younger self: since his youth Haeckel had longed to travel to the tropics, and now the urge became as acute as that which had impelled Goethe to Italy. For both, their ventures promised a loosening of the grip of settled middle age,

4. See, for example, Bernard Spilka, Ralph Hood, Bruce Hunsberger, and Richard Gorsuch, *The Psychology of Religion: An Empirical Approach* (New York: Guilford Press, 2003), 179.

release from domestic malaise in a cold climate, and escape to warmer, more romantically enticing environs.

Through friends in Berlin and with the encouragement of Wilhelm Peters (1815–1883), professor of zoology and director of the Berlin Zoological Museum, Haeckel made application to the Humboldt-Stiftung for 12,000 to 15,000 marks to fund his research trip. At the meeting of the Berlin Academy of Sciences, which would decide the matter, his cause was advanced by Peters and Helmholtz but opposed by Virchow and Du Bois-Reymond. Despite the eminent Helmholtz's support, the funds were denied.[5] In some desperation, Haeckel appealed to Darwin, Huxley, and Lubbock to see if any British sources could be found. In his letter to Darwin, he struck the plaintive cord: "A few weeks ago, in the decisive meeting, the academy (with a bare majority) denied my application, since 'Professor Haeckel is the most enthusiastic and most dangerous apostle of the Darwinian error doctrine [*Irrlehre*] and since his zoological works are without true merit!!'"[6] His English friends were certainly sympathetic to his request—and Darwin offered him a personal check for 100 pounds[7]—but the British learned societies had no significant funds for such purpose.[8] Haeckel was yet determined to leave Germany, if he could secure at least minimal funding. He drew on his own accounts and, exercising delicate solicitation, on monies from friends, as well as on a small stipend from the Weimar administration.[9]

The scientific results of his trip to Ceylon would be modest. He would have great difficulty in preserving specimens in the tropical heat, yet his stay at the southern tip of this exotic land would prove to be the "most

5. The meeting of the Academy of Sciences occurred on 5 May 1881. Helmholtz seemed to think the opposition to Haeckel was based mainly on personality differences. Du Bois-Reymond wrote Helmholtz to say that the rejection of Haeckel reached above personality to such doctrines as the *Plastidule* theory—the very theory that Virchow had derided in his Munich lecture. See Du Bois-Reymond to Helmholtz (7 May 1881), in *Dokumente einer Freundschaft: Briefwechsel zwischen Hermann von Helmholtz und Emil du Bois-Reymond, 1846–1894*, ed. Christa Kirsten (Berlin: Akademie, 1986), 264.

6. Haeckel to Darwin (21 June 1881), in the Darwin Papers, DAR 166.1, Special Collections, Cambridge University Library.

7. Haeckel to Darwin (1 July 1881), in ibid. In this note Haeckel thanked Darwin for the offer but said he would try other sources first.

8. See also Haeckel to Thomas Henry Huxley (21 June 1881) and Huxley to Haeckel (1 July 1881), in "Der Briefwechsel zwischen Thomas Henry Huxley und Ernst Haeckel," ed. Georg Uschmann and Ilse Jahn, *Wissenschaftliche Zeitschrift der Friedrich-Schiller Universität Jena* (Mathematisch-Naturwissenschaftliche Reihe) 9 (1959–60): 26–27.

9. Erika Krauße, *Ernst Haeckel* (Leipzig: Teubner, 1984), 98.

interesting and happiest six weeks of my life."[10] The excursion would yield a large travel book of great charm and continuing interest, his *Indische Reisebriefe* (Indian travel letters, 1882), and over two hundred watercolors displaying a vibrancy of hue and depth of saturation that indicated a new artistic and personal sensibility.[11] After some difficulty, he managed to have a sampling of these watercolors (and complementary photographs) reproduced in a volume in 1904, at a time when the paintings from tropical Tahiti by the ex-banker Paul Gauguin were starting to arouse public interest.[12] During Haeckel's journey, the artistic and personal would submerge the scientific and professional.

Haeckel departed Jena for Trieste with sixteen large trunks on 8 October 1881. From Trieste, he took a Lloyd's Austrian steamer through the Suez Canal and on to Bombay, where he spent a week. On 21 November he arrived in Colombo, the capital of Ceylon. There he reveled in the great tumult of peoples—the Sinhalese and Tamils, who had been engaged in an age-old conflict, Arabs and Europeans—English, Dutch, Portuguese, Germans—and those of mixed descent. In Colombo he discovered what great havoc the hot, humid climate could produce: his dark evening coat came out of his trunk white with mildew; his books were blackened with mold; and the wood on his cameras had warped. The sultry air seemed to feed the mosquitoes, scorpions, and land leeches that besieged him, and to make the

10. Ernst Haeckel, *Indische Reisebriefe*, 2nd ed. (Berlin: Gebrüder Paetel, 1884), 276. Haeckel found that many delicate specimens, when transported from the ocean in glass jars to his research station, disintegrated in the warmer water of their containers.

11. Prior to the publication of his *Indische Reisebriefe* (1882), Haeckel gave an account of his journey in a series of articles for the *Deutsche Rundschau*. These reports were summarized by various European and American newspapers—for instance, the *London Daily News* and the *New York Times*. See, for example, the article in the latter: "Ways of Life in Ceylon: Professor Haeckel's Record of His Own Experiences There," *New York Times*, 30 September 1882, 2. Haeckel's trip had derivative consequences. He inspired several others to make comparable trips to the regions of India and Malaya. See Uwe Hoßfeld, "The Travels of Jena Zoologists in the Indo-Malayan Region," *Proceedings of the California Academy of Sciences* 55, supplement 2 (2004): 77–105.

12. Upon his return to Germany, Haeckel tried to get more than a dozen publishers interested in putting out an edition of his paintings—without success. However, after the turn of the century, he did have an audience, and his collection appeared as *Ernst Haeckels Wanderbilder* (Gera-Untermhaus: Koehler'sche Verlagsbuchhandlung, 1904). The well-known Berliner painter Ernst Körner (1846–1927) wrote Haeckel to say of the collection: "The view for the sublime in the whole, which indicates the artist, is united with the loving observation of the particulars, wherein the study of the researcher is evidenced." The letter is quoted by Walther May in his *Ernst Haeckel: Versuch einer Chronik seines Lebens und Wirkens* (Leipzig: Johann Ambrosius Barth, 1909), 246.

coverings of his bed a moist haven for snakes of every kind, including the cobra. Undeterred, he continued south, down the island, over roads roughly cut through the jungle. On 12 December he reached his destination, the village of Belligam, on the southwestern coast; and there he set up his research station.

Haeckel was aided in his work by divers who brought up corals, sea urchins, medusae, and other creatures from around the coastal reefs. These Sinhalese helpers performed extraordinary feats of endurance and dexterity, lifting fifty- and eighty-pound chunks of coral into their boats. Haeckel, enjoying a renewed, youthful exuberance, initially joined the divers. But he proved not as adept as his helpers in avoiding seductively beautiful but painfully dangerous creatures. Piercing nettles from stealthy jellyfish left him in pain for a week, and unobtrusive sea urchins cut up his feet.[13] Dissection and microscopic observation yielded safer modes of research. Despite the difficulties in preserving his specimens, he did solder away in tins enough radiolarians and medusae to enlarge substantially the number of new species for his *Challenger* reports.

Haeckel's casual observations of the Sinhalese led him to make the wistful colonial's kind of judgment: in contrast to "Europeans with our thousand superfluous requirements," they "content themselves to be simple men, natural men, who dwell in paradise and who enjoy this paradise."[14] Haeckel likewise thought he had arrived in Elysium, where adolescent boys and girls of singular beauty in face and form might frolic, wearing only the briefest of aprons. He noticed one especially:

> [There] was a girl of about sixteen years, a niece of the Aretschi [a friend of Haeckel], whose perfectly beautiful figure could have served as a model for a sculptor. Among the boys, several could have competed with Ganymede in beauty.[15]

If Somerset Maugham's own experience in Ceylon is suggestive, the adolescent girls might come to live with a European for a while, perhaps until he departed the country. The girls generally, however, married at ten or twelve and were grandmothers at twenty-five. In Haeckel's view the sensual bloom quickly faded, and age descended rapidly to blot out

13. Haeckel, *Indische Reisebriefe,* 193–94.
14. Ibid., 176.
15. Ibid., 270.

their ephemeral youth. But as he came to know some Sinhalese more intimately, his attitude about these natural men changed to one of deep appreciation of their interior qualities, appreciation that grew, perhaps, even into love.

Shortly after he arrived in Belligam, Haeckel met an older Sinhalese who helped him with his various needs. Because of this fellow's appearance and demeanor, Haeckel dubbed him Socrates. The man's wry intelligence and casual acquaintance with cleanliness also suggested the appositeness of the name. They became good friends. But he found a deeper intellectual companion in Abayavira, a member of the Aretschi tribe. This Sinhalese was about Haeckel's age, industrious, curious, and "spoke rather good English and carried himself with a natural dignity and a clear intelligence that often astounded me." He and Haeckel spent many evenings discussing problems of science and life.[16] The most memorable figure of Haeckel's travels, though, was a lad of about nineteen whom he hired as a personal valet. The young man was of the lowest and most despised caste but had a beauty of face and figure that singularly struck the Romantic German. When Haeckel first saw the quite naked, bronze boy with long dark hair, he believed a Greek statue had come to life. Since the adolescent was from the village of Gamameda ("Gama" = village and "meda" = middle), Haeckel called him Ganymede—a name of libidinous provenance, likely reflecting a recollection of Goethe's poem.[17] The servant became completely devoted to his master. And when Haeckel reached the end of his time in the village, Ganymede fell despondent and tearfully beseeched his employer to take him back to Europe.[18] Haeckel's later recounting of this event would leave his friend Allmers gasping for breath and for superlatives to express how sensually exquisite the scene struck him. He even wondered if it were not yet possible to secure the beautiful youth passage to Germany.[19]

After quitting Belligam, Haeckel traveled up to the hill country with the aim of climbing the mountain called Adam's Peak (Sri Pada), a modestly high (7,300 feet) cone that could be seen from several miles out at sea. He took an interest in the peak, since toward the top a giant footprint, some two and a half feet by five feet, had been impressed into the rock, or at least

16. Ibid., 217.

17. Ibid., 205–6.

18. Ibid., 277.

19. Allmers to Haeckel (6 September 1882), in *Haeckel und Allmers: Die Geschichte einer Freundschaft in Briefen der Freunde*, ed. Rudolph Koop (Bremen: Arthur Geist, 1941), 147.

the several religions of the region had so believed. The Sinhalese thought the god Saman had left the imprint. The Buddhists built a small temple around the place where the Buddha had left his mark preaching Nirvana. The Hindus were convinced that the print indicated the spot where Siva had leapt into heaven. The Christian Portuguese thought it a memento of Saint Thomas, the apostle to the Indies. The Muslims regarded it as the footprint of Adam, left when the angel had set him on the mountain after he had been driven from Paradise. Not to be outdone, the few Persians in the area argued that actually the print was from Alexander the Great, who had passed through the island on his fateful conquest of the East. On 12 February 1882, Haeckel climbed the mountain to see the acclaimed vestige. His own examination made him wonder about the powers of imagination that transformed the odd formation into a giant footprint. He suggested that the impression held a mixed message both about the credulity of the religiously minded, yet also about the remarkable toleration, particularly exemplified by the Buddhists, that allowed the several faiths mutually to venerate the spot. At the summit Haeckel took the occasion to juxtapose belief in the footprint with an account, delivered to his companions, of another revered occasion—the birth of Charles Darwin, which had occurred on that same February day in 1809.[20]

Shortly after coming down from the mountain, Haeckel began to make his way home. He steamed from Colombo on 4 March, reaching Egypt on 28 March. He spent ten days in Cairo, and then continued on to Alexandria, whence he departed for Trieste, arriving there 18 April. He reached Jena on 21 April, bearing twice the amount of luggage with which he had departed. At his return, he learned that his revered friend, Charles Darwin, the man whom he had commemorated on Adam's Peak, had died three days before, on 19 April.

During the summer of 1882, Haeckel busied himself with the construction of his new home, which would be called Villa Medusa, an ample dwelling whose rooms carried escutcheons of medusae on the ceilings (fig. 9.2). He also supervised construction of a new zoological institute. Both were completed by the fall of the following year. And in October he traveled to Eisenach, a morning's train ride away, to attend the fifty-fifth annual meeting of the Society of German Natural Scientists and Physicians. There he celebrated the contributions to science of the recently departed Darwin.

20. Haeckel described his climb and the religious significance of the mountain in ibid., 297–326.

Fig. 9.2. Haeckel's house, Villa Medusa. (Photograph by the author.)

"Science Has Nothing to Do with Christ"—Darwin

Darwin's Letter

In his plenary lecture, Haeckel chanted a hymn of devotion to Darwin's genius and to the extraordinary impact of his theory on all realms of human thought, emancipating that thought for a rational approach to life.[21] Haeckel argued that the Englishman followed upon the path first hacked through the jungle of religiously overgrown biology by the likes of Lessing, Herder, Goethe, and Kant. Indeed, Darwin had solved the great problem posed by Kant, namely, "how a purposively directed form of organization can arise without the aid of a purposively effective cause."[22] In his encomium Haeckel, like the devil, could appeal even to scripture—or at least to one who translated scripture in the very city of Eisenach: just as Martin Luther, who "with a mighty hand tore asunder the web of lies by the world-dominating papacy, so in our day, Charles Darwin, with comparable

21. Ernst Haeckel, "Ueber die Naturanschauung von Darwin, Göthe und Lamarck," in *Tageblatt der 55. Versammlung Deutscher Naturforscher und Aerzte in Eisenach, von 18. bis 22. September 1882* (Eisenach: Hofbuchdruckerei von H. Kahle, 1882), 81–91.

22. Ibid., 82.

overpowering might, has destroyed the ruling error doctrines of the mystical creation dogma and through his reform of developmental theory has elevated the whole sensibility, thought, and will of mankind onto a higher plane."[23]

Haeckel certainly advanced no new ideas in his lecture—something Allmers tactfully observed after reading the text[24]—but he did eloquently reinforce four points: that Darwin fulfilled the promise of higher German thought—especially that of Goethe; that the evolutionary theories of Goethe, Lamarck, and Darwin were as vital to modern culture and as substantial as the locomotive and the steamship, the telegraph and the photograph—and the thousand indispensable discoveries of physics and chemistry; that Darwinism yielded an ethics and social philosophy that balanced altruism against egoism; and, in summary, that Darwinian theory and its spread represented the triumph of reason over the benighted minions of the anti-progressive and the superstitious, particularly as shrouded in the black robes of the Catholic Church. In Haeckel's analysis, then, Darwinism was thoroughly modern, liberal, and decidedly opposed to religious dogmatism.

To drive his message home, Haeckel read to the audience a letter that Darwin had sent to a student of Haeckel, a young Russian nobleman who had confessed to the renowned scientist his bothersome doubts about evolutionary theory in relation to revelation. Darwin's response to the student read:

> Dear Sir:
> I am much engaged, an old man, and out of health, and I cannot spare time to answer your questions fully,—nor indeed can they be answered. Science has nothing to do with Christ, except in so far as the habit of scientific research makes a man cautious in admitting evidence. For myself, I do not believe that there ever has been any revelation. As for a future life, every man must judge for himself between conflicting vague probabilities.
>
> Wishing you happiness, I remain, dear Sir, Yours Faithfully,
> Charles Darwin[25]

23. Ibid., 81.
24. Allmers to Haeckel (January, 1883), in *Haeckel und Allmers*, 149–50.
25. Haeckel, "Ueber die Naturanschauung," 89. Haeckel translated the letter into German. A copy of the original, which I have used here, is held in the Manuscript Room of Cambridge University Library. The letter was addressed to Nicolai Alexandrovitch Mengden.

What Darwinism offered instead of traditional orthodoxy, Haeckel contended, was Goethe's religion: a "monistic religion of humanity grounded in pantheism."[26] This declaration of rationalistic faith would hardly be the recipe to satisfy those who yet hungered after the old-time convictions.

For those assembled at Eisenach—and for the many others who read the published text of Haeckel's lecture—the recitation of Darwin's letter seemed to drive a wedge into the soft wood of compatibility between science and traditional religion, splitting the two. The letter revealed that an aggressive, preacher-baiting German was not the only evolutionary enemy of faith but that the very founder of the theory had also utterly rejected the ancient beliefs. Jacob Moleschott (1822–1893), the Dutch physiologist and ardent materialist, wrote Haeckel immediately to say that the publication of Darwin's letter was "of incalculable importance."[27] But several English authorities complained that Haeckel had committed a great indiscretion in making public Darwin's private communication even before the earth had settled around his grave.[28] But indiscreet or not, the message could hardly be plainer: Darwinian theory was decidedly opposed to that old-time religion. And as Haeckel discovered during the next three decades (and as we are still quite aware), that old-time religion was decidedly opposed to modern Darwinian theory.

Monistic Religion

Haeckel had, over the course of a quarter of a century, expressed his own religious views both negatively and positively. The negative critique attacked orthodox religion, dismissing its belief in an anthropomorphic deity and deriding its view of an immaterial human soul. Haeckel was an equal-opportunity basher of all orthodox doctrines—that of Christianity, Judaism, Muslimism, and the faiths of the East. Yet he still thought of himself

26. Haeckel, "Ueber die Naturanschauung," 89.

27. Moleschott to Haeckel (23 October 1882), in *Carl Vogt, Jacob Moleschott, Ludwig Büchner, Ernst Haeckel: Briefwechsel*, ed. Christoph Kockerbeck (Marburg: Basilisken-Presse, 1999), 120.

28. Haeckel mentioned to Allmers the unfavorable response coming from England at the publication of Darwin's letter. See Haeckel to Allmers (26 December 1882), in *Ernst Haeckel: Sein Leben, Denken und Wirken. Eine Schriftenfolge für seine zahlreichen Freunde und Anhänger*, ed. Victor Franz, 2 vols. (Jena: Wilhelm Gronau und W. Agricola, 1943–44), 2:81. Edward Aveling, consort of Karl Marx's daughter and translator of *Das Kapital* into English, wrote Haeckel to describe the cowardly reaction of the British press to Haeckel's exposition of the letter. See Aveling to Haeckel (6 October 1882), in Ernst Haeckel, *Die Naturanschauung von Darwin, Goethe und Lamarck* (Jena: Gustav Fischer, 1882), 62–64.

as a religious person, though his was the religion of Spinoza and Goethe. He took opportunity to synthesize his negative and positive critiques when invited to Altenburg (thirty miles south of Leipzig) to help celebrate the seventy-fifth anniversary of the Naturforschende Gesellschaft des Osterlandes (The Osterland Natural History Society).[29] At the meeting on 9 October 1892, Haeckel was preceded by a speaker who said something rather irritating about the relationship between science and religion. Haeckel tossed aside his prepared text and gave a lecture extemporaneously, which he wrote down the next day from memory, augmenting where necessary. The lecture was published in the popular press and as a small monograph, *Der Monismus als Band zwischen Religion und Wissenschaft* (Monism as the bond between religion and science, 1892)—a book that would reach a seventeenth edition just after Haeckel's death. The broadside resonated with freethinkers all over the world and became the foundation for the even more successful *Die Welträthsel* (The world puzzles), which would be published in 1899.[30]

In his small tract, Haeckel argued for a universe in which homogeneous atoms of matter express various properties through the fundamental powers of attraction and repulsion. These atoms propagate their effects through vibrations set up in an ocean of ether. From the inorganic, through the simplest organisms, right up to man, no unbridgeable barriers arise separating one kind of substance from another; rather a continuous, law-governed unity runs through the whole. Even what might be called man's soul—his central nervous system—appeared over the course of ages by slow increments out of antecedents in the lower animals. Though Haeckel's enemies thought this cosmology the sheerest materialism, he yet maintained it was a strict monism: all matter had its mental side, just as all examples of mind displayed a material face. This meant that the elements of perception and thought could be traced right down to the simplest organisms—every one-celled protist could thus boast of a "soul," after a manner of speaking. This sort of conception gave the comparative psychologist, according to Haeckel, permission to discover the antecedents of human cognitive ability in animal life. The great unity pervading the universe, a universe

29. Osterland is the region around Altenburg and Gera.

30. Fritz Müller, from his small island home off the southern coast of South America, wrote Haeckel on receipt of the book: "That I agree with you completely in all essential points needs hardly be said—indeed, especially your judgment of Bismarck [which acknowledged his role in forming the German nation]." Müller to Haeckel (28 February 1893), in the Correspondence of Ernst Haeckel, the Haeckel Papers, Institut für Geschichte der Medizin, Naturwissenschaft und Technik, Ernst-Haeckel-Haus, Friedrich-Schiller-Universität, Jena.

governed by ineluctable law, could be understood materially as nature in her organized diversity and spiritually as God.

Haeckel wished to uproot all anthropomorphisms from religion, yet he thought something was worth preserving from the old dispensation. This was the ethical core of traditional orthodoxy, especially of Christianity:

> Doubtless, human culture today owes the greater part of its perfection to the spread and ennobling [effect] of Christian ethics, despite its higher worth being injured, often in a regrettable way, by its connection with untenable myths and so-called "revelation."[31]

Haeckel was no Nietzsche. He had no desire to replace the heart of traditional morality with some superseding ethical system. He believed that altruism, the selfless love of one's neighbor, functioned as the chief moral virtue. It could be explained, not by appeal to a divine hereafter—which reduced moral action to a selfish desire for reward—but by the natural selection and evolution of cooperative behavior.

Haeckel's *Monismus* had an immediate and, for the author, a surprising outcome: he was sued. This occurred because of a note that he appended to his discussion of anti-Darwinian scientists. He mentioned Louis Agassiz and, of course, Virchow. He added that more recently, his former student and assistant Otto Hamann (1857–1928) had taken a reactionary turn in his book *Entwicklungslehre und Darwinismus* (Evolutionary theory and Darwinism, 1892). Hamann went from being an enthusiastic supporter of Darwinian evolutionary theory, while he was Haeckel's student, to rejecting it for a more distinctively teleological and ultimately religious conception.

In his new book, Hamann variously argued: that the paleontological evidence indicated gaps in the fossil record; that von Baer had shown long ago that embryos were of consistent type, not passing from one type to another; and that the divide between the mental abilities of men and animals was absolute.[32] He maintained, in opposition to "Darwinian dogmatism," that one had to explain the goal-directed character (*Zielstrebigkeit*) of life as based on "inner causes" that produced macro-mutations responsive to altered environments. The great harmony in the natural system of coor-

31. Ernst Haeckel, *Der Monismus als Band zwischen Religion und Wissenschaft, Glaubensbekenntniss eines Naturforschers* (Bonn: Emil Strauss, 1892), 29.

32. Otto Hamann, *Entwicklungslehre und Darwinismus. Eine kritische Darstellung der modernen Entwicklungslehre* (Jena: Hermann Constenoble, 1892), 7–20, 21–26, 120.

dinated adaptations discovered by the naturalist was "the same as that unity and harmony which men prior to all scientific research feel and have sensed—a unity and limitlessness that goes by the name of God."[33]

Haeckel felt the sting of this apostasy. The argument of Hamann's volume, he remonstrated, was the very opposite of science; rather it was "from the beginning to the end a great lie."[34] Haeckel attributed the reversal in his onetime student's attitude not to the discovery of new truths about the failure of Darwinism but to his own failure to receive an academic appointment. Hamann had implored his former teacher to recommend him for a vacant chair (the Ritter Professor) in zoology at Jena. Haeckel did put him on a list of candidates submitted to the faculty senate but did not place his former student among the top contenders.[35] Hence, as Haeckel charged in his *Monismus,* Hamann took his revenge by going over to the dark side. Yet all that would be needed to bring him running back, Haeckel supposed, would be "the jingle of coins."[36]

Hamann sued Haeckel because of this characterization, contending loss of income and libel. He requested the court grant him a total of 7,500 marks, 6,000 for reduced income and 1,500 as punishment for the libel. Haeckel countersued, and the case was heard in the Schöffengericht (a lower court) in Jena. During the process, it came out that Hamann had misrepresented himself as a professor at Göttingen, whereas he was only a *Privatdozent* there. Haeckel put in evidence a series of obsequious letters from Hamann, in which the supplicant likened his former teacher to a god whom he revered.[37] The court concluded that Haeckel did slightly slander Hamann and fined him 200 marks; the judge also levied a fine of 30 marks against Hamann. Both were enjoined not to speak of the conflict again, and Haeckel complied by expunging his remarks from subsequent editions of

33. Ibid., 288.

34. Haeckel, *Der Monismus,* 42–43.

35. Haeckel to Hamann (18 August 1889). On the back of a letter in which Hamann made the request (16 August 1889), Haeckel sketched out a list of candidates, initially putting Hamann first, Richard Semon second, and Wilhelm Roux sixth. These letters are in the Haeckel Correspondence, Haeckel-Haus, Jena.

36. Haeckel, *Der Monismus,* 43.

37. Most of the letters to Haeckel were in the typical reverential tone accorded a teacher by a former student. So, for instance, Hamann took Haeckel's side in the continually sniping of the likes of Semper and Virchow. See Hamann to Haeckel (23 September 1887), in the Haeckel Correspondence, Haeckel-Haus, Jena. Haeckel had more than a half-dozen letters marked in which Hamman was unstinting in his reverence and praise for his former teacher; these were apparently included in his court brief against Hamann.

his *Monismus.* Most onlookers thought that Haeckel had won the moral victory, or so an anonymous account of the case reported.[38] This trial is probably the source of the rumor, one still bubbling in the heads of many creationists, that Haeckel had been brought before a "university court" by five of his colleagues where he was judged guilty of having committed scientific fraud. Though Jena had a student *Kerker,* a jail for misbehaving adolescents, a university court is an unknown entity and any talk of one could come only from brains on the boil.[39]

Erich Wasmann, a Jesuit Evolutionist

The Challenge of the Catholic Church

Ever since his medical school days in Bavaria, Haeckel had been both attracted and repelled by the Catholic Church, especially by its black-robed combat troops, the Jesuits. While in Rome, Haeckel felt his north German sensibilities continually assaulted—unlike Goethe, who rather enjoyed the pomp of papal celebrations. Protestant liberals like Haeckel had come to perceive the wars against Austria and France not only as political-social conflicts but also as struggles against an alien religious force. Intellectual and cultural threats from the church were codified for liberals in the series of condemnations listed in Pope Pius IX's *Syllabus errorum* (1864), his brief of particulars brought against the modern world. Condemned were such heretical tenets as pantheistic naturalism, the autonomy and sufficiency of reason to discover the truth, freedom of individuals to embrace any religion, civil control of education, and unbridled speech. The declaration by the Vatican Council (1870) of papal infallibility only heightened the cultural clash between the Vatican and liberal movements all over Europe—including those within the Catholic Church itself.[40]

38. Anonymous, *Der Ausgang des Prozesses Haeckel-Hamann* (Magdeburg: Listner & Drews, 1893). See also Georg Uschmann, *Geschichte der Zoologie und der zoologischen Anstalten in Jena 1779–1919* (Jena: Gustav Fischer, 1959), 119–20.

39. This mythical story can be found on a large number of creationist websites. The words "Haeckel" and "university court" in any search engine will disgorge thousands of sites on to an innocent computer.

40. For example, in Germany, Johann Joseph Ignaz von Döllinger (1799–1890) led the opposition to the declaration of papal infallibility. Döllinger, priest and professor of theology at Munich, was a leading liberal theologian who had been elected delegate to the Frankfurt National Assembly in 1848. During the Vatican Council, he published letters in the *Augsburger Allgemeine Zeitung* (1869) that decried the *Syllabus errorum* and the movement to define papal infallibility, describing them as contrary to scripture and the traditions of the church. The let-

Bismarck recognized that the negative reaction of liberals to the Roman Catholic Church made opportune a move to curb the growing power of the Catholic Center Party. He promoted what Virchow called a *Kulturkampf*—a cultural battle—but one fought with the forces not of persuasion but of legislation. At Bismarck's instigation, the Reichstag passed a series of laws, the so-called May Laws of 1872–75, that restricted the civil activities of the Catholic clergy, especially in performing state-recognized marriages and in education. In 1872 the Jesuits, perceived as the sinister agents of Pius IX, were expelled from Germany; and the next year all religious orders, except those directly concerned with care of the sick, had to disband. The suppression of the Catholic Church in Germany by the liberal-dominated Reichstag ran against the principles of those same liberals, who often acted out of religious intolerance and prejudice, and, as Gordon Craig has suggested, not a little out of the economic advantages accruing to those of a more materialistic taste.[41] Even among individuals differing on many other issues—Haeckel and Virchow, for instance—the exclusion of the Jesuits and the restrictions on the Catholic clergy found favor. By the end of the 1870s, however, the political situation began to flex as Bismarck's worries turned from Catholics to the growing socialist movements.

In 1878 a new pope, Leo XIII, ascended to the chair of Peter. Leo sought

---

ters roused widespread opposition amongst liberal Catholics in Germany to the ultramontane position. After the promulgation of the doctrine, he was formally excommunicated and became an adviser to the "Old Catholic" group in Germany, of which Gegenbaur was a member. Efforts to bring him back into the fold were unavailing. See Johann Ignaz von Döllinger, *Das Papstthum* (Munich: Beck, 1892). In England John Henry Cardinal Newman (1801–1890) attempted to dissipate the miasmic disdain of many of his countrymen for the *Syllabus* by employing two tactics: he noted that the list was not drawn up or signed by the pope, and thus the document itself had no authoritative power—certainly it was not declared ex cathedra as a matter of faith or morals; and he then argued that the specific condemnations looked quite different when replaced in the encyclicals, letters, and pronouncements whence they were drawn—they lacked the universal and absolute character they seemed to have when enumerated in isolation. See John Henry Newman, *Certain Difficulties Felt by Anglicans in Catholic Teaching Considered*, new impression, 2 vols. (London: Longmans, Green, 1900), 2:276–98. In the many discussions leading up to the declaration of papal infallibility, Newman argued for a very narrow definition, if the doctrine had to be defined at all: the pope had expressly to say he was so legislating, and it could only be on matters of faith or morals—not on science or any other subject outside of the limited purview; further the pope had to be referring to the apostolic deposit of faith as found in scripture or tradition. Others in the hierarchy wished the doctrine to be applicable virtually to any sphere of human concern. At the Vatican Council (1870), Newman preferred that the assembled bishops not define the doctrine at all. The doctrine was, however, promulgated, though in the narrower terms that Newman favored. Unlike Döllinger, Newman remained in the church.

41. Gordon A. Craig, *Germany, 1866–1945* (Oxford: Oxford University Press, 1980), 78–79.

accommodation with the German government; and with a lessening of tensions, the legal and extralegal opposition to the Catholic Church began to ease. The old *Kulturkampf* abated, but a new one, more personal, was turned against its original author as the young Emperor Wilhelm II (ruled 1888–1918) strove to take a greater hand in the social and foreign affairs of his government. Quickly relations with his aged chancellor deteriorated, until the exit became clearly marked and the door opened. Bismarck departed in 1890. Thereafter the Social Democrats and the Center Party continued to gain seats in the Reichstag, as a more accommodating head of state took command.[42]

The new dispensation stiffened Haeckel's spine and drove the old liberal to prepare an aggressive defense against the sudden rise of politically and culturally militant religious forces. In a move that angered many of his colleagues at Jena, he and several other professors, students, and townspeople met with the deposed Bismarck and invited him to visit Jena to be honored for his creation of and service to the empire. With this as something of a fait accompli, Haeckel then informed Archduke Carl Alexander of Saxe-Weimar-Eisenach (1818–1901), officially rector of the university, of the personal invitation. The archduke made the invitation official, and Bismarck accepted it. At the end of July 1892, the old chancellor addressed a cheering throng of students and townspeople gathered in the marketplace. Since he had already received honors from various law and medical faculties throughout the empire, his benefactor devised a new degree to be conferred on the former chancellor—the degree of doctor of phylogeny, *honoris causa*! The degree, of course, suggested more about the turn of the new government—with rumors spreading that the king might convert to Catholicism—than about any contributions Bismarck might have made to this special branch of biology.[43]

Throughout the next decade, the political and social situation, from the old liberal point of view, continued to deteriorate. In 1903 the newly elected pope, taking the ominous name of Pius X, cast a lengthening shadow up from the south. The threat of Catholic revanchism brought an invitation from friends in Berlin for Haeckel to sally forth and to take up arms against the newly resurgent church. The invitation, as Haeckel described it,

42. See James Sheehan, *German Liberalism in the Nineteenth Century* (Chicago: University of Chicago Press, 1978), 223.

43. See the brief account of Haeckel's involvement in the invitation to Bismarck by Else von Volkmann, granddaughter of Haeckel, in her "Ernst Haeckel veranlasste die Einladung Bismarck's," in *Ernst Haeckel: Sein Leben, Denken und Wirken*, 1:82–86; see also Haeckel's account of the invitation, in ibid., 2:119–22.

Fig. 9.3. Market square in Jena, with monument commemorating Bismarck's address in 1892. (Photochrom print courtesy of U.S. Historical Archive.)

> especially mentioned that the continually growing reaction in the leading circles, the overweening confidence of an intolerant orthodoxy, the shift in balance toward ultramontane papism, and the consequent threat to German spiritual freedom in our universities and schools—that all of this made an energetic defense a pressing necessity.[44]

Haeckel accepted the invitation and, in 1905, gave three lectures in the great hall of the Sing-Akademie in Berlin to over two thousand enthusiastic auditors on each of the succeeding days. Accounts of the combative lectures spread beyond Germany to other European nations and America—the *New York Times* titled its story "Haeckel Kills the Soul."[45] In the lectures he rehearsed, in a minor key, the indictment against old enemies, especially those who either rejected or hesitated to endorse evolutionary theory (of course, Du Bois-Reymond and Virchow). In addition, he orches-

44. Ernst Haeckel, *Der Kampf um den Entwickelungs-Gedanken: Drei Vorträge, gehalten am 14., 16., und 19. April 1905 im Salle der Sing-Akademie zu Berlin* (Berlin: Georg Reimer, 1905), 7.

45. "Haeckel Kills the Soul," *New York Times*, 8 May 1905, 8.

trated a thundering denunciation of a new and quite sinister foe. This was a group most conspicuously represented by an entomologist, a man who was chiefly responsible for bringing the old bear out of his cave.[46] This new enemy argued strongly for evolutionary theory, grounding his defense in extremely compelling empirical evidence; and he had just written a scientifically exemplary study, *Die moderne Biologie und die Entwicklungstheorie* (Modern biology and evolutionary theory, 1904). But the scientist was also a Jesuit priest, Father Erich Wasmann.

For the Jesuits to endorse evolution meant that subtle chicanery had to be afoot. Haeckel declared Wasmann's book "a masterpiece of Jesuitical confusion and sophistry."[47] Wasmann bears some extended consideration not only because of the vehemence of Haeckel's reaction but also because of this Jesuit's scientific acumen, which has preserved his name in the reference lists of modern entomological studies, and especially because he provides a telling case of an individual whose scientific observations trumped his initial dogmatic convictions.[48]

Wasmann's Entomological Studies

Erich Wasmann was born in the southern Tyrol, in the Austrian village of Meran, in the fateful year of 1859.[49] His father, an excessively religious convert, had a minor reputation as a landscape painter working in the late Romantic tradition. His mother, also a convert from a Protestant family, served as the disciplinarian. Wasmann showed some of his father's talent with a sketchpad but yielded more to the lure of nature and the investiga-

46. Haeckel mentioned to his biographer, Wilhelm Bölsche, that it was Wasmann who provoked what he thought would be his last public lectures. See Haeckel to Bölsche (3 April 1905), in *Ernst Haeckel–Wilhelm Bölsche, Briefwechsel 1887–1919*, ed. Rosemarie Nöthlich (Berlin: Verlag für Wissenschaft und Bildung, 2002), 173.

47. Haeckel, *Kampf um den Entwickelungs-Gedanken*, 32.

48. Of the hundreds of authors cited by Edward O. Wilson in his *Insect Societies* (Cambridge, MA: Harvard University Press, 1971), Wasmann has about the eighth largest number of citations, some fourteen (521).

49. I have drawn the facts of Wasmann's life from his autobiographical recollections "Jugenderinnerungen," *Stimmen der Zeit* 123 (1932): 110–19, 191–99, 259–68, 327–34, 407–13; and also from H. Schmitz, S.J., "P. Erich Wasmann S.J." *Tijdschrift voor Entomologie* 75 (1932): 1–57; and from Franz Heikertinger, "P. Erich Wasmann, S.J.: Ein Nachruf," *Koleopterologische Rundschau* 17 (1931): 88–96. Abigail Lustig has written an illuminating essay on Wasmann and colleagues. See her "Ants and the Nature of Nature in Auguste Forel, Erich Wasmann, and William Morton Wheeler," in *The Moral Authority of Nature*, ed. Lorraine Daston and Fernando Vidal (Chicago: University of Chicago Press, 2004): 282–307. Lustig also has published a comparison of the intellectual styles of Haeckel and Wasmann. See her "Erich Wasmann, Ernst Haeckel and the Limits of Science," *Theory in Biosciences* 121 (2002): 252–59.

tion of colorful insects. In 1875 he entered the novitiate of the German Jesuits in Limburg, the Netherlands, where the order had relocated after Bismarck's putsch.

During his years in the seminary studying philosophy—Thomas Aquinas in the fore—he examined more mundane subjects as well, namely, insects, especially beetles (a passion he shared with the young Darwin). He also started reading the expanding literature surrounding the new theories of animal and human evolution. His avocation spiked when he was twenty-four years old with the publication of his book on the very odd *Trichterwickler,* a small black beetle (*Rhynchites betulae*) that cuts a precise pattern in a leaf and rolls it into a funnel for the deposition of eggs. Wasmann argued that the kind of geometrical knowledge these insects evinced in cutting their leaves could not be mechanically caused, as the Darwinians must suppose, but ultimately required a divine geometer. "Is the highly purposeful activity of the *Trichterwickler,* which produces so intriguing an artificial product, also purposeful striving? Or can we explain this instinct without the assumption of a purposeful cause?"[50] The answers to these questions came foreordained, since Aristotle and Saint Thomas guided the pen of this bug-besotted seminarian and blotted out the more naturalistic considerations of the likes of Darwin and Haeckel.[51] Yet scholastic philosophical assumptions did not really despoil Wasmann's minute analyses of the anatomy and behavior of the leaf-rolling beetle. In his examination of the instincts of this creature and that of its several related species, Wasmann executed a thorough and exacting study on a subject that would occupy him through the rest of his career.

Because of a recurring lung infection, the young seminarian could not go to the missions or teach in a Jesuit school after finishing the philosophy curriculum. Instead he was allowed to engage in private theological study and to continue exercising his obvious talent for entomological research. His

50. Erich Wasmann, S.J., *Der Trichterwickler, eine naturwissenschaftliche Studie über den Thierinstinkt* (Münster: Aschendorff'schen Buchhandlung, 1894), 19.

51. At this time Wasmann did suggest that some evolution of instincts was possible within well-defined limits. But he thought that one would have to grant such evolution could occur only as guided by an internal law of divine origin. Many difficulties, he argued, opposed the Darwinian view—e.g., if care of young through more primitive instincts were sufficient, there would be no utility in the more elaborate instinct of the little geometer; very small variations, as proposed by Darwin, would have no utility, etc. When Darwin suggested a continuous development connecting animal instinct, animal intelligence, and human reason, the philosophy student was assured that "this man who has become so famous possessed no fundamental philosophical education." See Erich Wasmann, S.J., "Die Entstehung der Instincte nach Darwin," *Stimmen aus Maria-Laach* 28 (1885): 333–53; quotation from 343–44.

Fig. 9.4. Erich Wasmann (1859–1931) as a seminarian; painting by his father. (Courtesy of the Natural History Museum, Maastricht, the Netherlands.)

interest in this latter quickly turned to ants and a class of beetles that lives commensally in ant nests, the so-called "myrmecophile," the inquilines or "guests of ants." In the short period from 1884 to 1890, Wasmann had over sixty publications on ants, termites, and their guests. His meticulous study of slave-making behavior in ants of the new and old worlds culminated in a work that secured his reputation as a leading authority in entomology: *Die zusammengesetzten Nester und gemischten Kolonien der Ameisen* (The composite nests and mixed colonies in ants, 1891). He concluded that work with a consideration of its bearing on evolutionary theory. He argued that slave-making ants in the Americas and Europe, which displayed common instincts, had either to have been created originally with these behavioral traits or to have evolved in the two, widely separated locations in a strictly parallel fashion, which on Darwinian grounds seemed quite improbable. One had to acknowledge, therefore, that a higher intelligence had established internal laws of development and instilled their causal processes in the hereditary structure of these organisms.[52]

Wasmann's accomplishments marked him for higher studies; and over three terms, beginning in 1890, he heard lectures in zoology at the Charles University in Prague. He subsequently returned to the Netherlands and in 1894 became an editor of the Jesuit cultural journal *Stimmen aus Maria-Laach* (later *Stimmen der Zeit*).

While serving as editor for the journal, Wasmann continued publishing on ants and their behaviors, using his empirical studies to draw philosophical conclusions. During the last part of the 1890s, he saw published three provocative monographs on the instincts of ants: *Instinct und Intelligenz im Thierreich* (Instinct and intelligence in the animal kingdom, 1897), *Vergleichende Studien über das Seelenleben der Ameisen und höheren Thiere* (Comparative studies in the psychic life of ants and higher animals, 1897), and *Die psychischen Fähigkeiten der Ameisen* (The psychological faculties of ants, 1899).[53] In these tracts he wished to argue against the Darwinians, who held that animal instinct shaded into animal intelligence and that this latter differed only in degree from human intelligence and reason. Like Darwin, he maintained that instinct proper resulted from inherited patterns of behavior that the animal exhibited without any view of the purpose of the act: thus worker ants disposed of the dead bodies of their comrades without any notion of health requirements. Yet he dissented from the Darwinian view that ants or any other animal might exhibit "animal intelligence." Behavior so labeled could result only from sensory associations, and thus should also be classed with instinct, though of a quite flexible sort. Human intelligence strictly understood was, according to Wasmann, a quite different faculty. Human intelligence could abstract relationships and patterns, a feat beyond the ability of any animal.

Wasmann also attempted to disrupt notions of continuity by arguing that ants—often regarded as mere mechanical automata—were capable of the same modes of sensory learning as higher animals. With the observational acuity of the obsessive naturalist, he argued, for instance, that the military strategies displayed by slave-making ants (*Formica sanguineae*) rivaled the

52. Erich Wasmann, *Die zusammengesetzten Nester und gemischten Kolonien der Ameisen* (Münster: Aschendorff'schen Buchdruckerei, 1891), 252–53.

53. Erich Wasmann, *Instinct und Intelligenz im Thierreich* (Freiburg: Herder'sche, 1897); *Vergleichende Studien über das Seelenleben der Ameisen und höheren Tiere* (Freiburg: Herder'sche, 1897); *Die psychischen Fähigkeiten der Ameisen* (Stuttgart: E. Schweizerbartsche, 1899).

supposed intellectual accomplishments of higher animals—though, as he contended, in both cases only sensory cognition, as opposed to real intelligence or reason, would be exhibited. He even took up the challenge brought by another naturalist, Albrecht Bethe (1872–1954), who objected that dogs could be trained to do tricks in a short time but ants at best would take weeks or months, thus illustrating progressive cognitive development in the animal kingdom. Accepting the challenge, Wasmann trained his ants in a few days—a feat that won the approval and delighted surprise of his new correspondent, the American entomologist William Morton Wheeler (1865–1937).[54] For Wasmann, however, the point remained that so-called animal intelligence had no connection with human intellectual abilities.[55]

Wasmann's analyses of the differences between animal instinct and human intelligence, under the guidance of scholastic lights, brought a quite negative response even from those entomologists with whom he had a friendly and protracted correspondence, such as Wheeler and Auguste Forel (1848–1931).[56] But in an area that Wasmann had made his own and in

54. Erich Wasmann, *Vergleichende Studien über das Seelenleben der Ameisen und höheren Thiere*, 2nd ed. (Freiburg: Herder'sche Verlagshandlung, 1900), 44. Wasmann sent Wheeler some papers on which his tract was based. Wheeler, then professor at the University of Texas, responded that "you have refuted Bethe most excellently!" See Wheeler to Wasmann (10 January 1900), in the Papers of Father Erich Wasmann, Maastricht Natural History Museum, the Netherlands. Lustig discusses the philosophical and evolutionary views of Wheeler in "Ants and the Nature of Nature," 298–306.

55. Wasmann, *Instinct und Intelligenz*, 80–92.

56. One protracted exchange went this way: Auguste Forel, in his "Ueber die psychischen Eigenschaften der Ameisen und einiger anderer Insekten," *Verhandlungen des V. internationalen Zoologenkongresses* (1901): 141–69, opened with a salvo against Wasmann's dualism: if the immaterial human soul pumps energy into the brain to produce behavior, there would be a clear violation of the law of the conservation of energy (145–50). Wasmann responded with "Die monistische Identitätstheorie und die vergleichende Psychologie," *Biologisches Centralblatt* 23 (1903): 545–56, in which he distinguished between the material substrate of thought and its formal character, with the former doing mechanical work but the latter not; and it was the latter that served as the object of psychological science, which Forel must simply regard as a "contentless subjective illusion" (552). Forel rejoined with a two-part article entitled "Naturwissenschaft oder Köhlerglaube," *Biologisches Centralblatt* 25 (1905): 485–93; 519–27 (Natural science or the faith of an incense burner). While decrying his friend Wasmann's Jesuitical wordplay, he also took the opportunity to attack Wasmann's newly revealed version of evolutionary theory. As discussed below, Wasmann now asserted evolutionary transitions within original *Urspecies*—which primitive types may have been equivalent to the primitive archetypes indicated by both Darwin and Haeckel. In the final shot of this series, Wasmann responded that his objections to the monistic identity theory were sufficient, but that he took personal exception to Forel's questioning his professional integrity because of his "incense-burner's faith." Wasmann then outlined comparable objections to the monistic thesis by the philosopher Carl Stumpf, who certainly did not wear the Roman collar. See Erich Wasmann, "Wissenschaft-

which he was recognized as the world's leading authority, the biology of the myrmecophile—that class of beetles that live in ant nests—he drew some inferences that startled both opponents and coreligionists.[57]

The Guests of Ants—Evidence for Evolution

In a series of articles first appearing in *Biologisches Centralblatt* and in *Stimmen aus Maria-Laach*,[58] and then summarized in *Die moderne Biologie und die Entwicklungstheorie,* Wasmann presented extensive and detailed empirical evidence for a new and unexpected conclusion: namely, that the myrmecophile exhibited evolutionary transformations.[59]

Wasmann distinguished three kinds of ant guests according to their morphology and behavior: the aggressive type (*Trutztypus*), the symphilic type, and the mimetic type. Aggressive, tanklike beetles could be found in the genus *Dinarda.* These species displayed heavily armored, compact individuals that were impervious to ant attacks. Wasmann examined four species that were distributed over north-central Europe and showed that they varied in color and size depending on the color and size of the species of ant with which they lived. The similarity of color made the beetles less conspicuous in the nests; and appropriate size made them less vulnerable to attacks on their appendages. Wasmann asserted that "we have here, therefore, a case in which we can explain effortlessly and completely satisfactorily, by the simplest natural causes, the differentiation of similar species of the same genus from a common progenitor."[60] He further argued that the genus *Chitosa,* which inhabited southern Europe, had to be related to *Dinarda*

liche Beweisführung oder Intoleranz?" *Biologisches Centralblatt* 25 (1905): 621–24. Through all of these public duels, Forel and Wasmann continued to correspond, exchanging information about ants. Wasmann's side of the still-polite correspondence can be found in Forel's papers at the Medizinhistorisches Institut, University of Zürich. There is little doubt, however, that Wasmann was wounded by Forel's suggestion that being a Jesuit twisted his scientific intelligence to squeeze out doctrinaire pronouncements.

57. Wilson, *Insects Societies,* 390, simply says: "Erich Wasmann initiated the modern study of arthropod symbionts."

58. See Erich Wasmann, "Gibt es tatsächlich Arten, die heute noch in der Stammesentwicklung begriffen sind?" *Biologisches Zentralblatt* 21 (1901): 685–711, 737–52; and "Konstanztheorie oder Deszendenztheorie?" *Stimmen aus Maria-Laach* 56 (1903): 29–44, 149–63, 544–63.

59. Erich Wasmann, *Die moderne Biologie und die Entwicklungstheorie,* 2nd ed. (Freiburg: Herdersche Verlagshandlung, 1904), 210–45. The third edition (1906) was also published in English translation as *Modern Biology and the Theory of Evolution,* trans. A. M. Buchanan (St. Louis, MO: B. Herder, 1914).

60. Wasmann, "Gibt es tatsächlich Arten?," 694–95.

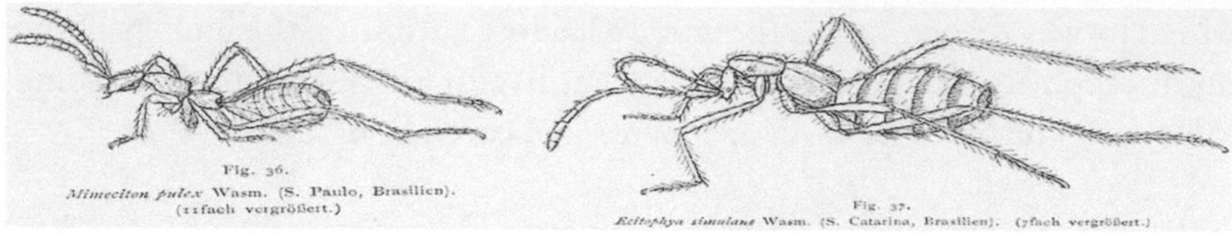

Fig. 9.5. Two species of the "guests of ants," beetles that have evolved to look like ants. (From Wasmann, *Die moderne Biologie und die Entwicklungstheorie,* 2nd ed., 1904.)

through a common ancestor. Thus, he concluded, evolutionary adaptations had been acquired in the descent of species. Moreover, inquilines found in termite nests in India suggested that beetle species in the genus *Doryloxenus,* typical of the myrmecophile dwelling with African wandering ants (*Dorylus*), had come to live with termites, quite different insects; moreover, one could trace alterations in the species of this genus as they evolved more effective adaptations for protecting themselves against termite attacks.

Wasmann deployed further evidence of evolutionary transformation in the symphilic group of myrmecophile, those that secreted a sweet exudate and were fed by the ants in return. He showed that species of the *Lomechusini* varied in features, depending on the species of ant with which they lived. The most startling evidence he produced, however, was within the mimetic group. These were beetles that had evolved to look like ants. Wasmann showed that myrmecophile of quite different genera, which inhabited nests of the same species of ant, had converged in their morphologies (see fig. 9.5). On the basis of such evidence, he affirmed that "we ought calmly accept the evolutionary doctrine insofar as it is scientifically founded on a definite class of structures with a sufficient degree of probability."[61]

While Wasmann thought his inquilines—and also various ant species—offered compelling empirical evidence for descent with modification, he would still not yield to Darwinian theory. He argued that several considerations precluded natural selection as the primary agent of change. First, selection could only eliminate possibilities once they arose, not create them initially—a common-enough objection (and a common-enough misunderstanding of Darwin's device). Second, he argued that most variations were neutral, so that selection would have no purchase on them. Third, though species of the *Lomechusini* evolve because the ants, as it were, selected those with the sweetest liquor—what Wasmann called "amical selection"—the beetles yet ate ant pupae and thus were positively harm-

61. Wasmann, *Moderne Biologie,* 219.

ful to the ant community, something natural selection should have prevented.[62] Finally, a gradual change, as Darwin would have it, in these inquiline species ought to take hundreds of thousands of years, exhausting, as Wasmann estimated, the geological time available.[63] Instead of Darwinian evolution, Wasmann proposed a theory of evolution that was a hybrid of ideas drawn from Hugo de Vries (1848–1935) and Hans Driesch. Like De Vries, he argued that alterations in species would come as macromutations; and like Driesch, he held that *Anlagen*—dispositions—in the hereditary structure of organisms would respond to external causal relationships in a teleologically directed way.

Wasmann maintained that since we had no evidence of spontaneous generation, we had to assume a divine act as the source of the several foundational species. These natural *Urspecies* formed the base of the stem-trees whose branches held the derived species of plants and animals. He regarded it an open question as to the number of original types—perhaps only a few, perhaps more. But one type, he vigorously insisted, was unique, namely, the human.

Wasmann rejected the possibility that human beings might have arisen out of the stock of lower animals.[64] Human reason simply bore no relationship to what passed as animal intellect—an argument that he retained from his earliest considerations of the question. He thus continued to reject Haeckel's monistic metaphysics as the proper foundation for understanding human beings or animals. While he allowed that man's body might have been prepared by an evolutionary process prior to the reception of the soul, the leading contenders for this kind of pre-adaptation—Neanderthal man and Dubois's Java man—were, he thought, both unlikely candidates as proto-humans. Neanderthals, as Virchow suggested, were quite within the range of human variation—so they were real human beings; and Dubois's discovery appeared to be only that of a giant ape unrelated to the human stock.

62. While E. O. Wilson cites Wasmann's work throughout his *Insect Societies,* he obviously did not penetrate Wasmann's German very deeply. Wilson believes that Wasmann did not recognize that symphilic beetles often preyed on ant pupae (390), something that Wasmann, in fact, emphasized as part of his argument against natural selection.

63. We now know that beetles were diversely proliferating during the Permian, 300 million years ago; and fossil ants of more than 90 million years old have recently been discovered. It is reasonable to suppose the symbiosis between the two has existed for many millions of years. See D. A. Grimaldi, D. Agosti, and J. M. Carpenter, "New and Rediscovered Primitive Ants (Hymenoptera: Formicidae) in Cretaceous Amber from New Jersey, and Their Phylogenetic Relationships," *American Museum Novitates,* no. 3208 (1997): 1–43.

64. Wasmann, *Moderne Biologie,* 273–304.

The Confrontation between Wasmann and the Monists

In his Berlin lectures, Haeckel took delight in referring to Wasmann as the "Darwinian Jesuit," an ironically intended designation that yet begrudgingly suggested some respect for this Jesuit's accomplishments in entomology.[65] But he simply derided Wasmann's rejection of a thoroughgoing evolutionism in the case of human beings: "If Wasmann assumes this introduction of the soul for the development of the type, then he must postulate in the phylogeny of the anthropoid apes a historical moment in which God descends and injects his spirit into this hitherto spiritually bereft ape soul."[66] Haeckel thought the whole assumption absurd, but not innocent of political consequence. He suspected that the conservative Prussian government would seek a union of "crown and altar" not for reasons of religious conviction but for reasons of practical advantage. He was convinced that this would be no even match; under the banner of reconciliation, the crown would become "the footstool of the altar," as the church bent the state to its own purposes.[67]

When Wasmann read of Haeckel's attack in the several newspapers that described the lectures, he penned a long open letter to his nemesis, which appeared on the front page of the morning edition of the *Kölnische Volkszeitung* (2 May 1905).[68] He complained that Haeckel too easily identified evolutionary theory with monism, and thus misleadingly suggested that the Jesuits and the church had come over to the Darwinian side. Wasmann rejected Haeckel's assumption of only one meaning for evolution, and he protested that his own theistic version had no official sanction from the church or the Jesuits. About this second point, Wasmann would eventually be proved mistaken: his view of evolution came to be widely accepted by the Catholic Church as a way of accommodating this latest scientific, though dangerous, advance. Under Wasmann's orchestration, the Vatican could at last admit the world actually moved—but not too much.

The drama of the evolution-religion conflict and a sense of its high-cultural entertainment value brought Wasmann, amidst a flurry of newspaper interpretations of the debate, an invitation in 1906 to reply to Haeckel at the Sing-Akademie. He declined the offer, but a short time later did accept a comparable invitation issued by a group of prominent scientists in

65. Haeckel, *Der Kampf um den Entwickelungs-Gedanken*, 75.

66. Ibid., 83.

67. Ibid., 84.

68. Erich Wasmann, "Offener Brief an Hrn. Professor Haeckel (Jena)," *Kölnische Volkszeitung* 46, no. 358 (2 May 1905): 1–2.

Ein Jesuit als freier Forscher.

Erich Waßmann vom Jesuiten=
orden, einer der wenigen katho=
lischen Geistlichen, die die Ent=
wicklungslehre zum Teil an=
erkennen, wird demnächst in
Berlin Vorträge halten.

Fig. 9.6. Father Erich Wasmann, S.J.: "A Jesuit as unfettered researcher. Erich Wasmann of the Jesuit order, one of the few Catholic spirituals who acknowledges descent theory in part, will soon hold lectures in Berlin." (From *Berliner Tageblatt,* 7 February 1907.)

Berlin. Initially he was to have addressed a meeting of the entomological society, but Ludwig Plate (1862–1937), a member of the inviting committee and eventual successor to Haeckel at Jena, insisted that the meeting be open to the public.[69] Wasmann agreed and he further allowed that after his three public lectures, his opponents could present their objections and he would respond. Initially some twenty-five critics requested time, but Wasmann left it up to the committee to pare down the list to something manageable.

69. Wasmann had already crossed pens with Plate in the pages of the *Biologisches Centralblatt* (1901), where he defended evolutionary descent in the guests of ants but not on the monist's terms. See Wasmann, "Gibt es tatsächlich Arten?"

On 13, 14, and 17 February 1907, Wasmann lectured in the Sing-Akademie each day to over one thousand people, who paid one mark for each occasion (two for reserved seating). He took as his subjects: the general theory of evolution and its support drawn from entomology; varieties of evolutionary theory—theistic and monistic (atheistic); and the problem of human evolution.[70] At 8:30 on the evening of 18 February, with the audience swelling to some two thousand men and women, eleven opponents confronted Wasmann in the auditorium of the Zoological Gardens. His objectors were allotted varying amounts of time, with Plate, the principal organizer, receiving the longest period at half an hour. Wasmann was granted thirty minutes to answer his eleven critics. He mounted the podium at 11:30 p.m., with the full complement of the audience still in their seats. He focused his response on Plate's objections and brought in others as his scant time permitted. He asserted that he would surrender to the idea of spontaneous generation if the scientific evidence demonstrated the likelihood, but he could not allow that the creation of matter and its laws were proper scientific subjects. These latter problems lay in the province of metaphysics, about which he would nonetheless be happy to argue. His own position on the purely scientific issues, he said, was close to that of Hans Driesch: one had to postulate internal vital laws to devise adequate explanations of species descent. Though Plate and others continued to attribute an interventionist theology to Wasmann, he claimed that his science did not require that—though he was philosophically committed to the belief that God had created matter and its laws, which laws might, he allowed, eventually include those governing spontaneous generation. And while the evolution of man's body from lower creatures had yet to be shown, he also allowed that as a possibility. But, he maintained, it was the natural science of psychology that absolutely distinguished human mentality from animal cognition, and therefore a gradual transition in mind from animals to man was precluded by science itself.

Wasmann's opponents shelled him not only with intellectual objections but also lobbed the occasional invective designed to dismember less substantial egos—Plate concluded that "Father Wasmann is not a genuine

70. Several accounts of Wasmann's lectures and the ensuing debate are extant. I have relied on two book-length descriptions given by Wasmann himself and by his principal opponent, Ludwig Plate. See Erich Wasmann, *Der Kampf um das Entwicklungsproblem in Berlin* (Freiburg: Herdersche, 1907); and Ludwig Plate, *Ultramontane Weltanschauung und moderne Lebenskunde, Orthodoxie und Monismus* (Jena: Gustav Fischer, 1907). Wasmann's book was also published in English as *The Berlin Discussion of the Problem of Evolution*, authorized translation (St. Louis, MO: Herder, 1909).

research scientist [*Naturforscher*], not a true scholar"; the anthropologist Hans Friedenthal (1870–1943) referred to Wasmann as a "dilettante in the area of human evolution."[71] Yet Wasmann met the overwrought responses with a calm professionalism made piquant by a "dry sense of humor" (as the *Berliner Morgenpost* characterized his lectures).[72] The *Deutsche Tageszeitung* judged that with the exception of Plate, Wasmann's opponents "seemed almost like pygmies."[73] After midnight, at the conclusion of the reply to his critics, Wasmann, according to the *Kölnische Volkszeitung*, received from the audience a "thunderous ovation."[74] It seems clear that if he did not always convince his auditors—some five hundred articles in the various German papers reported a variety of judgments—he at least charmed them. But from our historical perspective, he did more than that. Although evolutionary theory was rapidly achieving fundamental agreement among professionals of every philosophical persuasion, Wasmann showed that it had still not achieved consensus at the turn of the century. And his subtle arguments demonstrated that no necessary antagonism had to exist between evolutionary theory and a liberal, philosophically acute brand of theology. Not all objectors from the side of religion showed themselves as high-minded as Wasmann. Certainly Arnold Brass of the Protestant Keplerbund did not.

The Keplerbund vs. the Monistenbund

Haeckel's book *Die Welträthsel*, which I will more carefully consider in the next chapter, set off a conflagrational reaction from the many quarters that had already been incited by Haeckel's frequent attacks on religion. To the young, the book seemed like a torch lighting the way to emancipation from the heavy hands of orthodox science and religion; others, though, thought it a flaming faggot set at the root of Christian civilization. Many of those for whom it illuminated the path to freedom joined the Monistenbund, originally a union of scientists and educated citizens who subscribed to Haeckel's program of monistic philosophy. Haeckel had harbored the idea of such an organization for several years. While attending the International Free-Thinkers Conference in Rome in 1904, where he was celebrated as the

71. Plate, "Ultramontane Weltanschauung," 77, 93.

72. "Pater Wasmanns Berliner Vorträge," *Berliner Morgenpost*, 14 February 1907.

73. *Deutsche Tageszeitung*, 19 February 1907, as quoted by Wasmann, *Kampf um das Entwicklungsproblem in Berlin*, 148.

74. "Pater Wasmann," *Kölnische Volkszeitung* (morning edition), no. 149 (20 February 1907), 2.

anti-pope, he thought it might then spontaneously form. When that failed, he took practical steps to bring it into existence.[75]

The planning began in the wake of his Berlin lectures against Wasmann, and the initial meeting took place on 11 January 1906, in Jena. The first president selected was the radical Protestant pastor Albert Kalthoff (1850–1906), who died just a few months after taking up his position. Haeckel quickly thereafter importuned the noted naturalist Auguste Forel to assume leadership.[76] Eventually the Nobel Prize winner Wilhelm Ostwald (1853–1932) would occupy the chair (1911). The organization would grow to some six thousand members before disbanding in 1933 rather than be taken over by the Nazis. While the league was initially guided by Haeckel's declarations of monistic philosophy—especially its anti-dualism, anticlericalism, and notions of scientific management of the state—it became a more heterogeneous alliance, embodying, as one of its early presidents maintained, the principles of the Enlightenment further elevated through modern science. It continued to stress scientific epistemology, world peace, international cooperation, and eugenic principles of forming a healthy society. While some of its members—Wilhelm Schallmayer (1857–1919), for instance—would preach race hygiene, others—like Magnus Hirschfeld (1868–1935)—would urge tolerance for homosexuals, a stand that Haeckel endorsed.[77] After the Great War, the Monistenbund became decidedly more pacifistic and socialistic. The society spread to most European countries, as well as America, where the journal the *Monist*, edited by Paul Carus (1852–1919), published Haeckel and many other like-minded philosophers and scientists.[78]

75. Haeckel to Bölsche (15 October 1905), in *Ernst Haeckel–Wilhelm Bölsche*, 180–81.

76. Heiko Weber, "Der Monismus als Theorie einer einheitlichen Weltanschauung am Beispiel der Positionen von Ernst Haeckel und August Forel," in *Monismus um 1900: Wissenschaftskultur und Weltanschauung*, ed. Paul Ziche (Berlin: Verlag für Wissenschaft und Bildung, 2000), 81–127.

77. Hirschfeld asked Haeckel (21 February 1912) if he might dedicate his book *Naturgesetze der Liebe* (Natural laws of love) to him. Haeckel acquiesced after reading the page proofs that Hirschfeld sent him. Hirschfeld visited Haeckel in Jena in the late spring of 1912 and thereafter sent him several works on various studies of sexuality, including hermaphroditism, transvestitism, and bisexuality. See Hirschfeld to Haeckel (6 June 1912), in the Haeckel Correspondence, Haeckel-Haus, Jena.

78. See Niles Holt, "Monists and Nazis: A Question of Scientific Responsibility," *Hastings Center Report* 5 (1975): 37–43. See also Richard Weikart, "Evolutionäre Aufklärung? Zur Geschichte des Monistenbundes," in *Wissenschaft, Politik und Öffentlichkeit*, ed. Mitchell Ash and Christian Stifter (Vienna: Universitätsverlag, 2002), 131–48. For a contrasting picture of the Monist League, see Daniel Gasman, *The Scientific Origins of National Socialism: Social Darwinism in Ernst Haeckel and the German Monist League* (New York: Science History Publications, 1971), especially 31–54. Paul Carus, founder of both the *Monist* and the Open Court

Fig. 9.7. Meeting of the Monistenbund on 6 May 1906; *from right:* Arthur Schwarz, Ernst Haeckel, Wilhelm Breitenbach, F. Siebert, C. H. Thiele, Johannes Unold. (Courtesy of Ernst-Haeckel-Haus, Jena.)

In 1907, the year after the founding of the Monistenbund, Eberhard Dennert (1861–1942), a botanist and teacher in the Evangelical Pädagogium in Bad Godesberg, called into existence the Keplerbund for the Advance of Natural Knowledge. This was an organization of Protestant scientists and laymen dedicated, as their initial call declared, to the conviction that

> truth encompasses the harmony of natural scientific facts with philosophical knowledge and religious experience. Accordingly, the Kepler-

Press in Chicago (later LaSalle, Illinois), did have some reservations about Haeckel's very strident version of monism. He made his criticisms plain, if indirect. He sent a letter to Haeckel's acquaintance Edgar Ashcroft, when both men were in Algiers. He asked Ashcroft to indicate to Haeckel that he did not agree with his friend's "monistic confession of faith." The difference, he explained, was "mainly a difference of attitude not as to the subject of his [Haeckel's] contentions." He wanted Haeckel to know what his U.S. followers felt: "There is no need of saying that we do not believe for that antagonizes without any necessity and there is no need of negating all other creeds in the world." He thought that "it would be better to formulate a kind of confession of faith in positive terms which at the same time would be acceptable to many people who for some reason or other do not wish to cut themselves off from church life." Carus to Ashcroft (6 February 1905), in the Haeckel Correspondence, Haeckel-Haus, Jena.

Fig. 9.8. Eberhard Dennert (1861–1942). (From Dennert, *Vom Sterbelager des Darwinismus,* 1905.)

bund is expressly distinguished from the materialistic dogma of biased Monism and struggles against the thoroughly atheistic propaganda of this latter, which falsely claims to be grounded on natural science.[79]

The founder of the *Bund,* Dennert, had trained in the *Realschule* at Lippstadt under the Darwinian enthusiast Hermann Müller (1829–1883), who was the brother of the more famous Fritz Müller. The schoolmaster sent his best pupils to Jena. Dennert went to Marburg, where, under the strongly anti-Darwinian Albert Wigand (1821–1886), he cultivated a distaste for evolutionary doctrine.

Dennert had reacted like an overwound spring to Haeckel's *Welträthsel,* immediately flinging off a broadside—*Die Wahrheit über Ernst Haeckel und seine "Welträtsel"* (The truth about Ernst Haeckel and his "World Puzzles," 1901)—and, before he wound down, over ninety books

79. Eberhard Dennert, *Die Naturwissenschaft und der Kampf um die Weltanschauung,* Schriften des Keplerbundes, Heft 1 (Godesberg: Naturwissenschaftlicher, 1910), 29. Uwe Hoßfeld describes the membership of the Keplerbund in his *Geschichte der biologischen Anthropologie in Deutschland von den Anfängen bis in die Nachkriegszeit* (Stuttgart: Franz Steiner, 2005), 248–52.

and pamphlets venting multiple religious enthusiasms had sprung from his pen.[80] Through his many scattered tracts, he sought the reconciliation of religion and science—by draining the blood from the one and emasculating the other. Religion, he asserted, was not a matter of understanding, of intellectual demonstration, but a matter of feeling. He thought it manifest from his own surveys of the faith of past scientists that "natural scientific research [*Naturforschung*] does not exclude simple biblical faith, and that religious belief and religious life do not draw their proof from the intellect, but entirely from other factors. These factors [feelings of the heart] are available to every person."[81] But the heart froze when affronted by Darwinian theory, especially the version coldly contrived by Haeckel. Thus, as a second requirement for reconciliation, Darwinian evolution had to be rejected. Typical of Dennert's effort was the often-reprinted and translated tract *Vom Sterbelager des Darwinismus* (On the deathbed of Darwinism, 1902), which cursorily examined the work of several biologists (e.g., Kölliker, Oscar Hertwig, Gustav Eimer) who proposed alternative descent theories to that of Darwin and Haeckel.[82] The argument seems to be that all of these different variations on transmutation theory somehow prove Darwin and Haeckel's version to be moribund. The heterogeneity of proposals concerning evolution and the ultimately inadequate efforts to substantiate it suggested to Dennert that the very doctrine of descent itself must also be quite doubtful. At least we could have no "clear and exact demonstration of evolutionary theory [*Entwicklungslehre*]," and thus the mode of transmutation, if it occurred at all, would remain forever hidden.[83]

Dennert found a particularly aggressive and paranoid ally in another hapless naturalist, Arnold Brass (b. 1854). Brass had failed to start his academic career in a way that would lead to a professorship: he wanted to

80. Eberhard Dennert, *Die Wahrheit über Ernst Haeckel und seine "Welträtsel," nach dem Urteil seiner Fachgenossen,* 2nd ed. (Halle: C. Ed. Müller, 1905). The book is mostly a compilation of the positions of the various objectors to Haeckel, beginning with Rütimeyer's charge of fraud.

81. Eberhard Dennert, *Bibel und Naturwissenschaft* (Halle: Richard Mühlmann, 1911), 312–20.

82. Eberhard Dennert's book was translated into English as *At the Death Bed of Darwinism,* trans. E. V. O'Harra and John H. Peschges (Burlington, IA: German Literary Board, 1904). It became a staple of the fundamentalist movement in the first half of the twentieth century.

83. Eberhard Dennert, *Vom Sterbelager des Darwinismus,* neue Folge (Halle: Richard Mühlmann, 1905), 6. Dennert rather liked Peter Kropotkin's emphasis on cooperation in nature but thought it militated against the Russian's retention of Darwinian selection theory (123–34). But in sum, he thought transformation might occur, but we would never have any proof of it nor could we ever discover its mode. If we yet postulated it, we would have to assume internal driving forces (*Triebkräften*) as responsible (6).

work at Dohrn's Naples Zoological Station but was not chosen; and his application for recognition of his habilitation was rejected at Marburg. He had to fall back on itinerant work in zoology, usually producing drawings for various books and articles in anatomy. After the turn of the century, as he reflected on the derailment of his academic career two decades before, Brass began to suspect the hand of Ernst Haeckel.[84] Haeckel would later deny any such connivance, since he barely knew the man. In 1906 Brass published a tract that came to the defense of Dennert, who had been dismissed by Plate and Haeckel as an inept Christian apologist. In the booklet *Ernst Haeckel als Biologe und die Wahrheit* (Ernst Haeckel as biologist and the truth, 1906), Brass remained fairly polite, actually rather sycophantic. He acknowledged Haeckel's "genius" and the latter's command of vast areas of zoology—far superior, he thought, to Darwin's own in this respect. He, nonetheless, felt himself able to meet the Jena lion on common ground. He expended most of his effort in the book describing the presumed deficiencies of Darwinian theory and arguing for the compatibility of reliable science with evangelical theology. After this publication, Brass began to lecture on Haeckel's monism, for which he received some financial support from the Keplerbund.[85] In these lectures his opposition to monism in general and Haeckel in particular grew in stridency.

On 10 April 1908, Brass delivered a lecture in Berlin to a meeting of the Christian Social Party at which he claimed that Haeckel had illustrated a recent talk in an "erroneous" fashion.[86] As reported in the Berlin *Staats-*

84. Naively Brass let slip his various failures to obtain desired academic positions, and increasingly he became convinced that Haeckel was the culprit in his downward spiral. See Arnold Brass, *Ernst Haeckel als Biologe und die Wahrheit* (Halle: Richard Mühlmann, 1906), 10–11. See also the second edition of Brass's *Affen-Problem* (1909) as quoted by Gursch, *Die Illustrationen Ernst Haeckels,* 89: "In 1886, I had submitted a habilitation work on the systematics of the mammals, etc. at Marburg for the first and only time. This audacity had angered Haeckel and others at the time. To exclude the possibility of my again attempting a habilitation in Marburg, Plate, a student of Haeckel, was admitted to the position of docent."

85. Brass later denied he received any money from the Keplerbund—and maybe he did not. But the business director of the Keplerbund, Wilhelm Teudt, reported that Brass did receive financial guarantees from the society for his lectures in the winter of 1807–8. Haeckel would use Teudt's avowal as evidence of connivance. See Wilhelm Teudt, *Im Interesse der Wissenschaft! Haeckel's "Fälschungen" und die 46 Zoologen,* Schriften des Keplerbundes, Heft 3 (Godesberg: Naturwissenschaftlicher, 1909), 7.

86. I have reconstructed the course of these debates from two opposing sources, from the account of the Keplerbund's general business manager, Wilhelm Teudt, and from that of the secretary of the Monistenbund, Heinrich Schmidt. Both quote verbatim from newspaper articles and other sources, and both, of course, offer their particular interpretations of the events. See Teudt, *Im Interesse der Wissenschaft;* and Heinrich Schmidt, *Haeckels Embryonenbilder: Dokumente zum Kampf um die Weltanschauung in der Gegenwart* (Frankfurt: Neuer Frank-

bürgerzeitung, Brass asserted that in arguing for the biogenetic law, Haeckel had made a "mistake" (*Missgeschick*) by depicting an ape embryo sporting the head of a human embryo and a human embryo with an ape head. The newspaper reported that "the lecturer could speak here from the most exact personal knowledge, since he himself had presented to Haeckel the correct illustrations."[87] The supposedly "mistaken" illustration was from Haeckel's Jena lecture on the occasion of the two hundredth anniversary of Linnaeus's birth. The lecture was published as *Das Menschen-Problem und die Herrentiere von Linné* (The problem of man and the anthropoid animals of Linnaeus, 1907), and it had several illustrations appended to it. In the illustration that compared the embryos of a bat, gibbon, and human being, Brass claimed that Haeckel had switched the heads of the gibbon and human being that were depicted in the second row (see fig. 9.9, GII and MII).[88]

When Haeckel learned of Brass's lecture, he explosively responded in an open letter to a colleague that the charge was a "barefaced lie" (*freche Lüge*). He retorted that he did not make the alleged "mistake" and that Brass certainly never prepared any illustrations for him. In a fury, he had his lawyer contact several newspapers threatening suit if they perpetuated this "brazen invention."[89] Brass immediately modified his charge in two newspaper articles (*Staatsbürgerzeitung* and *Volk,* Berlin, 25 April 1908), now saying that the head of the gibbon in the illustration bore "more than the usual similarity to the human embryo at a similar developmental stage, which I have repeatedly sketched and illustrated from a preparation."[90] Haeckel quickly wrote to the same newspapers saying that he himself had not drawn the illustrations but had a designer do so relying on figures taken from well-known authors: the ape embryo, which he called a "hylobates" (a genus of gibbon), he said he took from Emil Selenka (1842–1902) and the human embryo was based on the work of a couple of authors, including Wilhelm His.[91] A comparison of Selenka's and His's images with those of

furter, 1909). In 1900 Schmidt had become Haeckel's assistant and protégé. See Uwe Hoßfeld, "Haeckels 'Eckermann': Heinrich Schmidt (1874–1935)," in *Klassische Universität und akademische Provinz: Die Universität Jena von der Mitte des 19. bis in die 30er Jahre des 20. Jahrhunderts,* ed. Matthias Steinbach and Stefan Gerber (Jena: Bussert & Stadeler, 2005), 270–88.

87. Schmidt, *Haeckels Embryonenbilder,* 8.

88. Ernst Haeckel, *Das Menschen-Problem und die Herrentiere von Linné: Vortrag, gehalten am 17. Juni 1907 im Volkshause zu Jena* (Frankfurt: Neuer Frankfurter, 1907), table 3. This is the same illustration Haeckel had used in his *Der Kampf um den Entwickelungs-Gedanken* two years earlier.

89. Schmidt, *Haeckels Embryonenbilder,* 8; Teudt, *Im Interesse der Wissenschaft,* 13.

90. Teudt, *Im Interesse der Wissenschaft,* 14.

91. Ibid., 14–15; Schmidt, *Haeckels Embryonenbilder,* 9.

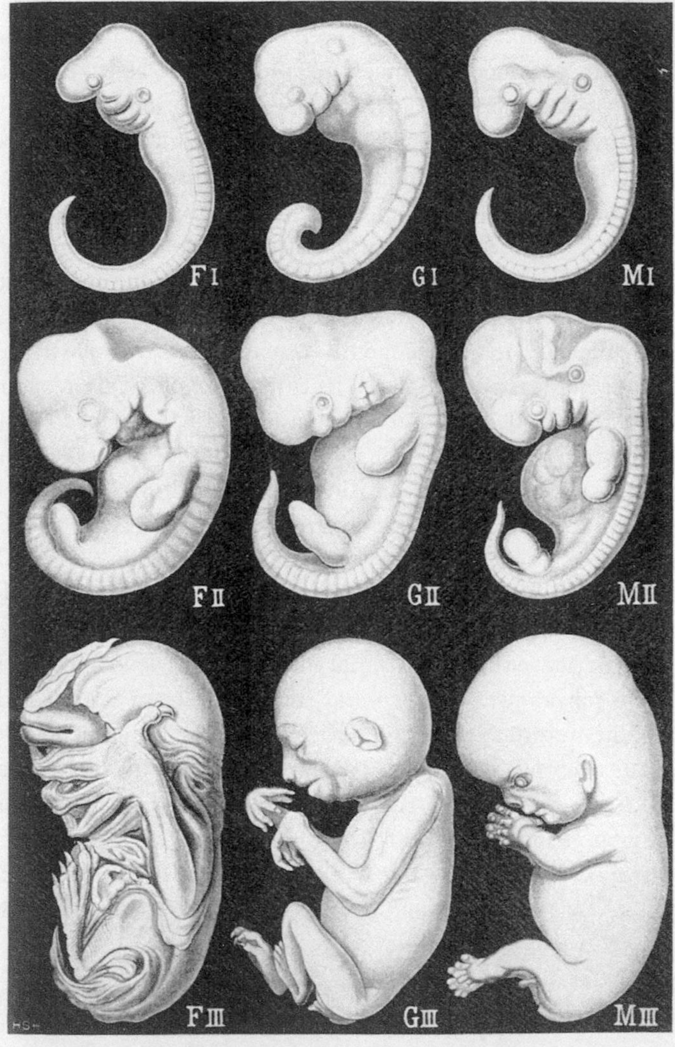

Fig. 9.9. Illustration of the biogenetic law: comparison of bat, gibbon, and human embryos at three stages of development. (From Haeckel, *Das Menschen-Problem*, 1907.)

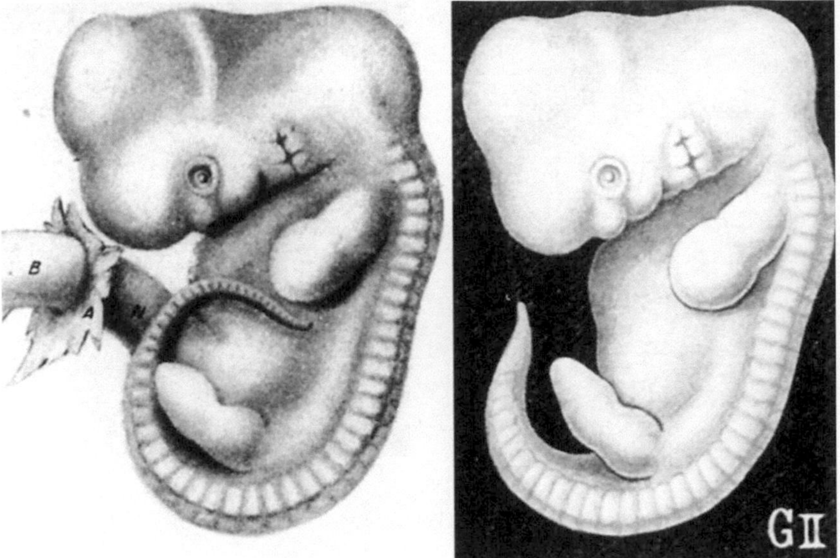

Fig. 9.10. Ape embryos at comparable stages. (From Selenka, *Menschenaffen*, 1903; and Haeckel, *Das Menschen-Problem*, 1907.)

Haeckel's lecture shows, indeed, a close similarity (see figs 9.10 and 9.11).[92] It is quite clear that Haeckel did not switch the heads of the embryos as Brass had asserted.

Brass, nonetheless, quickly escalated his charges in further lectures: "Haeckel has not only falsely represented the developmental condition of the human, ape, and other mammals, in order to be able to sustain his hypothesis; he took from the scientific store of a researcher the figure of a macaque, cut off its tail, and made a gibbon out of it."[93] Haeckel in fact did use a macaque embryo with a shortened tail instead of a gibbon embryo. In Selenka's volume, the illustrations of gibbon embryos immediately follow those of macaques, without, however, any gibbon embryo at the stage that Haeckel needed.[94] In the fifth edition (1903) of the *Anthropogenie*, Haeckel

92. For their respective depiction of a macaque embryo and a human embryo, see Emil Selenka, *Menschenaffen (Anthropomorphae): Studien über Entwickelung und Schädelbau*, vol. 5 of *Zur vergleichenden Keimesgeschichte der Primaten* (Wiesbaden: C. W. Kreidel, 1903), 357; and Wilhelm His, *Anatomie menschlicher Embryonen*, 3 vols. with 3 atlases (Leipzig: F. C. W. Vogel, 1880–85), III atlas, table 10.

93. Schmidt, *Haeckels Embryonenbilder*, 9–10; Teudt, *Im Interesse der Wissenschaft*, 15.

94. Selenka, *Menschenaffen*, 353–63. Haeckel had used the macaque embryo in the fifth edition (1903) of the *Anthropogonie* also to represent a gibbon at the midstage of development. But in that depiction, the embryo retains the long tail of the macaque embryo. In the sixth edi-

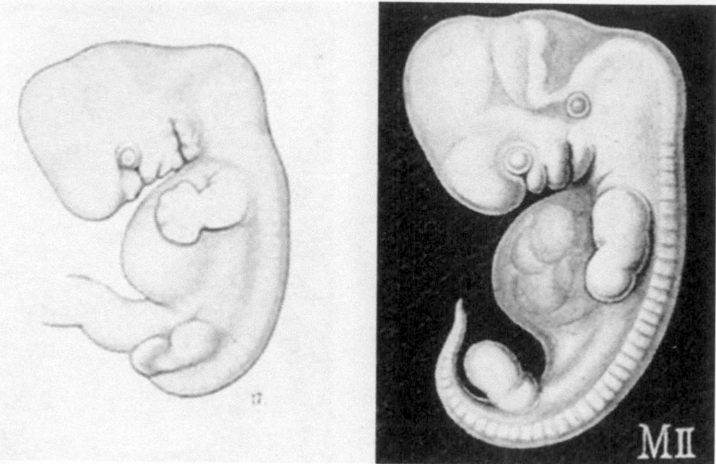

Fig. 9.11. Human embryos at comparable stages. (From His, *Anatomie menschlicher Embryonen,* 1880–85; and Haeckel, *Das Menschen-Problem,* 1907.)

had depicted a similar series of "gibbon" embryos; but at the middle stage he had placed a macaque embryo with the proper length of tail (see fig. 8.16, NII). The similarity of macaque and human embryos would, actually, seem to make Haeckel's case even stronger. But there is no doubt that his use of the macaque embryo instead of a gibbon embryo rendered him vulnerable. Brass promised that Haeckel's malfeasance would be extensively demonstrated in a little book he was preparing. Haeckel perceived the forthcoming tract as another repetition of the ancient charge, a creature he had slain over and over, which was now returning to seek vengeance against an old man.

Brass's book appeared as *Das Affen-Problem* in late 1908.[95] In the tract he expanded his indictment by enumerating several trivial particulars and at the same time deflating what had been his initial, quite serious charge. The first plate of Haeckel's *Das Menschen-Problem* depicted a representation of four ape skeletons and a human skeleton assuming poses similar to those in a famous illustration by Huxley (see fig. 9.12). Brass contended that Haeckel had made the human too stooped, the gorilla too erect, the apes with their feet flat on the ground, and the gorilla displaying his teeth in an

tion (1910), the macaque embryo is still used to represent the gibbon, though in that edition it has a shorter tail.

95. Arnold Brass, Das *Affen-Problem: Prof. E. Haeckel's Darstellungs- u. Kampfesweise sachlich dargelegt nebst Bemerkungen über Atmungsorgane u. Körperform d. Wirbeltier-Embryonen* (Leipzig: Biologischer, 1908).

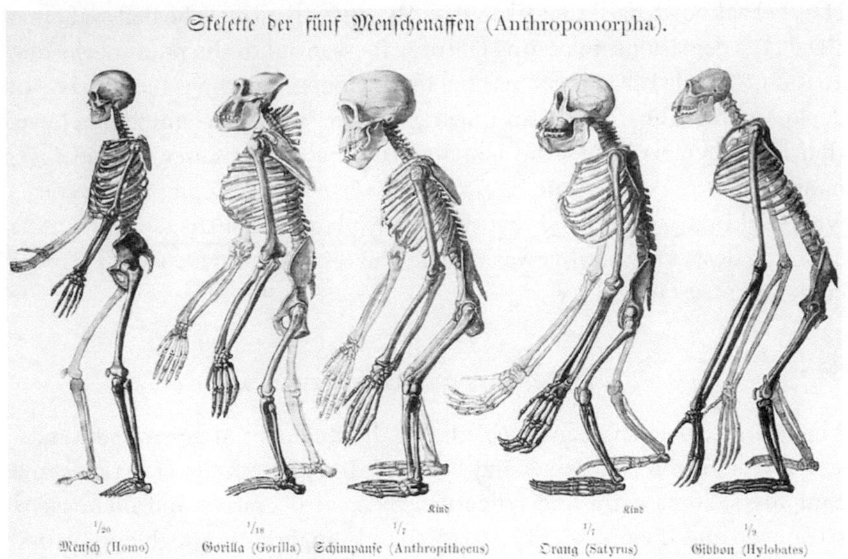

Fig. 9.12. Comparison of four ape skeletons (normalized for size) with that of a human. (From Haeckel, *Das Menschen-Problem,* 1907.)

all-too-human grin.[96] Concerning the second plate, which shows embryos of a pig, rabbit, and human being at three very early "sandal" stages, Brass mostly suggested they lacked other surrounding features (e.g., yolk) and that they were too symmetrical.[97] Finally, concerning the third plate of the embryonic stages of the bat, gibbon, and human being (fig. 9.9), Brass simply dropped his original charge that Haeckel had swapped the heads of the gibbon and human embryos. He found other falsifications, however: the bat was the common bat (*Vespertilio murinus*) instead of the horseshoe nosed bat (*Rhinolophus*) that Haeckel claimed; the human embryo in MII was represented with forty-six vertebrae instead of the thirty-three to thirty-five normally present; and the so-called gibbon at GII was really a macaque that had its tail removed.[98]

Haeckel responded to Brass's new charges in the 29 December 1908 issue of the *Berliner Volkszeitung* in a long article that recounted the activities of the Keplerbund and its opposition to Darwinian theory and monism.

96. Ibid., 8.
97. Ibid., 8–10.
98. Ibid., 15–21.

Haeckel acknowledged that like virtually every illustrator he had "schematized" his depictions, removing features inessential to the point of the discussion.[99] While Haeckel often acted injudiciously—perhaps recklessly—in deploying his illustrations, an impartial judge would recognize, I believe, that his schematizations did not materially alter his essential message, namely, that the embryonic structures of vertebrates at comparable stages were strikingly similar and that the best explanation of the similarity was common descent. And this was the view of forty-six of Haeckel's eminent contemporaries in biology.

The Response of the Forty-six

The contretemps between Haeckel and the Keplerbund generated a massive reaction from scientists and laymen alike. Hundreds of articles and pamphlets, some calm and reflective, most vituperative and dismissive, streamed from the presses. The Keplerbund sought a thorough condemnation of Haeckel, and to that end they sent a letter to many famous anatomists and embryologists seeking their support. They did get a response, but not precisely the one they had hoped for. In mid-February 1909 the following letter, signed by some of the most distinguished researchers in biology, appeared in a number of German newspapers:

> The undersigned professors of anatomy and zoology, directors of anatomical and zoological institutes and natural history museums, and so on, herewith declare that they certainly [*zwar*] do not approve [*nicht gutheissen*] of the few instances in which Haeckel practiced a kind of schematization but that in the interest of science and the freedom to teach they condemn in the sharpest way the battle that Brass and the Keplerbund have waged against him. They further declare that the developmental concept, as it is expressed in descent theory, can suffer no injury from a few inappropriately repeated embryo illustrations.[100]

The letter was signed by forty-six biologists, including Theodor Boveri, Karl Escherich, Max Fürbringer, Alexander Goette, Richard Hertwig, Karl Kraepelin, Arnold Lang, Ludwig Plate, Carl Rabl, Gustav Schwalbe, and August Weismann. Lest their meaning be unclear about their mild re-

99. Teudt, *Im Interesse der Wissenschaft*, 28; Schmidt, *Haeckels Embryonenbilder*, 16–17.
100. Schmidt, *Haeckels Embryonenbilder*, 50; Teudt, *Im Interesse der Wissenschaft*, 49.

proof of Haeckel, Carl Rabl (1853–1917), the great Leipzig cytologist, published in the *Frankfurter Zeitung* a clarification of what they meant by "schematization":

> Concerning the schematizations that went a bit too far, this is not a question of falsification or betrayal. The mild form in which the objection was clothed has been dictated by the great regard the zoologists and anatomists feel for Haeckel. They know very well how to appreciate how much they owe Haeckel and they know also that the few schemata of lesser value are hardly of consequence, as opposed to the numerous first-rate ones that Haeckel has produced and that have become the common property of science.[101]

Rabl and most of the signers had known Haeckel personally. Some, like Goette, had done battle with him; others, like Weismann, deeply disagreed with some of his theories. They understood from the inside the practices, the craft of science, in ways closed to outsiders, especially to those fearful of the great changes in society wrought by modernist intellectuals. Rabl and his colleagues provided, I think, a just evaluation of the old warrior's protracted dispute with the Keplerbund.

Conclusion

"Darwin's *Origin of Species* had come into the theological world like a plough into an ant-hill," wrote Andrew Dixon White in 1894. "Everywhere," he remarked, "those thus rudely awakened from their old comfort and repose had swarmed forth angry and confused."[102] None were more angry and confused than the theologians and theologians manqué who saw in Haeckel the embodiment of the Antichrist. From sophisticated German theologians who found his scientific worldview an appropriate challenge to Christianity to English preachers who feared "the depth of degradation and despair into which the teachings of Haeckel will plunge mankind," the

101. Schmidt, *Haeckels Embryonenbilder,* 63.

102. Andrew Dixon White, *A History of the Warfare of Science with Theology in Christendom,* 2 vols. (1894; repr., New York: George Braziller, 1955), 1:70. Michael Ruse delivers a spiked account of the reaction of contemporary religious sects to evolutionary theory in *The Evolution-Creation Struggle* (Cambridge, MA: Harvard University Press, 2005). Ronald Numbers provides a scholarly treatment of the American fundamentalist response to evolution in the early part of the twentieth century in *The Creationists* (New York: Knopf, 1992).

German Darwinian came to symbolize evolution militant.[103] Certainly that was the designation spread in the popular press, and with good reason.[104] Moreover, the complex relations of religion with political parties and revolutionary social movements, especially the Marxists, made even more hyperbolic the reactions of the lower-minded orthodox to a doctrine that seemed to deny the hand of the Creator in shaping the living world. To what shoals did that dangerous doctrine lead? "Primitive barbarism, Sun worship, Mohammedanism, self-love: these are the awful rapids to which Haeckel would steer the ship of humanity"—so warned the preacher of the Hampstead Congregationalist Church.[105]

I have argued that the strident and militantly anti-religious face that evolution has turned to the broader culture during the last century and a quarter has been shaped largely by Ernst Haeckel. Of course, I cannot produce a conclusive demonstration of this contention, but I think the evidence strongly supports it, as exemplified by the tracts footnoted in the previous paragraph. But there is also the mundane fact that more people learned of evolutionary theory through Haeckel's popular works than through any other source, including Darwin's own writings. Darwin once complained to his friend Asa Gray that he "had no intention to write atheistically."[106] Haeckel had every intention to do so. The twelve editions of his *Natürliche Schöpfungsgeschichte* and the six editions of his *Anthropogenie* were riddled with anti-religious declamations, asides, and arguments. *Die Welträthsel,* which sold in the hundreds of thousands of copies in over twenty-five languages, made the anti-religious theme central to its considerations. Then, of course, Hackelianism became the faith of the Monist League, whose members spread across several continents. Darwin had taken measures to blunt any religious backlash, even suggesting in the *Origin of Spe-*

103. For examples of calm and sophisticated responses to Haeckel's attacks on religion, see, for example, Friedrich Loofs, "Offener Brief an Herrn Professor Dr. Ernst Haeckel in Jena," *Die Christliche Welt* 13 (1899): 1067–72; and Georg Wobbermin, *Ernst Haeckel im Kampf gegen die christliche Weltanschauung* (Leipzig: J. C. Hinrichs'sche Buchhandlung, 1906). The analytic and reflective consideration was not the strong suite of the English preacher R. F. Horton; see his "Ernst Haeckel's 'Riddle of the Universe,'" *Christian World Pulpit* 63 (10 June 1903): 353–56; quotation from 353.

104. The *New York Times,* for instance, covered the fight that Haeckel waged with the Keplerbund in several long articles. See "Prof. Ernst Haeckel Accused of Faking," *New York Times,* 3 January 1909, C3; "'Man-Apes' the Subject of a War of Science," *New York Times,* 7 February 1909, SM9; and "Accused of Fraud, Haeckel Leaves the Church," *New York Times,* 27 November 1910, SM11.

105. Horton, "Ernst Haeckel's 'Riddle of the Universe,'" 355.

106. Darwin to Gray (22 May 1860), in *The Correspondence of Charles Darwin,* ed. Frederick Burkhardt et al., 15 vols. to date (Cambridge: Cambridge University Press, 1985–), 9:224.

cies that the Creator was behind it all. And as James Moore and Jon Roberts have shown, mainline Protestants in England and the United States came to terms with Darwin's conciliatory message.[107] No religious thinkers politely received Haeckel's vitriolic message. His publications evoked from religiously minded individuals, in Germany, in England, and in the United States, a bristling defense against atheistic evolutionism.

In the early part of the twentieth century, the character of scientists' religious attitudes suggests the impact of Haeckel's campaign. Anecdotally, many elite biologists of the period—such as Richard Goldschmidt and Ernst Mayr (see chapter 1)—testified to the transformations wrought on their psyches by their countryman. More representative evidence comes from James Leuba, a psychologist at Bryn Mawr, who in 1914 conducted several surveys of scientists and college students regarding their religious beliefs.[108] Among scientists generally, 41.8 percent indicated they accepted a personal God (defined as a being to whom one could pray and expect a response). However, among elite biologists, only 16.9 percent were believers; while among elite physical scientists twice that number were.[109] Today the percentage of elite biologists professing faith in a personal God is, at 5.5 percent, the lowest among the different kinds of scientist. These figures are in contrast to the numbers of the entire class of those American scientists expressing belief in God, 39.3 percent, about the same level obtaining a century ago.[110] The discrepancy between attitudes of elite biologists and those of the run of American citizens is even more brutal: a recent Gallup poll (1999) indicates 86 percent of the public accepts a personal God—and more when belief in a "higher power" counts as religious conviction (94 percent). If prominent biologists have rejected the deity in such numbers, then there must be something more than the sheer logic of biological theory—since that logic of itself cannot tell against religion. But Haeckel certainly did tell against religion, and the force of his argument had considerable momentum.

Evolutionary theory was not in any necessary conflict with sophisticated theology; and Erich Wasmann's construction of evolution makes this clear. Today few philosophers—or even theologians of cultivated taste—

107. See James Moore, *The Post-Darwinian Controversies* (Cambridge: Cambridge University Press, 1979); and Jon Roberts, *Darwinism and the Divine in America: Protestant Intellectuals and Organic Evolution, 1859–1900* (Madison: University of Wisconsin Press, 1988).

108. The surveys were published in James Leuba, *The Belief in God and Immortality: A Psychological, Anthropological and Statistical Study* (Boston: Sherman, French & Co., 1916).

109. Ibid., 255. Elite scientists were the ones receiving a star designation in James McKeen Cattell's survey, *American Men of Science* (Lancaster, PA: Science Press, 1906).

110. See Edward Larson and Larry Witham, "Scientists Are Still Keeping the Faith," *Nature* 386 (1997): 435–36; and "Leading Scientists Still Reject God," *Nature* 394 (1998): 313.

would be ready to endorse Wasmann's Thomistic dualism. Yet his readiness to reflect on articulate scientific theory and accept striking empirical evidence indicate the kind of flexible mind that is not saturated with dank ideology—a mind that in a later day might be ready to conceive sensory cognition (which he thought the provenance of animals) and human reason as more dynamically related, one that might interpret the "soul" not as an entity but as an achievement.

Wasmann stands as a case of an individual for whom empirical truth triumphed over recalcitrant dogmatism. By contrast, the crude opposition of individuals like Brass would not have stirred Haeckel to wrath, except for that failed academic's mendacity. Wasmann's scientific intelligence and sophisticated acumen created for Haeckel a much more dangerous situation: this Jesuit showed how one could be both an intelligent evolutionist and a sophisticated religious thinker. This was the deeper problem for the Monist position. Of course, it did not take much to discharge Haeckel's long-term suspicion and disdain for the church of Rome. Even when the more acidic and personally damaging dispute with the Keplerbund broke out, he still thought of that Protestant group as somehow allied with Wasmann's Jesuits, so intellectually pernicious did he regard the latter. In 1910 Haeckel brought out a small tract entitled *Sandalion: Eine offene Antwort auf die Fälschungs-Anklagen der Jesuiten* (Sandalion: an open answer to the charges of falsification by the Jesuits).[111] "Sandalion" referred to the sandal-shaped embryos of vertebrates. But by "Jesuits" he meant not only the Catholic religious order but also Protestant religious thinkers of a low, Jesuitical type. Protestant Jesuits! He saw those dark shapes looming everywhere. That part of the world soul where Haeckel now dwells must be even more chagrined and suspicious of Jesuit intrigue after eavesdropping on the meeting of the Pontifical Academy of Sciences in 1996, where Pope John Paul II declared that "fresh knowledge leads to recognition of the theory of evolution as more than just a hypothesis."[112] The pope, in stating the church's position, however, hardly broke new theological ground.

111. Ernst Haeckel, *Sandalion: Eine offene Antwort auf die Fälschungs-Anklagen der Jesuiten* (Frankfurt: Neuer Frankfurther, 1910).

112. John Tagliabue, "Pope Bolsters Church's Support for Scientific View of Evolution," *New York Times*, 25 October 1996, A1. This is a report of Pope John Paul II's address to the Pontifical Academy of Sciences. The current pope, Benedict XVI, may be having second thoughts. His friend, the cardinal archbishop of Vienna, Christoph Schönborn, has asserted: "Evolution in the sense of common ancestry might be true, but evolution in the neo-Darwinian sense—an unguided, unplanned process of random variation and natural selection—is not. Any system of thought that denies or seeks to explain away the overwhelming evidence for design in biology

He essentially reiterated the resolution that Wasmann had worked out a century before.

Haeckel had lost his taste for any orthodox religion after his habilitation work in Italy and Sicily. The wonderful excesses of southern Catholicism should, perhaps, have amused him; instead he took them as a personal affront. The death of his first wife, Anna, not only caused him to abandon formal observance; the soul-searing event turned him against the kind of superstition that would tolerate such malevolence. Yet because of his second wife, his children, and their social life in Jena, Haeckel retained nominal membership in the Evangelical Church. The attacks of the Keplerbund, however, finally drove him out. In December 1910 he formally declared, in a published account of his religious trajectory, that he was leaving the Evangelical Church.[113] Those who read the article were undoubtedly surprised to learn that he had been, up to that time, still a member of the church.[114]

Coda: "The Rape of the Ants"

After his encounter with Haeckel and the Monists, Wasmann continued his research on inquilines and their hosts. His correspondence network of important ant-men—Forel, Wheeler, and Hugo von Buttel-Reepen (1860–1933)—continued apace, with the exchange of many ant species among them. Wasmann built up the largest entomological collection of ants in the world, some thirty-five hundred different species. He also strove unremittingly against Haeckelian evolutionary theory and its cultural spread, which he believed to be rife during the first decades of the new century. He lectured and wrote on the dangers to German culture of monistic thought, especially that connection about which Virchow had warned, namely, its alliance with the Social Democratic Party and the Communists. Wasmann thought this danger particularly acute after the Great War, with German institutions and society in shambles and with their need of

is ideology, not science." His essay appeared as an op-ed in the *New York Times:* Christoph Schönborn, "Finding Design in Nature," *New York Times*, 7 July 2005, A27.

113. Ernst Haeckel, "Mein Kirchenaustritt," *Das freie Wort* 10 (1910): 714–17.

114. The first sentence in a long article in the *New York Times*, 27 November 1910, SM11—"Accused of Fraud, Haeckel Leaves the Church"—expresses this same response: "Any surprise that was felt over the news that Prof. Ernst Haeckel had withdrawn from the Evangelical Church in Germany must have been tempered by wonder that he had not withdrawn from it long before."

reconstruction. In a lecture delivered to the Catholic Union in Aachen on 28 January 1921, Wasmann asked, rhetorically, about the direction to take in the wake of the destruction of German cultural and social life.

> Our answer can only be shouted: back to Christianity and away with Haeckelian monism! For the impregnation of anti-Christian ideas of this neopaganism into our social networks bears the chief responsibility for not only the material collapse of our Fatherland but also its ethical and religious orientation. For that reason we say: Haeckel's monism is a cultural danger [*Kulturgefahr*].[115]

During Wasmann's last years, he saw the beginning of a transformation in German society in a direction that confirmed his dark forebodings. Wasmann died in 1931. His ants, however, were fated to have a curious connection with the Nazi regime.[116]

After his death Wasmann's large collection of ants and beetles, along with his personal library, were donated to the Natural History Museum of Maastricht, where they were to be available to all researchers. In October 1942 Dr. Hans Bischoff, curator of the Berlin Zoological Museum, received an urgent order from Heinrich Himmler, head of the Schutzstaffel (SS) and himself an amateur entomologist. Bischoff was to go to Holland and get Wasmann's ants. Bischoff first traveled to the Jesuit house in Limburg looking for the collection. He was told it had been transferred to the Natural History Museum in Maastricht. The museum personnel and other citizens learned of Bischoff's mission; and, with the connivance of even the Quisling mayor, they hid the ants in the basement of the city hall. Only temporarily foiled, Bischoff returned to Maastricht the next spring with a contingent of SS troops. He stated quite formally that the ants were being repatriated. They were German ants! The burgomaster retorted that Wasmann was born in the Tyrol. They were Italian ants. The Dutch, needless to say, did not win the argument. The ants and Wasmann's book and reprint collection were carted off to Berlin. A *Time* magazine article of 1944,

115. The lecture is in the Nachlass of Erich Wasmann held in the Natural Museum of Maastricht.

116. The outline of the following story was told to me by Dr. Fokeline Dingemans of the Natural History Museum of Maastricht. For other details, I have relied on a report, "Ants Rescued by Richmonder," in the *Richmond Times-Dispatch*, 10 February 1946. I am grateful to David Leary (University of Richmond) for providing information on John Wendell Bailey.

Lieutenant-Colonel J. W. Bailey, who recovered the loot, looks on while Dr. Van de Geijn, curator of the Maastricht Museum, shows a tray to Dr. W. M. Van Kessenich, chairman of the Maastricht City Council.

Fig. 9.13. John Wendell Bailey returning Wasmann's ants to the Natural History Museum in Maastricht. (From the *Richmond Times Dispatch,* 10 February 1946.)

entitled "The Rape of the Ants," stood aghast at the perfidy of the SS, who stooped so low as even to steal ants.[117]

After the Normandy invasion, Colonel John Wendell Bailey (1895–1986), head of typhus control in Europe, made his way to Maastricht in the fall of 1945 to examine Wasmann's collection. Bailey was a professor of entomology at the University of Richmond and a former student of Harvard professor William Morton Wheeler, Wasmann's old friend. When he got to the museum, he learned about the fate of the ants. He decided to chance it and traveled the six hundred miles to Berlin and the Zoologisches Museum. Like most of Berlin, the museum lay in rubble. Bailey did manage to locate

117. "The Rape of the Ants," *Time* 44, no. 21 (20 November 1944).

Bischoff and with some strategic threats discovered that Wasmann's ants and books had been stored in the deep vaults of a bank. The bank lay in ruins, but the vaults were still secure. Miraculously the entire collection of ant species and the library had survived. Since the bank was in the Russian sector, Bailey had to negotiate with a Russian general, whom he befriended with many cartons of American cigarettes and several bottles of whiskey. After the proper papers were signed, Bailey and several GIs loaded the ants and books—some 160 insect trays, 150 small boxes, 100 bottles of specimens in alcohol, and 50,000 books and reprints—on two trucks and three jeeps and took them to the American sector. Bailey discovered, however, that some of the insects were missing, which he later found in Himmler's country home in Waischenfeld, just over the Swiss border. Bailey shipped the ants and books back to the Maastricht Natural History Museum, where today they are still used in research.

CHAPTER TEN

Love in a Time of War

In 1896 Haeckel completed the last major scientific work of his career, his *Systematische Phylogenie.* That stolid, three-volume account of the kingdoms of protists, invertebrates, and vertebrates stands like a snow-capped volcano, only vaguely reminiscent of the fire-belching *Generelle Morphologie* that threatened the orthodox three decades earlier. Completion of this work on systematic phylogeny seems an accomplishment that might have initiated a period of rest and repose, a gentle decline during which accolades could be enjoyed at the end of a career—and through the turn of the century, Haeckel's honorary degrees and awards from learned societies accumulated at an accelerated rate.[1] Yet his tranquillity lasted only for a moment. The last two decades of his life exploded with awakened passion and ferocious combat. Three major events brought on the troubles: a new love, which grew in frustrating intensity; his book *Die Welträthsel* (The world puzzles, 1899), which ignited intellectual war on all fronts; and the Great War, a real war, which produced catastrophic cultural chaos and untold death and misery.

At Long Last Love

In 1927 a book appeared with the title *Franziska von Altenhausen: Ein Roman aus dem Leben eines berühmten Mannes in Briefen aus den Jahren 1898–1903* (Franziska von Altenhausen, a novel of the life of a famous man in letters from the years 1898–1903).[2] The editor, Johannes Werner,

1. See chapter 8 for a list of his honorary degrees and awards.

2. *Franziska von Altenhausen: Ein Roman aus dem Leben eines berühmten Mannes in Briefen aus den Jahren 1898–1903*, ed. Johannes Werner (Leipzig: Koehler & Amelang, 1927).

indicated that though the collection had the qualities of a romantic novel, the letters were nonetheless authentic. The names of the two protagonists, Paul Kämpfer and Franziska von Altenhausen, were pseudonyms used to protect the parties involved; place-names had also been changed. And, as it later became clear, the editing of the letters was quite imaginative: only a small portion of the original collection was included, crucial passages were cut, and completely fictitious lines were inserted. Shortly after the book appeared, Haeckel was easily identified as "the famous man"; and in the English edition published three years later, the veil partly fell from the title: *The Love Letters of Ernst Haeckel Written between 1898 and 1903.*[3] For years the woman remained a mysterious figure. Her identity subsequently came to public light as a result of evidence preserved at Haeckel-Haus, the main repository of Haeckel's manuscripts and letters. She was Frida von Uslar-Gleichen, a young woman of the minor nobility living in Hanover. The remains of the letters themselves—over five times the number in the original volume—had been deposited in the State Library in Berlin.[4] Haeckel's son, Walter, moved by what seems an anti-Oedipal impulse, arranged for the publication of the highly edited 1927 book.[5] That first publication went through many translations and editions—in Germany alone sales reached 140,000 copies by 1943. Despite the extensive editing, the letters revealed a poignant relationship between two people quite in love

3. *The Love Letters of Ernst Haeckel Written between 1898 and 1903*, ed. Johannes Werner, trans. Ida Zeitlin (London: Methuen, 1930).

4. After the highly selected publication in 1927, the bulk of the letters (over nine hundred of them) remained in the family of the publisher. Just before the Russian takeover of eastern Germany, the publisher's family fled to the west and finally deposited the letters in the Staatsbibliothek zu Berlin in 1968. This original correspondence (or that which has survived) and letters of Haeckel's son and nephew, as well as relevant letters of the Uslar-Gleichen family, have recently been published. See *Das ungelöste Welträtsel: Frida von Uslar-Gleichen und Ernst Haeckel, Briefe und Tagebücher 1898–1900*, ed. Norbert Elsner, 3 vols. (Göttingen: Wallstein, 2000). Prior to this publication, I had used the collection (Nachlass Ernst Haeckel) in the Staatsbibliothek (Preussischer Kulturbesitz); and now, after checking originals against the published versions, I have relied on Elsner's three-volume edition.

5. After the death of Frida von Uslar-Gleichen, her family returned Haeckel's letters to him. Haeckel systematically arranged the correspondence; and just before he died, he left them with his nephew Heinrich, presumably because of their autobiographical interest. Heinrich Haeckel went through the correspondence, apparently with an eye to publication, but he died two years after his uncle. The letters were then sent to Haeckel's son, Walter, who arranged for the highly redacted publication. The 1927 book surprised and upset both families, but Walter justified it because his father's work, except for *Die Welträthsel* (which had been adopted by the "socialists and communists"), had fallen from public view. He hoped the publication would stimulate new interest in his father's legacy. And as an artist and devoted son, he undoubtedly thought the artistic quality of the correspondence deserved public appreciation. See the letters describing these events in *Das ungelöste Welträtsel*, 3:1205–10.

but whose consciences apparently restrained their behavior.[6] As a review in the *Times Literary Supplement* (11 September 1930) put it: "Nobody who reads these letters can doubt either the spiritual fire of their emotion or their suffering." Their suffering resulted from "the moral scrupulousness and the stern sense of duty which animated each of them."[7]

In the six years of their correspondence, over nine hundred letters passed between Haeckel and Frida, about one every two and a half days. Their bond grew more intimate than the published letters suggested. They certainly had moral scruples about their affair, but their relationship moved, nonetheless, beyond the platonic phase represented in the initial publication. In their letters the lovers would endlessly recount to one another their meetings in out-of-the-way hotels, their embraces, their strolls through gardens and parks, and their fugitive plans. Haeckel began to think of Frida as a reincarnation of his first wife—significantly Frida was born in 1864, the year of Anna's death. Beyond the lasting relation with his first wife, no other attachment affected Haeckel in so profound a way. At a time when attacks on him mounted because of *Die Welträthsel* and when his own wife Agnes and his daughter Emma had both withdrawn into the deep depressions and invalid valleys of the nineteenth-century neurasthenic, Frida provided the emotional escape that probably stayed his hand from taking his own life, which he had seriously contemplated on several occasions prior to their meeting. She also served as an intellectual and cultural confidante. She urged him to reduce the force of his assaults on religion and other orthodoxies; she advised him on the selection of illustrations for inclusion in his *Kunstformen der Natur* (Art forms of Nature, 1899–1904); she recited poetry to him, discussed music, and generally encouraged him in his work. She elevated his life when it threatened to plunge into the recesses of bitter despair and extinguished hope. Their story, though, does not have a happy ending.

Frida von Uslar-Gleichen was the eldest of five children of Bernhard von Uslar-Gleichen (1830–1873) and his wife, Anna (1833–1915). Branches of the family had been vassals to the elector of Hanover and, later, kings of Hanover and England. Her father had fought on the Austrian side in the Austro-Prussian War (1866) and died shortly after the peace was concluded, leaving the family with only a small income and a modest estate

6. As I initiated this study, I showed photocopies of the correspondence to a graduate student. He later remarked that he had never really been in love, but now, having read a portion of the correspondence, he thought he knew how love must feel.

7. [R. D. Charques], "Haeckel's Love Letters," *Times Literary Supplement*, no. 1493 (11 September 1930): 714.

Fig. 10.1. Frida von Uslar-Gleichen (about age twenty) and her brother Bernhard. (Courtesy of Georg Freiherr von Uslar-Gleichen.)

at Gelliehausen near Göttingen. Frida was raised by her mother and helped with the care and education of the smaller children. She had tutors to age sixteen and thereafter saw to her own education. She was a cultivated woman who read generously; she frequently attended musical concerts, preferring Beethoven; and she painted tolerably well. She could write fluently and critically, as Haeckel would discover. She was also quite an attractive woman, slim, blond, and handsome. She was barely thirty-four and unmarried when she first corresponded with Haeckel; he was sixty-five.

Like many young women in her position during the nineteenth century, she felt smothered in layers of duties and expectations, with diversion relegated to occasional teas with maiden aunts. She spent her days among "pedestrian people [*alltäglichen Leuten*] and listening only to pedestrian people."[8] Ernst Haeckel arose in her eyes as a modern titan of science and something of a dangerous man; he opened the possibility of flight from her Biedermeier cocoon.

Their relationship began inauspiciously enough. In January 1898 she wrote him a fan letter about his *Natürliche Schöpfungsgeschichte* and asked if she could pursue a few questions with him. She obviously had considerable intelligence, and Haeckel sent her some books, including his *Reisebriefe von Ceylon,* to whet her appetite. They ritually exchanged photographs; and she, rather forwardly, remarked: "It is not a question that you please me, but from a purely artistic standpoint, you are a handsome man [*ein schöner Mann*], and I'm quite happy about that."[9] He undoubtedly appreciated her beauty as well, and periodically she would send him updated photographs. He, in turn, would keep her supplied with books, both scientific and literary, the kind that he thought would further her self-education. When he learned that she had a great love for Goethe, a certain set of feelings fell into place for him.[10]

The letters through late winter and early spring of 1898 became increasingly more personal—with Frida detailing her hopes for a life with larger horizons and Haeckel emphasizing his miserable existence at a home in which the miasma of depression and recrimination hung heavy in the air. In July she tentatively suggested that he stop at Göttingen on his way to England, where he planned to attend an international conference and would receive an honorary doctorate from Cambridge University. The meeting did not take place. During the following weeks and months, each would continue to make suggestions for a rendezvous, which for one reason or another never occurred. Through the spring of 1899, Haeckel nailed himself to his desk to finish his book *Die Welträthsel.* As the page proofs appeared, he sent them to Frida, who would mark various passages in an effort to dilute the acid with which he etched his condemnations of orthodoxy; and later she would expend much ink consoling him for the scorn the book evoked from enemies and even friends. Finally, Haeckel felt he had to meet the

8. Frida to Haeckel (21 October 1901), in *Das ungelöste Welträtsel,* 2:706.
9. Frida to Haeckel (24 March 1898), in ibid., 1:68.
10. Frida to Haeckel (15 July, 1898), in ibid., 1:80.

woman who was becoming so dear, so necessary to his existence. In June 1899 the university planned an academic festival, which offered him the opportunity to invite her to a rather safe, public event.

Frida and Haeckel spent virtually the entire day of Saturday, 17 June, and the next day in deep conversation in his office at the Zoological Institute and in walks through the city. Immediately after she returned to her hotel on Sunday evening, she wrote to assure him that "what we have spoken about will remain only for you and me alone; and that through our conversation you have become still more beloved and dear."[11] It was not only conversation they shared, but also a kiss. Later he recalled for her that kiss, when through his body arced "a shiver of desire of the sort I experienced with my dear first wife, Anna S., 40 years ago, and which I was never granted with my poor unhappy second wife—a born Vestal virgin."[12]

After their encounter the salutations of their letters moved from the very formal "Most Honored Professor" and "Dear Honored Young Lady" to "My Dear Teacher" and "My Dear Frida." Finally, her letters addressed him variously as "My Dear Ernst," "My Sweetheart," even "My Silver Bunny." His became more simplified, from "My Loveliest, Dearest Friend" to just "L. F." (*Liebe Freundin* or *Liebste* Frida). In the course of a letter, when his keenest desires seized his pen, he might refer to her as "bride of my soul."

Their first meeting allayed their mutual anxieties about face-to-face contact and altered their relationship dramatically. They excitedly agreed to another visit on her return from her sister's home, this time for three days (14 to 16 July 1899). Haeckel planned the event meticulously. He had her get off the train just outside of Jena at Papiermühle on Friday afternoon, 14 July. He met her there, and they lingered until nine o'clock in the garden restaurant, still a romantic setting today. The next day they traveled to Dornburg (a bit north of Jena) to visit an art gallery; on Sunday they stole time at the Zoological Institute; and on Monday they traveled to Weimar to walk through Goethe's house, and then on to Eisenach and the Wartburg, where Luther translated the New Testament. During these intimate sojourns, they often embraced and kissed.[13] Thereafter in his letters, Haeckel would refer to this three-day excursion as their "honeymoon" (*Brautfahrt*).[14]

Throughout the course of their relationship, they remained laced up in

11. Frida to Haeckel (18 June 1899), in ibid., 1:142.

12. Haeckel to Frida (5 August 1899), in ibid., 1:200.

13. Recounted by Frida in her diary (17 July 1899), in ibid., 1:180.

14. For example, Haeckel to Frida (18 July 1899), in ibid., 1:162.

Fig. 10.2. Agnes Haeckel in later years. (Courtesy of Ernst-Haeckel-Haus, Jena.)

a fraying Victorian morality. He wrote: "Is it not a tragedy that two highly gifted children of the earth, who are so completely made for each other, seem to be kept so far apart by reason of age and position, of standing and propriety?"[15] She replied that it was not age, position, or propriety that kept them apart, but only duty to his wife.[16] She wanted him to tell his wife, if not all the details of their relationship, at least that they were friends.[17] When these requests were made, Haeckel always demurred, saying that

15. Haeckel to Frida (19 July 1899), in ibid., 1:164.
16. Frida to Haeckel (21 July 1899), in ibid., 1:165–66.
17. Frida to Haeckel (8 August 1899), in ibid., 1:210.

his wife could not stand the shock, especially with her weak heart. On Frida's side, despite a stated desire for candor, she hid the extent of their relationship from her mother, who disapproved of her writing and visiting a married man—especially an infamous man like Ernst Haeckel.

As their relationship progressed, Haeckel constantly devised plans for their future. As a first possibility, they could simply maintain their relationship as that of friends—but he knew he could not keep the friendship nonphysical: "even against our judgment and will, it will be hand in hand, then arm in arm, and then mouth to mouth." In the past, he confessed, many beautiful women had flung themselves at him; yet, he said, he never permitted himself any "sexual dissipation" (*geschlechtliche Ausschweifung*)—a tenuous claim, perhaps. With Frida, however, he would not be able to restrain himself. There was a second possibility. They could wait for his "unhappy wife to have her wish fulfilled to be freed from her difficult suffering of many years by an easy death"—but who knew how long that would take. Or, with the money he had amassed from his numerous publications, they could run away to an exotic island, while leaving sufficient funds for his wife and daughter. He concluded the first possibility would destroy his spirit and the latter two were unrealistic. Over the first year and a half of their relationship, their plight gradually scored in the souls of each wounds of deep melancholy, bleak pessimism, and unremitting desire.[18]

The World Puzzles

In September 1899 Haeckel's *Welträthsel* debuted in the bookshops. Almost immediately his publisher had to bring out a second edition; and then by mid-November, a third was readied. Haeckel wrote his friend Allmers the next April to report that "the success of *Die Welträthsel* surpasses all of my expectations; the fourth (unchanged) edition (8 to 10 thousandth) has now already appeared. Correspondence about it has occupied my whole winter."[19] During its first year, some 40,000 copies had been produced. The publication of the "people's edition" (1903), selling for one mark, helped boost the total sales in Germany to 400,000 before the First World War. And letters responding to the book flooded his offices. In 1903 alone, he had received over three thousand letters, both commendatory and condemnatory

18. Haeckel to Frida (5, 15, 18 August and 2 September 1899), in ibid., 1:204, 234, 244, 268.

19. Haeckel to Allmers (12 April 1900), in *Haeckel und Allmers: Die Geschichte einer Freundschaft in Briefen der Freunde*, ed. Rudolph Koop (Bremen: Arthur Geist, 1941), 205.

(the majority).[20] It was an extraordinary succès de scandale. And scandal it was. A *New York Times* reviewer, evaluating the quickly published English translation (1900), epitomized what for many readers was the essence of the book:

> One of the objects of Dr. Haeckel—it would not be unfair to say the chief object—is to prove that the immortality of the human soul and the existence of a Creator, designer, and ruler of the universe are simply impossible. He is not at all an agnostic. Far from it. He knows that there can be no immortality and no God.[21]

There was, of course, more to the book than that.

The book took its title from Du Bois-Reymond's conceit that seven world enigmas existed: (1) the nature of matter and force; (2) the initiation of motion; (3) the beginning of life; (4) the design of nature; (5) the appearance of sensibility; (6) the origin of consciousness and speech; and (7) the problem of free will.[22] Du Bois-Reymond contended that the first, second, and fifth were transcendental problems for which there could be no solution, while the third, fourth, and sixth had yet to be solved. He was not sure into which category freedom of the human will fell. That Haeckel should have chosen as his title "The World Puzzles" was a bit like Darwin taking

20. Haeckel to Max Fürbringer (12 August 1903), in *Ernst Haeckel: Biographie in Briefen mit Erläuterungen*, ed. Georg Uschmann (Leipzig: Prisma, 1984), 282. While most of the objections concerned Haeckel's attacks on religion, some complained about the political considerations. John Lubbock (Lord Avery) and Haeckel began their friendship in the early 1870s and continued to be in communication through the first decade of the new century, with Haeckel supplying his English friend with copies of his various publications. Lubbock complained about Haeckel's not-so-subtle attacks on English political policy: "I have read your Riddle of the Universe [the English translation of *Die Welträthsel*] with interest, but am surprised at the unjust attack on England in 362 [*sic*, 354]. When did we take any colonies from Germany? . . . We are accustomed to unfounded attacks in some of the German newspapers, but surely a Philosopher should not attempt to sow dissension between two great and cognate peoples." See John Lubbock to Haeckel (12 December 1901), in the Correspondence of Ernst Haeckel, in the Haeckel Papers, Institut für Geschichte der Medizin, Naturwissenschaft und Technik, Ernst-Haeckel-Haus, Friedrich-Schiller-Universität, Jena. In his book Haeckel made the remark in passing that Christianity emphasized unrealistically love of neighbor at expense of self. He observed that when the injunction was translated into modern politics, it would suggest: "When the pious English take from you simple Germans one after another of your new and valuable colonies in Africa, let them have all the rest of your colonies also—or, best of all, give them Germany itself." See Ernst Haeckel, *The Riddle of the Universe*, trans. Joseph McCabe (New York: Harper & Brothers, 1900), 354.

21. "A Little Riddle of the Universe," *New York Times*, 27 July 1901.

22. Emil Du Bois-Reymond, "Die sieben Welträtsel," in his *Vorträge über Philosophie und Gesellschaft*, ed. Siegfried Wollgast (Hamburg: Felix Meiner, 1974), 159–86.

"Origin of Species" as his: both books denied the existence of their subject. Darwin argued that "species" served only as a term of convenience; in the course of nature, only similarity and variability existed. Haeckel believed that modern science, in its monistic version, had solved, at least in principle, all the world puzzles that Du Bois-Reymond had discriminated. They were no longer real conundrums. The framework of "world puzzles," however, allowed Haeckel to sketch the advances made by modern science, the weight of which had extinguished, as he never tired of proclaiming, the old dispensation of a religiously infected science.

Haeckel's view of the accomplishments of modern science, in broadest outline, is the one widely shared by scientists today. The details of the physical theory he described, then at the leading edge of science, have been greatly modified during the last hundred years. But the idea of continuity between the nonliving and living worlds; the application of natural law to account for all physical phenomena; the ultimate resources of observation, experiment, and logical analyses in the discovery of new knowledge; the validity of evolution by natural selection—all of these have been sanctioned by scientists in the modern day. The watchtowers that Haeckel erected around science to prevent the ingressions of supernatural entities continue to be manned by alert contemporary scientists, while in the plains below creationists and intelligent designers marshal the forces of an increasingly bellicose and politically armored religious fundamentalism.

In conformity to the physics of his day, Haeckel asserted that the universe consisted of congregations of atomic elements swimming in a sea of ether; the behavior of the elements and the sea itself ran in currents strictly governed by what he called the laws of substance—that is, the conservation of matter and the conservation of energy. The known elements—about seventy in Haeckel's day—exhibited chemical affinities that formed larger molecules, the very stuff of macroscopic physical bodies. In Haeckel's monistic reading, physical objects—even down to elemental atoms—had a quasi-mental side, which was displayed at the lowest level by bonding inclinations among constituents, their elective affinities. Among larger complexes, as found in living organisms, these fundamental forces were expressed in sensation, volition, and ultimately consciousness. What this monistic image precluded was an independent, nonphysical soul or distinct mental entity.[23]

Had Haeckel's depiction remained at the level of an abstract scientific

23. Ernst Haeckel, *Die Welträthsel, gemeinverständliche Studien über Monistische Philosophie* (Bonn: Emil Strauss, 1899), chap. 12.

materialism, of the sort just indicated, the book would not have caused the gorge to rise in the throats of any but the most theologically sensitive. But he relentlessly applied this monistic view to discuss the "nature of the soul" (an expression of forces of matter);[24] the "embryology of the soul" (from the amoeba-like movements of spermatozoa and egg to conscious functions of brain);[25] the "phylogeny of soul" (the continuity of psychic life from protists to invertebrates and then to vertebrates, as the evolutionary doctrine maintained);[26] and the "immortality of the soul" (persistence of elemental forces, while higher souls evanesced with their complex bodies).[27] Against this scientific image, Haeckel cracked the many myths of Western and Eastern theology like so many goose eggs.[28] And, of course, he applied this monistic worldview to the question of the deity. As he had already suggested thirty years earlier in his *Generelle Morphologie*, the only God that a thoroughgoing monism might tolerate is the God of Spinoza: *Deus sive natura*.[29]

Haeckel did not wish to advance a Nietzschean ethics of a superior morality to replace the shards of the old morality.[30] The foundations of orthodoxy, whether derived from Christianity or from the more austere considerations of Kant's practical reason, had to be rejected in light of modern science; but the code of conduct that they supported—Haeckel wished to leave those principles substantially intact. The new foundation for the Golden Rule and the biblical injunction to love one's neighbor as oneself, he found in the Darwinian doctrines of self-preservation and social instinct: by reason of selection, we are designed to preserve our own ego's integrity but also to cooperate in promoting the welfare of our community. Five years later, in *Die Lebenswunder* (The wonder of life, 1904), Haeckel made clear his rejection—for "personal reasons," as well as for good biological reasons—of the one-sided ethics of the "modern prophets of pure egoism, Friedrich Nietzsche, Max Stirner and the like." They committed, he argued, a fundamental biological error:

> Indeed, the natural commandments of sympathy and altruism not only arose in human society millennia before Christ; they were to be found

24. Ibid., chap. 6.
25. Ibid., chap. 8.
26. Ibid., chap. 9.
27. Ibid., chap. 11.
28. Ibid., chaps. 16–17.
29. Ibid., chap. 15.
30. Haeckel explicitly rejected Nietzsche's extreme egoism. See ibid., 463.

> characterizing those higher animals that live in herds and social groups. These traits have their oldest phylogenetic roots in the formation of the sexes in the lower animals and in the sexual love and parental care upon which the preservation of the species rests.[31]

For Haeckel, this kind of evolutionary foundation gave theoretical substance to those native impulses that he himself felt, particularly in his relationship with Frida.

Haeckel was not insensitive to the contemplative repose and aesthetic satisfaction that Christian art—and especially medieval churches—offered the reflective individual. Such experiences had real value for creating a cohesive society and for producing a feeling of communal solidarity. He thought a continued social and educational evolution would transform the worship of a supernatural deity gradually into the enjoyment of a spiritually enriching nature. He even imagined—though he buried his musings discreetly in the fine print of the "notes and remarks" of *Die Welträthsel*—that something like Comte's church of science might eventually be instituted:

> In place of the mystical faith in supernatural wonders clear knowledge of the true wonders of nature would be introduced. The houses of God, as contemplative places, would not be decorated with holy pictures and crucifixes but with representations of the uncreated realm of natural beauty and the lives of men. Between the high pillars of the Gothic dome, entwined by liana vines, slender palms and ferns, delicate banana and bamboo would remind us of the creative power of the tropics. In large aquariums beneath the church windows, charming medusae and siphonophores, colorful corals and starfish would exemplify the "art forms" characterizing the life in the sea. At the high altar, a "Urania" would step forth to explain the omnipotence of the laws of substance governing the motions of the planets. Indeed, there are now numerous educated people who find their real edification, not in listening to prolix and meaningless sermons but in attending public lectures on science and art, in the pleasures of the limitless beauty that flows in inexhaustible streams from the womb of our Mother Nature.[32]

31. Ernst Haeckel, *Die Lebenswunder: Gemeinverständliche Studien über Biologische Philosophie, Ergänzungsband zu dem Buche über die Welträthsel* (Stuttgart: Alfred Kröner, 1904), 131.

32. Haeckel, *Welträthsel,* 463. This passage was omitted from the English edition.

Though Haeckel presented to the world the face of a coldly rational man of the modern scientific age, the mask occasionally slipped and the Goethean Romantic looked back on a world still resonant of a deeply spiritual, not to say mystical, core.

A large portion of the educated public reacted to Haeckel's *Welträthsel* with cataclysmic furor. Those of an orthodox religious temper execrated the book (as I have indicated in the previous chapter). Beyond those with theological concerns, the attitude was still virulent enough that Haeckel feared some reprisals from the government. Even, as he said, in freethinking Jena, the response to the book had been highly negative.[33] The deepest wound, however, came from his dear friend Gegenbaur. He anticipated his old colleague's reaction—or nonreaction: "Gegenbaur (like many other close friends!) has not written me one word about it! He shares my views completely, from beginning to end!—He has always been of the opinion, however, that esoteric secrets are not to be revealed to the larger public—and, besides, he disapproves of my sharply aggressive mode of expression."[34] Haeckel found his fears realized when he traveled to Heidelberg in August 1900 to celebrate his friend's seventy-fourth birthday. Gegenbaur received him coolly. His onetime colleague had not read the book but had read the review written by their old friend Kuno Fischer, who called the book a "wretched effort" (*Machwerk*). Though Gegenbaur in the past invited his colleague to stay in his home, this time he did not. Haeckel walked to the Necker Bridge in the rain, stood there, and wept.[35] When Gegenbaur died three years later (1903), Haeckel felt remorse anew that his *Wurstbuch*, as his onetime colleague called it, had destroyed their forty-six-year friendship.[36]

The Consolations of Love

During the time when his book jolted the intellectual public to reaction—and certainly it was not all negative; the sales and congratulatory letters confirm that—Haeckel enjoyed the frustrating consolation of his affair with Frida. Though qualms of conscience had kept them apart since their second meeting in July 1899, she agreed to see him at the end of the follow-

33. Haeckel to Frida (19 October 1899), in *Das ungelöste Welträtsel*, 1:315.

34. Haeckel to Frida (1 March 1900), in ibid., 1:405–6.

35. Haeckel to Frida (31 August 1900), in ibid., 2:554.

36. Haeckel to Max Fürbringer (12 August 1903), in *Ernst Haeckel: Biographie in Briefen*, 282.

ing March. She initially proposed that they rendezvous in Naumburg, taking a hotel suite with two bedrooms and a sitting room between, and that she register as his daughter.[37] They finally met on 31 March in Magdeburg. The meeting confirmed Haeckel in his love for Frida, as he recounted to her the next day:

> Dearest, best, truest wife! So I might now call you, you who after some considerable worry opened to me the entire depths of your marvelous soul and unfolded the entire magic of your ideal person! You tell me and write, my dear Frida, that I should not idealize your person. Love, I cannot do that—since you are my ideal—the real ideal of a living wife, who with me finds the true religion in the cult of the true, the good, and the beautiful. . . . After I waved the last good-by at your departure this morning at 6:15, I remained another two hours in our romantic hotel!! Your "great mad child" committed all sorts of foolishness—washed himself yet again "from top to bottom" out of your washbasin, celebrated solemn memories in each of the two magical rooms—numbers 17 and 16—and delighted in yours, etc., with a princely tip [to the staff]. Two hours later, as I traveled from Magdeburg to Berlin, I read in Goethe's letters to Charlotte von Stein only your dedication [she gave the volume to him] and the few sentences you underlined. The entire remaining time (two and a half hours) I reveled in the sweetest memories.[38]

Frida had her own memories of their night together: "You write that the touch of my hand has benefited you. The moment when you had permitted me to lay my hand gently on your body—that remains for me an unforgettable time."[39]

After this one-day excursion, they planned another tryst, in view of a long journey Haeckel was planning for the late summer and winter of 1900–1901. On 1 June they met at Plauen (about fifty miles south east of Jena) and then traveled to Munich the next day. They spent five days there in the Hotel Grünwald, leaving 7 June, and then on to Erfurt and Sangehausen (about fifty miles north of Erfurt). On 9 June they journeyed to Bad Frankenhausen, a cure resort. They departed from one another on

37. The letters in which Frida suggests this meeting have not survived. However, Haeckel's nephew Heinrich Haeckel made extracts of the letters. See *Das ungelöste Welträtsel*, 3:1140–41.

38. Haeckel to Frida (1 April 1900), in ibid., 1:410–11.

39. This is from Heinrich Haeckel's extract of a no longer extant letter. See ibid., 3:1174.

11 June. Frida's diary, pages of which she sent to her beloved, memorializes their time together:

> In the cave of Barbarossa [a large tourist attraction near Frankenhausen] you were completely bewitched. You felt my power more than usual, or otherwise how were you drawn so strongly to my lips? Our charming trip in the one-horse carriage.—Our sweet union in the small, quiet rooms. Can I tell you how gladly I stroked your lovely body and how often I now do it in my imagination! Marriage = belonging together soul and body—that is the sweetest that union can bestow here below.—If I were your legal wife, then you would lay your lovely head on my breast and with your hand press my little electric buttons [*elektrische Knöpfchen drückst*], while I would caress you softly and sweetly, my sacred one. Amen.[40]

One of the marvelous attractions of the World's Fair Exposition in Paris, which received wide publicity and which Haeckel would shortly visit, was the electrification of the buildings; with a press of an electric button, a room would glow with warmth and brilliance. One century ended and a new had begun, with hope and possibilities for a new kind of life.

Second Journey to the Tropics—Java and Sumatra

Haeckel had been planning a second voyage to the tropics for a while. Some admirers thought he intended to build on the work of his protégé, Eugène Dubois, by finding further evidence of the missing link. He dismissed that notion, though perhaps not completely, since he would engage in some protracted study of the apes of Malaya, the anatomical features of which he believed provided surer evidence of descent than scattered paleontological remains.[41] His stated reason for the journey was to complete his plankton studies and to gather more interesting exhibits for his *Kunstformen der Natur,* which began appearing in a folio series in 1899.[42]

Haeckel had planned some ten installments in the series, which would then be published as a whole in a large folio volume. Each installment

40. From Heinrich Haeckel's extract of a no longer extant diary entry. See ibid., 3:1177.

41. Ernst Haeckel, *Aus Insulinde: Malayische Reisebriefe* (Bonn: Emil Strauss, 1901), 218–19. Other rumors sprung up. Haeckel heard that Cornelius Vanderbilt and Jay Gould had funded a rival expedition to find evidence of the missing link (ibid.).

42. Ibid., 3–6.

would have ten beautifully lithographed plates by Adolf Giltsch, the printer with whom Haeckel worked on many of the atlases for his systematic investigations. With the help of Frida, he carefully chose illustrations from his previous volumes on marine invertebrates for inclusion in the fascicles. The journey to Malaya would supply material for several new paintings of exotic creatures observed in the jungles and pulled up from the crowded seas around the islands. All of the illustrations would be reproduced in lithographs of vibrant color or stark black and white. Haeckel expressed the premise of the series in the introduction to the first installment: "Nature generates from her womb an inexhaustible plethora of wonderful forms, the beauty and variety of which far exceed the crafted art forms produced by human beings." But because creatures displaying these wondrous structures lay hidden in the depths of the ocean or camouflaged in the jungle, they remained inaccessible to the lay public. Haeckel thus wished to make visible to a wider audience the extraordinary artistry of nature that the science of the nineteenth century had uncovered. He also hoped his series would provide "a rich cornucopia of newer and more beautiful motifs" for modern artists.[43] This hope would be realized during the next several decades as his *Kunstformen der Natur* (1899–1904) had a decided impact on the movement of Jungenstil (Art Nouveau) in Europe.[44] Even today selections from his *Kunstformen* continue to be reproduced as aesthetic exemplars.

So Haeckel had his professional and artistic justifications for setting out on an extensive journey to the tropics. But he revealed a more personal, underlying motive in his letters to Frida: he simply could not abide the thought of spending another winter confined to his own gloomy home and depressive family. He obviously betrayed his feelings to his wife, since Agnes thought he would never return to their home.[45] He left Germany with regret because of the distance between him and his "true wife." The memories of his last rendezvous with her in June, though, would carry him sweetly along for a while. And, of course, even old men dream of native girls bringing breadfruit.

43. Ernst Haeckel, "Vorwort," in *Kunstformen der Natur* (Leipzig: Bibliographisches Instituts, 1904).

44. Christoph Kockerbeck traces some of the lines of Haeckel's aesthetic influence in *Ernst Haeckel's 'Kunstformen der Natur' und ihr Einfluß auf die deutsche Bildende Kunst der Jahrhunderwende* (Frankfurt: Peter Lang, 1986). Kockerbeck focuses on the work of the Munich sculptor and painter Hermann Obrist (1863–1927) and his friend the architect August Endell (1871–1925), but seems unaware of the impact on René Binet (1866–1911), architect and designer of the Paris Exposition—see below.

45. Haeckel to Frida (31 August 1900), in *Das ungelöste Welträtsel*, 2:555.

Haeckel departed Jena on 21 August 1900 but initially headed for Paris to spend a few days at the World's Fair with his nephew Heinrich Haeckel (son of his brother, Karl), who was chief of hospital in Stettin. He undoubtedly walked through René Binet's extraordinary gate that opened off the Champs-Elysées onto the midway of the fair (see fig. 10.3). That gate rose up like some giant radiolarian, and not by accident. Binet explained to Haeckel that the gate and various ornamental features of the fair's buildings had been inspired by the scientist's radiolarian work.[46] From Paris, Haeckel traveled to Basel to confer with Paul von Ritter, whose foundation supported a professorship in Haeckel's honor at Jena and who was planning to commission a statue of Haeckel (which was never produced). Finally on 4 September, he boarded the North German Lloyd steamer *Oldenburg* in Genoa. As the ship entered the bay of Naples on 5 September, Haeckel's thoughts traveled back to 1859, when he roamed the island of Capri with Allmers. The Neapolitan melodies that drifted over from the island "made my heart heavy in thought of the loved ones left at home to whom I said good-by for nine months."[47] The ship passed through the Suez Canal on 9 and 10 September, and then took twelve more days to reach Ceylon. While gazing out on the Indian Ocean from the ship's rail, Haeckel made many observations about the abundant life of the sea: myriads of siphonophores floating just below the glasslike surface of the water and squalls of medusae and jellyfish. But here, too, his ruminations took him back to Frida and their plight. He repeated to himself the couplet: "Resignation, the most sere word of release, / Only this opens for us the gates of peace."[48] On 2 September the ship sailed into the harbor at Colombo, which rekindled memories of his earlier travels to Ceylon. After a brief visit, the ship sailed for five more days, passing through the Straits of Malaka to Singapore, where it dropped anchor for several days. While in Singapore, Haeckel spent his time in the Raffles Museum and Garden, where he examined the exotic

46. Binet initiated a correspondence with Haeckel in 1899, when he indicated to him that he had read the *Challenger* volumes on radiolarians; he was especially interested in their artistic features, their "architectural and ornamental" qualities. See Binet to Haeckel (21 March 1899), in the Haeckel Correspondence, Haeckel-Haus, Jena. As his *Kunstformen der Natur* was published in fascicles, Haeckel would send Binet copies. Robert Proctor illuminates the relationship between Binet's architectural designs and Haeckel's biological depictions in his "Architecture from the Cell-Soul: René Binet and Ernst Haeckel," *Journal of Architecture* 11 (2006): 407–24.

47. Haeckel, *Aus Insulinde,* 14–15.

48. Ibid., 22: "Resignation, dies herbste aller Worte, / Eröffnet uns allein des Friedens Pforte!"

Fig. 10.3. René Binet's Porte Monumentale at the Paris Exhibition of 1900. (From the author's collection.)

plant life and took to strolling along with a young orangutan from the zoo.[49] Finally, on 13 October, he loaded his fourteen cases onto the steamer *Stettin* and made for Java in the Dutch East Indies, arriving in the harbor of the principle city, Batavia (now Jakarta), on 15 October.

Not fifteen minutes after he disembarked in the port city, Haeckel had his pocket picked. He lost a wallet that Frida had made for him, as well as his passport.[50] He left those troublesome environs rather quickly, traveling some fifty miles outside of the city to the gardens of Buitenzorg ("without worry," now Bogor), where he was hosted by the director of the Botanical Institute, Melchior Treub (1851–1910). Haeckel spent two and a half months at the institute, his stay prolonged by the aggravation of an old knee injury compounded by arthritis in the joint. Despite his generally vigorous health, this kind of travel adventure proved arduous for a man of sixty-six years.

Haeckel's convalescence offered opportunity to study the exotic plant life of the gardens—including fossil plants—and various organisms brought to him by neighboring children, invertebrates such as horseshoe crabs (which he regarded as the living descendents of trilobites) and a slew of lower vertebrates. He had become convinced by his experience of Dar-

49. Sir Thomas Stamford Raffles (1781–1826) had bought the island of Singapore from the sultan of Johor and established there a natural history museum and botanical garden. Later he founded the Zoological Society of London.

50. Haeckel to Frida (21 October 1900), in *Das ungelöste Welträtsel*, 2:573.

win's English garden that one had to investigate organisms in light of their "ecology" (*Oekologie*), that is, "the relation of plants and animals to their environment."[51] One particular feature of the environment that seemed quite significant was the climate—Java hardly had any seasons, only a kind of endless summer. That gave some promise that the phylogenetic history of many of the region's plants might be read off their individual development, since adaptations to the seasons seemed not to be a factor.[52] He also had the leisure to investigate various embryos—fish, amphibians, reptiles, and mammals—which he sealed up in tubes for the return.[53] These specimens likely led him to the further comparative displays of the biogenetic law in the fifth (1903) and sixth (1910) editions of his *Anthropogenie* (see conclusion to the previous chapter).

While at Buitenzorg and in the highlands of Java, Haeckel both painted in oils (see plate 8) and took photographs of local scenes, particularly of native groups. This set him to considering the comparative advantages of the painterly eye over the mechanical eye for rendering the true character of the vegetation that lay in the complex weave of the tropical forest:

> In the colorful confusion produced by the mass of tangled plants, the eye vainly seeks a resting place. Either the light is reduced and distorts the thousands of crisscrossed branches, twigs, and leaf surfaces—themselves covered with a chaos of epiphytes—or the light of the overhead sun shines brightly through the gaps of the tree crowns and produces on the mirrored surface of the leather-like leaves thousands of glancing reflections and harsh lights, which allow no unified impression to be gathered. In the depths of the primitive forest, the various complexes of light are extraordinary and cannot be simply reproduced by means of photographs. . . . Only the carefully wrought sketch can bring out the true character of the primeval forest. . . . A good landscape painter—especially when he possesses botanical knowledge, is able in a larger oil painting to place before the eye of the viewer the fantastic, magical world of the primeval forest in a realistic way.[54]

For the representation of plants—and animals rapidly passing through increasingly complex stages of development—the steady painterly eye of the

51. Haeckel, *Aus Insulinde,* 75.
52. Ibid., 77–80.
53. Ibid., 92.
54. Ibid., 106–8.

artist-scientist captured a truer, more precise rendering of living organisms than the shuddering, light-perplexed eye of the camera. Haeckel, like all competent illustrators, recognized that in photographs lighting posed serious problems—natural light and shadow might obscure some structures and distort others. The botanical or anatomical illustrator, by contrast, is able to manipulate light and produce shadings impossible in a natural setting, so as to render structures as they "really" are.[55] (See the previous chapter for further considerations about the contrast of photography to illustration.)

In mid-January 1901 Haeckel traveled through the south-central part of Java, mostly by train. His excursion convinced him that the island was the most beautiful he had ever visited. But it was not only the scenes of exotic plants and animals that captured his attention; his naturalist's eye also alighted on the peoples of the region and their customs and habits. He became completely enamored of the colorful life of cities, like Djokja, in the mid-part of Java. Here the camera could be used to best advantage; and he filled the travel book that came out of this trip, his *Aus Insulinde: Malayische Reisebriefe* (From the Islands: Malayan travel letters, 1901), with photographs of villagers and townspeople in their bright costumes and in their bare-breasted beauty.

Haeckel returned to Batavia at the end of January to gather his materials. On 23 January he boarded the steamer *Princess Amalia,* sailed past Krakatau (which had exploded in a mighty eruption in 1883, producing the loudest sound ever experienced by human beings), and landed in the harbor of Padang, Sumatra, two days later. He was shown hospitality by the chief engineer of the Dutch rail system on the island, a Mr. Deiprat. Shortly after arriving, he again injured his left knee, which laid him up for some four weeks. During that time he amused himself by giving biological instruction to Deiprat's two daughters, one fourteen and the other sixteen years old. He used his *Natürliche Schöpfungsgeschichte* and *Kunstformen* as his texts.[56] Despite an immobile convalescence, he did get a bit of work done; he had native divers at his disposal, who furnished him specimens from the waters around Padang. And the German consul on the island supplied him with apes, large land turtles, and reptiles for his study.

55. I have discussed these problems of light and shadow with Alta Buden, anatomical illustrator, who pointed out that in her renderings of structures, the shadings and bright areas could never occur in nature—or even under artificial light—though the illustrations were designed to prove true to nature. Helmholtz considered these and other problems that painters faced in rendering scenes true to life. See Hermann von Helmholtz, "Optisches über Malerei," in *Vorträge und Reden,* 3 vols. (Braunschweig: Friedrich Vieweg und Sohn, 1884), 2:97–137.

56. Haeckel, *Aus Insulinde,* 184–85.

Fig. 10.4. Discomedusa *Rhopilema Frida* (*center*): "This magnificent new species of the genus *Rhopilema,* one of the most beautiful of the medusae, was captured on 10 March 1901 under the equator in the Malaccan Straits. It bears its name as a remembrance of Fräulein Frida von Uslar-Gleichen, the artistic friend of nature, who has advanced the 'Kunstformen der Natur' in numerous ways by her exquisite judgment." (From Haeckel, *Kunstformen der Natur,* 1904.)

When his knee healed sufficiently, Haeckel spent the last two weeks on the island traveling by train to various towns and villages. Again, the sociological and anthropological features of the population continually stimulated his interest. He noted, for instance, that the Islamic religion's usual restrictions on women had to accommodate the matriarchal structure of

Fig. 10.5. Hamburg German-American liner *Kiautschou* (launched 1900); renamed *Prinzess Alice* (1904); seized by Americans during World War I (1917) and renamed *Princess Matoika;* the ship in 1917. (Courtesy of Navel Historical Center, Department of the U.S. Navy.)

social life in the Sumatran villages: women were not sequestered at home; they were not compelled to hide behind the veil; and divorce was rather easy for both sexes.[57] Haeckel thought Sumatran matriarchy produced in the people a more forceful, independent nature than could be found among the Javanese. That prideful spirit, he observed, caused the Dutch colonial power many more difficulties than they encountered elsewhere. The Dutch, in their turn, seemed to have absorbed many local traits, including wild superstitions. Haeckel was rather astonished that quite well-educated planters and businessmen might try to convince him that, for instance, putting a pearl in a sack of rice and burying the sack would yield several small pearls as offspring. The Dutch also yielded to stories of ghosts and sprits. Haeckel concluded that one should not be surprised that native peoples harbored strange convictions since more civilized individuals could be brought to comparable credulity.[58] While he did not think the Malayan people had achieved much by way of civilization—at least compared to the other branches of the Mongolian family (e.g., Japanese and Chinese)—he concluded that their customs and ways of life ought not be suppressed by the occupying colonial powers nor should the natives be subjected to missionary efforts at conversion.[59] The same kind of intricate ecological relationships Haeckel found

57. Ibid., 242–43.
58. Ibid., 210–11.
59. Ibid., 241.

amidst plants and animals, he also discovered characterizing the peoples of these tropical islands, both the native and the European.

On 5 March 1901 Haeckel boarded the Dutch steamer *Soembing* with all his specimen crates, diary notes, sketchpads, and canvases. And after transfer to the 2,000-passenger Hamburg-American luxury ship *Kiautschou*—outfitted with electric lights, fans, and refrigeration for good Munich beer—he enjoyed a very pleasant journey back to Genoa, where he disembarked on 2 April. He could not quite bring himself to rush back to Jena. With the excuse of a hobbled knee, he stopped at Baden-Baden for three weeks to take the waters and decompress. He finally returned home on 28 April. Though the whole eight-month excursion really yielded little by way of scientific results, he did accomplish his primary mission, which was to avoid spending a thoroughly miserable winter at home.

Growth in Love and Despair

Haeckel dedicated his travel book, *Aus Insulinde,* to "his true life's partner [*Lebensgefährtin*], Frau Agnes Haeckel"; but during the trip he kept in constant communication with his "true bride," Frida von Uslar-Gleichen. As soon as he arrived in Italy, he wrote Frida, begging her to join him at Baden-Baden; since, after all, she too had injured her knee. She responded that because she was neither too old nor too ugly, no one would believe their friendship was only platonic.[60] And she herself certainly did not believe that they could be merely spiritual friends: "The reunion you desire is not to exchange spiritual thoughts unhampered by distance but only to have me physically, to kiss and embrace me, and I'm too weak and love you too much to deny you this poor consolation since I know how much you hunger for it."[61] Though she expressed irritation at his presumption, she nonetheless finally yielded to the plan for another meeting. They rendezvoused at a clinic near Göttingen, where she was receiving some therapy for her own knee; and on 26 April, they traveled ten miles to Münden, where they took adjoining rooms at the Hotel Tivoli and enjoyed two days that, as Haeckel later wrote her, "filled my heart with new joy and will be forever unforgettable."[62]

During the course of their relationship, Frida had pressed to meet Haeckel's beautiful daughter Elisabeth and his son Walter, and Agnes as well.

60. Frida to Haeckel (4 April 1901), in *Das ungelöste Welträtsel,* 2:618.
61. Frida to Haeckel (14 April 1901), in ibid., 2:626.
62. Haeckel to Frida (27 April 1901), in ibid., 2:639.

Fig. 10.6. Haeckel's letter to Frida (Jena, 12 December 1901): "To Charlotte from the Wartburg: 'If I were a little bird / And thus had two wings, / I would fly to you! / But that cannot be / And I must be alone, / I remain alone.'" Here Haeckel recites a verse from a traditional folk song. (Photo courtesy of Preussischer Kulturbesitz, Berlin.)

Haeckel dodged these requests because, as he would weakly protest, bourgeois morals and a superstitious church would not understand their relationship.[63] Frida quietly responded that she believed Haeckel to be ashamed of her. But then on one of his trips in October 1902, he picked up Frida in Leipzig, and they went to visit Elisabeth and her family.[64] Quite obviously, both his daughter and—as the publication of the correspondence would later show—his son had considerable sympathy for their father's plight.

As Haeckel became more aware of how precarious were the finances of Frida and her mother and the other children in her family, he would periodically send money, usually about 300 marks at a time for various domestic expenses. Then as Frida required increasing medical attention, he made ever more generous contributions. So, for instance, in June 1901 he paid for

63. Haeckel to Frida (17 July 1902), in ibid., 2:777–78.
64. Frida, diary entry for mid-October, in ibid., 2:795–97.

a five-week stay for Frida at Baden-Baden, where she sought therapy for her knee. He took care of the costs by forwarding to her half the 2,000-mark prize money he received with the Darwin Medal from the Royal Society of London.[65] And when she entered a sanatorium in 1903, Haeckel paid for most of her expenses. The money seemed to him demanded by his love and by the solace Frida provided, as he made so clear in a letter of 4 August 1901:

> My dear best Frida!
>
> What you have been to me during these last two years! How can I thank you? When I think about the small miseries of my daily domestic life, when I think about the meaningless complaints about the smallest things with which this poor being [Agnes] wastes her time, when I think about the trivial attacks daily made by my many opponents that drag my best work through the mud—then your wonderful form arises before me, offering consolation and encouragement—in reality, *a being of a higher kind* that lifts me up to a purer, nobler height! "The eternal feminine draws us on."[66]

Unbeknownst to Agnes, Haeckel set up a reserve fund of 30,000 marks for Frida, to be executed after his death by his nephew Heinrich.[67] The fund, however, would ultimately prove unnecessary.

Haeckel did not see how his relationship with Frida could survive, especially since she was reluctant to step too far over the decaying boundaries of a bourgeois marriage and he remained restrained therein by duty and guilt. In the somber intimacy of their letters, they frequently spoke of suicide, conversations that fed on Frida's declining health and Haeckel's miserable home situation, as well as on the personal attacks unleashed by his *Welträthsel.* She requested that Haeckel, a medical doctor, send her morphine, "as strong as possible," for the occasion in which she would need "self-help."[68] He quite reluctantly sent her some on several occasions, but just enough to ameliorate the pain in her joints and her debilitating menstrual cramps.[69] Finally, however, he stopped sending the prescriptions altogether in fear of what she might do. During the summer of 1903, she began to suffer cardiac symptoms—shortness of breath, angina, exhaustion—

65. Haeckel to Frida (18 June 1901), in ibid., 2:649–50.
66. Haeckel to Frida (4 August 1901), in ibid., 2:688.
67. Frida to Haeckel (13 December 1903), in ibid., 2:930.
68. Frida to Haeckel (mid-February 1902), in ibid., 2:742.
69. Frida to Haeckel (mid-June 1902), in ibid., 2:764.

and again pressured him to prescribe enough morphine so that, should it come to it, she could put an end to her suffering:

> My love and trust will be in you. I ask with all my heart for morphine to release me. The thought won't leave me that I will at some point choose this easy death. So long as your lovely eyes stand watch, you have nothing to fear from me. But for how long will you still be my true protector?[70]

Haeckel did send her sufficient morphine for that final occasion but extracted from her a promise that she would not use it before his own death, and then only "in the most extreme necessity."[71]

Frida's heart problems continued to worsen. During the fall of 1903, she became exhausted with even the smallest exertion. In mid-December she pleaded for more morphine in a letter that Haeckel's wife intercepted. Agnes became quite upset, both because of the request for morphine and because Frida had used the familiar form of "Du" when addressing her husband.[72] Then on 15 December 1903, Haeckel received a letter from Luise von Uslar-Gleichen, Frida's sister, who knew of their relationship. She wrote that Frida had been sick for the last few days and then suffered an "accident" from taking too much morphine:

> Then yesterday afternoon, she took to bed, though that evening, as I prepared her for bed, she spoke of domestic matters and was very quiet. We didn't know that she had taken several pills of codeine [prescribed by her doctor] and also wine, which her doctor had expressly warned against. Later she became very restless. Mother was with her, and she then got up and put on her ring [which Haeckel had given her]. About the middle of the night, she came suddenly up the stairs to my sister and me and said she felt very ill. She thought she would die. I immediately laid her in my bed, which was very difficult; she had collapsed and her breathing was very bad, and after a short time she passed away.[73]

A few days later Luise asked Haeckel to keep from her mother that Frida had taken her own life.[74] Frida's death, though not a surprise, seemed like a reprise of the earlier tragedy of Anna's last hours. He scored the event

70. Frida to Haeckel (28 August, 1903), in ibid., 2:877.
71. Haeckel to Frida (28 August 1903), in ibid., 2:878.
72. Haeckel mentioned this in a letter to Frida (14 December 1903), in ibid., 2:927.
73. Luise von Uslar-Gleichen to Haeckel (15 December 1903), in ibid., 3:953–54.
74. Luise to Haeckel (18 December 1903), in ibid., 3:954–55.

Fig. 10.7. Frida von Uslar-Gleichen, with leaves from her graveside pressed against her photo. (Courtesy of Ernst-Haeckel-Haus, Jena.)

into his memory with delicate leaves, taken from her grave site, which he pressed against her picture (fig. 10.7).

For the six years of their relationship, Frida had brought him back into the living presence of Anna. She had discussed with him in intimate and warm detail those matters of art and science so close to his own vital concerns—and so far from the concerns of Agnes. Frida had enlivened the pulse of love that had been missing from his life. He wanted to commemorate his spiritual rebirth in a way that would represent the transcendent nature of her significance. He had in mind an artistic-scientific tableaux that would crown his *Kunstformen der Natur,* the last installment of which was being readied at the time that Frida's health became ever more precarious. He had planned the hundredth and final illustration to be that of the eternal feminine, the ideal of Eve and Mary, of Helen and Aspasia—and of

Fig. 10.8. "Apotheosis of Evolutionary Thought," illustration by Ernst Haeckel and Gabriel von Max. (From Supplement to Haeckel, *Wanderbilder,* 1905.)

Frida. He sketched out a scene that would portray the ideal of the female surrounded by a company of anthropoid apes, the forms of which would indicate human phylogenetic history. Since he did not think he could do artistic justice to the plan, he asked his friend the Munich painter Gabriel von Max to execute the final version. His publisher, however, balked, thinking that it would cause too much of a scandal.[75] To the publisher's refusal, Haeckel reacted with dismissive pique: as the hundredth illustration in the *Kunstformen*, he inserted a depiction of a community of antelope! He would, however, seek another venue for the symbol of his transformed passion, namely, in a supplement to his *Wanderbilder* (1905).

In the supplement, a slim addition to the photography and paintings that came from his trip to Indonesia, he aesthetized his evolutionary convictions about human beings and, it would seem, about one individual in particular. He maintained that through millennia of sexual selection within a race, males and females co-formed one another according to their respective standards of beauty. In the tableaux included in the supplement, he sought to display "the apotheosis of evolutionary thought, the human female as the most perfect 'art form of nature'" (fig. 10.8).[76] For Haeckel, Frida had been lifted into a symbol of his science and his art. Frida had been given the only kind of immortality he could imagine.

Lear on the Heath

From 1899 through the next decade, religious opponents constantly attacked the redoubtable author of *Die Welträthsel* and the titular leader of the Monist League.[77] And the old lion struck back, supported by faithful students, old friends, and biologists of considerable intellectual mettle. Shortly after Frida's death, he also received a momentary lift from that freethinking cultural spirit and modern Terpsichore, Isadora Duncan

75. Frida mentioned this in a letter of consolation, though she agreed that the illustration would cause a scandal. See Frida to Haeckel (27 September 1903), in ibid., 2:905.

76. Ernst Haeckel, "Apotheose des Entwickelungs-Gedanken nach Ernst Haeckel und Gabriel Max," erstes Beiheft zu der Sammlung Wanderbilder, bound with *Wanderbilder: Nach eigenen Aquarellen und Oelgemälden* (Gera-Untermhaus: W. Koehler, 1905), plate 1. Haeckel was clear that the European, the Mongolian, the Dravidian, etc., each had specific standards of physical beauty. He aimed only to represent the European female.

77. The variety and extent of the negative reactions to Haeckel's *Welträthsel*, just during the first year of its publication, are surveyed by his assistant Heinrich Schmidt in *Der Kampf um die "Welträtsel": Ernst Haeckel, die "Welträtsel" und die Kritik* (Bonn: Emil Strauss, 1900).

(1877–1927), who wore as native dress Yeats's line about fine women eating a crazy salad with their meat.

Duncan was performing in Berlin when she had the theater manager write to inform Haeckel that she would celebrate his seventieth birthday with a dance that would be based on "Darwin and Haeckel." She prompted the manager to tell her new master that she initially had no religion, but "that when she read your *Natürliche Schöpfungsgeschichte* she had religion and was happy."[78] After some exchange of correspondence, Duncan invited Haeckel to Bayreuth, where she would be staying that summer.[79] She met him at the train station and "was immediately enfolded in his great arms, and found my face buried in his beard. His whole being gave forth a fine perfume of health and intelligence, if one can speak of the perfume of intelligence."[80] Cosima Wagner (1837–1930), however, did not find the perfume of the notorious scientist quite so alluring; she was aghast that he had come into the sanctuary dedicated to the art of her husband. Duncan took Wagner's disdain as her cue to make an entrance. At intermission during the performance of *Parsifal,* "before the astonished audience, I promenaded . . . in my Greek tunic, bare legs and bare feet, hand-in-hand with Ernst Haeckel, his white head towering above the multitude" (see fig. 10.9).[81] When Duncan later gave birth to a baby boy, she wrote her master to proclaim: "This Boy will be a Monist and who knows but some of your great and Beautiful Spirit may be in him."[82]

Despite the aid of friends and the glamorous touch offered by the hand of Isadora Duncan, the religious wars that Haeckel unleashed, and the death of Frida, along with the inevitable consequences of age—these psychic and physical strains eroded his powerful constitution. He spent most of the winter of 1908–9 confined to his quarters because of crippling arthritis. At Easter the seventy-five-year-old Darwinian stalwart officially tendered his resignation as professor of zoology and director of the Zoological Institute. He had spent almost a half century at Jena advancing the evolutionary cause through his research, teaching, and lecturing; but now he sought to remove himself from the most taxing arenas of struggle.

78. Otto Borngräber to Haeckel (9 February 1904), in the Haeckel Correspondence, Haeckel-Haus, Jena. Borngräber was also chair of the Richard Wagner Society for Dramatic Art and Culture. He had planned an elaborate celebration of Haeckel's birthday at his theater in Berlin.

79. Isadora Duncan to Haeckel (7 April 1904), in the Haeckel Correspondence, Haeckel-Haus, Jena.

80. Isadora Duncan, *My Life* (New York: Liveright, 1927), 112–13.

81. Ibid., 113.

82. Duncan to Haeckel (8 May 1910), in the Haeckel Correspondence, Haeckel-Haus, Jena.

Fig. 10.9. Isadora Duncan walking with Haeckel at Bayreuth: "This fugitive remembrance of Bayreuth, with best wishes for the new year." (Photo courtesy of Ernst-Haeckel-Haus, Jena.)

As a more permanent legacy of his work, Haeckel, just prior to his retirement, began planning a museum that would promote research and education in evolutionary theory. The museum would not only display exhibits of embryology, anthropology, comparative anatomy, and various specimens exemplifying phylogenetic relationships; it would also include a library, a manuscript room, and a gallery for his paintings and artifacts from his numerous excursions. He conceived of the museum as a "temple to the philosophy of nature."[83] For its construction, he organized a building fund, which quickly reached over 100,000 marks; about half the monies came from Haeckel himself, mostly from royalties on his *Welträthsel*. On 30 July 1908 the Phyletic Museum was dedicated, a modernist building displaying motifs of medusae, bivalves, and other sea creatures (fig. 10.10).

Lest his efforts either lose momentum or go astray, Haeckel also sought to have appointed a successor of like mind and comparable energies. He believed he found these qualities in Ludwig Plate, a former student and director of the Museum für Meereskunde in Berlin. Plate had authored a comprehensive text on Darwinism and had organized the critics' response

83. Georg Uschmann, *Geschichte der Zoologie und der zoologischen Anstalten in Jena 1779–1919* (Jena: Gustav Fischer, 1959), 173.

Fig. 10.10. Phyletic Museum, Jena. (Photo by the author.)

to Wasmann's brand of evolutionism (see the previous chapter).[84] Supported by Haeckel, the ducal court appointed Plate professor in the university as well as director of both the Zoological Institute and the new Phyletic Museum. In gratitude for Haeckel's support, Plate pledged that he would arrange three rooms in the upper level of the museum for his former teacher's use as archive, library, and workroom and that he would "direct the museum with you and according to your intention."[85] Despite such avowals, the appointment marked the beginning of another source of tribulation for the old man.

The troubles cascaded as respect and honor fell to bureaucracy and officiousness. Shortly after his appointment, Plate required that Haeckel clean out his workroom in the Zoological Institute and move everything to the space in the Phyletic Museum, and simultaneously he informed Haeckel that a room in the museum, which had been designed for assistants, would now be used to hold some eighty-four cages of mice used in heredity experiments. Haeckel objected that the dirt and smell would be intolerable

84. Ludwig Plate, *Selectionsprinzip und Probleme der Artbildung: Ein Handbuch des Darwinismus*, 3rd ed. (Leipzig: Wilhelm Engelmann, 1908).

85. Plate to Haeckel (9 January 1909), quoted by Adolf Heilborn, *Die Leartragödie Ernst Haeckels* (Hamburg: Hoffmann & Campe, 1920), 12.

and asked whether there were not some other place in Jena for the animals. Plate responded that "since 1 April, I am the sole director of the Phyletic Museum and you must obey unconditionally all my directives."[86]

As the relationship between the two rapidly deteriorated, Haeckel sought to confirm an understanding about the disposition of the three rooms in the upper floor of the museum. He stipulated that while the museum would be run by the director of the Zoological Institute (Plate), he (Haeckel) would retain use of the three rooms during his lifetime. Though Haeckel initially sought to have control of the rooms pass to his son and son-in-law after his death, he dropped that condition. Plate acknowledged the understanding but continued to put pressure on the increasingly infirm older scientist. He had another set of keys made to the rooms under Haeckel's control, a move that the old man regarded as patently giving the lie to the understanding they had reached. Further, the director demanded the return to the institute's library of books that Haeckel held in his museum rooms and at home. Many of these books Haeckel himself had contributed to the library—books from his own private collection and books given to him by many authors; and a large portion of the institute's library he had purchased out of a fund over which he had proprietorship. It was not so much the demand that galled him, but the venomous suggestion that he had virtually stolen the books. Plate's contumelious assessment of his onetime teacher's character became explicit just after the latter's death, as judgments sympathetic to the renowned Darwinist came rolling in. In a suppurating defense, Plate declared:

> Haeckel was a crass materialist and atheist and had ridiculed Christianity in numerous ways. For that reason he was celebrated by the Social Democrats and the Jews as the world-famous light of true science. I, on the other hand, am an idealist, freethinking Christian, German populist, and anti-Semite.[87]

86. Haeckel described Plate's "brutal" remark to his friend and biographer Wilhelm Bölsche. See Haeckel to Bölsche (27 July 1909), in *Ernst Haeckel–Wilhelm Bölsche: Briefwechsel 1887–1919*, ed. Rosemarie Nöthlich (Berlin: Verlag für Wissenschaft und Bildung, 2002), 281.

87. Ludwig Plate, *Deutsch-völkische Monatshefte* (1921), 33, as quoted by Heinrich Schmidt, *Ernst Haeckel und sein Nachfolger Prof. Dr. Ludwig Plate* (Jena: Volksbuchhandlung, 1921), 19. Plate was later accused by student organizations of preaching anti-Semitism in the classroom and of slandering the Social Democratic Party. See George Levit and Uwe Hoßfeld, "The Forgotten 'Old-Darwinian' Synthesis: The Evolutionary Theory of Ludwig H. Plate (1862–1937)," *Internationale Zeitschrift für Geschichte und Ethik der Naturwissenschaft, Technik und Medizin* (NTM), n.s. 14 (2006): 9–25.

Fig. 10.11. Haeckel flanked by his granddaughter Else Meyer and his son Walter, in 1915. (Photo courtesy of Ernst-Haeckel-Haus, Jena.)

At the end of 1909, Haeckel finally decided to avoid what he regarded as the petty pomposity and the stinging ingratitude of his successor. He simply abandoned his rooms in the museum and confined his work to his own home. And with that, the ambit of his movements became ever more circumscribed. In the spring of 1911, he fell and broke his leg and had to remain confined to an even smaller geographical space—only imaginatively expanded by the many guests he continued to receive. He recovered some mobility, but until the end he had to use a crutch. After the death of his wife Agnes in the spring of 1915, his twenty-year-old granddaughter, Else Meyer, came to live with him; she helped in the running of the household and, to his delight, aided him in preparing manuscripts and correcting page proofs for new editions of earlier works (see fig. 10.11). Though this beautiful and intelligent woman brightened the last years of her grandfather's life, darkness descended on him whenever he passed the museum or the Zoological Institute. However, in the summer of 1918, the cloud of rejection dissipated when a plan was executed to transform his own house into a museum dedicated to evolutionary theory and to his role in establishing that doctrine. With the help of the Carl-Zeiss-Stiftung, Villa Medusa was to be purchased from the heirs and given to the university.[88] His many paintings, research materials, manuscripts, and letters would be preserved—as, indeed, they are today—in Haeckel-Haus, Friedrich-Schiller-Universität.

The Great War

In the city of Sarajevo on 28 June 1914, Gavrilo Princip (1894–1918), a Bosnian Serb, fired two shots into the car carrying the heir to the Austro-Hungarian throne, Archduke Franz Ferdinand (1863–1914), and his wife, Sophia (1868–1914). Both were mortally wounded and the assassination set off a catastrophic chain of events. During the previous several years, the great and not-so-great powers of Europe had been edging closer toward the abyss that became World War I, the Great War.[89]

The War, 1914–1918

The events that seemed to lead inexorably to war built slowly from the late 1890s to the outbreak two decades later. It was like a freight train picking up cars at various stations and an engineer in a drunken stupor increasing the speed with each new coupling. One car slipping the rails would cause the entire line to hurdle into oblivion.

The perceptible beginning of the tragedy occurred when the German military command convinced Emperor Wilhelm II that the country required a superior navy in order to achieve equality among the great powers of Europe. To that end, Alfred von Tirpitz (1849–1930), secretary of state of the Imperial Naval Office, persuaded the Reichstag in the late 1890s to increase by significant measure the size of the naval budget. The Brit-

88. Heilborn, *Leartragödie*, 54–55.

89. In characterizing the buildup to and events of World War I, I have relied principally on Gordon A, Craig, *Germany: 1866–1945* (Oxford: Oxford University Press, 1980); William Carr, *A History of Germany: 1815–1990*, 4th ed. (London: Edward Arnold, 1991); and John Keegan, *The First World War* (London: Hutchinson, 1998).

ish, who had by far the largest navy in the world, feared the possible consequences of German expansion, especially because the emperor had the most powerful standing army among the European states. Russia became wary of Germany's growing power in the East, as it made inroads in China and strengthened economic ties with the Ottoman Empire. Reich's Chancellor Bernard Prince von Bülow (1849–1929) counseled Wilhelm that he should maintain good relations with Britain and Russia while the fleet was being built up and calmly await "the future development of elemental events."[90]

From the period of the Napoleonic invasions to the Franco-Prussian War of 1870–71, relations between France and Germany had oscillated between active hostility and uneasy peace. When France thwarted Wilhelm's efforts to gain an economic foothold in Morocco, the German High Command began to plan for a preventive war with their neighbor. The time for action was ripe, since Russia, on the eastern flank, was weakened by a punitive struggle with Japan. Only Wilhelm's lack of resolve momentarily disengaged the throttle of the careening train. In October 1908 Austria-Hungary, Germany's partner in the empire, annexed the territory of Bosnia and Herzegovina, which it had administered since 1878. The move struck at the ambition of the Serbs, who believed the territory formed part of Greater Serbia—a territorial desire that still lingers. This meant that if Austria went to war with Serbia over Bosnia, Germany would be forced to confront the Serbian ally, Russia. In 1912, with international tensions rapidly increasing, Winston Churchill (1874–1965), first Lord of the British Admiralty, urged a naval-building holiday in order to give contending powers time to assess their true interests. The emperor and his new Reich's chancellor, Theobald von Bethmann-Hollweg (1856–1921), rebuffed the proposal; they supposed that the British were simply temporizing in view of Germany's new might. In reaction, Britain moved to strengthen its entente cordiale with France; and British prime minister Herbert Asquith (1852–1928) informed the Germans that his country would not remain idle if Austria attacked Serbia. Then in June 1914 Gavrilo Princip and seven co-conspirators, members of the anarchistic Young Bosnia movement, attempted revolution to extirpate the Austrian tyrant holding their land. The train hurtling down the tracks derailed in Sarajevo, and each car pulled the next into the dark void: Austria declared war on Serbia, Russia mobilized, and Germany preemptively attacked France through Belgium, which brought in the English to defend

90. Von Bülow as quoted by Craig, *Germany,* 311.

their ally and to face off in the Dardanelles against the Turks, who had joined the Central Powers.

All sides entertained the idylls of their kings, believing war would be a patriotic and glorious endeavor—and would be quickly concluded. Moreover, hot steel would cut through the knot of treacherously entangled relations, allowing the establishment of a natural hierarchy of states. The Germans in particular, fully confident of their superior power and military acumen, planned for a very short war. Both industrialists and intellectuals anticipated reparations in the form of annexed territories—Belgium, Luxembourg, Lithuania, Estonia, eastern Poland; these lands would receive the gift of German culture, of which they stood in sore need.

In early September 1914, the emperor's armies met with stiffer resistance from the French than had been expected. The French general Ferdinand Foch (1851–1929) repulsed an initially brutal attack along the Marne River, a hundred miles east of Paris. Reputedly Foch sent Commander in Chief Joseph Joffre (1852–1931) a note that read: "My right is driven in; my center is giving way, the situation is excellent, I shall attack."[91] Initial bravura would quickly drain away in the trenches. The Russians, urged by the French, invaded East Prussia and momentarily gained the upper hand, until, that is, generals Paul von Hindenburg (1847–1934) and Erich Ludendorff (1865–1937) were brought into the fray. These shrewd military strategists skillfully routed the ill-prepared Russians and drove them back across the border. But another contingent of the Russian army moved into Galicia and Serbia, crushing the Austrians, who suffered over 200,000 casualties. The new chief of the German General Staff, Erich von Falkenhayn (1861–1922), an aristocratic West Prussian, decided that the war had to be quickly ended, since the reserves of munitions and material resources were being exhausted at a rate not contemplated. He commanded the Fourth and Sixth armies to end the war with a decisive battle in Flanders against the British, French, and Belgian forces. The effort at Ypres failed miserably due to weather and Allied reinforcements, with the Germans losing some 130,000 men in a few weeks' time. Meanwhile on the eastern front, the Austrian and German armies crushed the Russians, who suffered 300,000 casualties.

In the forests of northeastern France and in Flanders Fields, trench warfare broke out with spectacular ferocity. The very names of the major battles—Verdun, the Somme, Passchendaele—ring with horror and un-

91. Quoted by S. L. A. Marshall, *World War I* (New York: Houghton Mifflin, 2001), 91.

Fig. 10.12. French trenches at Verdun, 1914.
(Courtesy of Heritage of the Great War, the Netherlands.)

remitting sadness. Falkenhayn thought his armies at Verdun could bleed white Joffre's forces. Instead the million-man German army ground to a halt despite some ten months of brutal bombardment of the Allies and the introduction of their new weapon, phosgene gas—colorless and hard to detect, it dissolved lung tissue into a frothy slime. The trenches on both sides ran with maroon ooze, composed of mud, waste, and blood. From the end of February to the middle of December 1916, when the Germans withdrew from Verdun, the combined armies lost a total of over a million men, roughly in equal numbers. The halt of the German advance resulted partly from the necessity of shifting forces both to the east, because of a new Russian front, and to the area along the river Somme in northeastern France, where the British began a new offensive in July under their commander in chief, Sir Douglas Haig (1861–1928).

Haig's initial bombardment of German positions was fierce but with little strategic consequence, merely producing mists of blood over the trenches. When the British Expeditionary Force went over the top into no-man's-land, company after company danced a quick jig of death played by German machine guns. On the first day of the Battle of the Somme, 1 July 1916, the British lost some 58,000 men—the most casualties of any single day of combat before or since. The various engagements along the Somme led to wholesale slaughter, while moving the lines only a few feet.

In September the British rolled out a new weapon, the tank, which caught the Germans by surprise and initially pushed them back. But tanks were unwieldy; and when the advance was halted in November because of driving snow, the Allies had gained less than ten miles and had lost a half million men, with comparable losses suffered by the Germans.

Part of the German strategy at the outbreak of hostilities was to use its small fleet of U-boats to disrupt British shipping, even targeting passenger ships, most famously the Cunard liner *Lusitania,* which went down in May 1915 with the loss of 1,198 lives, including 128 United States citizens. To lift the siege of the sea lanes, Haig sought to break through the German lines in Flanders and attack U-boat bases along the Belgium coast. The third battle of Ypres—known by the British as the Battle of Passchendaele, because of the small village in the vicinity—began 31 July 1917 and continued through November. Heavy rains at the beginning of August stopped the English advance, mired tanks in mud, and flooded the trenches. Siegfried Sassoon knew the worst:

> The place was rotten with dead; green clumsy legs
> And trunks, face downward, in the sucking mud,
> Wallowed like trodden sand-bags loosely filled;
> Bulged, clotted heads slept in the plastering slime,
> And then the rains began.[92]

During the previous March, the Bolshevik Revolution caused the Russian armies to pull back from battle, thus allowing German forces from the eastern front to pour into Belgium. With reinforcements against him mounting, Haig called off the attack at Passchendaele. British casualties in the fruitless effort climbed to over 300,000.

With an apparent stalemate in the offing and the disastrous cost of the war in terms of lives and economic resources, the Reichstag, on 19 July 1917, passed a nonbinding peace resolution, suggesting that if all parties gave up territorial claims and kept the sea lanes free, then the German people would negotiate a peace. The resolution infuriated the German High Command and brought the downfall of Prime Minister Bethmann-Hollweg, who had rightly predicted that German U-boat attacks on American supply ships and passenger vessels would pull the United States into the war.

After the sinking of the *Lusitania* and the revelation that Germany had

92. Siegfried Sassoon, from his "Counter-Attack," in *The War Poems* (London: Faber and Faber, 1983), 105.

Fig. 10.13. German postcard with "Argonnerwald-Lied" by Hermann Albert von Gordon. (Courtesy of Archives Universität Osnabrück.)

encouraged Mexico to attack the United States if it should enter the conflict, President Woodrow Wilson (1856–1924) asked Congress for a declaration of war on 2 April 1917 and received it four days later. American troops came slowly but steadily into battle. Under General John "Black Jack" Pershing (1860–1948), the American Expeditionary Force pushed into the French Argonne Forest, driving the Germans back. Hermann Albert von Gordon, a minor German poet, prophesied correctly: "Argonne Wood, Argonne Wood, soon, / You will be a quiet tomb."[93] Inspired by the American success, the other Allies intensified their attacks, and the Central Powers began to collapse. On 9 November 1918 Emperor Wilhelm abdicated, and two days later the German High Command sued for peace.

The total casualties on all sides amounted to over eight and a half million men dead and twenty-one million wounded, with Germany alone suffering over a million and a half soldiers dead and four million severely wounded. Such numbers gave the lie to the oft-quoted Horatian epigram: "dulce et decorum est, pro patria mori." The surrender terms of the Versailles Treaty (7 May 1919) left Germany shorn of its colonies and stripped of almost 15 percent of its territory. War reparations crippled the German

93. "Argonnerwald, Argonnerwald, / Ein stiller Friedhof wirst du bald." Hermann von Gordon, "Argonnerwald-Lied," ballad inscribed on a German postcard, 1915. See fig. 10.13.

economy for the next decade, leading to political instability and eventually to the rise of the Nazi Party.

Haeckel during the Time of the Great War

In February 1914, on the eve of the European disaster, Haeckel experienced a burst of happiness. On the occasion of his eightieth birthday, his former students and colleagues invited him to Leipzig to celebrate his work and accomplishments. He was presented with a two-volume festschrift, *Was Wir Ernst Haeckel Verdanken* (What we owe to Ernst Haeckel, 1914), that contained reminiscences and tributes from over 120 contributors, including such individuals as Wilhelm Ostwald, Richard Semon, Auguste Forel, Carl Rabl, Paul Kammerer, Jacques Loeb, Richard Hertwig, Max Verworn, and Robert Lowie. Social reformers also saluted Haeckel, notably Magnus Hirschfeld and Helene Stöcker, both of whom worked to change attitudes about sexual relations (especially about homosexuality).[94] His daughter Elisabeth, his son, Walter, his nephew Heinrich, and his grandchildren were all present. His wife Agnes remained in Jena, though she, too, sent him congratulations. She did, however, add that "all the newspapers give an account of your life and even mention the most intimate family affairs, but of your second wife, the daughter of an excellent scholar and one who shared with you almost fifty years of your combative life, there is not one word, and she belongs truly to your life's course."[95] But she never really wished to share what was central to his life's course; rather, she seemed to detest it.

In his later years, Haeckel did achieve a kind of reconciliation with his wife, apparently abetted by his own experience as an invalid. On 21 April 1915, she passed away at age seventy-two. A year later Haeckel wrote to her nephew Konrad Huschke a note indicating his mutedly mixed feelings:

> This morning it is now already a year since my faithful wife has been freed from her years of suffering by a gentle death. After forty-eight

94. Both Hirschfeld and Stöcker indicated that Haeckel's conceptions were central to their own concerns. Hirschfeld made the personal acquaintance of Haeckel when the latter invited him to Villa Medusa; and Haeckel had, in 1909, attended Stöcker's lecture on the reform of marriage. Both of these individuals were persecuted by the Nazis. See Heinrich Schmidt, ed., *Was Wir Ernst Haeckel Verdanken: Ein Buch der Verehrung und Dankbarkeit*, 2 vols. (Leipzig: Unesma, 1914), 2:282–84, 325–28.

95. Agnes Haeckel to Ernst Haeckel (15 February 1914), in *Ernst und Agnes Haeckel: Ein Briefwechsel*, ed. Konrad Huschke (Jena: Urania, 1950), 213.

> years of a happy marriage, I feel her loss daily, so much that I must always console myself with the thought that with her delicate health we could no longer expect a lasting recovery. I am also consoled that she would be spared the further experience of the awful losses of this terrible world war. She suffered greatly both in sympathetic response for the unspeakable sacrifices that this incessant human murdering has cost and in the worry about the dark future that lies before us.[96]

Haeckel shared his wife's anxieties—though, in a greatly intensified form as he quite early perceived the disaster lowering on the horizon. When the apocalyptic hoofbeats grew louder, he took action. He joined with the French socialist and educational reformer Henriette Meyer in founding a league for international peace: the French-German Institute of Reconciliation to Prepare for a Perpetual Peace between Our Two Countries. The first meeting of the league was held in August 1913; and the cofounders issued a journal—*La Réconciliation*—in an effort to appeal to the remaining fragment of rational individuals who might work for a reduction of tensions and a peaceful resolution of territorial disputes.[97] In the lead article ("Vernunft und Krieg"—Reason and war) of the first number (October 1913), Haeckel identified trouble spots in the Balkans and China, and especially the stockpiling of armaments, as creating a momentum for war that might not be stoppable. He condemned the pathological chauvinism gripping France, Britain, and Germany. He ended with the urgent injunction: "Pacifism is a duty of humanity." The journal lasted less than a year.

After the war broke out in the summer of 1914, Haeckel cast his lot with his own nation, a nation he initially thought no different in its bellicose enthusiasms than members of the Triple Entente—Britain, France, and Russia. After Germany preemptively invaded Belgium to attack France, Britain—to the shock of many German intellectuals—declared war on the invader. The consensus quickly formed that England took the occasion of a two-front war to stanch Germany's growing economic and intellectual

96. Haeckel to Huschke, in ibid., 215. Haeckel expressed the same sentiments to his friend and biographer Wilhelm Bölsche: see Haeckel to Bölsche (30 April 1915), in *Ernst Haeckel–Wilhelm Bölsche: Briefwechsel 1887–1919*, 261.

97. Haeckel had first been contacted by Henriette Meyer when she was secretary of the International League for Rational Education in France in 1908. In spring 1913 they began their efforts to form L'Institut Franco-Allemand de la Réconciliation. As the planning reached its critical phase, she wrote Haeckel in the spring to herald the new, last chance: "The hour of la Réconciliation has arrived." She declared their movement to be "for the triumph of civilization." See Meyer to Haeckel (12 June 1913), in the Haeckel Correspondence, Haeckel-Haus, Jena. See also Ernst Haeckel, "Vernunft und Krieg," *La Réconciliation* 1 (October 1913): 1–10.

Fig. 10.14. Haeckel in his study, 1914. (Courtesy of Ernst-Haeckel-Haus, Jena.)

influence. In August Haeckel composed a tract—*Englands Blutschuld am Weltkriege* (England's blood guilt for world war, 1914)—that repeated the emperor's explanation for the war: that members of the Triple Entente had grown insanely jealous of Germany's economic success, its loyal citizens, and its growing might. Who was chiefly responsible for the war? Not Czar Nicholas, who was but a spineless instrument in the hands of his officers; not the French people, who wanted peace while their leaders wanted revenge for earlier defeats; but "Perfidious Albion," who could not stand any restraint on the spread of British rule over the entire world. Despite his patriotic cant, at the end of his tract Haeckel confessed that

> it is with bleeding heart, and only because of the compulsion of my patriotic feeling that I, an eighty-year-old German citizen, have composed this complaint against brother England. For more than sixty years I have belonged to the group of those scholars who have held the mighty cultural work of Great Britain in the highest respect.[98]

98. Ernst Haeckel, *Englands Blutschuld am Weltkriege* (1914), reprinted in *Ernst Haeckel: Sein Leben, Denken und Wirken. Eine Schriftenfolge für seine zahlreichen Freunde und Anhänger*, ed. Victor Franz, 2 vols. (Jena: Wilhelm Gronau und W. Agricola, 1943–44), 1:81.

Haeckel then went on to recount his intimate ties of old with the scientists and intellectuals of England. He naively supposed that were those scientists of his acquaintance still alive, they would condemn the malicious aggression of their own nation. He was at a loss to explain the perfidy of the British in what he perceived as its aggression against Germany. He speculated that the pathology of the English might be due to their isolation on an island: "The same influence of geographical isolation that affects island selection is that which separates the British island realm from the neighboring continent and promotes its peculiar nationalistic egoism."[99] Not very likely, of course. But his stretch for an explanation suggests his conflict not only over England's role in the war but Germany's as well.

Haeckel's defensiveness over his nation's actions reflected that of the scholars and writers who signed the declaration printed in Germany's leading newspapers in October 1914, the call "An die Kulturwelt!" (To the cultural world).[100] In a repetitive mantra, the document asserted "it is not true": that Germany instigated the war, that it injured the neutrality of Belgium, that its soldiers pillaged the cities of Belgium and murdered its citizens, that they brutalized the city of Louvain or violated "the laws of human rights." Nor was it true that German militarism had subsumed German culture; on the contrary, the German military, the declaration asserted, had preserved the treasure of German culture. The proclamation was signed by ninety-three intellectuals, writers, and scientists, including the physicists Wilhelm Conrad Röntgen and Max Planck; the chemists Wilhelm Ostwald, Fritz Haber, and Walther Nernst; the mathematician Felix Klein; the philosophers Rudolf Eucken and Wilhelm Windelband; the writer Gerhart Hauptmann; the musician Engelbert Humperdinck—and the biologist Ernst Haeckel.

During the course of the war, Haeckel had some twelve nephews and grandnephews on the field of battle. Within a year half of them were dead or severely wounded.[101] The losses at the University of Jena were comparably staggering. By the autumn of 1917, some three hundred onetime students had fallen.[102] What could be the sense of this slaughter? What good might come out of it? Haeckel tried to answer these questions—questions

99. Ernst Haeckel, "England als Feind," *Deutsche Montags-Zeitung,* 2 November 1914, 2.

100 The document was drafted by writer Ludwig Fulda. See "An die Kulturwelt!" in *Aufrufe und Reden deutscher Professoren im Ersten Weltkrieg,* ed. Klaus Bohme (Stuttgart: Reclam, 1975), 47–49.

101. Haeckel to Max Fürbringer (1915), in *Ernst Haeckel: Biographie in Briefen,* 296.

102. Haeckel to Richard Hertwig (19 September 1917), in *Ernst Haeckel: Sein Leben, Denken und Wirken,* 1:64.

urged upon him by many correspondents sympathetic to monism—in a little book he composed in the early fall of 1915. It was entitled *Ewigkeit: Weltkriegsgedanken über Leben und Tod, Religion und Entwicklungslehre* (Eternity: thoughts on life and death, religion and evolutionary doctrine prompted by the world war).[103]

In the tract Haeckel ticked off examples of provocative British "egoism": the suppression of peoples in India and the Middle East, the Boer Wars in South Africa, the opium trade in China.[104] Yet the conflict with the British, he suggested, was not fated; there was a certain chanciness to the whole event, just as in ordinary life "the mad play" of chance often overturned plans and people.[105] He was sure that the losses suffered during the conflict affected the Germans more than their enemies, since "a single finely educated German fighter—who has fallen, so sadly now, in massive numbers—possesses a higher intellectual and moral worth than a hundred of the raw, natural men whom England and France, Russia and Italy have brought to the front." Yet the losses of millions on the killing fields of Europe, where "intelligent and cultured peoples attempt to extirpate each other"—this gross tragedy could only be balanced in this awful equation by a faint hope. Perhaps the Great War would be a turning point, after which education and scientific advance would "open the eyes of men and in clear sunlight manifest the true worth of life on this earth."[106]

After America entered the war and Germany stumbled back toward its borders, Haeckel became less sure of his country's singular probity in the awful slaughter. He disdained the sort of dogmatic certainty that drove the German people to follow their political and church leaders without question; but after the outbreak of war, he likewise wandered along in confusion. In April 1918 he made a small contribution to the *Süddeutsche Monatshefte*, in an issued devoted to "German dreamers." In a letter dated 28 March 1918, he indicated that he was troubled by the lack of reality and "false cosmopolitan idealism" of the German people in their desire for peace; he cautioned that the need for a strong German state not be neglected:

> I fear that German dreamers will be strangled by political cant and utopian phantasms of freedom. The Reichstag, in its notorious decision of 19 July 1917, has provided new, shocking evidence of the politically

103. Ernst Haeckel, *Ewigkeit: Weltkriegsgedanken über Leben und Tod, Religion und Entwicklungslehre* (Berlin: Georg Reimer, 1915).

104. Ibid., 62.

105. Ibid., 24.

106. Ibid., 36, 127.

> clichéd inability of the "German Michael" [i.e., Germany itself]. Concerning a peaceful conclusion [to this war], we must, unfortunately, be wary of the triumph of an ideal desire for international compromise over an indispensable real, national augmentation of power [*nationale reale Machtvergröszerung*].[107]

The logic of Haeckel's letter fractured against hard, desiccated memories of earlier times, distorting even recollection of his own efforts at cosmopolitan reconciliation; its refractions yet spelled out an ironic message. A week before he penned his note, Ludendorff launched the so-called Michael offensive along the Somme in an effort to break through the Allied lines and to move on Paris before American troop numbers grew too large. Initially the German advance penetrated more deeply than any side had in the trench warfare up to that time; but Allied reinforcements caused the advance to founder after about ten days. Several other efforts were made in the spring offensive, but the Allied counterattacks finally pushed the Germans inexorably back. By November no peaceful compromise was at hand, only unconditional surrender.

That Christmas Eve of 1918, Haeckel, now severely incapacitated, reflected on the four years of tragedy:

> At the conclusion of the calamitous four years of this disastrous world war, we stand almost without hope on the ruins of a ravaged culture. The false idealism of the German people, the fractured unity of the German heritage, the lack of a naturally unified worldview—all of this threatens our fatherland with complete collapse. We must yet seek hope for a better future in a more solidly realistic school education, in a true knowledge of actual nature founded on a monistic evolutionary doctrine. State and school must remain free of the traditional restraints of the church. Pure reason must overcome the governing superstition.[108]

Haeckel preached enlightenment and freedom of thought all of his life. He believed, not without foundation, that the collapse of his nation resulted from a failure of rational considerations. He revered and celebrated, as did much of the world in the late nineteenth century, the cultural wonders of Germany's best artists and scientists. He felt true surprise and shock that these accomplishments not only had not prevailed against an

107. Ernst Haeckel, [Letter, 28 March 1918], *Süddeutsche Monatshefte* 15 (April 1918), 11.
108. From a loose note in the Haeckel Papers, Haeckel-Haus, Jena.

Fig. 10.15. Haeckel's grave in the backyard of his home, Villa Medusa. The marble bust was originally executed in 1908 by Gustav Herold (b. 1839). (Photo by the author.)

enemy but were in grave danger of dissolution. Yet the disaster of the war, the imbecilic slaughter of a generation, was not simply due to the failure of cultural values and enlightened reason. There was also the tragic failure of human nature itself, which in this instance manifested a disposition to act aggressively in the pursuit of goods and territory, to respond reflexively to any perceived opponents of desire, and to enshroud judgment with nationalistic fervor. These traits, useful perhaps in human prehistory, threatened and continue to threaten the other side of our slowly evolving nature, the side harboring rational ideals and universal compassion.

At the beginning of the new year of 1919, Haeckel started to suffer bouts of faintness, likely due to heart failure. In the spring he fell unconscious and reinjured his left hip, which left him in constant pain. Unable to negotiate the stairs, he was confined to his bedroom and study on the second floor of his house. In the summer he was visited by colleagues and friends. Ludwig Lange, who had worked in Wilhelm Wundt's psychology laboratory at Leipzig, reported to his former professor that he had had the good fortune to see "the old Häckel, who knew exactly how it stood with him; he faced his own death with exemplary calmness and serenity [*Heiterkeit*]."[109] On 4 August Haeckel wrote his onetime student and close friend Richard Hertwig that he "desired entrance to eternal rest, which would occur probably at the latest by October."[110] The next day he fainted and, falling against his desk, broke his left arm. His son, Walter, rushed to him. A few days afterward, on the morning of 9 August, Ernst Haeckel died in his sleep. A year later, on 30 October 1920, his ashes were buried in the backyard of Villa Medusa, with a small bust overlooking his grave.

109. Lange to Wundt (9 September 1919), in Wundt Nachlass (nr. 434d), Universitäts Archiv, Leipzig. I am grateful to Gabriel Finkelstein for drawing this note to my attention.

110. Haeckel to Hertwig (4 August 1919), in *Ernst Haeckel: Sein Leben, Denken und Wirken*, 1:70.

CHAPTER ELEVEN

Conclusion: The Tragic Sense of Ernst Haeckel

Cautious historians of science are reluctant to use the word "genius" in referring to their subjects of study. But if the term refers to an individual of high intelligence, extreme creativity, powerful expression, and extensive influence on the thought of his and succeeding generations, then Ernst Haeckel was, undeniably, a scientific and even artistic genius. His over twenty technical monographs and hundreds of articles manifest that genius through their scope, theoretical sophistication, artistry, and iconoclasm. A simple count of new organisms described in systematic detail—thousands of them—indicate an empirical curiosity and investigative energy of vast proportions. His numerous theoretical conceptions, from the biogenetic law and theory of gastrulation to his location of the hereditary substance in the cell nucleus, to his graphic innovations, to his experimental procedures demonstrating the reality of evolutionary change—these all testify to a scientific mind of extraordinary proportions. His talent with the artist's brush and his ability to depict evocatively the "art forms of nature" continue to be exhibited in the countless reproductions of his works. His freethinking attitudes and execration of religious dogmatism anticipated the prevailing views in the contemporary scientific community.

These various gifts and attitudes made him a magnet for young scientists wishing to train in the most advanced biology of the day. Students came in droves to that little outpost of evolutionary thought in the Thüringen forests. Though his raw personality irritated some of his contemporaries, he could rely on colleagues of stature for quite personal support: Darwin, Huxley, Weismann, Gegenbaur. Even those not entirely disposed to his point of view would not deny his scientific acumen—certainly the doyen of German science in the second half of the nineteenth century, Hermann von Helmholtz, did not. The puzzle then becomes acute: Why, in the present period, has this individual been so maligned, not simply by conservative religious critics—who continue to batter away at his science and at evolution more generally—but by historians and scientists of considerable standing? The puzzle deepens insofar as Haeckel's evolutionary views were hardly different from those of Darwin, whose virtues contemporary scholars, including this historian, have apotheosized.

Early Assessments of Haeckel Outside of Germany

Two early historical assessments, which still resonate, began the warping of Haeckel's scientific achievements—the evaluations of E. S. Russell (1887–1954), a British fisheries expert turned historian, and Erik Nordenskiöld (1872–1933), a Finnish-Swedish zoologist, who also found his vocation in history. Russell's masterly *Form and Function: A Contribution to the History of Animal Morphology* (1916) examined the course of theorizing about the structure of vertebrates that occurred during the late eighteenth through the early twentieth centuries. Aside from Lamarck, Russell's hand stayed pressed on Haeckel longer than any other figure, since this German materialist, in Russell's judgment, wielded more influence in evolutionary morphology than even Darwin himself.[1] More precisely, Russell pressed his foot against Haeckel's neck. The judgment was harsh. Russell thought Haeckel's *Generelle Morphologie* "a medley of dogmatic materialism, idealistic morphology, and evolutionary theory." He particularly execrated Haeckel's materialism, which exhibited "the most intransigent character."[2]

Russell treated his countryman Darwin more gently but, nonetheless, dismissed Darwinian theory through subtle manipulation of the grammar of his historical analysis: he juxtaposed Darwin's conception against a series of objections brought by von Baer, Kölliker, Owen, and others.[3] The objections were all to the same effect and, indeed, were Russell's own: Darwinian chance and blind mechanism could not explain the purposive features of evolutionary development and its "orderly tendency towards perfection."[4] Russell was an Aristotelian, who rejected the possibility of mechanism explaining vital phenomena. In other of his works, he argued

1. E. S. Russell, *Form and Function: A Contribution to the History of Animal Morphology* (1916; repr., Chicago: University of Chicago Press, 1982), 252.

2. Ibid., 248.

3. I discuss the features of "historical grammar" in the second appendix.

4. Russell, *Form and Function*, 242.

against the Morgan school's postulation of a material gene to explain the transmission of traits. He believed some property like Lamarck's inner striving, a nonmaterialistic "hormé," was required to account for development.[5] Prognosticating on the future of biology in 1916, he predicted a turning away from "dogmatic materialism" and "dogmatic evolutionism," and a return to a general Aristotelian perspective.[6] Obviously, a talent for history does not harbor a comparable talent for prophesy.

Nordenskiöld, like Russell, devoted more space in his *History of Biology* to Haeckel than to any other figure. He thought few personalities "have so powerfully influenced the development of human culture" as the German Darwinian. During the 1880s Haeckel and Gegenbaur's "ideas universally prevailed without opposition." Haeckel's tireless industry and aggressive, powerful personality made his *Natürliche Schöpfungsgeschichte* "the chief source of the world's knowledge of Darwinism."[7]

Nordenskiöld's justly famous three-volume history began as a series of lectures in 1916–17 at the University of Helsingfors in Finland; his work was first published in Swedish (1920–24), with an English translation arriving shortly thereafter in 1928 and followed by many reprints.[8] In his history he construed Haeckel's science differently from Russell. The biology was not truly mechanistic; rather its deep structure arose out of something like Schelling's idealistic philosophy. Nordenskiöld thought the pull of idealism so powerful on Haeckel that "the mere observation of natural phenomena is deeply despised" by him. The attribution of idealism notwithstanding, Nordenskiöld condemned his subject's "blind faith in the power of 'mechanical causality' to explain anything whatever."[9] The wellspring of that faith flowed from Haeckel's "unbounded enthusiasm for Darwin's theory."[10] Nordenskiöld contended that by the beginning of the new century the waters of that spring had run completely dry. Darwinism was dead.

Nordenskiöld contended that in his own day Darwin's theory of the

5. E. S. Russell, *The Study of Living Things: Prolegomena to a Functional Biology* (London: Methuen, 1924), 56–64. See also the perceptive study of Nils Roll-Hansen, "E. S. Russell and J. H. Woodger: The Failure of Two Twentieth-Century Opponents of Mechanistic Biology," *Journal of the History of Biology* 17 (1984): 399–428.

6. Russell, *Form and Function*, 364.

7. Erik Nordenskiöld, *The History of Biology: A Survey* (1920–24), trans. Leonard Eyre, 2nd ed. (New York: Tudor, 1936), 505, 522, 515.

8. Nils Hofsten, "Obituary Notice: Erik Nordenskiöld (1872–1933)," *Isis* 38 (1947): 103–6.

9. Nordenskiöld, *History of Biology*, 513.

10. Ibid., 513–14.

origin of species had simply failed and was being replaced by real laboratory science in the form of genetics. What had initially persuaded people and led to the dominance of the Darwinian-Haeckelian view was the political liberalism that it supported. In an analysis of the faux success of the theory, one that would be repeated by social constructionists much later, Nordenskiöld claimed:

> From the beginning Darwin's theory was an obvious ally to liberalism; it was at once a means of elevating the doctrine of free competition, which had been one of the most vital corner-stones of the movement of progress, to the rank of a natural law, and similarly the leading principle of liberalism, progress, was confirmed by the new theory—the deeper down the origin of human culture was placed, the higher were the hopes that could be entertained for its future possibilities. It was no wonder, then, that the liberal-minded were enthusiastic; Darwinism must be true, nothing else was possible.[11]

For Russell and Nordenskiöld, Haeckel's creative genius and powerful personality constituted both his strength and ultimately his fatal weakness. His creative enthusiasms thrashed about in a blind acceptance of Darwinian science, which both of these historian-scientists rejected as speculative and philosophically misbegotten. Thus for these early scientists and historians, the deep objection to Haeckel rested on a rejection of that theory that now serves as the binding web of all biological science.

Haeckel in the English-Speaking World at Midcentury

Two histories of biology at midcentury had, after the development of the modern synthesis of evolution and genetics (1930s–1940s), a much more sanguine view about Darwin and his theory, but little better estimate of Haeckel. These are the histories by the Briton Charles Singer (1876–1960) and the American Jane Oppenheimer (1911–1996). Singer was a longtime professor of biology at London University and lecturer in the history of biology at Oxford. His *Story of Living Things* (1931) first appeared in the interwar period and was subsequently revised twice. The final edition, under the title *A History of Biology to about the Year 1900: A General Introduction to the Study of Living Things,* came out during the centennial

11. Ibid., 477.

year of the *Origin of Species*, 1959.[12] Singer published many other works in the history of biology; especially noteworthy are his editions of and commentaries on Galen and Vesalius.[13] He served as a medical officer during the First World War and lived in London during the blitz of the Second. He shared with Russell and Nordenskiöld the experience of the Great War and with it the aggressive polemics and real consequences of German bellicosity. During the lead up to the Second World War, he aided refugee scientists from Germany.[14] It is hard to imagine that the following characterization of Haeckel did not have its particular vehemence charged by Singer's war experiences:

> It is difficult to estimate Haeckel's place in the history of biology. His faults are not hard to see. For a generation and more he purveyed to the semi-educated public a system of the crudest philosophy—if a mass of contradictions can be called by that name. He founded something that wore the habiliments of a religion, of which he was at once the high priest and the congregation. A large part of his insatiable energy was devoted to propaganda for the great liberal intellectual movement of his time, the essential nature of which he misunderstood. In science, his peculiar employment of hypothesis had close affinities with that of the scholasticism that he denounced and that vitiated alike his observations and his inferences.[15]

Despite such characterization, Singer had grudgingly to admit that Haeckel's works "contain contributions which are still fundamental to the fabric of scientific thought." He judged the leading accomplishment to be none other than the biogenetic law, which through Haeckel's efforts "initiated the modern movement which has, in effect, transformed comparative anatomy into comparative embryology." [16] With bile expended, Singer went

12. Charles Singer, *A History of Biology to about the Year 1900: A General Introduction to the Study of Living Things* (London: Abelard-Schuman, 1959). The book has been reprinted several times.

13. See, for example, *Galen on Anatomical Procedures: De anatomicis administrationibus*, trans. with notes by Charles Singer (London: Oxford University Press, 1956); and *Vesalius on the Human Brain*, trans with notes by Charles Singer (London: Oxford University Press, 1952).

14. Edwin Clark, "Charles Joseph Singer" (obituary), *Journal of the History of Medicine and Allied Sciences* 16 (1961): 411–19.

15. Singer, *History of Biology*, 487–88.

16. Ibid., 488, 489.

on to laud Haeckel's work on radiolaria, sponges, and hydrozoa, and especially his *Challenger* volumes.[17] As he warmed to his subject, he also enumerated other of Haeckel's significant enhancements to biological science: his distinction between protozoa and metazoa (and their developmental continuity), his exploration of the germ layers, and his theory of gastrulation. These important studies, Singer observed, were further developed by Francis Balfour (1851–1882) and Edwin Ray Lankester (1847–1929) in England and the Hertwig brothers in Germany. Singer, having experienced the triumph of Darwinian theory at midcentury, quite obviously became caught up in the many innovative and startling advances in biology made by Haeckel that heralded the triumph of evolutionary theory in the later period.

Jane Oppenheimer, a fish embryologist at Bryn Mawr turned historian, had a dimmer view of Haeckel's science than did Singer. She thought his powerful personality, his "fervency," had two effects: a positive one, in that he drew many young men into biology and embryology (e.g., Wilhelm Roux); and a distinctively negative one, in that he "delayed rather than accelerated the course of embryological progress"—a view obviously in opposition to that of Singer.[18] According to Oppenheimer, Haeckel shared the responsibility for retarding the science of embryology with Darwin himself. She argued that Haeckel's proselytizing for the principle of recapitulation occurred because of Darwin's endorsement of it. She maintained, correctly, that Darwin had formulated the principle in rough terms before he encountered Haeckel's work and that they mutually encouraged each other after their interaction began.[19] In virtually all of her historical essays, Oppenheimer found her heroes among the professionalizers of embryology—such as von Baer, His, and Roux. They, and Oppenheimer taking up their cause, thought the science of embryology needed no help from speculative theories outside the discipline.

Haeckel Scholarship in Germany (1900–Present)

In the German-speaking world, the assessment of Haeckel's science and character went through four phases. The first phase, which extended to the early 1930s, consisted of various celebrations, collections of letters and

17. Ibid., 489, 342–43, 489–92.

18. Jane Oppenheimer, "Analysis of Development: Problems, Concepts and Their History" (1955), in her *Essays in the History of Embryology and Biology* (Cambridge, MA: MIT Press, 1967), 117–72; quotation from 151 and 166; see also her "An Embryological Enigma in the *Origin of Species*" (1959), in ibid., 221–55.

19. Oppenheimer, "Embryological Enigma," 254.

works, and defenses of Haeckelian ideas. Heinrich Schmidt (1874–1935), Haeckel's former student who became director of the Haeckel archives, was especially vigorous in producing Haeckeliana—several small books on Haeckel's work, editions of his letters and monographs, and a biography.[20]

The second phase of Haeckel's historiographic existence in Germany occurred during the mid-1930s and through the war years. During this period, he was initially recruited to the side of National Socialism but then quickly rejected by party functionaries. During the recruitment phase, Haeckel-Haus, under the directorship of party member Victor Franz (1883–1950), published collections of his letters as well as essays on his contributions to biological science.[21] Haeckel also became the subject of a new kind of biography, one that assessed his bloodlines as a measure of the purity of his ideas. Heinz Brücher's *Ernst Haeckels Bluts- und Geistes-Erbe* (Ernst Haeckel's racial and spiritual legacy, 1936) attempted to show that Haeckel's racial ideas were consistent with those of Hitler (a matter discussed in the second appendix). Notably, however, Brücher attempted to shatter the widely held belief that Haeckel was a friend of Jews.[22]

20. Among the small books by Heinrich Schmidt on Haeckel are *Der Kampf um die "Welträtsel": Ernst Haeckel, die "Welträtsel" und die Kritik* (Bonn: Emil Strauss, 1900); *Haeckels Embryonenbilder: Dokumente zum Kampf um die Weltanschauung in der Gegenwart* (Frankfurt: Neuer Frankfurter, 1909); and *Ernst Haeckel und sein Nachfolger Prof. Dr. Ludwig Plate* (Jena: Volksbuchhandlung, 1921). Schmidt edited the following collections of Haeckel's letters: *Entwicklungsgeschichte einer Jugend: Briefe an die Eltern, 1852–1856* (Leipzig: K. F. Koehler, 1921); *Italienfahrt: Briefe an die Braut, 1859–1860* (Leipzig: K. F. Koehler, 1921); and *Himmelhoch Jauchzend: Erinnerungen und Briefe der Liebe* (Dresden: Carl Reissner, 1927). Schmidt was also the editor of *Was Wir Ernst Haeckel Verdanken: Ein Buch der Verehrung und Dankbarkeit*, 2 vols. (Leipzig: Unesma, 1914), which includes a biography. See also Uwe Hoßfeld, "Haeckels 'Eckermann': Heinrich Schmidt (1874–1935)," in *Klassische Universität und akademische Provinz: Die Universität Jena von der Mitte des 19. bis in die 30er Jahre des 20. Jahrhunderts*, ed. Matthias Steinbach and Stefan Gerber (Jena: Bussert & Stadeler, 2005), 270–88.

21. Franz, while director of Ernst-Haeckel-Haus (1935–45), was an active Nazi; after the war he was dismissed from his professorship. He edited a two-volume collection of Haeckel's letters to Oscar and Richard Hertwig and to Hermann Allmers; the collection includes a miscellany of essays on Haeckel as well as then-contemporary evolutionary theory. It also contains a reprint of Haeckel's *Englands Blutschuld am Weltkriege.* See *Ernst Haeckel: Sein Leben, Denken und Wirken. Eine Schriftenfolge für seine zahlreichen Freunde und Anhänger*, ed. Victor Franz, 2 vols. (Jena: Wilhelm Gronau und W. Agricola, 1943–44). The Hermann-Allmers-Gesellschaft published correspondence between Haeckel and Allmers: *Haeckel und Allmers: Die Geschichte einer Freundschaft in Briefen der Freunde*, ed. Rudolph Koop (Bremen: Arthur Geist, 1941). Uwe Hoßfeld describes Franz's career in "Staatsbiologie, Rassenkunde und Moderne Synthese in Deutschland während der NS-Zeit," in *Evolutionsbiologie von Darwin bis heute*, ed. R. Brömer, U. Hoßfeld, and N.A. Rupke (Berlin: Verlag für Wissenschaft und Bildung, 2000), 249–305.

22. Heinz Brücher, *Ernst Haeckels Bluts- und Geistes-Erbe: Eine kulturbiologische Monographie* (Munich: Lehmanns, 1936), 118. Other nineteenth-century German scientists

The efforts to recruit the author of *Die Welträthsel* to the Nazi cause foundered almost immediately because of a quasi-official monitum issued by Günther Hecht, who represented the National Socialist Party's Department of Race Politics (Rassenpolitischen Amt der NSDAP). Hecht, also a member of the Zoological Institute in Berlin, explicitly rejected the suggestion that Haeckel's materialistic conceptions should be regarded as having contributed to the doctrine of the party:

> *The common position of materialistic monism is philosophically rejected completely by the völkisch-biological view of National Socialism.* Any further or continuing scientific-philosophic disputes concerning this belong exclusively to the area of scientific research. The party and its representatives must not only reject a part of the Haeckelian conception—other parts of it have occasionally been advanced—but, more generally, every internal party dispute that involves the particulars of research and the teachings of Haeckel must cease.[23]

Another functionary writing in the same party organ seconded the warning issued by Hecht. Kurt Hildebrandt, a political philosopher at Kiel, maintained it was simply an "illusion" for Haeckel to have believed that "philosophy reached its pinnacle in the mechanistic solution to the world puzzles through Darwin's descent theory."[24] Neither Hecht nor Hildebrandt thought compatible with Nazi doctrine a scientific-philosophical conception that had been embraced by the likes of such socialists and Marxists as August Bebel, Karl Kautsky, and Eduard Bernstein—not to mention V. I. Lenin. The warnings of Hecht and Hildebrandt were enforced by an official edict of the Saxon ministry for bookstores and libraries condemning material inappropriate for "National Socialist formation and education in the

---

were also recruited to the Nazi cause. Alexander von Humboldt, for instance—cosmopolitan, friend of Jews, and homosexual—was declared by Alfred Rosenberg, chief party propagandist, to be a supporter of the ideals of the party. Like Haeckel, Humboldt had his Aryan pedigree established, indicating the purity of his ideas. See Nicolaas Rupke, *Alexander von Humboldt: A Metabiography* (Frankfurt: Peter Lang, 2005), 81–104.

23. Günther Hecht, "Biologie und Nationalsozialismus," *Zeitschrift für die Gesamte Naturwissenschaft* 3 (1937–38): 285. This journal bore the subtitle: *Organ der Reichsfachgruppe Naturwissenschaft der Reichsstudentenführung* (Organ of the Reich's section natural science of the Reich's students administration). For a discussion of this journal's role in the National Socialist Party, see Uwe Hoßfeld, *Geschichte der biologischen Anthropologie in Deutschland von den Anfängen bis in die Nachkriegszeit* (Stuttgart: Franz Steiner, 2005), 329–34.

24. Kurt Hildebrandt, "Die Bedeutung der Abstammungslehre für die Weltanschauung," *Zeitschrift für die Gesamte Naturwissenschaft* 3 (1937–38): 17.

Third Reich." Among the works to be expunged were those by "traitors," such as Albert Einstein; by "liberal democrats," such as Heinrich Mann; by "sexologists," such as Magnus Hirschfeld; by "all Jewish authors no matter what their sphere"; and by individuals advocating "the superficial scientific enlightenment of a primitive Darwinism and monism," such as Ernst Haeckel.[25]

The third phase of Haeckel scholarship came during the time of the Soviet occupation of East Germany. For most of this period, Georg Uschmann (1913–1986), director of Haeckel-Haus (1959–79), produced numerous collections of Haeckel's letters, issued reprints of his works, wrote essays on various aspects of his theories, and published a splendid biography in letters. Though Haeckel's materialism and disdain for religion fell comfortably in line with official East German philosophy, Uschmann's scholarship—factual, intelligent, and comprehensive—did not bend Haeckel to the political views of the new dispensation. Uschmann himself had been a member of Haeckel-Haus during the previous dispensation, when the institute was under Nazi purview. Other East German scholars failed to show the same restraint; they forcefully impressed Haeckel to join the Marxist cause.[26] Following in the steps of Uschmann, Erika Kraußе (1935–2003) became archivist at Haeckel-Haus; and she continued the sober approach of her predecessor. In 1984 she produced a small biography of Haeckel that provided the essentials of his life and work without praise or blame; after the Berlin Wall was breached, her scholarship blossomed with a new enthusiasm.[27] The contemporary successors to Uschmann and Krauße at Haeckel-Haus have continued to issues studies of Haeckel and the period of his activity.[28] Especially noteworthy are the several reproductions of Haeckel's paintings and illustrations (with explanatory essays), which place his artistic accom-

25. "Richtilinien für die Bestandsprüfung in den Volksbüchereien Sachsens," *Die Bücherei* 2 (1935): 279–80.

26. The East German philosopher Reinhard Mocek, typically, found Haeckel to be in essential agreement with Marx and Engels, though lacking their kind of dialectical considerations. See Reinhard Mocek, "Ernst Haeckel als Philosoph," in *Leben und Evolution*, ed. Bernd Wilhelmi (Jena: Friedrich-Schiller-Universität, 1985), 86–95.

27. Erika Krauße, *Ernst Haeckel* (Leipzig: Teubner, 1984). During the last dozen years of her life, Krauße was particularly productive, writing essays and taking part in symposia on Haeckel. Especially significant was her participation in coordinating the volume of essays and reproductions *Haeckel e L'Italia: La vita come scienza e come storia*, ed. Giampiero Bozzolato and Rüdiger Stolz (Brugine: Edizioni Centro Internazionale di Storia dello Spazio e del Tempo, 1993).

28. Uwe Hoßfeld, a scholar at Haeckel-Haus whose many titles are scattered through the footnotes of this book, has been a leader in reconstructing evolutionary theory in Germany from the mid-nineteenth century through the Nazi period.

plishments in the context of other graphic developments of the time.[29] The interest in Haeckel and his works was equally keen in the English-speaking world from the 1970s to the present but with a rather different valence.

The Contemporary Evaluation: Haeckel and the Nazis Again

The fortunes of Haeckel's postmortem history took another sharp turn with the publication of Daniel Gasman's indictment in his *Scientific Origins of National Socialism* (1971), which was based on his dissertation at the University of Chicago. Gasman set out to show that Haeckel's ideas wallowed in romanticism, not materialism, and that

> the content of the writings of Haeckel and the ideas of his followers—their general political, philosophical, scientific, and social orientation—were proto-Nazi in character and that the Darwinist movement which he created, one of the most powerful forces in nineteenth- and twentieth-century German intellectual history, may be fully understood as a prelude to the doctrine of National Socialism.[30]

Gasman adduced Haeckel's "racism" as evidence of his crucial role in forming the biological concepts that oriented Nazi doctrine, especially what he took, with negligible evidence, to be Haeckel's anti-Semitism (a matter discussed above, in chapter 7).[31] In the introduction to a recent reprint of his book, Gasman has elevated Haeckel not only to the singular cause of Nazism but of fascism more generally:

> On a basic level the history of National Socialism in Germany, and fascism in other countries like Italy and France, should be viewed largely from the perspective of the scientific culture rooted in evolutionary biology that emerged under the sway of Haeckelian Monism during the second half of the nineteenth and the beginning of the twentieth century.[32]

29. Especially important is the magnificent volume issued by Olaf Breidbach, *Ernst Haeckel, Bild Welten der Natur* (Munich: Prestel, 2006).

30. Daniel Gasman, *The Scientific Origins of National Socialism: Social Darwinism in Ernst Haeckel and the German Monist League* (New York: Science History Publications, 1971), xiv.

31. Ibid., 157–59.

32. Daniel Gasman, "Introduction to the Transaction Edition," in *The Scientific Origins of National Socialism* (1971; repr., New Brunswick: Transaction, 2004), xiii.

Gasman's efforts exemplify what might be called the fallacy of mono-causality: the attempt to explain complex historical phenomena by appeal to one simple cause. Richard Weikart provides a more recent example of this fallacy, as the title to his book might suggest: *From Darwin to Hitler* (2004). Both Gasman's and Weikart's analyses (further discussed in chapter 7 and in the second appendix) impute moral responsibility to Haeckel for the crimes of Hitler and the Nazis, injudiciously ignoring the tangle of social, political, religious, and economic causes that snaked through the interwar period to foster the rise of Hitler and his party. They also appear to have been completely unaware that Nazi Party officials rejected Haeckelian mechanistic materialism as having anything to do with *völkische Biologie*.

Despite the causal complexity that fostered the advent of National Socialism, Gasman's mono-causal claims have been eagerly accepted by religious fundamentalists, who take the presumed connection between Haeckel and the Nazis as an indictment of evolutionary theory more generally. Additionally—and surprisingly—many influential scientists and historians also received Gasman's argument without scholarly scruple.[33] Stephen Jay Gould immediately endorsed it.

33. The number of historians that have unquestioningly adopted Gasman's thesis is quite large. Here are just a few authors who recently have: J. W. Burrow, *The Crisis of Reason: European Thought, 1848–1914* (New Haven, CT: Yale University Press, 2000), 256, 258; Nicholas Goodrick-Clarke, *The Occult Roots of Nazism: Secret Aryan Cults and Their Influence on Nazi Ideology* (New York: New York University Press, 2004), 13; Scott Gordon, *History and Philosophy of Social Science: An Introduction* (London: Routledge, 1991), 528; Joseph L. Graves, *The Emperor's New Clothes: Biological Theories of Race at the Millennium* (New Brunswick, NJ: Rutgers University Press, 2001), 130–31; Robert Jay Lifton, *The Nazi Doctors: Medical Killing and the Psychology of Genocide*, 2nd ed. (New York: Basic Books, 2000), 125, 441; Richard M. Lerner, *Final Solutions: Biology, Prejudice, and Genocide* (University Park: Pennsylvania State University Press, 1992), 33; Daniel Pick, *Faces of Degeneration: A European Disorder, c. 1848–1918* (Cambridge: Cambridge University Press, 1989), 28; Pat Shipman, *The Evolution of Racism* (Cambridge, MA: Harvard University Press, 1994), 134–135; and Milford Wolpoff and Rachel Caspari, *Race and Human Evolution: A Fatal Attraction* (Boulder, CO: Westview Press, 1998), 136. Gasman's thesis, however, has indeed been critically scrutinized and disputed by several historians: Robert Bannister, *Social Darwinism: Science and Myth in Anglo-American Social Thought* (Philadelphia: University of Pennsylvania Press, 1979), 133; Richard J. Evans, "In Search of German Social Darwinism: The History and Historiography of a Concept," in *Medicine and Modernity: Public Health and Medical Care in Nineteenth- and Twentieth-Century Germany*, ed. Manfred Berg and Geoffrey Cocks (Cambridge: Cambridge University Press, 1997), 55–89 (see 64); and even Richard Weikart believes Gasman has gone too far—see his *From Darwin to Hitler: Evolutionary Ethics, Eugenics, and Racism in Germany* (New York: Palgrave Macmillan, 2004), 116–17. Paul Weindling offers the most balanced assessment of Haeckel and other German biologists in the context of the Nazi movement; see his *Health, Race and German Politics between National Unification and Nazism, 1870–1945* (Cambridge: Cambridge University Press, 1989).

Gould's first book, *Ontogeny and Phylogeny,* on which he worked during the first half of the 1970s, took as its theme Haeckel's principle of recapitulation, both its historical fortunes and its then-contemporary status.[34] As part of his scientific endeavor, he wished to show the importance of the concept of heterochrony (i.e., differential regulation of the several phases of embryological development), which Haeckel conceived as part of the cenogenetic alteration of a potentially perfect recapitulation. Gould credited Haeckel with introducing the concept, but he then expended considerable energy to show that this German never coherently analyzed or pursued it. Indeed, Gould struggled to show that Haeckel's contributions to biological science were dogmatic, unfounded, and distinctly non-Darwinian. I say "struggled," since his effort to show a distinction between Haeckel's conception of recapitulation and Darwin's was like cracking granite with a baseball bat—if you had enough bats and time, you might get somewhere. Gould had neither. After citing passages from Darwin that appeared to be identical to Haeckel's principle, he pounded away: "These two views imply radically different concepts of variation, heredity, and adaptation—the fundamental components of any evolutionary mechanism."[35] What he failed to show was that Haeckel's concepts of variation, heredity, and adaptation essentially differed from Darwin's (see chapter 5).[36] Gould then traced out the mostly unsavory uses of the principle in the works of Lombroso, Freud, and Piaget; and he even hinted that William Westmoreland, commanding general of American forces in Vietnam, had his thinking infected by the principle![37]

The wedge that seemed quite effective in prying Darwin away from Haeckel came from Gasman, to whose thesis Gould unhesitatingly subscribed:

> But as Gasman argues, Haeckel's greatest influence was, ultimately, in another tragic direction—national socialism. His evolutionary racism; his call to the German people for racial purity and unflinching devo-

34. Stephen Jay Gould, *Ontogeny and Phylogeny* (Cambridge, MA: Harvard University Press, 1977). Gould introduced his subject with what would become a familiar trope, namely, mentioning his own childhood experience in the public schools, where he first learned of Haeckel's principle (1).

35. Ibid., 73.

36. I have discussed at length Darwin's use of the principle of recapitulation and Gould's fruitless effort to distinguish Darwin's position from Haeckel's in my *Meaning of Evolution: The Morphological Construction and Ideological Reconstruction of Darwin's Theory* (Chicago: University of Chicago Press, 1992), chap. 4.

37. About Westmoreland, see Gould, *Ontogeny and Phylogeny,* 126.

> tion to a "just" state; his belief that harsh, inexorable laws of evolution ruled human civilization and nature alike, conferring upon favored races the right to dominate others; the irrational mysticism that had always stood in strange communion with his grave words about objective science—all contributed to the rise of Nazism. . . . Our narrow subject impinges upon these wider implications of Haeckel's beliefs, for Haeckel buttressed many of his political claims by reference to recapitulation.[38]

Actually, Haeckel made the most pointed political use of his principle when he observed with amusement that those of royal blood would be chagrined to learn that during the first months of development, the human embryo of a noble or middle-class individual could not be distinguished from that of a dog.[39]

Gasman's historical thesis obviously made a strong impact on the young Gould. Prior to reading Gasman, Gould actually confessed admiration for Haeckel. He liked the kind of pluralism that characterized Haeckel's phylogenetic theorizing.[40] After reading Gasman, however, he felt compelled to denigrate the German's science at almost every turn.[41] Gasman's thesis had greater sting because of another event that occurred two years before the publication of Gould's book.

38. Ibid., 77–78.

39. Ernst Haeckel, *Natürliche Schöpfungsgeschichte* (Berlin: Georg Reimer, 1868), 240.

40. In 1973 Gould wrote a response to Gary Nelson's own positive portrait of Haeckel. He thought Nelson was not appreciative enough of the quite sound ways by which Haeckel established phylogenetic relationships. See Stephen Jay Gould, "Systematic Pluralism and the Uses of History," *Systematic Zoology* 22 (1973): 322–24. After he read Gasman, Gould had almost nothing good to say about Haeckel's science.

41. Gould took frequent occasion to denigrate Haeckel's science. So among the works in which a disparaging word was nearly always heard are *Ever Since Darwin* (New York: Norton, 1977), 215–17 (Haeckel as racist); *The Panda's Thumb* (New York: Norton, 1980), 237–41 (metaphysics of the *Urschleim*), 246–47 (fruitless results of recapitulation); *The Flamingo's Smile* (New York: Norton, 1985), 90 (colonial organisms as inappropriate models for human society); 412–13 (discredited law of recapitulation), 439 (good imagination in predicting *Homo erectus*—actually a positive evaluation); *Wonderful Life: The Burgess Shale and the Nature of History* (New York: Norton, 1989), 263–67 (progressivist bias of tree graphs); "A Developmental Constraint in *Cerion*, with Comments on the Definition and Interpretation of Constraint in Evolution," *Evolution* 43 (1989): 516–39 (ignoring natural selection in phyletic theory); "Redrafting the Tree of Life," *Proceedings of the American Philosophical Society* 141 (March 1997): 30–54 (progressivist bias of tree graphs); *Living Stones of Marrakech* (New York: Three Rivers Press, 2000), 330 (prescient anticipation of Precambrian fossil—mildly positive); *The Hedgehog, the Fox, and the Magister's Pox* (New York: Harmony Books, 2003), 157–62 (unjustified self-justification in representations of organisms in *Art Forms of Nature*); *I Have Landed* (New York: Three Rivers Press, 2003), 305–20 (Haeckel's fraudulent illustrations).

In 1975 E. O. Wilson's *Sociobiology* appeared with great fanfare. A year later Gould along with thirty-four other biologists, members of Science for the People, countered the initial favorable reception of Wilson's book with a screed entitled "Sociobiology—Another Biological Determinism."[42] The battle against this presumed perversion of Darwinian theory would occupy Gould for many years. Ernst Haeckel would be drawn retrospectively into that contest and treated as if he were an errant contemporary instead of a child of the nineteenth century. Gould occasionally became fascinated by some features of Haeckel's science, but his negative judgment persisted to the end. In a book that appeared during the last year of his life (2003), he reprinted an essay that bore the bludgeoning title: "*Abscheulich!* (Atrocious!): Haeckel's Distortions Did Not Help Darwin."[43]

In his essay Gould echoed Nordenskiöld's observation that Haeckel's books "surely exerted more influence than the works of any other scientist, including Darwin and Huxley (by Huxley's own frank admission), in convincing people throughout the world about the validity of evolution."[44] But the influence gave off a malign odor. Gould again mentioned Gasman's by-then twenty-nine-year-old thesis and immediately turned to join Haeckel's contemporary Louis Agassiz in outrage over the illustrations in *Natürliche Schöpfungsgeschichte*—"*Abscheulich!*" is the expostulation that Agassiz penned at the time in his copy of the book. Gould rehearsed the charges of fraud brought back in 1868 and buttressed them with Michael Richardson's more recent analyses of Haeckel's embryo illustrations (see chapter 8). The collective efforts of Gasman, Gould, and Richardson have continued to keep Haeckel's accomplishments in the sulfurous regions of sinister thought.

A telling instance of their effect is the impact on Scott Gilbert, a justly admired developmental biologist. Gilbert concurred in Gould's rejection of Haeckel's biogenetic law, yet at the same time he used Haeckel's illustration of embryo similarity when he thought the image to be drawn by a different author (see chapter 8). When Gilbert was later interviewed about Richardson's new evidence, he quickly rejected Haeckel's illustration as the result of indulgent license.[45] That is to say: to the expert eye of this

42. Garland Allen et al., "Sociobiology—Another Scientific Determinism," *BioScience* 26 (1976): 182, 184–86.

43. Stephen Jay Gould, "*Abscheulich!* (Atrocious!): Haeckel's Distortions Did Not Help Darwin," *Natural History* 190, no. 2 (2000): 42–49; reprinted in *I Have Landed*, 305–20. Further references are to the original article.

44. Gould, "*Abscheulich*," 42.

45. Quoted by Elizabeth Pennisi in her article on Richardson's analysis; see her "Haeckel's Embryos: Fraud Rediscovered," *Science* 277 (1997): 1435.

developmental biologist, Haeckel's illustration of embryos at different stages initially seemed perfectly acceptable when Gilbert thought it was by someone else; only when he learned that the depiction came from the tainted hand of Haeckel did he find it objectionable. The lava of destructive opinion seems inexorably to advance, continuously eating away Haeckel's reputation as a scientist and human being.

The Tragedy of Haeckel's Life and Science

What is the deeper source of the eruptive rejection of Haeckel's work? That he got things wrong? So has every scientist since yesterday. That he failed to advance many innovative ideas and greatly expand the borders of his discipline? That his empirical work was circumscribed and small? The evidence against these presumptions, I have argued, is large. That his influence on the science of his time—and later—failed for lack of energy? The aforementioned scholars agree that his impact was more significant than even Darwin's. That he was a proto-Nazi, as Gasman, Gould, and others claim? That charge, of course, has functioned in the repudiation. Yet Haeckel, I believe, would have rejected the vulgar and dogma-driven Nazis, just as they rejected him. The sustained hostile reaction to Haeckel over the years has stemmed, I believe, from his passionately driven personality and the reckless abandon with which he pursued his Darwinian modernist convictions. And the blood, once let, fed the avenging response. Tacitus reckoned such cases well: "Proprium humani ingenii est odisse quem laeseris."[46]

Miguel de Unamuno, in his *Del Sentimiento Trágico de la Vide* (1913)—*The Tragic Sense of Life*—described in quietly intense language the tragedy that he believed to lie at the very foundation of Western thought:

> The tragic history of human thought is simply the history of a struggle between reason and life—reason bent on rationalizing life and forcing it to submit to the inevitable, to mortality; life bent on vitalizing reason and forcing it to serve as a support for its own vital desires.[47]

Unamumo argued that the desire for immortality, the longing to unite with eternal, divine nature, could not overcome the skepticism of grounded rea-

46. Tacitus, *Agricola*, 42: "It is characteristic of human nature to hate the individual whom it has injured."

47. Miguel de Unamuno, *The Tragic Sense of Life*, trans. J. E. Crawford Flitch (London: Macmillan, 1921), 115.

son, but neither could science extinguish the passion to be united with nature and nature's God. These contending forces of a fully human existence, he thought, were made flesh through sexual love.

The iconic figure of this tragedy for Unamuno was Don Quixote, who epitomized the contradictions of reason and passion, especially in his quest after the imaginary Dulcinea. Unamuno perceived that same predicament in the tragic hero of Goethe's *Faust,* the twin-souled figure who sought after the legendary Helen and found her in the visage of the sweet innocent Gretchen. Unamuno's quixotic existentialism—so modernist Catholic as to send his book onto the Index librorum prohibitorum and him, during the Spanish upheavals of the 1920s and 1930s, into exile and then arrest—might also represent what I take to be the tragedy of Ernst Haeckel. Torn by the two souls in his breast—the deeply feeling spirit and the aggressively rational mind—Haeckel quested after his own Dulcinea, his Helen. She initially appeared when he was a young man in the form of a beautiful, natural girl with whom he fell so deeply in love; and then she momentarily reappeared in his old age as a more sophisticated but no less passionately desired figure. The human incarnations perished untimely and absurdly young. With these deaths, the image of the eternal feminine more forcefully led him on, though staying just beyond the grasp of his evolutionary science. That science, impelled by and bearing the mark of his frustrated desires, lashed out against the false promises of ancient religion and then fell into the dark shadows of modern thought.

APPENDIX ONE

A Brief History of Morphology

Haeckel's *Generelle Morphologie* reconstructs the study of organic forms in light of Darwinian evolutionary theory. In order to appreciate what this reformulation entailed, I will provide here a sketch of the history of morphology during the late eighteenth and first half of the nineteenth centuries.[1] Haeckel drew on this history in complex ways and forged from it his own style in evolutionary thought. He adapted and rekeyed several themes characteristic of the morphological ideas of the earlier period: the unity of type—that creatures can be divided into groups, or archetypes, displaying common patterns; the principle of recapitulation—the assertion that the embryo goes through developmental stages replicating the hierarchy of more primitive species forms; the attendant idea that individuals are constructed of more elemental parts that themselves are creature-like; the assumption that all transmutation of forms can be captured in law-like propositions; and the conviction that the aesthetic character of nature provides an avenue into such law-governed relationships. These features of earlier morphology allowed two further notions to float to the surface of nineteenth-century thought as transforming possibilities: that divine superintendence proved unnecessary for scientific comprehension; and that human beings suffered the same deterministic forces as the rest of nature.[2]

1. For a sketch of the development of morphological theory in the wake of Haeckel, see Uwe Hoßfeld and Lennart Olsson, "The Road from Haeckel: The Jena Tradition in Evolutionary Morphology and the Origin of 'Evo-Devo,'" *Biology & Philosophy* (2003): 285–307.

2. In this brief history, I have omitted treatment of the views of Alexander von Humboldt and Matthias Schleiden, both of whose works were considered in chapter 2. I have, though, expanded the analysis of Goethe's morphological thought—so important to Haeckel—which was also touched on in that chapter.

Johann Wolfgang von Goethe (1749–1832)

The anatomist Karl Friedrich Burdach first publicly used the word *Morphologie* in a medical handbook in 1800, where he confined his investigations to the human form and its pathological deviations.[3] Goethe, however, had already employed the term by 1796 in correspondence with the poet Friedrich Schiller (1759–1805). He informed his friend that he was working on his "famous *morphology* [*berühmte Morphologie*]," which would encompass the study of all natural forms.[4] The essays constituting his intended work were written over a period of some thirty years, but only began publicly appearing in 1817 with the first fascicles of a two-volume collection, which he titled *Zur Morphologie* (1817–24). In the first number of the first volume, he indicated that by "morphology" he meant especially the study of organic forms, both external and internal, and the development of such structures during the life of the organism:

> Scientific men at all times have displayed a drive to comprehend living forms as such, to grasp the connections of their external visible and tangible parts as indicative of the internal parts, and so to control the whole, to a certain extent, in an intuitive perception [*Anschauung*]. How close this scientific urge is connected to the artistic and imitative drive we need not go into. One finds thus in the course of art, of knowledge, and of science several attempts to ground and develop a doctrine, which we would like to call morphology.[5]

The forms that animals exhibit, Goethe maintained, never remain constant; they transmute over time:

3. Karl Friedrich Burdach, *Propädeutik zum Studium der gesammten Heilkunst* (Leipzig: Breitkopf und Härtel, 1800).

4. Goethe to Schiller (12 November 1796), in *Briefwechsel zwischen Schiller und Goethe in den Jahren 1794 bis 1805*, in *Johann Wolfgang von Goethe, Sämtliche Werke nach Epochen seines Schaffens* (Münchner Ausgabe), ed. Karl Richter et al., 21 vols. (Munich: Carl Hanser, 1985–98), 8.1:268.

5. Johann Wolfgang von Goethe, "Die Absicht Eingeleitet" (1807), in his *Zur Morphologie*, vol. 1, no. 1 (1817), as collected in *Goethe, Die Schriften zur Naturwissenschaft*, 1st division, vol. 9: *Morphologische Hefte*, ed. Dorothea Kuhn (Weimar: Böhlaus Nachfolger, 1954), 7. See also Robert J. Richards, *The Meaning of Evolution: The Morphological Construction and Ideological Reconstruction of Darwin's Theory* (Chicago: University of Chicago Press, 1992), 29–39. For a more extensive consideration of Goethe's morphology, see Robert J. Richards, *The Romantic Conception of Life: Science and Philosophy in the Age of Goethe* (Chicago: University of Chicago Press, 2002), chap. 11.

Fig. app. 1.1. Johann Wolfgang von Goethe (1749–1832). Portrait (1787) by Angelika Kauffmann. (Courtesy Stifting Weimarer Klassik.)

When we consider all forms, especially the organic, we never find them stationary, at rest, fixed; rather they all shimmer in constant movement. That is why our language customarily employs, quite fittingly and appropriately, the word *formation* [*Bildung*] for that which has been produced as well as for that which is in the process of being produced.[6]

6. Goethe, "Die Absicht Eingeleitet," in *Morphologische Hefte*, 7.

Goethe's morphology thus attempted to understand external and internal organic structures in their dynamic development.

As Goethe's delineation of morphology hints, he initially took up the study of organic form for artistic reasons. In 1781 he secured time from his court duties at Weimar to study anatomy with J. C. Loder (1753–1832) at Jena. He quickly extended his investigations beyond anatomical structures to consider embryological development. In early 1784 he claimed to have found an osteological formation in the upper jaw of the human fetus that previously had been thought characteristic only of lower vertebrates.[7] This was the *Zwischenkiefer,* or the intermaxillary bone (also called the "premaxillary"). The discovery confirmed for Goethe that a common plan ran through all the vertebrates. He developed a comparable view in botany.

During his Italian journey (1786–88), Goethe pursued the study of art and took every opportunity to continue his instruction in anatomy, the indispensable aid to his aesthetic education. This man of many wiles spent the later part of his sojourn in a search for the *Urpflanze,* the primitive form that lay at the foundation of all plants. As he later confessed, he had naively assumed that this original *Bauplan* could be found perfectly realized somewhere in the land that harbored such a rich profusion of life. After his return from Italy, Goethe undertook a more protracted examination of plant life, utilizing his own gardens and those of the university at Jena. This study resulted in his first published scientific monograph, his *Metamorphose der Pflanzen* (Metamorphosis of plants, 1790).[8] In this work he argued that each part of a plant had an underlying common structure—represented by the ideal leaf—that underwent transformation during individual development. According to this view, the stem, leaves, petals, sexual organs, and seeds could be understood as transformations of the elemental type. Through the mid-1790s, Goethe pursued his morphological conviction that organisms could best be understood by extracting, through comparative analyses, the common plan—or *Urbild*—that underlay its diverse expressions. In the case of vertebrates, the primitive plan would consist of an abstract pattern of bones that retained their topological relationships to each other, despite individual differences within the various species being

7. Goethe announced his discovery in a pamphlet he wrote in 1786 and circulated to friends and professional anatomists. The work, "Dem Menschen wie den Tieren ist ein Zwischenknochen der obern Kinnlade zuzuschreiben," was published in 1820 in the first volume, second number of *Zur Morphologie.* See Goethe, *Morphologische Hefte,* 154–86.

8. Johann Wolfgang von Goethe, *Die Metamorphose der Pflanzen,* in *Sämtliche Werke,* 12:29–68.

compared. So, for example, the bones in the arm and hand of a human being, those in the paddle of a porpoise, and those in the wing of a bat, while differing from species to species, yet displayed the same pattern of relationships across species. (Later Richard Owen would christen these relationships "special homologies.") As Goethe put it in an essay in 1796:

> Metamorphosis operates in two ways in the more perfect animals. In the first, . . . identical parts, according to a certain plan, become differently formed in constant fashion through the formative power [*die bildende Kraft*]. In this way the type in general becomes possible. In the second way, the particular parts composing the type become constantly altered through all the animal groups and species without losing their character.[9]

In 1790 Goethe undertook a second trip to Italy, where he made a potent discovery. While walking along the Lido in Venice, near a Jewish cemetery, his secretary tossed him the skull of a sheep, feigning it was a Jewish head. In later essays (1820), published in his *Zur Morphologie,* Goethe claimed that he had perceived in the fused bones of that battered skull transformed vertebrae. He elaborated his initial insight, proposing that the vertebrate skull could thus be understood as a construction out of more elemental parts, represented by vertebrae. Earlier in 1807 Lorenz Oken had advanced a similar claim, which in time would ignite a priority dispute that proved an irritant to Goethe and a festering wound to Oken.[10]

Goethe's theory that the constantly changing forms of plants and animals could best be understood as transformations of simpler structures—namely, the leaf or the vertebra—received unexpected confirmation in a conception developed in Kant's *Kritik der Urteilskraft* (Critique of the power of judgment), which was published in 1790, the same year as Goethe's *Metamorphose der Pflanzen.* Kant argued that the harmony of organic structures found in a creature had to be comprehended—an epistemological requirement—*as if* they expressed a fundamental plan that gave the harmonious relationships their teleological orientation, *as if* the

9. Johann Wolfgang von Goethe, "Vorträge, über die drei ersten Kapitel des Entwurfs einer allgemeinen Einleitung in die vergleichende Anatomie" (1796), in *Zur Morphologie,* in *Sämtliche Werke,* 12:211.

10. I have discussed this dispute in my *Romantic Conception of Life,* 491–502. There I argue that Goethe only retrospectively "discovered" the vertebral theory of the skull and that Oken did have priority in the discovery.

structures and their interactions had been produced by an *intellectus archetypus.*[11] Goethe later reformulated Kant's conception: instead of regarding the archetype, or *Urbild,* as a symbol of the mind of a transcendent Creator, he gave that notion a deeply Romantic twist:

> To be sure, the author [Kant] here seems to indicate a divine understanding. Morally, we should raise ourselves through belief in God, virtue, and immortality into a higher region and thus approach the first being. But it should be the same in the intellectual sphere as well. We ought to be worthy of mentally taking part, through an intuitive perception [*Anschauen*] into an ever creative nature, in her productions. It was fortunate that, initially and unconsciously, but incessantly pressed by an inner drive, I arrived at that *Urbild,* that type which allowed me to construct a natural representation. So nothing could constrain me from intrepidly embarking on that adventure of reason, as the elder of Königsberg himself calls it.[12]

For Goethe's Romantic Spinozism, the comprehension of nature's archetypal forms—a *scientia intuitiva*—constituted participation in the very creative process of nature herself and formed the intellectual love of *Deus sive natura.* His Spinozistic monism—the conviction that nature exhibited properties of mind and matter, which were but features of an underlying *Urstoff* that could not be identified with either of its salient traits—this kind of monism would anchor Haeckel's loose metaphysics and serve as the foundation for his evolutionary conception of nature.

There is another feature of Goethe's morphology hinted at in the above quotation. The "adventure of reason" that he believed himself entitled to pursue was that of the transmutation of species. In the third *Critique,* Kant considered the possibility that an organism of a type exhibiting a less purposive structure might give birth to other, more purposively structured types "that would be better adapted to the place where they arose and to the relationships formed with each other." This possibility Kant denominated a "daring adventure of reason."[13] Yet he rejected the possibility because he found no empirical evidence to support such transformations. Quite obviously Goethe believed he had the requisite evidence. Later

11. See chapter 2.

12. Johann Wolfgang von Goethe, "Anschauende Urteilskraft," in *Zur Morphologie,* in *Sämtliche Werke,* 12:98–99.

13. Immanuel Kant, *Kritik der Urteilskraft,* in *Werke in sechs Bänden,* ed. Wilhelm Weischedel, 6 vols. (Wiesbaden: Insel), 5:549 (A366, B370–71).

Haeckel would suggest to Darwin that Goethe was a true forerunner of the Englishman.

Karl Friedrich Burdach (1776–1847)

Karl Friedrich Burdach, as mentioned above, first used the term *Morphologie* in 1800. Initially he confined the study of form to human structures in the context of medicine. He believed that the physician had to appreciate normal structures in the human body in order to understand pathological alterations. His first extensive elaboration of this idea came in 1814 as propaedeutic to a diverting study of the human sexual organs, especially the spectrum of their varieties parading between the poles of the normal male and the normal female. After a Schellingian discussion of the nature of human existence in the cosmos,[14] Burdach specified the task of morphology:

> This is no other than to represent, both in their appearance and meaning, the mechanical qualities [i.e., texture, thickness, cohesion, size, form, continuity] of the particular structures and their spatial relationships with one another and with the whole of the human body. The purpose of this representation is partly to ground the entire natural theory of man and partly to put the physician in a position to recognize abnormal spatial relationships so that he can, in respect of their consequences and effects, evaluate and correct them and can maintain unimpaired, so far as it is possible, the normal forms.[15]

Burdach, like Kant, Schelling, and Goethe, understood the various structures of organs constituting a creature to be consequences of forces—expressed in the laws of chemical activity—that were ultimately governed by the "idea that lies at the foundation of the organism and reveals its true essence."[16] The researcher, Burdach thought, would emotionally respond to these organic forms—especially as they indicated either the wondrous

14. Burdach confessed to being intrigued with Schelling's philosophy of nature, though he attempted to distinguish his own independent considerations, especially as regards method. See Karl Friedrich Burdach, *Rückblick auf mein Leben* (Leipzig: Bosz, 1848), 150. Uwe Hoßfeld places Burdach in a more general context in his *Geschichte der biologischen Anthropologie in Deutschland von den Anfängen bis in die Nachkriegszeit* (Stuttgart: Franz Steiner, 2005), 84–87.

15. Karl Friedrich Burdach, *Anatomische Untersuchungen bezogen auf Naturwissenschaft und Heilkunst*, first number (Leipzig: Hartmannschen Buchhandlung, 1814), 8.

16. Ibid., 11.

Fig. app. 1.2. Karl Friedrich Burdach (1776–1847). Lithograph in 1832. (Courtesy of the National Library of Medicine.)

ideas of nature lying behind them or the pathological ways those ideas failed to be realized; but the researcher must not stop there, lest he be beguiled by the higher mysticism. He must rather follow through with exact, rational investigation into the formational laws (*Bildungsgesetze*) governing the internal and external connections between force and form.

In 1817 Burdach sent Goethe a pamphlet that expanded a lecture he had given on the occasions of the opening of the new Royal Anatomical Institute in Königsberg, a construction for which he had lobbied the Prussian government. His tract, *Ueber die Aufgabe der Morphologie* (On the task of morphology),[17] extended his considerations of morphology along lines that Goethe found complementary to his own views.[18] Burdach distinguished three levels in the development of the science of morphology: as a branch of medicine, as a branch of natural history (*Naturkunde*), and as a branch of natural science (*Naturwissenschaft*). The task of morphology in medicine

17. Karl Friedrich Burdach, *Ueber die Aufgabe der Morphologie: Bey Eröffnung der Königlichen anatomischen Anstalt in Königsberg geschrieben und mit Nachrichten über diese Anstalt begleitet* (Leipzig: Dyk'schen Buchhandlung, 1817).

18. Goethe acknowledged the receipt of the *Aufgabe* and expressed his agreement with Burdach's program. See Goethe to Burdach (25 January 1818), in *Goethes Briefe und Briefe an Goethe*, 6 vols., 3rd ed. (Hamburger Ausgabe), ed. Karl Mandelkow (Munich: C. H. Beck, 1988), 3:414–16.

remained what he had previously suggested, namely, as a guide in the detection of pathology. This pragmatic employment of morphology, though, lacked the dignity of disinterested knowledge, the kind of knowledge to be found in natural history and natural science.

The natural historian, in contrast to the practicing physician, sought to investigate organic forms by conducting comparative analyses throughout the animal kingdom. The knowledge thus acquired would certainly exceed what a physician strictly needed for his practice, yet such study might strengthen his confidence and provide the security that an exhaustive understanding of an area would usually bring.[19] Morphology in the context of disinterested comparative analyses allowed the researcher to distinguish properties that united whole groups of animals, those that constituted "the foundational form and mother-pattern of the species" from those that characterized the particular features of individual organisms.[20] In addition to its medical and scientific uses, this kind of knowledge, Burdach urged, should also be cultivated for the pleasures and civilized education it provided; it was simply another stage of *Bildung* for the refined thinker. In this respect, it had an effect comparable to poetry:

> Just as the essence of poetry consists in grasping through feeling the internal content and variety of existence, and in the proclamation of the warm feeling of the eternally active, self-forming displays of life in all of its splendor, so also the discovery of the brightly colored fabric of natural phenomena has it particular pleasures.[21]

Indeed, organic forms are the "poetic products of nature," which the comparative anatomist comprehends in "their purity through artistic experience [*Kunstverfahren*]."[22]

Burdach's Romantic conception of morphological knowledge—that it was conveyed not only through rational analyses but also through artistic experiences—seems to have sprung principally from Schelling.[23] Burdach's consideration of morphology as a *natural science* displayed a comparable

19. Burdach, *Ueber die Aufgabe der Morphologie*, 14.
20. Ibid., 8.
21. Ibid., 8–9.
22. Ibid., 9.
23. Burdach echoes Schelling's aesthetic metaphysics, as developed in the philosopher's *System des transsendentalen Idealismus*. "The objective world," Schelling asserted, "is only the original, though unconscious, poetry of the mind [*Geist*]." See Friedrich Schelling, *System des transscendentalen Idealismus* (1800), in *Schellings Werke* (Münchner Jubiläumsdruck), ed. Manfred Schröter, 12 vols. (Munich: C. H. Beck, 1927–59), 2:349.

debt to the philosopher's idealism. Natural history, he argued, provided descriptions of particulars, while natural science plunged below phenomenal relationships to causes, and finally to the structure of universal subjectivity whence arose the natural world of experience. "Pure science," he intoned in Schellingian terms, "raises itself to an intuition of that which is directly given in self-consciousness; it grasps the inner connections of that which presents itself with necessity, that which arises out of the core and true essence of our being, out of the eternal laws of thought."[24] Every organic form, from this perspective, is revealed as expressing a power that has its origin ultimately within subjective mind and "can only be understood as thought," though in external manifestation it "merely occupies time." Every particular power behind an organic form is simply "a modification of subjective being, and thus a part of world-thought [*des Weltgedankens*], of the idea which is the internal concept of all powers."[25]

Like Schelling, therefore, Burdach believed that objective organic structures and their lawful relationships could ultimately be understood as expressing universal and necessary structures of thought. This kind of thought, however, was a product of mind itself (not merely the mind of a particular individual); and the correlate of universal mind was nature. Schelling explicitly and Burdach implicitly endorsed the Spinozistic conception of reality as *Deus sive natura.* The particular finite forms of nature, which the morphologist might come to know, were the specifications of infinite mind: nature was simply absolute mind in its finite though infinitely variable expression, and such variegated finite manifestation provided partial empirical realization of eternal mind.

Lorenz Oken (1779–1851)

Like Burdach, Lorenz Oken drew inspiration from Goethe and Schelling. He advanced several morphological conceptions that would become staples in the study of organisms during the early nineteenth century. In his first major publication after his doctorate at Freiburg in 1804, Oken proposed that the entire animal kingdom could itself be conceived as a complete animal. In that first book, *Die Zeugung* (1805), he maintained that each component animal—from infusoria to fish, birds, and mammals—would supply, as it were, the graded hierarchy of organs for this larger beast; this is because each animal had an organ system that dominated its life (for

24. Burdach, *Ueber die Aufgabe der Morphologie,* 20.
25. Ibid., 27.

Fig. app. 1.3. Lorenz Oken (1779–1851). (Lithograph courtesy of the National Library of Medicine.)

example, worms were principally epidermal systems, birds bony systems, fish liver systems, and so on). Moreover, according to Oken, each individual within the hierarchy of animals repeated in its embryological development the structures of creatures lower in the hierarchy. Thus the mammalian fetus would begin as something like a polyp, and subsequently pass through the morphological stages of insect, worm, fish, and amphibian, finally assuming the form of the mammal.[26]

This principle of recapitulation would take on an added dimension when united with evolutionary theory. Haeckel would argue that the embryo recapitulated the morphological stages not of extant creatures of more primitive form but of the forms passed through in the phylogenetic history of the organism. He would make this principle of parallelism between onto-

26. Lorenz Oken, *Die Zeugung* (Bamberg: Goebhardt, 1805), 146–47. Oken was not the first to propose the general principle of recapitulation. He was preceded by the Englishman John Hunter (1728–1793) and the Germans Carl Friedrich Kielmeyer (1765–1844) and Johann Heinrich Autenrieth (1772–1835). See Richards, *Meaning of Evolution*, 18–20.

genetic development and phylogenetic development the linchpin of his own evolutionary morphology, but the principle enjoyed currency long before Haeckel employed it. As I will discuss in a moment, Friedrich Tiedemann had already joined the basic notion of species evolution, of a Lamarckian flavor, with the proposition that higher animals in their embryological development recapitulated the morphological stages of those lower in the scale.

In his inaugural lecture at Jena in 1807, Oken proposed another morphological idea that would resonate in German and English biological works for the next half century. This was the vertebral theory of the skull—the assertion that the various bones of the vertebrate skull were really modified vertebrae. It was an idea that Goethe himself would later advance, though contending that he had originally discovered this morphological fact while in Italy in 1790 (see above). As differences between Oken and Goethe heated to a boil, each accused the other of stealing priority for the theory.[27] In his inaugural lecture, Oken did extend certain ideas already suggested in Goethe's *Metamorphose der Pflanzen* and applied them to animals. Just as the entire plant had parts that were transformations of one element, the ideal leaf, so the parts of the vertebrate skeleton could be understood as modifications of one ideal part, the vertebra. As Oken succinctly put it in his lecture, "The entire human being is only a vertebra."[28] German morphologists like Oken found a universe in a grain of sand.

Friedrich Tiedemann (1781–1861)

Friedrich Tiedemann, who had studied with Schelling at Würzburg and Cuvier in Paris, joined paleontological evidence with the recapitulational hypothesis to formulate the principle that became foundational for Haeckel. In the first volume of his *Zoologie* (1808), Tiedemann concluded:

27. Lorenz Oken published his version of the theory of the vertebrate skull in *Ueber die Bedeutung der Schädelknochen* (Bamberg: Göbhardt, 1807). He held that the bones of the skull were only elaborations of three vertebrae (6). Goethe announced his own theory in 1820, though indicated that his original discovery had been made before the turn of the century. Oken reacted to this account in high dudgeon. He believed Goethe was attempting to steal his originality. He was especially offended when several writers, including Hegel, accused him of pilfering Goethe's idea. Oken's wounded ego never healed. Even some quarter of a century later, in 1847, he wrote Richard Owen: "Göthe had the audacity [*Keckheit*] to maintain that he had already discovered [the vertebrae of the skull] twenty years before I had. One has only to read his other osteological treatises to recognize that he had no idea of it. So it goes in the world." See Oken to Owen (12 January 1847), OC62: 20/362a–b, in the Correspondence of Richard Owen, Natural History Museum, London. I have discussed this dispute in my *Romantic Conception of Life*, 491–502.

28. Oken, *Bedeutung der Schädelknochen*, 1.

> It is clear from the previous propositions that from the oldest strata of the earth to the most recent, there appears a graduated series of fossil remains, from the most simply organized animals, the polyps, to the most complex, the mammals. It is evident too that the entire animal kingdom has its evolutionary periods [*Entwickelungsperioden*], similar to the periods that are expressed in individual organisms. The organs of those animal species and genera that have undergone an evolution [*Entwickelung*] can be compared with the organs that in the course of the evolution of each animal have vanished—for example, after birth the vascular system of the navel, the thymus gland, etc. have disappeared in human beings; with frogs, the tail finally vanishes. Just as these parts, the organs in the evolutionary periods of the individual organism, have vanished, so have animals, the organisms of the evolutionary periods of the animal kingdom.[29]

The basic notion that ontogeny recapitulates phylogeny was thus not original with Darwin or Haeckel, but had a preexistence in the biological literature of the early part of the century.

The recapitulation idea fit snugly into the folds of evolutionary theory as it passed from the early advocates of species descent to Darwin and Haeckel. It also supported what would be a major proposition of developmental theory, namely, progressive species evolution. Here was a secular mirror held to the biblical story of decline, giving the reverse image, that of progressive development. That species and varieties could be arranged in a progressive hierarchy was a conclusion seemingly confirmed by recapitulation and evolutionary theory—and by common sense. Who could doubt that dogs were a higher species than frogs or that European man outstripped the Negro in mental ability? The anatomists Pieter Camper (1722–1789), Samuel Thomas Soemmerring (1755–1830), and Georges Cuvier (1769–1832) held that the Ethiopian brain resembled the ape's and displayed comparable intellectual ability.[30] Darwin, though a man of humane sensibility, could hardly believe that the wild Fuegian Indians living at the tip of South America were of the same species as himself. And later in *The Descent of Man*, he expressed comparable reservations about the Irish.

Tiedemann, however, had the considerable acumen not simply to ac-

29. Friedrich Tiedemann, *Zoologie, zu seinen Vorlesungen enworfen*, 3 vols. (Landshut: Weber, 1808–14), 1:73–74.

30. See Robert J. Richards, "Race," in *Oxford Companion to the History of Modern Science*, ed. John Heilbron (Oxford: University of Oxford Press, 2002).

Fig. app. 1.4. Friedrich Tiedemann (1781–1861). Engraving (1830) from a portrait. (Courtesy of the Wellcome Institute Library, London.)

cept the conclusion that the Negro race fell short of the European in raw brainpower. Stimulated by debates in the British Parliament over slavery, he undertook comprehensive measurements of the brains and skulls of hundreds of individuals from the several human groups. In *Das Hirn des Negers* (The brain of the Negro, 1837), he found no significant differences among the groups.[31]

In his study Tiedemann used two kinds of measurement: weight of brains and capacity of skulls (determined by the weight of millet seeds an empty cranium could hold). He took account of the absolute differences between men and women, as well as the differences among infants, children, and adults. He also measured the weight of brains relative to gross body weight, showing thereby that infants had proportionately much big-

31. Friedrich Tiedemann, *Das Hirn des Negers mit dem des Europäers und Orang-outangs verglichen* (Heidelberg: Karl Winter, 1837), 47. I have considered this work in the context of a general discussion of racial hierarchies in "Race." See also the brief discussion in chapter 8.

ger brains than adults. He also noted that though women had, on average, smaller brains than men, they also had lighter bodies; but in proportion to their weight, women's brains were just as large as those of men.[32] Human beings, he observed, had larger brains than most animals—though bested by elephants and whales; relative to body weight humans still stood above most animals, except at the other end of the scale—some small apes, songbirds, and rodents displayed proportionately larger brains.[33] These calculations suggested to Tiedemann that though brain size mattered in intelligence, it was hardly a certain marker.

After tabulating a large number of instances (430 males of all races and 56 females), Tiedemann found that among individual adults of a given race and sex, considerable variability existed; but within a midrange of sizes, no significant differences could be discerned among Caucasians, Mongolians (Chinese), American Indians, Malays, and Ethiopians (Negroes).[34] He completed his study with accounts of Negroes who had received an education and had made important contributions to the sciences and to literature.[35] Tiedemann's conclusions could not, however, overcome the deeply ingrained assumptions of European superiority, especially as those assumptions seemed to be confirmed by theories of evolutionary progress.

32. Tiedemann, *Das Hirn des Negers,* 18: "This reveals [i.e., measurements of body weight in relation to brain weight] that the brain of a women, although absolutely smaller than that of a man, relative to the body is not smaller than his."

33. Ibid., 14–16.

34. Tiedemann's measures were by weight of millet seed that each skull could hold. The extreme of the range was between 59 ounces (for an American Indian) and 13 ounces (for a Mongolian). Among males, he took the range between 42 ounces and 32 ounces as his standard. Within that range fell 64 of 70 Ethiopians; 144 of 186 Caucasians; 29 of 45 Mongolians; 20 of 31 American Indians; and 63 of 98 Malays. Above 42 ounces were 5 Ethiopians, 42 Caucasians, 10 Mongolians, 7 American Indians, and 21 Malays. Below 32 ounces were 1 Ethiopian, 1 Caucasian, 3 American Indians, 7 Mongolians, and 13 Malays. (Note: these figures seem to be off by one or two; e.g., of 186 Caucasians, he has 144 in the midrange plus 42 in the high range and 1 in the low range, which equals 187.) These figures indicate that Tiedemann computed the average brain size differently than by simply dividing total weight of brains by the total number of individuals in each group. He concluded the following: "From the principal results of our investigation, it is undeniable that those anatomists and natural historians have been caught in a mistake who have subscribed to the proposition that the skull capacity and the size of the brain of the Negro are less than those of Europeans and the peoples of other human races. The skull capacity and brain of all the human races show a comparable mid-state [*mittlere*], within certain limits of the range in size. According to the foregoing facts, one can at best say that a few individuals of the Caucasian and Malayan races have reached a considerable size more often than peoples of the other races" (ibid., 47). Tiedemann knew that brain size varied with body size but could not really take advantage of the fact in measuring the brain sizes of the different human races, since body size ranged too outrageously to provide a good measure (from 100 pounds to 800 pounds in the adult males he surveyed).

35. Ibid., 64–82.

Fig. app. 1.5. Carl Gustav Carus (1789–1869).
(Photograph courtesy of the Smithsonian Institution.)

Carl Gustav Carus (1789–1869)

Carl Gustav Carus synthesized the ideas of Goethe, Burdach, and Oken in his widely influential *Von den Ur-theilen des Knochen- und Schalengerüstes* (On the primitive parts of the bony and external frame, 1828).[36] There he portrayed in graphic form the archetype of the vertebrate skeleton and its elemental part, the vertebra (see fig. app. 1.6). Indeed, he supposed

36. Carl Gustav Carus, *Von den Ur-theilen des Knochen- und Schalengerüstes* (Leipzig: Fleischer, 1828); see especially his historical sketch, vii–xii.

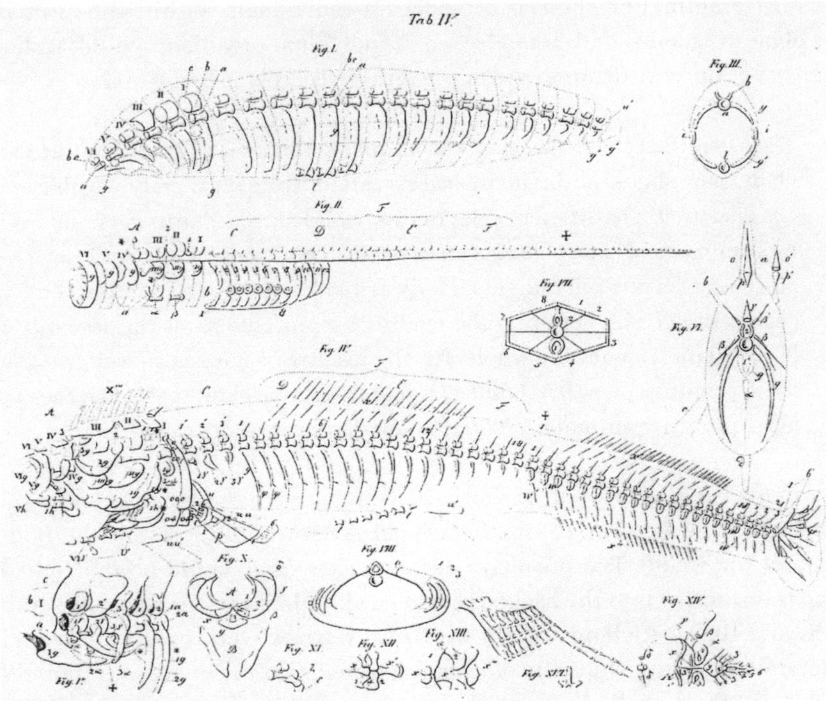

Fig. app. 1.6. Illustration of the vertebrate archetype (fig. 1) and an ideal vertebra (fig. 3). (From Carl Gustav Carus, *Von den Ur-theilen des Knochen- und Schalengerüstes*, 1828.)

that the architecture of the vertebra held the key to the plan of vertebrate organization. He also highlighted what he called the "idea of parallelism between the development of the higher animal forms—yes, even man himself—and the development of the particular classes and species in the animal kingdom."[37] This was the idea to which Oken and Tiedemann had earlier given currency, namely, that the development of higher organisms went through stages comparable morphologically to the permanent states of lower organisms—for example, that the human fetus sequentially took the forms of a simple invertebrate, a fish, an amphibian, a mammal, a primate, and finally displaying the particular structure of a human being.[38]

The idealistic and Romantic features of Carus's analysis echo those of his predecessors, especially Schelling and Oken. He regarded the most

37. Ibid., vii.

38. I have discussed the origin of the principle of recapitulation and Darwin's use of it in my *Meaning of Evolution*.

primitive animal or plant as a model for individuals within the animal or plant kingdoms; and, likewise, each individual organism would, in his construction, constitute a part of the entire ideal plant or animal:

> The scientific consideration of the animal realm as well as that of the plant realm should be distinguished essentially from the merely sensible consideration. The difference lies in this: while the sensible understands each animal and each plant only as a particular whole, the scientific approach conceives each individual only as a part or member; and from the perspective of a higher unity the scientific approach regards the assembly of all individuals as a whole. For the scientific approach as well, the ideal, primitive plant [*Ur-Pflanze*] is just this whole plant realm and the ideal, primitive animal [*Ur-Thier*] is just this entire animal realm.[39]

In his *Von den Ur-theilen,* Carus also proposed that the most elemental part of a plant or animal—its archetypal part—would represent the *Bauplan* of the whole. The primitive vertebra (*Urwirbel*) could be multiplied and transformed into the backbone and head, and its various processes into ribs and limbs. Richard Owen, who read Carus's work carefully, would elevate the relationships the German depicted into the concept of *homology.*[40] He would call the repetition of parts within the same animal (for instance, repetition of the vertebrae) "serial homology"; the repetition of the same parts in different but related species (for instance, the digits of the ape and those of the human being), "special homology"; and the repetition of parts in relationship to the archetype or *Bauplan,* "general homology."[41] When Darwin interpreted these relations as products of descent, he was only deepening the conception of development cultivated by the likes of Goethe and Carus.

Throughout a long scientific life, Carus sought to establish morphology as a science in a strict sense: there had to be laws of transformation of form such that by comprehending these laws one would rationally understand the developmental structure of life. The formation of mathematical laws had long since brought the physical universe to rational order; and German morphologists of the nineteenth century attempted something comparable

39. Carus, *Von den Ur-theilen,* 7.

40. Owen left reading notes on Carus's *Von den Ur-theilen des Knochen- und Schalengerüstes.* These are held in the London Natural History Museum, O.C. 90.2, fols. 37, 50, 150–53.

41. Richard Owen, "Report on the Archetype and Homologies of the Vertebrate Skeleton," in *Report of the Sixteenth Meeting of the British Association for the Advancement of Science: Held at Southampton in September 1846* (London: Murray, 1847), 175–76.

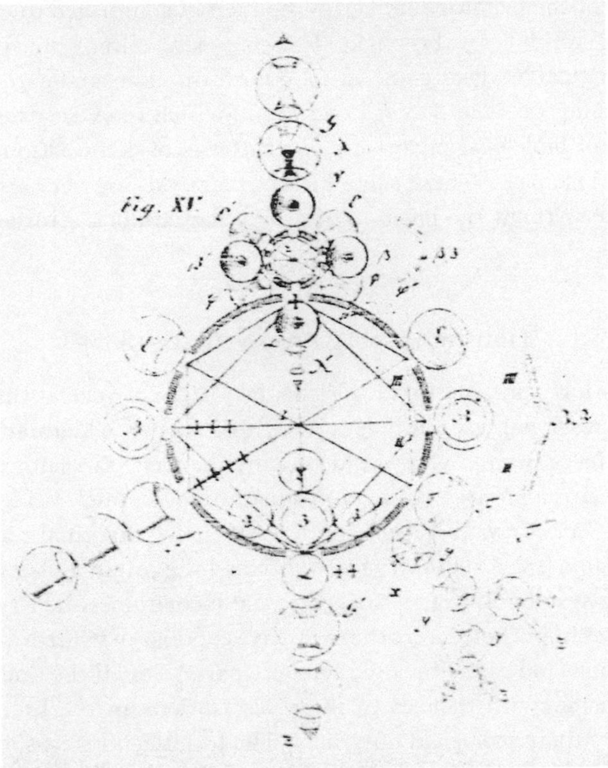

Fig. app. 1.7. A vertebra as decomposable into ideal spherical forms. (From Carl Gustav Carus, *Von den Ur-theilen des Knochen- und Schalengerüstes*, 1828.)

in the life sciences. Carus maintained that comparative analyses of animal skeletons demonstrated that the elemental figure out of which they could all be geometrically derived was the hollow sphere (*Hohlkugel*).[42] By duplication and deformation, the sphere could become a double sphere and then a cylinder, and with the repetition of these forms we could rationally understand the construction of the skeletons of radiate, articulate, molluscate, and vertebrate animals. So, for instance, the elemental vertebra itself can be decomposed into a central sphere and a series of smaller spheres radiating from its periphery (fig. app. 1.7). The vertebra of a temporally existing animal, of course, would display the impact of empirical circumstances, though would generally conform to the rational archetype. When Ernst Haeckel chose the *Heliosphaera actinota* as the Ur-radiolarian, the

42. Ibid., 35.

progenitor of the evolutionary series, he merely conformed to established patterns of morphological thought. The staggering climax of this tradition of mathematical analysis came in 1942 with the thousand-page edition of D'Arcy Thompson's *On Growth and Form,* which likewise examined the structures of biological organisms on the basis of deformation of simple geometrical shapes.[43] These same kinds of analysis are yet carried on today, though without the presumption that they unlock all the secrets of morphology.[44]

Heinrich Georg Bronn (1800–1862)

Of the over forty presentation copies of the *Origin of Species* that Darwin sent to professional acquaintances, only one was to a German scientist: Heinrich Georg Bronn, who would become the first translator of the volume into German.[45] Darwin read Bronn's three-volume *Handbuch einer Geschichte der Natur* (1841–49) and left extensive marginal notes in the second volume (1843).[46] Bronn's history traced the origin and development of the solar system, the earth, its geological formations, and its creatures. Of particular importance were the extensive catalogs in volume 3—totaling over two thousand pages (in three separate parts)—of all the known fossilized organisms, with indexes to their descriptions in the literature and estimates of their geological durations. The fossilized forms showed variable temporal spans, with some lasting but a geological moment, others for vast periods of time. Bronn considered at length, in the second volume of the *Handbuch,* the question of whether primitive animals arose spontaneously and whether the great variety of living species descended from a few original types. Did species change form in the manner similar to that of the embryo in its development? Bronn believed that there was little doubt that in the past certain whole groups of organisms went extinct and were replaced by others of a more modern character. He argued that the intro-

43. D'Arcy Wentworth Thompson, *On Growth and Form: A New Edition* (Cambridge: Cambridge University Press, 1942). Thompson wrote the first edition of this extraordinary book during World War I and completed the greatly expanded revised edition during the next world war.

44. See, for instance, Jorge Wagensberg, "On the Emergence of Forms in Nature," lecture delivered at the symposium *Forma i Funció* (6–7 March 2003), Barcelona.

45. See "Appendix VIII," in *Correspondence of Charles Darwin,* ed. Frederick Burkhardt et al., 15 vols. to date (Cambridge: Cambridge University Press, 1985–), 7:533–36.

46. Heinrich G. Bronn, *Handbuch einer Geschichte der Natur,* 3 vols. in 5 (Stuttgart: E. Schweizerbart'sche, 1841–49). Darwin's well-marked copy of the second volume is held in the manuscript room of Cambridge University Library.

Fig. app. 1.8. Heinrich Georg Bronn (1800–1862). (Lithograph courtesy of the University of Kansas Library.)

duction of new species followed definite laws, which related environmental conditions during past geological periods to the adaptational structures of the introduced species. As geological and environmental conditions changed, new kinds of organisms appeared, achieving a balance among themselves, but ever arching toward those more highly developed forms recognized today.[47]

Darwin especially profited from the large literature that Bronn summarized concerning the modifiability of organisms, especially the details of hybridization experiments. He was particularly interested in Bronn's biogeographical analyses: if an *Urtypus* arose on an island after the original ocean began to recede, then that type would have spread to other regions

47. Rupke has described the work of Bronn and others who attempted to provide a developmental history of species without, however, endorsing common descent. See Nicolaas Rupke, "Neither Creation nor Evolution: The Third Way in Mid-Nineteenth-Century Thinking about the Origin of Species," *Annals of the History and Philosophy of Biology* 10 (2005): 143–72.

as more land appeared. Despite what to Darwin's eyes, and ours, were the mountains of evidence supporting the transformation hypothesis, Bronn concluded against it—at least in the form that the Englishman would propose. Bronn declared that experiments on spontaneous generation, when carefully conducted, failed to produce evidence of the organic developing out of the inorganic. Moreover, alterations of organisms through the effects of climate, nutrition, and other natural forces—while extensive—did not produce changes in species; indeed, original forms reappeared when the altering causes were removed. Even hybridized organisms, he observed, reverted to the forms of the original progenitors after several generations. Nature gave no evidence of genealogical transition in species.

Through the extensive thickets of empirical studies that Bronn amassed, he detected a strict, lawful regularity in natural phenomena; natural forces were proximately responsible for the patterns that nature displayed beneath the cover of abundance. The patterns, though, raised the viewer's mind beyond secondary relationships to ultimate causes. Those patterns, in the last instance, had to be "the results of an unlimited, all-mighty power as well as the ordering of an unlimited wisdom."[48] Development of species did occur; huge numbers gradually went extinct and were replaced by forms better adapted to a changing environment. But replacement was not transformation: one form did not arise out of another in a natural, genealogical fashion. The patterns exhibited by fossils recorded the sequence of replacements orchestrated by unknown natural forces but according to a divine plan and executed ultimately by divine power.[49]

Bronn's conclusion would not have been utterly inimical to the young Darwin, who could have tolerated the notion that the Divinity was behind it all, as long as one supposed that proximate causes were natural forces acting in a regular manner. And though Bronn's was not a genealogical theory of transmutation, it would take but a few twists to turn his conception into a thoroughgoing evolutionary scheme, which is apparently one reason why Darwin scrutinized the *Handbuch* so carefully.

In 1850 the Académie des science in Paris announced a prize for an essay that gave the most convincing answers to a series of questions concerning organic development: What accounts for the appearance and disappear-

48. Bronn, *Handbuch einer Geschichte der Natur*, 3.3:746.

49. See also Lynn Nyhart's neat summary of Bronn's developmental theory in her *Biology Takes Form: Animal Morphology and the German Universities, 1800–1900* (Chicago: University of Chicago Press, 1995), 112–15.

ance of organisms in the fossil record? What is the relationship between the present state of the organic world and its previous state? Since no entry was deemed worthy, the Académie again offered the prize in 1854. This time Bronn submitted a tract, which took the French award. In 1858 he published his German translation of the submission as *Untersuchungen über die Entwickelungs-Gesetze der organischen Welt* (Investigations into the developmental laws of the organic world).[50] The monograph compressed the analyses and evidence of the multivolume *Handbuch* into about five hundred pages. He reiterated the principal conclusions of that previous work: "We believe," he asserted, "that all plants and animal species have originally been formed through an unknown force of nature [*Natur-Kraft*]—though they have not arisen through the restructuring of a few primitive forms. We believe that this force stands in the closest and most necessary relation to the forces and events that have built up the surface of the earth." He thus allowed natural causes to do the real work of production but did not deny the hand of the Creator ultimately orchestrating the whole. He also added some new emphases. He was especially keen to insist that "the most basic law [of organic phenomena] is the law of adaptation [*Gesetz der Anpassung*] of the successive populations of the earth to the external conditions of existence extant at those times."[51] This insistence of the close relationship of adaptational structures to particular environmental features, while well within the Cuverian tradition, yet made the evolutionary move tantalizing.

Bronn thus proposed a progressive force driving organisms to higher levels of perfection and a corresponding restricting "law of adaptation," which constrained and articulated organic development. His fundamental assumption of two antagonistic forces harkened back to Schelling and the *Naturphilosophen,* who analyzed biological life and all of nature in light of antithetic powers of expansion and contraction. In transmogrified form, this kind of dual-force theory would reappear in Richard Owen's own analyses of the progressive and adaptational development of life on earth. In a greatly attenuated fashion, it would be Darwin's conception as well.[52] The similarity of Bronn's theory to Darwin's would be signaled—at least to our modern eyes—when Bronn bedecked his ideas in English dress.

50. Heinrich G. Bronn, *Untersuchungen über die Entwickelungs-Gesetze der organischen Welt* (Stuttgart: E. Schweizerbart'sche, 1858).

51. Ibid., 82, 499.

52. I have discussed these ideas in *The Meaning of Evolution,* chap. 5.

An English translation of the last chapter of Bronn's monograph was published in August 1859 in the *Annals and Magazine of Natural History,* two months before the appearance of Darwin's book. It bore the startling title "On the Laws of Evolution of the Organic World during the Formation of the Crust of the Earth." Little wonder that Darwin acceded to Bronn's suggestion to have the *Origin of Species* translated into German, a task Bronn himself would undertake.[53] And anyone thoughtfully approaching the *Untersuchungen*—"this excellent work" of the "distinguished Bronn," as Haeckel put it—would be standing on the edge of the easy slide into Darwinism.[54]

If Haeckel became conceptually prepared for evolutionary theory by reading Bronn, he may also have received some advanced help in graphically interpreting the new theory. Bronn had included in his prize monograph a phylogenetic tree of the kind Haeckel would later make famous. This illustration, apparently the very first of its kind, depicts the appearance of organisms and their morphological relationships (see fig. 5.7). The large boughs *A* through *G* represent such groups as the invertebrates, fish, reptiles, birds, mammals, and man. The lowercase letters from *a* to *m* indicate species at different levels of development and time of appearance in the geological record. Bronn especially wanted to show by this graphic device that a main bough, which was temporally earlier in first appearance, might exhibit species that arose later than some on a temporally subsequent bough. In constructing his own phylogenetic trees in the 1860s, Haeckel would have Bronn's model as an inspiration—though not the only one, as I have indicated in chapter 5.

Karl Ernst von Baer (1792–1876)

Karl Ernst von Baer was perhaps the most eminent morphologist and embryologist of his day. He pursued postdoctoral study at Würzburg, where the residual influence of Schelling could still be felt; and in 1817 he moved to Königsberg to become assistant to Burdach. His sober experimental approach to anatomy led him to shed much of his earlier enthusiasm for a more Romantic and aesthetic interpretation of natural phenomena. In

53. Darwin to Bronn (4 February 1860), in *Correspondence of Charles Darwin,* 8:70.

54. In his *Natürliche Schöpfungsgeschichte* (Berlin: Georg Reimer, 1868), which was a series of lectures based on the *Generelle Morphologie,* Haeckel observed that "the distinguished Bronn" had provided evidence of the earth's early transformations in his "excellent *Untersuchungen über die Entwickelungsgesetze der organischen Welt*" (263).

his great work on embryology, *Über Entwickelungsgeschichte der Thiere* (On the developmental history of animals, 1828–37), he argued against recapitulation theory.

Like Georges Cuvier, von Baer held that the animal kingdom could be separated into four distinct archetypes: the Radiata (for instance, starfish and sea urchins), the Mollusca (for instance, clams and octopus), the Articulata (for instance, insects and crabs), and the Vertebrata (for instance, fish and human beings). He denied, however, that the embryos of more complex animals passed through morphological stages comparable to those of the adult forms of organisms lower in the hierarchy of life. He maintained that the embryo of an animal exemplified from the beginning of its gestation only the archetype or *Urform* of that particular organic group. "The embryo of the vertebrate," he asserted, "is already at the beginning a vertebrate."[55] So a human fetus, he held, would move through stages in which it would take on the form of a generalized vertebrate, a generalized mammal, a generalized primate, and finally a particularized human being. The form of the growing fetus moved from the general to the specific. The morphological forms of the developing fetus never passed from one archetypal form to another. Nor did the fetus advance through the adult forms of lower organisms even within the same archetype: the human embryo, in von Baer's view, thus never exhibited morphological stages in which it appeared as a mature form of a definite lower species. Von Baer's negative conclusion was simply that the human fetus never assumed the form of an invertebrate or of an adult fish.

Though he became wary of the Romantic *Naturphilosophie* exhibited by many German morphologists, von Baer yet retained certain metaphysical ideas that connected him with the tradition of transcendental idealism that was inaugurated by Schelling. He seems never to have abandoned the conviction, for example, that the archetype of the organism, as a kind of extra-physical entity, guided the creature's morphological development. "The type of every animal," he declared, "both becomes fixed in the embryo at the beginning and governs its entire development."[56]

Von Baer objected to the transmutational theory that was becoming quickly rooted in German biology. He understood that ideas of recapitula-

55. Karl Ernst von Baer, *Über Entwickelungsgeschichte der Thiere: Beobachtung und Reflexion*, 2 vols. (Königsberg: Bornträger, 1828–37), 1:220.

56. Ibid. Von Baer thought the doctrine of the type—"the Ideas according to the new school"—required the abandonment of any purely materialistic conception of embryogenesis. See ibid., 148.

Fig. app. 1.9. Karl Ernst von Baer (1792–1876). (Lithograph from the author's collection.)

tion and of species transmutation gave seductive succor to each other—and he firmly opposed them both:

> One gradually learned to think of the different animal forms as evolving [*entwickelt sich*] out of one another—and then shortly to forget that this metamorphosis was only a mode of conception. Fortified by the fact that in the oldest layers of the earth no remains from vertebrates were to be found, naturalists believed they could prove that such unfolding of the different animal forms was historically grounded. They then related with complete seriousness and in detail how such forms arose from one another. Nothing was easier. A fish that swam up on the land wished to go for a walk, but could not use its fins. The fins shrunk in breadth from want of exercise and grew in length. This went

on through generations for a couple of centuries. So it is no wonder that out of fins feet have finally emerged.[57]

The immediate target of this lampoon was, of course, Lamarck.

The community of German zoologists lived along several conceptual fissures. Some, like the redoubtable von Baer, rejected the notion of recapitulation and the supportive doctrine of species descent. Bronn, who thought species progressively appeared on the earth over vast periods of time according to a divine plan, recognized that embryonic development bore strong analogy to the morphological development of species. But fearing, perhaps, the consequences of the recapitulational idea, Bronn stressed the analogical character of the parallel and reaffirmed the boundaries to embryogenesis that von Baer had established.[58] Others, like the Romantic Oken and the aesthetically driven Carus, unhesitatingly advanced the theory of the archetype and its attendant notion of recapitulation, while embryologists like Tiedemann moved even further toward the doctrine of species transformation.[59] In England during the first half of the nineteenth century, German morphological theory received its most influential interpreter in the person of Richard Owen. Owen understood the close connection between ideas about embryological recapitulation and species transformation.

Richard Owen (1804–1892)

Richard Owen made morphological theory the framework for his many comparative studies of organisms, ranging from invertebrates to humans.[60]

57. Ibid., 200.

58. Bronn carefully remained on the side of orthodoxy: "Kielmeyer, Serres and others have thus assumed that each animal of a higher level must first pass through a lower level in order to reach the organizational apex of its mature age—so that the human being must first have become an infusorium, a worm, a fish, a reptile, a cetacean, and an ape before it can become a human being. Although there is a certain unmistakable analogy between an embryo's earlier condition and that of the lower stages of the animal realm, yet it is only an analogy. Even in its earlier appearances, one can discern that the beginning embryo goes through stages characteristic of its own species. The passage of the architecture of an animal from one kingdom or under-kingdom to another is, as we already know, impossible." See Heinrich Georg Bronn, *Morphologische Studien über die Gestaltungs-Gesetze der Naturkörper überhaupt und der organischen insbesondere* (Leipzig: E. F. Winter'sche Verlagshandlung, 1858), 145.

59. See my *Meaning of Evolution*, 39–55, for a more detailed account of the views of Oken, Carus, and Tiedemann.

60. For a comprehensive and exemplary account of Owen's life and work, see Nicolaas Rupke, *Richard Owen, Victorian Naturalist* (New Haven, CT: Yale University Press, 1994).

Fig. app. 1.10. Richard Owen (1804–1892). (Engraving courtesy of the Wellcome Institute Library, London.)

He derived much of his inspiration from such German authors as Johann Wolfgang von Goethe, Johann Friedrich Meckel, Johannes Müller, Karl Ernst von Baer, and Carl Gustav Carus, whose *Von den Ur-theilen* he read most assiduously.[61] In his "Report on the Archetype" (1847) and *On the Nature of Limbs* (1849), Owen, following Carus, depicted the vertebrate archetype as essentially a string of vertebrae.[62] The bones of various vertebrate species—head, ribs, limbs, pelvis—could be construed as developments of processes or features of the single vertebra; and all backboned animals, he argued, displayed structures that were modifications of this basic plan (see fig. app. 1.11). So, for instance, the forelimb of the mole, the wing of the bat,

61. Owen's reading notes on Carus's *Von den Ur-theilen* are held in the London Natural History Museum (O.C. 90.2).

62. Owen, "Report on the Archetype and Homologies of the Vertebrate Skeleton"; and *On the Nature of Limbs* (London: Van Voorst, 1849).

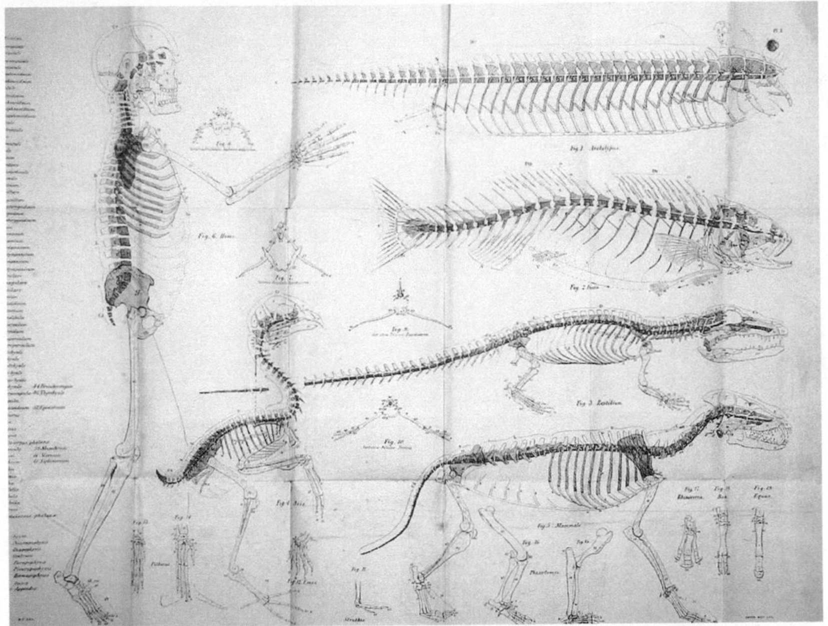

Fig. app. 1.11. Vertebrate archetype (*top right*). (From Richard Owen, *On the Nature of Limbs*, 1849.)

and the hand of a man exhibited the same bones in comparable topological arrangement—they would have, as Owen put it in his Germanophilic way, the same *Bedeutung*, the same meaning.[63] The component bones might be adapted to different uses (for instance, digging, flying, tool manipulation), but they nonetheless preserved a common pattern that underlay their specific adaptations.

Like von Baer, whose work he had carefully read, Owen was quite wary of the notion of the natural development of organisms, at least in its recapitulational form. He perceived that the doctrine of ontogenic recapitulation supported the doctrine of phylogenic transmutation. In his Hunterian Lectures of 1837, he expressly denied that "the Human Embryo repeats in its development that structure of any part of another animal; or that it passes through the forms of the lower classes." He understood that proposition to give comfort to the pernicious idea of organic evolution: "The doctrine of transmutation of forms during the Embryonal phases," he warned,

63. Owen, *On the Nature of Limbs*, 1.

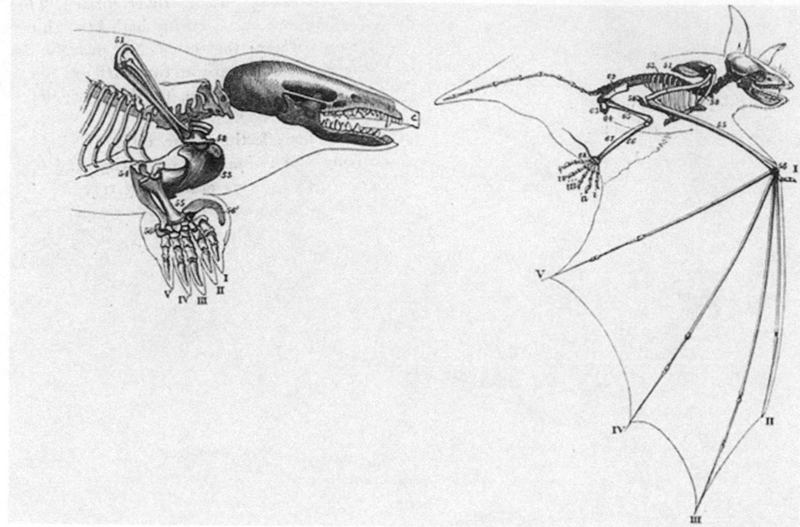

Fig. app. 1.12. Vertebrate limbs (mole and bat) displaying bones in the same topological arrangement, though adjusted to different environments. (From Richard Owen, *On the Nature of Limbs*, 1849.)

"is closely allied to that still more objectionable one, the transmutation of Species."[64] One of Owen's early colleagues would also perceive the connection between recapitulation and species evolution—but Charles Darwin was ready to argue for both.

Charles Darwin (1809–1882)

In his initial theorizing about species origins, Darwin employed the recapitulational principle as central to his conception of descent.[65] In 1837, on the very first page of his "Notebook B," in which he began to develop his species theory, he recognized recapitulation as the index of species transformation. The embryo, he wrote, passes "through the several stages (typical, <of the> or shortened repetition of what the original molecule has done)."[66] The "original molecule" referred to Erasmus Darwin's conception of the

64. Richard Owen, *The Hunterian Lectures in Comparative Anatomy, Mayand June 1837*, ed. Phillip Sloan (Chicago: University of Chicago Press, 1992), MS 98–99 (192).

65. I have discussed Darwin's usage of the concept of recapitulation in my *Meaning of Evolution*, 91–166; see also chapter 5, above.

66. Charles Darwin, "Notebook B," MS 1, in *Charles Darwin's Notebooks, 1836–1844*, ed. Paul Barrett et al. (Ithaca, NY: Cornell University Press, 1987), 170. Double-wedge quotes indicate Darwin's insertions; single wedges indicate his deletions.

primal "filament" that God had created and that subsequently developed into the myriad of species now populating the earth. Later in his notebooks and in the *Origin of Species,* Darwin situated the idea of recapitulation within his consideration of the unity of morphological types. Unlike von Baer and Owen, however, he interpreted the unity of plan exhibited by, say, the vertebrates, not as a result of a creative idea in the mind of God, but as the consequence of descent. He seems to have first expressed this fundamental notion on the back flyleaf of his copy of Owen's *On the Nature of Limbs,* where he jotted: "I look at Owen's Archetypes as more than idea, as a real representation as far as the most consummate skill & loftiest generalization can represent the parent form of the Vertebrata."[67] He thus suggested that the generalized vertebrate—the vertebrate archetype—was an actual creature that established the parental stock of all the vertebrates. The similarity of vertebrate structures, then, stemmed from common descent; and embryogenesis recapitulated the morphological phases of that descent. The archetype had thus been made historical.

In his notebooks Darwin assumed that the various archetypes—from Mollusca to Vertebrata—arose independently of one another and that in the course of phylogenetic history developed into the various species classified under a particular archetype. Even in the *Origin of Species,* Darwin did not propose common descent of *all* creatures to be a necessary postulate of his theory: "I believe that animals have descended from at most only four or five progenitors, and plants from an equal or lesser number." Analogy did suggest, however, that "probably all the organic beings which have ever lived on this earth have descended from some one primordial form, into which life was first breathed."[68]

In the *Origin of Species,* embryonic recapitulation became implicated in a set of nesting considerations about species evolution, from the theory of adaptation, to inheritance, to the progressive development of life. Darwin conceived that in the womb, little variability would be expressed or selected. Selection would occur after birth as animals manifested new traits that might adapt them to variously different environments—though in the course of ages, these adaptational differences would be shoved back into earlier stages of embryonic development. Such terminal additions of traits allowed the embryo to preserve, in Darwin's estimate, a record of the history of the species: "Thus the embryo comes to be left as a sort of picture,

67. Darwin's copy of Owen's *On the Nature of Limbs* is held in the Manuscript Room of Cambridge University Library.

68. Charles Darwin, *On the Origin of Species* (London: Murray, 1859), 484.

Fig. app. 1.13. Charles Darwin (1809–1882). Photo taken about 1857. (Courtesy of the Syndics of Cambridge University Library.)

preserved by nature, of the ancient and less modified condition of each animal."[69]

Darwin's theory of heredity, as can be detected from this account, supposed that environmental stimulation might alter individuals and that these alterations could be inherited. Like Haeckel after him, Darwin assumed that acquired characters could be inherited. He believed that heri-

69. Ibid., 338. My historical reconstruction differs considerably from that of Gould, who denies that Darwin endorsed recapitulation. See Stephen Jay Gould, *Ontogeny and Phylogeny* (Cambridge, MA: Harvard University Press, 1977), 70. Many other authors have followed Gould's interpretation. I have discussed at length Darwin's adoption of recapitulation in my *Meaning of Evolution*, 111–43.

table changes could be introduced at any time during the life of a creature, but most often as the result of the environment acting on the body of an organism or, in the case of animals, as a consequence of the creature adopting a habit that would gradually change its structure. Alterations of this sort might then be naturally selected if they provided the organism an advantage. Darwin, from the beginning of his theorizing to the end, allowed for the inheritance of acquired characteristics. He did argue, however, that the most complex adaptations would likely result from the environment randomly altering the sexual organs of parents, with changes passing to offspring and selected if advantageous. Only with the work of August Weismann and the reintroduction of Mendelian genetics at the turn of the century would notions of the inheritance of acquired characters be gradually rooted out of evolutionary theory.

In the German translation of Darwin's *Origin*, Bronn added, as we have seen (chapter 3), an appendix in which the problems of the theory were highlighted. Darwin actually encouraged such a critical response in his correspondence with "the old Bronn."[70] One of the challenges issued stuck to evolutionary theory like a large barnacle, namely, that while Darwin had shown the possibility of genealogical descent, the naturalist had not demonstrated the actuality of such descent. This became the challenge that Haeckel thought he was meeting in his monographs on radiolaria and sponges, and in his *Generelle Morphologie*. In that latter work, he believed he accomplished what was necessary for the success of any empirical theory: it had to be encased in ironclad law. And that law to which others would be subordinated was the biogenetic principle that ontogeny recapitulates phylogeny. Empirical evidence and fixed law would successfully answer the challenge issued by Bronn.

70. Darwin to Bronn (4 February 1860), in *Correspondence of Charles Darwin*, 8:70.

APPENDIX TWO

The Moral Grammar of Narratives in the History of Biology—the Case of Haeckel and Nazi Biology

Introduction: Scientific History

In his inaugural lecture as Regius Professor of Modern History at Cambridge University in 1895, Lord John Acton urged that the historian deliver moral judgments on the men and women under scrutiny. Acton declaimed:

> I exhort you never to debase the moral currency or to lower the standard of rectitude, but to try others by the final maxim that governs your own lives and to suffer no man and no cause to escape the undying penalty which history has the power to inflict on wrong.[1]

In 1902, the year Acton died, the president of thc American Historical Association, Henry Lea, in dubious celebration of his British colleague, responded to the exordium with a contrary claim about the historian's obligation, namely, objectively to render the facts of history without subjective moralizing. Referring to Acton's lecture, Lea declared:

> I must confess that to me all this seems to be based on false premises and to lead to unfortunate conclusions as to the objects and purposes of history, however much it may serve to give point and piquancy to a narrative, to stimulate the interests of the casual reader by heightening

This appendix is based on the Ryerson Lecture, which I delivered at the University of Chicago on April 12, 2005. The lecture was intended for inclusion in this volume. To keep this appendix independently comprehensible, I have repeated some points made in previous chapters.

1. John Acton, "On the Study of History," in *Lectures on Modern History* (London: Macmillan, 1906), 23.

lights and deepening shadows, and to subserve the purpose of propagating the opinions of the writer.[2]

As Peter Novick has detailed in his account of the American historical profession, by the turn of the century historians in the United States had begun their quest for scientific status, which for most seemed to preclude the leakage of moral opinion into the objective recovery of the past—at least in an overt way. Novick also catalogs the stumbling failures of this noble dream, when political partisanship and rampant nationalism sullied the ideal.[3]

Historians in our own time continue to be wary of rendering explicit moral pronouncements, thinking it a derogation of their obligations. On occasion, some scholars have been moved to embrace the opposite attitude, especially when considering the horrendous events of the twentieth century—the Holocaust, for instance. It would seem inhumane to describe such events in morally neutral terms. Yet even about occurrences of this kind, most historians assume that any moral judgments ought to be delivered as obiter dicta, not really part of the objective account of events. Lea thought a clean depiction of despicable individuals and actions would naturally provoke readers into making their own moral judgments about the past, without the historian coercing their opinions.

Of course, few would deny the historian the right—rather, the obligation—to establish the general values of a given historical period and measure the actions of individuals living in those times against the standards of the society. The great German philosopher of history Heinrich Rickert believed this task obviously incumbent on the historian. He distinguished, however, this kind of assessment from any intrusive moral evaluations. The historian should render an objective judgment of the conformity of the actions of an individual to the prevailing standards of the individual's times, but the scientific historian should refrain from making an evaluation of probity according to his or her own personal norms.[4]

2. Henry Lea, "Ethical Values in History," *American Historical Review* 9 (1904): 234.

3. See Peter Novick, *That Noble Dream: The "Objectivity Question" and the American Historical Profession* (Cambridge: Cambridge University Press, 1988).

4. Heinrich Rickert, *Die Grenzen der naturwissenschaftlichen Begriffsbildung, eine logische Einleitung in die historischen Wissenschaften* (Tübingen: J. C. B. Mohr, 1902), 356, 363: "In contrast to the person making a volitional judgment, the historian is not judging practically but theoretically, and thus always holds himself only as representing and not as evaluating—that is, he has in common with the practical man only the considering of a relationship, but not the willing and evaluating itself. . . . It is certainly not the place of the historical sciences to declare whether the individual reality, which the science represents, is valuable or malign.

This attitude of studied, objective neutrality has become codified in the commandments handed down by the National Center for History in the Schools, whose committee has recently proclaimed: "Teachers should not use historical events to hammer home their own favorite moral lesson."[5] Presumably this goes as well for the historian as teacher, whose texts the students study. And one might suppose that when the narrative describes episodes in the history of science, occasion for intrusive moral judgments would be quite limited.

The demand that historians disavow moral evaluations neglects a crucial aspect of the writing of history, whether it be political history or the history of science: the deep grammar of narrative history requires that moral judgments be rendered. And that is the thesis I will argue in this appendix, namely, that the historian's narrative must make moral assessments, not only in respect of the times in which the subject of the history operated but in respect of the historian's own moral standards. I will be especially concerned with an assessment that might be called that of "historical responsibility."

The role of moral judgment about past historical characters has been brought to eruptive boil recently in one area of the history of biology—that of nineteenth- and early twentieth-century evolutionary thought in Germany. The individual about whom considerable historical and moral controversy swirls is Ernst Haeckel, who offers a test case for my thesis. I will discuss the indictment brought against him in a moment. Now, I will simply point out that Haeckel, more than any other individual, was responsible for the warfare that broke out in the second half of the nineteenth century between evolutionary theorists and religiously minded thinkers, a warfare that continues unabated in the contemporary cultural struggle between advocates of intelligent design and those defending real biological science.

My motivation for considering the moral structure of narratives is encapsulated in the main title of a book that was published not long ago: *From Darwin to Hitler.*[6] The pivotal character in this historical descent, accord-

---

If that were the case, how should the science achieve a judgment that all would accept? On the contrary, what we understand under the rubric 'relation of an individual to a value' [in the common domain of his time] must be carefully distinguished from the historian's direct evaluation."

5. National Center for History in the Schools (UCLA), "Standard 5: Historical Issues." See http://nchs.ucla.edu/standards/thinking5-12-5.html.

6. Richard Weikart, *From Darwin to Hitler: Evolutionary Ethics, Eugenics, and Racism in Germany* (New York: Palgrave Macmillan, 2004).

ing to the author, is Ernst Haeckel. He and Darwin are implicitly charged with historical responsibility for acts that occurred after they themselves died. I don't think judgments of this kind, those attributing moral responsibility across decades, are unwarranted in principle. The warrant lies in the grammar of historical narrative. Whether this particular condemnation of Darwin and Haeckel is appropriate remains quite another matter.

The Temporal and Causal Grammar of Narrative History

Let me focus, for a moment, on two features of narrative history as a prelude to my argument and as an illustration of what I mean by the grammar of narrative. This concerns the ways time and causality are represented in narrative histories. Each seeps into narratives in at least four different ways.[7] Let me first consider, quite briefly, the temporal dimensions of narrative.

Embedded in the deep structure of narrative is the time during which events occur; that sort of time flows equitably on into the future, with each temporal unit having equal duration. Narratives project events as occurring in a Newtonian time. This kind of time, which we might call the *time of events,* allows the historian to place events in a chronology, to compare their durations, and to locate them in respect to one another as antecedent, simultaneous, or successive.[8]

The structuring of events in a narrative also exhibits *narrative time,* and this constitutes a different sort of temporal modality. Consider, for

7. I have discussed the temporal and causal structure of narrative history in "The Structure of Narrative Explanation in History and Biology," in *History and Evolution,* ed. Matthew Nitecki and Doris Nitecki (Albany: State University of New York Press, 1992), 19–53.

8. Under the rubric of Newtonian time (or time of events), one might distinguish longer periods of time, intermediate periods, and shorter periods—the kind of time intervals described by Fernand Braudel and the *Annales* school. The longer periods, *la longue durée,* encompass the slow movements of geological and environmental processes, the changing course of rivers and the alterations of coasts. These slow processes determine, according to Braudel, the prosperity (or lack thereof) of nations, the success of agriculture, the movements of armies, and the distribution of land. The intermediate periods, "conjunctures" in Braudel's terms, are the times of larger social phenomena: price cycles, demographic changes, spread of technological innovations. These social time scales will be affected by the slower geological and environmental changes, and in their turn have significant influence on the short times of momentary happenings: wars, discoveries, revolutions, elections, etc. These temporal distinctions, however, make sense only if one assumes, as Braudel does, that they occur through a time that flows equitably on, a Newtonian time. See Fernand Braudel, "History and the Social Sciences: The *Longue Durée,*" in his *On History,* trans. Sarah Matthews (Chicago: University of Chicago Press, 1980), 25–54.

instance, Harold Pinter's play *Betrayal*. The first scene is set temporally toward the end of the Newtonian sequence dramatized, with the next scene going in the right direction, occurring a few days later. But the third scene falls back to two years before, and the fourth a year before that, with subsequent scenes taking us back finally to a period six years before the final days with which the play opens. The audience, however, never loses its temporal bearings or believes that time staggers along, weaving back and forth like an undergraduate leaving the local college pub.

The historian might structure his or her narrative in a roughly comparable way, when one aspect of the history is related, but then the historian returns to an earlier period to follow out another thread of the story. Or the historian might have the narrative jump into the future to highlight the significance of some antecedent event. Again, when done with modest dexterity, the reader is never confused about the Newtonian flow of events. Narrative time is thus layered over the time of events.

The *time of narration* is a less familiar device by which historians restructure real time as well as narrative time. One of the several modes by which they construct this kind of time is through contraction and expansion of the number of sentences deployed. Let me illustrate what I mean by reference to a history with which most readers will be familiar: Thucydides' *History of the Peloponnesian War.* At the beginning of his story, Thucydides—a founder, along with Herodotus, of the genre of narrative history—expends a few paragraphs on events occurring in the earliest period of Cretan hegemony through the time of the Trojan War to just before the outbreak of the war between the two great powers of ancient Greece, Athens and Sparta. The temporal span he so economically describes in a few paragraphs extends, in Newtonian time, for about two thousand years. But Thucydides then devotes several hundred pages to the relatively brief twenty-year period of the war, at least that part of the war he recorded. Sentence duration is an indication of the importance historians place on the events to which they refer. Sentence expansion or contraction, however, may have other sustaining causes.

Simply the pacing or rhythm of a historian's prose might be one. The great French scientist and historian Bernard de Fontenelle said that if the cadences of his sentences demanded it, the Thirty Years' War would have turned out differently. Some historians will linger over an episode, not because it fills in a sequence vital to the tale, but because the characters involved are intrinsically interesting. Maybe some humorous event is inserted in the story simply to keep the reader turning the pages. In histories, centuries may be contracted into the space of a sentence, while moments may be expanded through dozens of paragraphs.

A fourth temporal dimension of narrative is the *time of narrative construction.* This is a temporal feature especially relevant to considerations of the moral structure of histories. A narrative will be temporally layered by reason of its construction, displaying, as it were, both temporal depth and a temporal horizon. The temporal horizon is more pertinent for my concerns, so let me speak of that. Thucydides wrote the first part of his history toward the end of the war that he described, when the awful later events allowed him to pick out those earlier, antecedent events of explanatory relevance—earlier events that would be epistemologically tinged with Athenian folly yet to come. Only the benefits of hindsight, for example, could have allowed him to put into the mouth of the Spartan messenger Melesipus, who was sent on a last desperate peace mission just before the first engagement of the war—to put into the mouth of this messenger the prophetic regret: "This day will be the beginning of great misfortunes to Hellas." By the horizontal ordering of time, the historian can describe events in ways that the actors participating in the events could not: Melesipus's prophecy was possible only because Thucydides had already lived through it. This temporal perspective is crucial for the historian. Only from the vantage of the future, can the historian pick out from an infinity of antecedent events just those deemed necessary for the explanation of the consequent events of interest.

Different causal structures of narratives correspond to these four temporal modalities. I will refrain from detailing all of their aspects, but let me quickly rehearse their dominant modes. The most fundamental causal feature of narratives is the *causality of events.* This is simply the causality ascribed to the events about which the historian writes. Typically the historian will describe events so as to depict and align their causal sequences, sequences in which the main antecedent causes are indicated so as to explain subsequent events, ultimately the central events that the history was designed to explain.

Events in a narrative, however, display a different causal grammar from events in nature. We may thus speak of the *causality of narrated events.* When in 433 B.C., the Athenians of Thucydides' history interfered in an internal affair of Corinth, a Spartan ally, they could not have predicted that war would result—though they might have suspected; they certainly could not have predicted their ignominious defeat in the Sicilian campaign twenty years later. From inside of the scene that Thucydides has set, the future appears open; all things are possible, or at least unforeseeable. Yet each of Thucydides' scenes moves inevitably and inexorably to that climax, namely, to the destruction of the fleet at Syracuse, the central event of his history. The historian, by reason of his or her temporal horizon, arranges antecedent events to make their outcome, the central event of interest, something the reader can expect—something, in the ideal case, that would be regarded as inevitable given the antecedent events, all the while keeping his actors in the dark until the last minute.

This is a view about the grammar of narration that some historians would not share. Some try assiduously to avoid surface terms redolent of causality in their narratives. But I think this is to be unaware of the deeper grammar of narrative. The antecedent events are chosen by the historian to make, as far as he or she is able, the consequent events a causal inevitability. That's what it is to explain events historically. To the degree this kind of causal structure is missing, to that degree the history will fail to explain how it is that the subsequent events of interests occurred or took the shape they did. Without a tight causal grammar, the narrative will begin to loosen to mere chronicle.

This grammatical feature of narrative has bearing on any moral characterization of the actions of the individuals about whom the historian writes. And this is done in two ways. First, we do think that when we morally evaluate an action, we assume the individual could have chosen otherwise. Good histories will evince a tension between the actor's belief that the future is open, full of possibilities, and the historian's knowledge that the future is closed and the possibilities restricted to what actually occurred. The actors did what they did because of the narrated events, events carrying those individuals to their appointed destiny. Yet when the historian looks back, he or she will also regard the actions in question from the agent's point of view, as behavior for which the agent is responsible. The grammar of narrative thus parses human events in two distinctive ways: one in which behavior is regarded as having antecedent causes that determine choices and fix outcomes—the explanatory mode; the other, in which the historian reflects on choices as freely flowing from individuals who are morally responsible for their actions—the normative mode. The grammar of narrative gives poignant expression to what might be called the Kantian dilemma: the critical requirement that we attempt to understand human events from the theoretical perspective and also from the moral perspective. In this latter instance, we reflect on the events from the agent's point of view, namely, as acts for which the agent is morally culpable. When histories relate the course of nonhuman events—for example, the evolutionary history of the rise of mammals—the grammatical structure is simpler, lacking as it does the moral dimension.

The second way the causality of narrated events bears on moral assessment has to do with the manner in which the historian renders a moral judgment: namely, by the construction of the sequence of events and their causal connections. The historian will arrange the sequence of events in which the character's actions are placed so as either morally to indict the individual or morally to exculpate the individual, or, what is more frequently the case, to locate the individual's action in a morally neutral plane. I'll say more about this feature of the grammar of narrative in a moment.

A third causal modality deployed by historians may be called the *causality of narration.* This aspect of causality has several features, but I will mention only one: this is the location in a narrative of various scenes. So, for example, Thucydides will place one scene before another to indicate what he presumes is an important condition or cause for a subsequent scene, even though the events described in the scenes may be at some temporal distance from one another. A speech made to motivate an action might be placed immediately before the scene in which the action is described, even though the speech and the resultant action may be separated by a fair amount of time, Newtonian time. Such juxtaposition can be used for other purposes as well—for example, to plumb the deeper dispositions of the actors. Immediately after Thucydides relates Pericles' great funeral oration, which extols the virtues of Athenian democracy and the glories of its laws, he shoves in a dramatic description of the Athenian plague, when citizens ignored the laws and each sought his own pleasure, thinking it might be his last. Yet Pericles' oration and the plague were separated by six months. This kind of causality effectively allows the reader to glimpse below the surface glitter of Athenian society.

Finally, there is the *causality of narrative construction.* There are two quite different causal features that would fall under this rubric. The first concerns what might be called the *final cause* in narrative construction. Most histories aim to explain some central event—the outbreak of the American Civil War, Darwin's discovery of natural selection, or the racial attitudes of Hitler. The antecedent events in the history provide the causal explanation of the central event, which latter might be thought of as the final cause, that is, the goal of the construction. Historians in their research use this final cause as the beacon in light of which they select out from an infinity of antecedent events just those that might explain the central event. No historian begins, as it were, at the beginning, rather at the end.

Without the final cause as guide—a guide that may alter, of course, during the research—the historian could not even start to lay out those antecedent causes that he or she will finally regard as the explanation for the end point of the historical sequence.

A related feature of the causality of narrative construction fixes on the motives guiding the historian, of which there may be several. The proclaimed motivation of the great nineteenth-century historian Leopold von Ranke was to describe an event "wie es eigentlich gewesen," how it actually was; and, insofar as *how it was* becomes the central event that needs explanation, that event—the final cause—furnishes the motive for constructing the history. Ranke's general standard must be that of every historian. Good historians will want to weigh purported causes of events and emphasize the most important, while reducing narrative time spent on the less important. Yet often other motivations, perhaps hardly conscious even to the historian, may give structure to his or her work. In his suspicious little book *What Is History?* E. H. Carr urged that "when we take up a work of history, our first concern should be not with the facts which it contains but with the historian who wrote it."[9] If the reader knows in advance that the historian is of a certain doctrinal persuasion, then a judicious skepticism may well be in order. After all, a historian may select events that have real but minor causal connections with central events of concern, while ignoring even more important antecedent causes. The history would then have a certain verisimilitude, yet be a changeling. Motivations of authors are often revealed by the moral grammar of narratives, another structural feature that lies in the syntactic depths of historical accounts.

The Moral Grammar of Narrative History

I am going to turn now specifically to the moral grammar of histories and then illustrate some of the ways this structure characterizes Ernst Haeckel's story. If narratives have a moral grammar, then it would be well for historians to become reflectively conscious of this and to formulate their reconstructions in light of some explicit principles. And in a moment I will suggest what those principles ought to be by which we morally judge the behavior of individuals who lived in the past and by which we assess their culpability for the future actions of others.

But let me first pose the question: Do historians make normative judg-

9. E. H. Carr, *What Is History?* (New York: Vintage, 1961), 24.

ments in their histories and should they? I will argue that not only should they, they must by reason of narrative grammar. At one level, it is obvious that historians, of necessity, do make normative judgments. Historical narratives are constructed on the basis of evidence, usually written documents—letters, diaries, published works—as well as oral statements. At times they also use artifacts as evidence, for instance, archaeological findings; and increasingly high-tech instruments, such as DNA analysis, are coming into play. Historians attribute modes of behavior to actors on the basis of inference from such evidence and in recognition of certain standards. Even when doing something apparently as innocuous as selecting a verb to characterize a proposition attributed to an actor, the historian must employ a norm or standard. For example, Thucydides could have had Melesipus *think* that disaster was in the offing, *believe* that disaster was in the offing, *be convinced* that disaster was in the offing, *suspect* that disaster was in the offing, *assume* that disaster was in the offing, or *prophesy* that disaster was in the offing. Whatever verb the historian selects, he or she will do so because the actor's behavior, as suggested by the evidence, has met a certain standard for such-and-such modal description—say, being in a state of firm conviction as opposed to vague supposition. All descriptions require measurement against standards or norms—which is not to say that in a given instance, the standard and consequent description would be the most appropriate. The better the historian, the more appropriate the norms employed in rendering the narrative.

Virtually all of the historian's choices of descriptive terms must be normative in this sense. But must some of these norms also be moral norms? I believe they must. The argument is fairly straightforward—at least as straightforward as arguments of this sort ever get. Human history is about res gestae, things done by human beings, human actions. Actions are not mere physiological behaviors, but behaviors that are intended and motivated. Inevitably these actions impinge on other individuals, immediate or remote. But intentional behavior impinging on others is precisely the moral context. The historian, therefore, in order to assign motives and intentions to individuals whose behaviors affect others and to describe those motives and intentions adequately—that historian must employ norms governing such intentional behaviors, that is, behaviors in the moral context.

Certainly the assessment of motives and intentions may yield only morally neutral descriptions. But even deciding that an intended behavior is morally neutral requires the historian, implicitly at least, also to judge it against standards of positive or negative moral valence and to decide that it conforms to neither. Even a morally neutral assessment is a moral assessment. I make no claim here, of course, that historians usually render such evaluations self-consciously. Mostly these evaluations occur reflexively, instead of reflectively. And they usually exist, not explicitly on the surface of the narrative, but in the interstices.

Let me offer a concrete example of the case that I am arguing, and this from a historian whom no one would accuse of cheap moralizing—his moralizings are anything but cheap. His descriptions reveal a rainbow of shaded moral evaluations, which range subtly between the polar categories of shining virtue and darkling vice. Byron called him the Lord of Irony, and it is often that trope through which he makes his moral assessments. I'm speaking, of course, of Edward Gibbon.

Let me quote just a short passage from the *Decline and Fall of the Roman Empire,* in which Gibbon describes what might have been the motives of Julian, as his soldiers were clamoring for his elevation to emperor even while Constantius still occupied the throne. Julian protested that he could not take the diadem, even as he reluctantly and sadly accepted it. Gibbon writes:

> The grief of Julian could proceed only from his innocence; but his innocence must appear extremely doubtful in the eyes of those who have learned to suspect the motives and the professions of princes. His lively and active mind was susceptible of the various impressions of hope and fear, of gratitude and revenge, of duty and of ambition, of the love of fame and of the fear of reproach. But it is impossible for us to ascertain the principles of action which might escape the observation, while they guided, or rather impelled, the steps of Julian himself. . . . He solemnly declares, in the presence of Jupiter, of the Sun, of Mars, of Minerva, and of all the other deities, that till the close of the evening which preceded his elevation he was utterly ignorant of the designs of the soldiers; and it may seem ungenerous to distrust the honour of a hero, and the truth of a philosopher. Yet the superstitious confidence that Constantius was the enemy, and that he himself was the favourite, of the gods, might prompt him to desire, to solicit, and even to hasten the auspicious moment of his reign, which was predestined to restore the ancient religion of mankind.[10]

10. Edward Gibbon, *The History of the Decline and Fall of the Roman Empire,* 6 vols. (London: Strahan and Cadell, 1777–88), 2:319–20.

In the cascade of rhetorical devices at play—zeugma, antithesis, irony—Gibbon explicitly refuses to attribute morally demeaning motives to Julian, and, of course, at the same time implicitly does precisely that.

There is another element of judgment that Gibbon evinces here, which is also an important feature of the moral grammar of historical narrative. Narratives explain action by allowing us to understand character, in this case Julian's character. Gibbon, however, has led us to comprehend Julian's action, not only by analytically diagnosing what the motives of a prince might be but also by shaping our emotional response to Julian's character and thus producing in us a feeling about his behavior. We morally evaluate individuals, partly at least, through feelings about them. The historian can orchestrate outrage—as some dealing with Haeckel have—by cutting quotations into certain vicious shapes, selecting those that appear damning while neglecting those that might be exculpating. Or, like Gibbon, the historian can evoke feelings of moral disdain with little more than the mist of antithetic possibilities. As a result, readers will have, as it were, a sensible, an olfactory understanding: the invisible air of the narrative will carry the sweet smell of virtue, the acrid stench of turpitude, or simply the bitter sweet of irony. These feelings will become part of the delicate moral assessment rendered by the artistry of the historian.

This is just one small example of the way moral judgment exists in the interstitial spaces of a narrative, instead of lying right on the surface. But sometimes such judgments do lie closer to the skin of the narrative. Let me now focus precisely on an instance of this and consider the principles that, I believe, should be operative in making moral judgments of historical figures. This is in the case of Ernst Haeckel.

The Case of Ernst Haeckel

Haeckel was Darwin's great champion of evolutionary theory in Germany; he was a principal in the theory's introduction there and a forceful defender of it from the mid-1860s until 1919, when he died. Haeckel's work on evolution reached far beyond the borders of the German lands. His popular accounts of evolutionary theory were translated into all the known and unknown languages—at least unknown to the West—including Armenian, Chinese, Hebrew, Sanskrit, and Esperanto. More people learned of evolutionary theory through Haeckel's voluminous writings during this period than from any other source, including Darwin's own work.

Haeckel achieved many popular successes and, as well, produced more than twenty large, technical monographs on various aspects of systematic biology and evolutionary theory. In these works he described many hitherto unknown species, established the science of ecology, gave currency to the idea of the missing link—which one of his protégés (Eugène Dubois) actually found—and promulgated the biogenetic law that ontogeny recapitulates phylogeny. Most of the promising young biologists of the next generation came to study with him at Jena. His artistic ability was considerable; and at the beginning of the twentieth century, he influenced the movement in art called Jugendstil. Haeckel became a greatly celebrated intellectual figure, often mentioned for a Nobel Prize. He was also the scourge of religionists, smiting the preachers at every turn with the jawbone of evolutionary doctrine. He advocated what he called a "monistic religion" as a substitute for the traditional orthodoxies, a religion based on science.

As a young student, trying to find a subject for his habilitation, Haeckel roamed along the coasts of Italy and Sicily in some despair. He thought of giving up biology for the life of a bohemian, spending his time in painting and poetizing with other German expatriates on the island of Ischia. But he felt that he had to accomplish something in biology, so that he could become a professor and marry the woman with whom he had fallen deeply in love—his love letters sent back to his fiancée in Berlin are something delicious to read. He finally hit upon a topic: a systematic description of a little known creature that populated the seas, the one-celled protist called a radiolarian. It was while writing his habilitation on these creatures in 1861 that he happened to read Darwin's *Origin of Species* and became a convert. Haeckel produced a magnificent two-volume tome on the radiolaria (see plate 1 and fig. 3.4), which he himself illustrated with extraordinary artistic and scientific acumen.[11] Later in the century, his illustrations of radiolaria would influence such artistic designs as René Benet's gateway to the Paris World's Fair of 1900 (see fig. 10.3). But the radiolarian monograph's most immediate and significant effect was to secure Haeckel a professorship at Jena, thus allowing him to marry his beloved cousin, Anna Sethe.

On his thirtieth birthday in 1864, Haeckel learned he had won a prestigious prize for his radiolarian work. And on that same day, a day that should have been of great celebration, his wife of eighteen months tragically died. Haeckel was devastated. His family feared he might commit suicide. As he related to his parents, this heart-crushing blow led him to reject all religion and replace it with something more substantial, some-

11. Ernst Haeckel, *Die Radiolarien (Rhizopoda Radiaria). Eine Monographie*, 2 vols. (Berlin: Georg Reimer, 1862). See chapter 3, above.

thing that promised a kind of progressive transcendence, namely, Darwinian theory.

In the years following this upheaval, Haeckel became a zealous missionary for his new faith, and his own volatile and combative personality made him a crusader whose demeanor was in striking contrast to that of the modest and retiring English master whom he would serve. This outsize personality has continued to irritate historians of smaller imagination.

The Moral Indictment of Haeckel

In 1868 Haeckel produced a popular volume on the new theory of evolution entitled *Natürliche Schöpfungsgeschichte* (Natural history of creation).[12] It would go through twelve editions up to the time of his death in 1919 and prove to be the most successful work of popular science in the nineteenth and early twentieth centuries. There are two features of that book that incited some of the fiercest intellectual battles of the last part of the nineteenth century and have more recently led not a few theologians, philosophers, and historians to comparably fierce judgments of Haeckel's moral probity.

The first has to do with what became the cardinal principle of his evolutionary demonstrations, namely, the biogenetic law that ontogeny recapitulates phylogeny. This principle holds that the embryo of a developing organism goes through the same morphological stages as the phylum went through in its evolutionary history: so, for example, the human embryo begins as a one-celled creature, just as we presume life on this earth began in a one-celled mode; it then goes through a stage of gastrulation to produce a cuplike form, similar, Haeckel believed, to a primitive ancestor that plied the ancient seas; then the embryo takes on the structure of an archaic fish, with gill arches; then of a primate; then of a specific human being.

The corollary to the law is that closely related creatures—vertebrates, for example—will pass through early embryological stages that are quite similar to one another. Some of Haeckel's enemies charged that he exaggerated the similarity in his graphic illustrations by lengthening the tail of the human embryo to make it more animal-like—a controversy that became known as *Die Schwanzfrage*. But the deeper, more damaging fight came with Haeckel's depiction of quite young embryos, those at the sandal stage, when they look like the sole of a sandal (see fig. 7.12). In the accom-

12. Ernst Haeckel, *Natürliche Schöpfungsgeschichte* (Berlin: Georg Reimer, 1868). See chapter 7, above.

panying text to his illustration, Haeckel remarked: "If you compare the young embryos of the dog, chicken, and turtle . . . , you won't be in a position to perceive a difference."[13] One of the very first reviewers of Haeckel's book, an embryologist who became a sworn enemy, pointed out that one certainly wouldn't be able to distinguish these embryos, since Haeckel's printer had used the same woodcut three times. The evolutionist had, in the terms of the reviewer, committed a grave sin against science and the public's trust in science.[14]

In the second edition of his book, Haeckel retained only one illustration of an embryo at the sandal stage and remarked in the text: It might as well be the embryo of a dog, chicken, or turtle, since you cannot tell the difference. The damage, however, had been inflicted, and the indictment of fraud haunted Haeckel for the rest of his life. The charge has been used by creationists in our own day as part of a brief not only against Haeckel but against evolutionary theory generally. But the creationists are not alone; several historians have employed it in their own moral evaluation of Haeckel and his science.

The second feature of Haeckel's work on which I would like to focus really did not create a stir in his own time but has become a central moral issue in ours. This has to do with the assumption of progress in evolution, an assumption that Haeckel certainly made. That assumption is forcefully displayed in the tree diagram appended to his *Natürliche Schöpfungsgeschichte*. The diagram displays the various species of humankind, with height on the vertical axis meant to represent more advanced types (fig. 7.13). Here the Caucasian group leads the pack, arching above the descending orders of the "lower species"—all rooted in the *Urmensch* or *Affenmensch*, the ape-man. A salient feature of the diagram should catch our attention: among the varieties of the Caucasian species, the Berbers and Jews were placed by Haeckel on the same vertical level of advancement as the Germans and southern Europeans. This classification should have had bearing on Haeckel's assignment by some historians to the ranks of the proto-Nazis.[15]

13. Ibid., 249.

14. Ludwig Rütimeyer, Review of "Ueber die Entstehung und den Stammbaum des Menschengeschlechts" and *Natürliche Schöpfungsgeschichte*, by Ernst Haeckel, *Archiv für Anthropologie* 3 (1868): 301–2. See chapter 8, above.

15. There is direct evidence for Haeckel's attitude about Jews, beyond his placement of them among the advanced races. In the early 1890s, he discussed the phenomenon of anti-Semitism with the Austrian novelist and journalist Hermann Bahr (1863–1934), who collected almost forty interviews with European notables on the issue, such individuals as August Bebel, Theodor Mommsen, Arthur James Balfour, and Henrik Ibsen. In his discussion with Bahr, Haeckel did think that lower-class Russian Jews would stand as an offense to the high stan-

Nazi Race Hygienists and Their Use of Haeckelian Ideas

Several Nazi race hygienists clearly appealed to Haeckel's work to justify their views. One pertinent example is Heinz Brücher, whose *Ernst Haeckels Bluts- und Geistes-Erbe* (Ernst Haeckel's racial and spiritual legacy, 1936) attempted to recruit Haeckel to the Nazi cause.[16] Not only did Brücher look to Haeckel's views of racial hierarchy as support for the policies of National Socialism; he gave full account of Haeckel's own impeccable pedigree as a sign of the integrity of the evolutionist's ideas. Included with the book was a five-foot chart laying out Haeckel's family tree. The implied argument was simply that only the best blood flowed through Haeckel's veins, so we may trust his ideas.

Brücher regarded Haeckel as a harbinger of "biological state-thinking" and made his case by citing *Die Lebenswunder* (The wonder of life, 1904), a book that flowed from the enraged pen of a man stricken by the tragic death of an intimate friend and suffering the loss of a daughter to mental illness.[17] From the turn of the century through the next two decades, Haeckel was besieged by the religiously orthodox; and, provocatively, he gave them cause. He suggested that the Spartans and other "natural peoples," who inspected infants at birth and killed those with debilitating diseases and monstrous physical deformities, had wisely solved a problem: "Is it not much more reasonable and better to remove at the beginning of the course of life the unavoidable suffering that their pitiable existence must bring to themselves and their family?" [18] Brücher thought Haeckel's attitudes quite in line with the sterilization policy of the Nazis.[19]

---

dards of German culture. But of educated German Jews, he remarked: "I hold these refined and noble Jews to be important elements in German culture. One should not forget that they have always stood bravely for enlightenment and freedom against the forces of reaction, inexhaustible opponents, as often as needed, against the obscurantists [*Dunkelmänner*]. And now in the dangers of these perilous times, when papism again rears up mightily everywhere, we cannot do without their tried-and-true courage." There is simply no reason to believe Haeckel to be racially anti-Semitic, as several historians have assumed. See Hermann Bahr, "Ernst Haeckel," in *Der Antisemitismus: Ein internationals Interview* (Berlin: S. Fischer, 1894), 62–69; quotation from 69. See also chapter 7, above.

16. Heinz Brücher, *Ernst Haeckels Bluts- und Geistes-Erbe: Eine kulturbiologische Monographie* (Munich: Lehmanns, 1936).

17. See chapter 10, above.

18. Ernst Haeckel, *Die Lebenswunder: Gemeinverständliche Studien über Biologische Philosophie, Ergänzungsband zu dem Buche über die Welträthsel* (Stuttgart: Alfred Kröner, 1904), 135–36.

19. Heinz Brücher, "Ernst Haeckel, ein Wegbereiter biologischen Staatsdenkens," *NS-Monatschrifte* 6 (1935): 1088–98; passage on 1098.

To tighten further the connection between the National Socialists and Haeckel, Brücher focused on a passage from the *Natürliche Schöpfungsgeschichte* that reads: "The difference in rationality between a Goethe, a Kant, a Lamarck, a Darwin and that of the lower natural men—a Veda, a Kaffer, an Australian and a Papuan is much greater than the graduated difference between the rationality of these latter and that of the intelligent vertebrates, for instance, the higher apes." Brücher then cited a quite similar remark by Hitler in his Nuremberg speech of 1933.[20] Through his several citations, he made Haeckel historically responsible, at least in part, for Hitler's racial attitudes.[21]

The Judgment of Historical Responsibility

Brücher's attribution of moral responsibility to Haeckel is of a type commonly found in history, though the structure of these kinds of judgments usually goes unnoticed, lying as it does in the deep grammar of narrative. For example, historians will often credit, say, Copernicus, who worked in the early sixteenth century, with the courage to have broken through the rigidity of Ptolemaic assumption and thus, by unshackling men's minds, to have initiated the scientific revolution of the late sixteenth and seventeenth centuries. This, too, is a moral appraisal of historical responsibility, though, needless to say, Copernicus himself never uttered: "I now intend to free men's minds and initiate the scientific revolution." Yet historians

20. Brücher, *Ernst Haeckels Bluts-*, 90–91.

21. Uwe Hoßfeld discusses Brücher's work and its intentions in his *Geschichte der biologischen Anthropologie in Deutschland von den Anfängen bis in die Nachkriegszeit* (Stuttgart: Franz Steiner, 2005), 312–16. See also Uwe Hoßfeld, "Nationalsozialistische Wissenschaftsinstrumentalisierung: Die Rolle von Karl Astel und Lothar Stengel von Rutkowski bei der Genese des Buches 'Ernst Haeckels Bluts- und Geistes-Erbe' (1936)," in *Der Brief als wissenschaftshistorische Quelle,* ed. Erika Krauße (Berlin: Verlag für Wissenschaft und Bildung, 2005), 171–94. Uwe Hoßfeld and Thomas Junker consider the degree to which some German evolutionists succumbed to Nazi racial policy in their "Anthropologie und synthetischer Darwinismus im Dritten Reich: Die Evolution der Organismen (1943)," *Anthropologischer Anzeiger* 61 (2003): 85–114. Brücher specialized in plant genetics. He fled Germany for Argentina after the war and in 1991 was assassinated either by drug smugglers or, perhaps, the Mossad. Both groups had reason to dispatch him: the smugglers because Brücher was developing an infectious pathogen to attack coca plants; the Mossad because during the war he was a member of the SS and had led raids on agricultural stations in the Ukraine. See the recollections of an American who met him in South America: Daniel W. Gade, "Converging Ethnobiology and Ethnobiography: Cultivated Plants, Heinz Brücher, and Nazi Ideology," *Journal of Ethnobiology* 26 (2006): 82–106. See also Ute Deichmann, *Biologists under Hitler,* trans. Thomas Dunlap (Cambridge, MA: Harvard University Press, 1996), 258–64.

do assign him credit for that, moral credit for giving successors the ability to think differently and productively.

The epistemological and historical justification for this type of judgment is simply that the meaning and value of an idea or set of ideas can be realized only in actions that themselves may take some long time to develop—this signals the ineluctable teleological feature of historical thinking and writing. While this type of judgment lies embedded in the moral grammar of history, this does not mean, of course, that every particular judgment of this sort is justified.

The Reaction of Contemporary Historians

How has Haeckel gone down with contemporary historians?[22] Not well. His ideas, mixed with his aggressive and combative personality, have lodged in the arteries feeding the critical faculties of many historians, causing sputtering convulsions. Daniel Gasman has argued that Haeckel's "social Darwinism became one of the most important formative causes for the rise of the Nazi movement."[23] Stephen Jay Gould and many others concur that Haeckel's biological theories—supported, as Gould contends, by an "irrational mysticism" and a penchant for casting all into inevitable laws—"contributed to the rise of Nazism."[24] And more recently, in *From Darwin to Hitler,* Richard Weikart traces the metastatic line his title describes, with the midcenter of that line encircling Ernst Haeckel.

Weikart offers his book as a disinterested historical analysis. In that objective fashion that bespeaks the scientific historian, he declares: "I will leave it the reader to decide how straight or twisted the path is from Darwinism to Hitler after reading my account."[25] Well, after reading his account, there can be little doubt not only of the direct causal path from Charles Darwin through Ernst Haeckel to Adolf Hitler but of Darwin's and Haeckel's complicity in the atrocities committed by Hitler and his party. The evolutionists bear, in Weikart's analysis, historical responsibility.

Taking E. H. Carr's advice to heart, we might initially be suspicious

22. See chapter 11 for an extended portrayal of Haeckel's treatment by twentieth-century historians.

23. Daniel Gasman, *The Scientific Origins of National Socialism: Social Darwinism in Ernst Haeckel and the German Monist League* (New York: Science History Publications, 1971), xxii. See also Daniel Gasman, *Haeckel's Monism and the Birth of Fascist Ideology* (New York: Peter Lang, 1998).

24. Stephen Jay Gould, *Ontogeny and Phylogeny* (Cambridge, MA: Harvard University Press, 1977), 77–81.

25. Weikart, *From Darwin to Hitler,* x.

Fig. app. 2.1. Plaque in the main university building at Jena: "The truth will make you free. To the victims of political repression at the Friedrich-Schiller University, 1933–1945, 1945–1989." (Photograph by the author.)

of Weikart's declaration of objectivity, coming as it does from a member of an organization having strong fundamentalist motivations.[26] Nonetheless, other historians have made similar suggestions, at least in the case of Haeckel—Gould and Gasman, for instance.

It is yet disingenuous for Weikart to pretend that most readers might come to their own conclusions despite the moral grammar of his history. Weikart, Gasman, Gould, and many other historians have created a historical narrative implicitly following—they could not do otherwise—the principles of narrative grammar: they have conceptualized an end point—Hitler's behavior (the final cause) regarded here as ethically horrendous—and have traced back causal lines to antecedent sources that might have given rise to those attitudes of the Nazi leader, tainting the sources along the way. It is like a spreading oil slick carried on an indifferent current and polluting everything it touches.

In the main text of this volume (especially in chapters 7 and 11), I have objected to the ways in which these historians have attempted to link Haeckel with the rise of the Nazis and the actions of Hitler in particular. They have not, for instance, properly weighed the significance of the many

26. Weikart is a member of the Discovery Institute, which supports the movement of intelligent design.

other causal lines that led to the doctrines of National Socialism—the social, political, cultural, and psychological strands that many other historians have emphasized. They have thus produced a mono-causal analysis that quite distorts the historical picture. They have ignored the many other nineteenth-century artists, scientists, and philosophers—Beethoven, Humboldt, Nietzsche—who had been recruited into the party by Nazi writers and propagandists. They have also failed to consider the explicit denials by National Socialist representatives of any contribution of Haeckel's Darwinian materialism to official *völkische Biologie.*[27]

While responsibility assigned to Darwin and Haeckel might be mitigated by a more realistic weighing of causal trajectories, some culpability might, nonetheless, remain. Yet are there any considerations that might make us sever, not the causal chain—say, from Darwin's writing, to Haeckel's, to Brücher's, to memos of high-ranking Nazis, and finally to Hitler's speeches—but the cord of moral responsibility? After all, Haeckel—and, of course, Darwin—had been dead decades before the rise of the Nazis.

Let me summarize at this juncture the different modal structures of moral judgment in historical narratives that I've tried to identify. First is the explicit appraisal of the historian, rendered when the historian overtly applies the language of moral assessment to some decision or action taken by a historical figure. This is both rare and runs against the grain of the cooler sensibilities of most historians, Lord Acton excepted.

Second is the appraisal of contemporaries (or later individuals). Part of the historian's task will often be to describe the judgments made on an actor by his or her own associates or temporally proximate individuals. This is the objective assessment of morals that, as I mentioned, Rickert would regard an obligation on the part of the historian. In the case of Haeckel, there were those who condemned him of malfeasance, as well as colleagues who defended him against the charge. The norms of a community—for example, German scientists of the second half of the nineteenth century—may be fairly uniform; but the evaluations in terms of those norms will likely be heterogeneous, as they certainly were in the case of Haeckel. In

27. See, for instance, Günther Hecht, "Biologie und Nationalsozialismus," *Zeitschrift für die Gesamte Naturwissenschaft* 3 (1937–38): 280–90: "*The common position of materialistic monism is philosophically rejected completely by the völkisch-biological view of National Socialism.* Any further or continuing scientific-philosophic disputes concerning this belong exclusively to the area of scientific research. The party and its representatives must not only reject a part of the Haeckelian conception—other parts of it have occasionally been advanced—but, more generally, every internal party dispute that involves the particulars of research and the teachings of Haeckel must cease" (285). Hecht represented Das Rassenpolitische Amt der NSDAP. See chapter 11, above.

such instances, the historian will highlight some evaluations and demote others; point out the uncertainty of some and the dubious motivations of others. In so doing, he or she must be employing covertly personal norms of probity.

Third is the appraisal by causal connection. This occurs when the historian joins the decisions of an actor with consequential behavior of moral import. The behavior itself might be that of the actor or behavior displaced at some temporal remove from the actor's overt intentions—Hitler's actions, for example, as supposedly promoted by Haeckel's conceptions. This latter is what I have called "historical responsibility." The causal trajectory moves from past to future, but the moral responsibility flows along the causal tracks from future back to past. It is the guiding hand of the historian—fueled by a complex of motives—that pushes this historical responsibility back along the causal rails to the past. And here a minor causal connection can be mistaken for a major moral bond. I will, in just a moment, indicate how I believe the historian ought reflectively to modulate the flow of responsibility.

Finally, there is appraisal by aesthetic charge. This occurs when the historian through artful design evokes a feeling of positive or negative regard for the actor. In the treatment of Haeckel by Gould, Gasman, and Weikart, the needle of regard has swung to the decidedly negative end of the scale.

Principles of Moral Judgment

This brings me to the final part of my argument, namely, the principles that ought to govern, in a reflective way, our moral judgments about historical figures, especially for actions that were at some temporal distance from their own historical positions. The same general principles I will discriminate are the ones that serve as standards for the assessments of our contemporaries, including ourselves. But much will depend on how those principles get specified when judging historically remote individuals.

First, there is the supreme principle of evaluation: it might be the golden rule, the greatest happiness, altruism, or the categorical imperative. Likely in the cases I have in mind, any of these presumptive first principles will yield a similar assessment of moral motives, since they express, I believe, the same moral core. Second, there is the intention of the actor: What did he or she attempt to do? What action did the individual desire to execute, to be distinguished, of course, from mere accidental behavior? Third is the motive for acting, the ground for that intention to act in a certain fashion. The motive will determine moral valence. Finally, in assessing moral behavior, we must examine the beliefs of the individual actor and try to determine whether they were reasonable beliefs—and this is the special provenance of the historian. Let me give an example.

When the Hippocratic physicians, during the great Athenian plague that Thucydides so dramatically described, purged and bled the afflicted, their treatment actually hastened the deaths of their patients. But we certainly do not think the physicians malign or malfeasant, since they had a reasonable belief in the curative power of their practices. Their intention was to apply the best therapeutic techniques. And their motive, we may presume, was altruistic, since they risked their own lives in caring for the sick. One should judge them, I believe, moral heroes, even though the consequence of their behavior was injury and even the death of their patients.

The case of Ernst Haeckel is decidedly more problematic. The historian quite properly and, in Rickert's sense, "objectively" tries to assess the actions of Haeckel against the deposit of values in the late nineteenth century. But the historian will also examine the several charges and efforts at exculpation made by Haeckel's contemporaries. In so doing, he or she will, inevitably, judge those nineteenth-century evaluations as adequate or as wanting; and that judgment will be made—must be made—in light of the historian's own standards. This rendering by the historian, though, will usually be covert. The covert judgments will occur in the historian's selection and assessment of those charges by Haeckel's colleagues, in the quotations deployed, and in the evaluation of the motives of those who made the charges. Again, the historian must ultimately make these assessments in light of his or her own standards of behavior. This would constitute a second order moral evaluation—a judgment of the moral adequacy of the then-contemporary moral assessments of Haeckel.

The historian would also have to explore the moral probity of Haeckel's replication of woodcuts, which would require the recovery of the scientist's intentions and motivations. Did Haeckel claim his woodcuts were evidence for the biogenetic law? If so, he must have been motivated to deceive, and we may be thus entitled to suspect his character. Or did he merely intend to provide an illustration of the law for a general audience, recognizing that indeed at an early stage the embryos cannot be distinguished? And thus at best, through a false economy, he committed a very minor infraction, one that would not rise to the level of fraud and moral condemnation.

Concerning Haeckel's conception of a racial hierarchy, the historian has the task of exploring two questions in particular: What did he intend to accomplish by his theory? And how reasonable were the beliefs he harbored about races? To take the first question: Could Haeckel's actions be understood as intending to set in motion something like the crimes of the Nazis? Or minimally, did he exhibit a careless disregard for the truth of his views about races, so that some malfeasant act could have been vaguely anticipated? It is in answering these questions that the grammar of narrative must be carefully observed. The historian may lay down the scenes of his or her history so as to lead causally to a central event, such as Hitler's racial beliefs and their results in the Holocaust; but the historian needs to keep the actors in the dark—insofar as it is reasonable to do so—about those future consequences. In this case, to keep Haeckel oblivious to the future use of his work. The historian may easily slip, since he or she knows the future outcome of the actor's decisions. It is easy to assume the actor also knew or could have anticipated those outcomes, at least in some vague way. More likely, though, the historian might simply fail to reflect on the crucial difference between his or her firm knowledge of the past and the actor's dim anticipation of the future.

In addition to carefully assessing intentions and motives, the historian must also consider the set of beliefs harbored by the actor. For example, was it reasonable for someone like Darwin or Haeckel to believe that evolutionary theory led to the postulation of a hierarchy of species within a genus or races within a species? Or did they hold these ideas in reckless disregard for the truth? To assess reasonableness of belief in this instance, the historian would have to know what the scientific consensus happened to be in the second half of the nineteenth century. And in this case, a modestly diligent historian would discover that the community of evolutionary theorists—as well as other biologists—did understand the human races to stand in a hierarchy, which might conveniently be represented by the graphic device of a branching tree. It was a scale comparable to that by which other animals were measured in regard to their evolved status. In the human case, the criterial traits establishing the level of ascent included those of intelligence, moral character, and beauty. Nineteenth-century evolutionary theory implied such a ranking, and all of the available evidence and methods supported it. We might recognize from our perspective certain social factors constraining the judgments of biologists; but it is safe to say they did not.[28]

Then the historian can further ask, in this particular instance, what does categorizing peoples as branches of a racial hierarchy mean for the treatment of those so classified? This question does not allow of a univer-

28. See chapter 7.

sal answer, but will depend more particularly on the individual scientist. Weikart, for instance, indicts Darwin for acceding to belief in a racial hierarchy, but neglects to mention that Darwin did not think any action should be taken to reduce the welfare of those lower in the scale. Haeckel's own attitudes about how one should treat those lower in the hierarchy is less explicit; but there is hardly room for moral condemnation, given that his own many interactions with peoples he regarded lower in the scale were, as far as we know, quite benign.

Conclusion

It can only be a tendentious and dogmatically driven assessment that would condemn Darwin for the crimes of the Nazis. And while some of Haeckel's conceptions were recruited by a few Nazi biologists, he hardly differed in that respect from Christian writers, whose disdain for Jews gave considerably more support to those dark forces. One might thus recognize in Haeckel a causal source for a few lines deployed by National Socialists, but hardly any moral connection exists by which to indict him.

A historian cannot write an extended account of the life of an individual without some measure of identification—always a possible source of less than cold, decisive judgment. The good historian will find in his or her own character something of the features of the individuals about whom he or she writes. That is a necessary source for understanding the actions of subjects and, of course, is one of the lures to take up a particular history initially. If one is going to recover the past with anything like verisimilitude, one must, as R. G. Collingwood has maintained, relive the ideas of the past, which is not only to unearth a long-interred intellectual corpus but also to feel again the pulse of its vitality, to sense its urgency, to admire its originality, and thus to empathize with its author. And yet one has to do all of this while retaining a reflective awareness of the moral structure in which an actor conceived those ideas and perceived their import.

BIBLIOGRAPHY

ARCHIVAL SOURCES

Correspondence and Papers of Charles Darwin, Manuscript Room, Cambridge University Library.

Correspondence and Papers of Ernst Haeckel, Ernst-Haeckel-Haus, Institut für Geschichte der Medizin, und der Naturwissenschaften und Technik, Friedrich-Schiller-Universität, Jena.

Correspondence and Papers of Richard Owen, Natural History Museum, London.

Maastricht Natural History Museum, the Netherlands.

Nachlaß of Ernst Haeckel, Staatsbibliothek Berlin (Preussischer Kulturbesitz).

Nachlaß of Rudolf Virchow, Archiv der Akademie der Wissenschaften, Berlin.

Nachlaß of Wilhelm Wundt, Universitäts Archiv, Leipzig.

Papers of Charles Lyell, Edinburgh University.

PRINTED SOURCES

Acton, John. "On the Study of History." In *Lectures on Modern History.* London: Macmillan, 1906.

Adickes, Erich. "The Philosophical Literature of Germany in the Years 1899 and 1900." *Philosophical Review* 10 (1901): 386–416.

Allen, Garland, et al. "Sociobiology—Another Scientific Determinism." *BioScience* 26 (1976): 182, 184–86.

Alter, Stephen. *Darwinism and the Linguistic Image.* Baltimore: Johns Hopkins University Press, 1999.

Anderson, O. Roger. *Radiolaria.* New York: Springer, 1983.

Anonymous. *Der Ausgang des Prozesses Haeckel-Hamann.* Magdeburg: Listner and Drews, 1893.

Baer, Karl Ernst von. *Über Entwickelungsgeschichte der Thiere: Beobachtung und Reflexion.* 2 vols. Königsberg: Bornträger, 1828–37.

Bahr, Hermann. "Ernst Haeckel." In *Der Antisemitismus: Ein internationals Interview*, 62–69. Berlin: S. Fischer, 1894.

Bannister, Robert. *Social Darwinism: Science and Myth in Anglo-American Social Thought*. Philadelphia: University of Pennsylvania Press, 1979.

Barrett, Paul, Donald Weinshank, and Timothy Gottleber, eds. *A Concordance to Darwin's Origin of Species, First Edition*. Ithaca, NY: Cornell University Press, 1981.

Bebel, August. *August Bebel Ausgewählte Reden und Schriften*. Edited by Horst Bartel, Rolf Dlubek, and Heinrich Gemkow. 10 vols. in 14 books. Berlin: Dietz, 1970–97.

Bender, Roland. "Der Streit um Ernst Haeckels Embryonenbilder." *Biologie in unserer Zeit* 28 (1998): 157–65.

Bergquist, Patricia. *Sponges*. Berkeley: University of California Press, 1978.

Bernard, Claude. *Introduction à l'étude de la médicine expérimentale*. 1865. Reprint, Paris: Éditions Garnier-Flammarion, 1966.

Bischoff, Theodor. *Entwicklungsgeschichte des Hunde-Eies*. Braunschweig: Friedrich Vieweg und Sohn, 1845.

Blavatsky, Helena Petrovna. *The Secret Doctrine*. 2 vols. London: Theosophical Society, 1888.

Blumenbach, Johann Friedrich. *De generis humani varietate nativa liber*. 3rd ed. Göttingen: Vandenhoek et Ruprecht, 1795.

Bölsche, Wilhelm. *Ernst Haeckel: Ein Lebensbild*. Berlin: Georg Bondi, 1909.

Bopp, Franz. *Vergleichende Grammatik des Sanskrit, Zend, Griechischen, Lateinischen, Littauischen, Gothischen und Deutschen*. Berlin: Königlichen Akademie der Wissenschaften, 1833.

Bowler, Peter. "A Bridge Too Far." *Biology and Philosophy* 8 (1993): 98–102.

———. *The Eclipse of Darwinism*. Baltimore: Johns Hopkins University Press, 1983.

———. *The Non-Darwinian Revolution*. Baltimore: Johns Hopkins University Press, 1988.

Brain, Robert. "Protoplasmania: The Vibratory Organism and 'Man's Glassy Essence' in the Later Nineteenth Century." In *Zeichen der Kraft: Wissensformationen 1800–1900*, ed. Thomas Brandstaetter and Christof Windgätter. Berlin: Kadmus, forthcoming.

Brass, Arnold. *Das Affen-Problem: Prof. E. Haeckel's Darstellungs- u. Kampfesweise sachlich dargelegt nebst Bemerkungen über Atmungsorgane u. Körperform d. Wirbeltier-Embryonen*. Leipzig: Biologischer, 1908.

———. *Ernst Haeckel als Biologie und die Wahrheit*. Halle: Richard Mühlmann, 1906.

Braudel, Fernand. "History and the Social Sciences: The *Longue Durée*." In *On History*, translated by Sarah Matthews, 25–54. Chicago: University of Chicago Press, 1980.

Braun, Alexander. "Das Individuum der Pflanze in seinem Verhältniss zur Species."

Abhandlungen der königlichen Akademie der Wissenschaften zu Berlin, aus dem Jahre 1853 (1854): 19–122.

Breidbach, Olaf. *Ernst Haeckel, Bild Welten der Natur.* Munich: Prestel, 2006.

———. *Ernst Haeckel: Kunstformen aus dem Meer.* Munich: Prestel, 2005.

Breitenbach, Wilhelm. *Ernst Haeckel. Ein Bild seines Lebens und seiner Arbeit.* Odenkirchen: Dr. W. Breitenbach, 1904.

Bronn, Heinrich G. *Handbuch einer Geschichte der Natur.* 3 vols. in 5 books. Stuttgart: E. Schweizerbart'sche Verlagshandlung, 1841–49.

———. *Morphologische Studien über die Gestaltungs-Gesetze der Naturkörper überhaupt und der organischen insbesondere.* Leipzig: E. F. Winter'sche Verlagshandlung, 1858.

———. "Schlusswort des Übersetzers." In *Über die Entstehung der Arten im Thier- und Pflanzen-Reich durch natürliche Züchtung; oder, Erhaltung der vervollkommneten Rassen in Kampfe um's Daseyn,* by Charles Darwin, translated by Heinrich Georg Bronn. Stuttgart: Schweizerbart'sche Verlagshandlung, 1860. Based on the 2nd edition of *On the Origin of Species by Means of Natural Selection; or, The Preservation of Favoured Races in the Struggle for Life.*

———. *Untersuchungen über die Entwickelungs-Gesetze der organischen Welt.* Stuttgart: E. Schweizerbart'sche Verlagshandlung, 1858.

Brücher, Heinz. "Ernst Haeckel, ein Wegbereiter biologischen Staatsdenkens." *NS-Monatschrifte* 6 (1935): 1088–98.

———. *Ernst Haeckels Bluts- und Geistes-Erbe: Eine kulturbiologische Monographie.* Munich: Lehmanns, 1936.

Bruhns, Karl. *Life of Alexander von Humboldt.* Translated by J. and C. Lassell. 2 vols. London: Longmans, Green, 1873.

Brusca, Richard C., and Gary J. Brusca. *Invertebrates.* Sunderland, MA: Sinauer Associates, 1990.

Buffon, Georges-Louis Leclerc. "Initial Discourse." In *From Natural History to the History of Nature.* Edited and translated by John Lyon and Phillip R. Sloan, 89–130. Notre Dame, IN: University of Notre Dame Press, 1981.

Burdach, Karl Friedrich. *Anatomische Untersuchungen bezogen auf Naturwissenschaft und Heilkunst.* First number. Leipzig: Hartmannschen Buchhandlung, 1814.

———. *Propädeutik zum Studium der gesammten Heilkunst.* Leipzig: Breitkopf und Härtel, 1800.

———. *Rückblick auf mein Leben.* Leipzig: Bosz, 1848.

———. *Ueber die Aufgabe der Morphologie. Bey Eröffnung der Königlichen anatomischen Anstalt in Königsberg geschrieben und mit Nachrichten über diese Anstalt begleitet.* Leipzig: Dyk'schen Buchhandlung, 1817.

Burrow, J. W. *The Crisis of Reason: European Thought, 1848–1914.* New Haven, CT: Yale University Press, 2000.

Carr, E. H. *What Is History?* New York: Vintage, 1961.

Carr, William. *A History of Germany, 1815–1990.* 4th ed. London: Edward Arnold, 1991.

Carus, Carl Gustav. *Von den Ur-theilen des Knochen- und Schalengerüstes.* Leipzig: Fleischer, 1828.

Cattell, James McKeen. *American Men of Science.* Lancaster, PA: Science Press, 1906.

[Chambers, Robert]. *Natürliche Geschichte der Schöpfung.* Translated by Carl Vogt. Braunschweig: Friedrich Vieweg und Sohn, 1851. Originally published as *Vestiges of the Natural History of Creation,* 1844.

Chandrasekhar, Subramanian. *Newton's Principia for the Common Reader.* Oxford: Oxford University Press, 1995.

Clark, Edwin. "Charles Joseph Singer" (obituary). *Journal of the History of Medicine and Allied Sciences* 16 (1961): 411–19.

Coleman, William. *Biology in the Nineteenth Century: Problems of Form, Function, and Transformation.* New York: John Wiley, 1971.

———. "Cell Nucleus and Inheritance: An Historical Study." *Proceedings of the American Philosophical Association* 109 (1965): 124–58.

Craig, Gordon A. *Germany, 1866–1945.* Oxford: Oxford University Press, 1980.

Cuvier, Georges. *Le Régne animal.* 2nd ed. 5 vols. Paris: Deterville Libraire, 1829–30.

Darwin, Charles. *The Correspondence of Charles Darwin.* Edited by Frederick Burkhardt, Sydney Smith, David Kohn, and William Montgomery. 15 vols. to date. Cambridge: Cambridge University Press, 1985–.

———. *The Descent of Man and Selection in Relation to Sex.* 2 vols. London: Murray, 1871.

———. *The Descent of Man and Selection in Relation to Sex.* 2nd ed. Edited with introduction by James Moore and Adrian Desmond. 1879. Reprint, London: Penguin, 2004.

———. *More Letters of Charles Darwin.* Edited by Francis Darwin. 2 vols. London: Murray, 1903.

———. *Naturwissenschaftliche Reisen.* [Journal of researches into the geology and natural history of the various countries visited by H.M.S. Beagle, 1834]. Translated by Ernst Dieffenbach. 2 parts. Braunschweig: Vieweg, 1844.

———. "Notebook B." In *Charles Darwin's Notebooks, 1836–1844.* Edited by Paul Barrett, Peter Gautrey, Sandra Herbert, David Kohn, and Sydney Smith. Ithaca, NY: Cornell University Press, 1987.

———. *On the Origin of Species.* London: Murray, 1859.

———. *The Origin of Species by Charles Darwin: A Variorum Text.* Edited by Morse Peckham. Philadelphia: University of Pennsylvania Press, 1959.

———. *Über die Entstehung der Arten im Thier- und Pflanzen-Reich durch natürliche Züchtung; oder, Erhaltung der vervollkommneten Rassen in Kampfe um's Daseyn.* [The origin of species by means of natural selection; or, The preservation of favored races in the struggle for life]. Translated by Heinrich Georg Bronn. Stuttgart: Schweizerbart'sche Verlagshandlung, 1860. Translation based on 2nd edition.

———. *Variation of Plants and Animals under Domestication.* 2nd ed. 2 vols. New York: D. Appleton, 1899. First published in 1868.

Daston, Lorraine. "Objectivity versus Truth." In *Wissenschaft als kulturelle Praxis, 1750–1900,* edited by Hans Bödeker, Peter Reill, and Jürgen Schlumbohm, 17–32. Göttingen: Vandenhoeck & Ruprecht, 1999.

Daston, Lorraine, and Peter Galison. "The Image of Objectivity." *Representations* 40 (1992): 81–128.

———. *Objectivity.* New York: Zone Books, 2007.

Dayrat, Benoit. "The Roots of Phylogeny: How Did Haeckel Build His Trees?" *Systematic Biology* 52 (2003): 515–27.

De Beer, Gavin R. *Embryos and Ancestors.* Oxford: Clarendon Press, 1940.

Deacon, Terrence. *The Symbolic Species: The Co-Evolution of Language and the Brain.* New York: Norton, 1997.

Deichmann, Ute. *Biologists under Hitler.* Translated by Thomas Dunlap. Cambridge, MA: Harvard University Press, 1996.

Dennert, Eberhard. *At the Death Bed of Darwinism* [*Vom Sterbelager des Darwinismus*]. Translated by E. V. O'Harra and John H. Peschges. Burlington, IA: German Literary Board, 1904.

———. *Bibel und Naturwissenschaft.* Halle: Richard Mühlmann, 1911.

———. *Die Naturwissenschaft und der Kampf um die Weltanschauung.* Schriften des Keplerbundes, Heft 1. Godesberg: Naturwissenschaftlicher, 1910.

———. *Vom Sterbelager des Darwinismus.* 2nd ed. Halle: Richard Mühlmann, 1905.

———. *Die Wahrheit über Ernst Haeckel und seine "Welträtsel," nach dem Urteil seiner Fachgenossen.* 2nd ed. Halle: C. Ed. Müller, 1905.

Desmond, Adrian, and James Moore. *Darwin: The Life of a Tormented Evolutionist.* New York: Norton, 1991.

Di Gregorio, Mario. *From Here to Eternity: Ernst Haeckel and Scientific Faith.* Göttingen: Vandenhoeck & Ruprecht, 2005.

———. "Reflections of a Nonpolitical Naturalist: Ernst Haeckel, Wilhelm Bleek, Friedrich Müller and the Meaning of Language." *Journal of the History of Biology* 35 (2002): 79–109.

Dilthey, Wilhelm. *Der Aufbau der geschichtlichen Welt in den Geisteswissenschaften.* Vol. 7: *Gesammelte Schriften.* 1883. Reprint, Göttingen: Vandenhoeck & Ruprecht, 1992.

Döllinger, Johann Joseph Ignaz von. *Das Papstthum.* Munich: Beck, 1892.

Driesch, Hans. "Entwicklungsmechanische Studien: I. Der Werthe der beiden ersten Furchungszellen in der Echinodermenentwicklung. Experimentelle Erzeugung von Theil- und Doppelbildungen. II. Über die Beziehungen des Lichtes zur ersten Etappe der thierischen Formbildung." *Zeitschrift für wissenschaftliche Zoologie* 53 (1891): 160–84.

———. *Die Lokalisation morphogenetischer Vorgänge, ein Beweis vitalistischen Geschehens.* Leipzig: Engelmann, 1899.

———. *Lebenserinnerungen: Aufzeichnungen eines Forschers und Denkers in entscheidender Zeit.* Munich: Ernst Reinhardt Verlag, 1951.

Du Bois-Reymond, Emil. "Gedächtnisrede auf Johannes Müller." *Abhandlungen der Königlichen Akademie der Wissenschaften zu Berlin* (1860): 25–190.

———. *Vorträge über Philosophie und Gesellschaft.* Edited by Reymond Siegfried Wollgast. Hamburg: Felix Meiner, 1974.

Dubois, Eugène. "Pithecanthropus Erectus—A Form from the Ancestral Stock of Mankind." *Annual Report, Smithsonian Institution* (1898): 445–59.

———. *Pithecanthropus erectus, eine menschenähnliche Übergangsform aus Java.* Batavia: Landesdruckerei, 1894.

Duncan, Isadora. *My Life.* New York: Liveright, 1927.

Ecker, Alexander. *Icones physiologicae: Erläuterungenstafeln zur Physiologie und Entwickelungsgeschichte.* Leipzig: L. Voss, 1851–59.

Eckermann, Johann Peter. *Gespräche mit Goethe in den letzten Jahren seines Lebens.* 3rd ed. Berlin: Aufbau, 1987.

Edwards, Samuel. *The Divine Mistress.* New York: McKay, 1970.

Ehrenberg, Christian Gottfried. "Über das Massenverhältnis der jetz lebenden Kiesel-Infusorien und über ein neues Infusorien-Conglomerat als Polirschiefer von Jastraba in Ungarn." *Abhandlungen der Königlichen Akademie der Wissenschaften zu Berlin* (1836): 109–36.

———. "Über die mikroskopischen kieselschaligen Polycystinen als mächtige Gebirgsmasse von Barbados." *Monatsbericht der Königlichen Akademie der Wissenschaften zu Berlin* (1847): 40–60.

Evans, Richard J. "In Search of German Social Darwinism: The History and Historiography of a Concept." In *Medicine and Modernity: Public Health and Medical Care in Nineteenth- and Twentieth-Century Germany,* edited by Manfred Berg and Geoffrey Cocks, 55–89. Cambridge: Cambridge University Press, 1997.

Farrar, Frederic. "Philology and Darwinism." *Nature* 1 (1870): 527–29.

Feest, Christian. "Germany's Indians in a European Perspective." In *Germans and Indians: Fantasies, Encounters, Projections,* edited by Colin Calloway, Gerd Gemünden, and Susanne Zantop. Lincoln: University of Nebraska Press, 2002.

Fischer, Kuno. *Geschichte der neuern Philosophie.* Book 6, *Friedrich Wilhelm Joseph Schelling.* 2 vols. Heidelberg: Carl Winter's Universitätsbuchhandlung, 1872–77.

———. *Geschichte der neuern Philosophie.* 10 vols. Mannheim: Basserman, 1855–.

———. *Kant's Leben und die Grundlagen seiner Lehre: 3 Vorträge.* Mannheim: Bassermann, 1860.

Forel, Auguste. "Naturwissenschaft oder Köhlerglaube." *Biologisches Centralblatt* 25 (1905): 485–93, 519–27.

———. "Ueber die psychischen Eigenschaften der Ameisen und einiger anderer Insekten." *Verhandlungen des V. internationalen Zoologenkongresses* (1901): 141–69.

Fulda, Ludwig. "An die Kulturwelt." In *Aufrufe und Reden deutscher Professoren im Ersten Weltkrieg,* edited by Klaus Bohme, 47–49. Stuttgart: Reclam, 1975.

Fürbringer, Max. "Wie Ich Ernst Haeckel Kennen Lernte und mit Ihm Verkehrte und wie Er mein Führer in den grössten Stunden meines Lebens Wurde." In *Was Wir Ernst Haeckel Verdanken,* edited by Heinrich Schmidt, 2:336–37. 2 vols. Leipzig: Unesma, 1914.

Gade, Daniel W. "Converging Ethnobiology and Ethnobiography: Cultivated Plants, Heinz Brücher, and Nazi Ideology." *Journal of Ethnobiology* 26 (2006): 82–106.

Galen. *Galen on Anatomical Procedures: De anatomicis administrationibus.* Translated with notes by Charles Singer. London: Oxford University Press, 1956.

Gandhi, Mahatma. *Ethical Religion.* Translated by A. Rama Lyer. 2nd ed. Madras: S. Ganesan, 1922.

Gasman, Daniel. *Haeckel's Monism and the Birth of Fascist Ideology.* New York: Peter Lang, 1998.

———. "Introduction to the Transaction Edition." In *The Scientific Origins of National Socialism: Social Darwinism in Ernst Haeckel and the German Monist League,* xi–xxxii. 1971. Reprint, New Brunswick: Transaction, 2004.

———. *The Scientific Origins of National Socialism: Social Darwinism in Ernst Haeckel and the German Monist League.* New York: Science History Publications, 1971.

Gegenbaur, Carl. *Beiträge zur näheren Kenntniss der Schwimmpolypen (Siphonophoren).* Leipzig: W. Engelmann, 1854.

———. *Erlebtes und Erstrebtes.* Leipzig: Wilhelm Engelmann, 1901.

———. *Gesammelte Abhandlungen von Carl Gegenbaur.* Edited by M. Fürbringer and H. Bluntschli. 3 vols. Leipzig: Wilhelm Engelmann, 1912.

———. *Grundzüge der vergleichenden Anatomie.* 2nd ed. Leipzig: Wilhelm Engelmann, 1870.

———. "Neue Beiträge zur näheren Kenntniss Der Siphonophoren." *Nova Acta Leopoldina* 28 (1859): 333–424.

———. *Untersuchungen zur vergleichenden Anatomie der Wirbelthiere.* 3 vols. Leipzig: Wilhelm Engelmann, 1864, 1865, 1872.

———. "Zur Lehre vom Generationswechsel und der Fortpflanzung bei Medusen und Polypen." *Verhandlungen der physikalisch-medicinischen Gesellschaft in Würzburg* 4 (1853): 154–221.

Gibbon, Edward. *The History of the Decline and Fall of the Roman Empire.* 6 vols. London: Strahan and Cadell, 1777–88.

Gilbert, Scott. *Developmental Biology.* 2nd ed. Sunderland, MA: Sinauer Associates, 1985.

———. *Developmental Biology.* 6th ed. Sunderland, MA: Sinauer Associates, 2000.
Goethe, Johann Wolfgang von. *Goethe-Briefe.* Edited by Philipp Stein. 8 vols. Berlin: Wertbuchhandel, 1924.
———. *Goethes Briefe und Briefe an Goethe.* Hamburger Ausgabe. Edited by Karl Mandelkow. 3rd ed. 6 vols. Munich: C. H. Beck, 1988.
———. *Johann Wolfgang von Goethe Werke.* Hamburger Ausgabe. 14 vols. Munich: Deutscher Taschenbuch, 1988.
———. *Sämtliche Werke nach Epochen seines Schaffens.* Vol. 2, *Die Natur.* Münchner Ausgabe. Edited by Karl Richter, Herbert Göpfert, Norbert Miller, and Gerhard Sauder. 21 vols. Munich: Carl Hanser, 1985–98.
———. *Sämtliche Werke nach Epochen seines Schaffens.* Vol. 6, *Faust, Eine Tragödie.* Münchner Ausgabe. Edited by Karl Richter, Herbert Göpfert, Norbert Miller, and Gerhard Sauder. 21 vols. Munich: Carl Hanser , 1985–98.
———. *Die Schriften zur Naturwissenschaft.* 1st division. Vol. 9, *Morphologische Hefte.* Edited by Dorothea Kuhn. Weimar: Böhlaus Nachfolger, 1954.
Goette, Alexander. *Die Entwickelungsgeschichte der Unke (Bombinator igneus).* 2 vols. Leipzig: Leopold Voss, 1875.
Goldschmidt, Richard. *Portraits from Memory: Recollections of a Zoologist.* Seattle: University of Washington Press, 1956.
Goodrick-Clarke, Nicholas. *The Occult Roots of Nazism: Secret Aryan Cults and Their Influence on Nazi Ideology.* New York: New York University Press, 2004.
Gordon, Scott. *History and Philosophy of Social Science: An Introduction.* London: Routledge, 1991.
Gould, Stephen Jay. "*Abscheulich!* (Atrocious!). Haeckel's Distortions Did Not Help Darwin." *Natural History* 109, no. 2 (2000): 42–49.
———. "A Developmental Constraint in *Cerion,* with Comments on the Definition and Interpretation of Constraint in Evolution." *Evolution* 43 (1989): 516–39.
———. "Eternal Metaphors of Palaeontology." In *Patterns of Evolution as Illustrated in the Fossil Record,* edited by A. Hallam, 1–26. New York: Elsevier, 1977.
———. *Ever Since Darwin.* New York: Norton, 1977.
———. *The Flamingo's Smile.* New York: Norton, 1985.
———. *The Hedgehog, the Fox, and the Magister's Pox.* New York: Harmony Books, 2003.
———. *I Have Landed.* New York: Three Rivers Press, 2003.
———. *Living Stones of Marrakech.* New York: Three Rivers Press, 2000.
———. *Ontogeny and Phylogeny.* Cambridge, MA: Harvard University Press, 1977.
———. *The Panda's Thumb.* New York: Norton, 1980.
———. "Redrafting the Tree of Life." *Proceedings of the American Philosophical Society* 141 (March 1997): 30–54.
———. "Systematic Pluralism and the Uses of History." *Systematic Zoology* 22 (1973): 322–24.
———. *Wonderful Life: The Burgess Shale and the Nature of History.* New York: Norton, 1989.

Grant, Robert. "Observations and Experiments on the Structure and Functions of the Sponge." *Edinburgh Philosophical Journal* 13 (1825): 94–107, 333–46; 14 (1826): 113–24, 336–41.

———. "Observations on the Structure and Functions of the Sponge." *Edinburgh Philosophical Journal,* n.s. 2 (1827): 121–41.

———. "Observations on the Structure of Some Silicous Sponges." *Edinburgh Philosophical Journal,* n.s. 1 (1826): 341–51.

———. "On the Structure and Nature of the Spongilla friabilis." *Edinburgh Philosophical Journal* 14 (1826): 270–84.

———. "Remarks on the Structure of Some Calcerous Sponges." *Edinburgh Philosophical Journal,* n.s. 1 (1826): 166–70.

Graves, Joseph L. *The Emperor's New Clothes: Biological Theories of Race at the Millennium.* New Brunswick, NJ: Rutgers University Press, 2001.

Greef, Richard. *Reise nach den Canarischen Inseln.* Bonn: Max Cohen & Sohn, 1868.

Grimaldi, D. A., D. Agnosti, and J. M. Carpenter. "New and Rediscovered Primitive Ants (Hymenoptera: Formicidae) in Cretaceous Amber from New Jersey, and Their Phylogenetic Relationships." *American Museum Novitates,* no. 3208 (1997): 1–43.

Hacking, Ian. *The Social Construction of What?* Cambridge, MA: Harvard University Press, 1999.

Haeckel, Ernst. *Anthropogenie; oder, Entwickelungsgeschichte des Menschen.* Leipzig: Wilhelm Engelmann, 1874.

———. *Anthropogenie; oder, Entwickelungsgeschichte des Menschen.* 3rd ed. Leipzig: Wilhelm Engelmann, 1877.

———. *Anthropogenie; oder, Entwickelungsgeschichte des Menschen.* 4th ed. 2 parts. Leipzig: Wilhelm Engelmann, 1891.

———. *Anthropogenie; oder, Entwickelungsgeschichte des Menschen.* 5th ed. 2 vols. Leipzig: Wilhelm Engelmann, 1903.

———. "Apotheose des Entwickelungs-Gedanken nach Ernst Haeckel und Gabriel Max." Erstes Beiheft zu der Sammlung Wanderbilder, bound with *Wanderbilder: Nach eigenen Aquarellen und Oelgemälden.* Gera-Untermhaus: W. Koehler, 1905.

———. *Arabische Korallen.* Berlin: Georg Reimer, 1876.

———. *Arbeitstheilung in Natur- und Menschenleben.* Berlin: C. G. Lüderitz'sche Verlagsbuchhandlung, 1869.

———. *Arbeitstheilung in Natur und Menschenleben.* 2nd ed. Leipzig: Alfred Kröner, 1910.

———. *Aus Insulinde: Malayische Reisebriefe.* Bonn: Emil Strauss, 1901.

———. *Berg- und Seefahrten: 1857/1883.* Edited by Heinrich Schmidt. Leipzig: K. F. Koehler, 1923.

———. "Beschreibung neuer craspedoter Medusen aus dem Golfe von Nizza." *Jenaische Zeitschrift für Naturwissenschaft* 1 (1864): 325–42.

———. "Eine Besteigung der Pik von Teneriffa." *Zeitschrift der Gesellschaft für Erdkunde* 5 (1870): 1–28.

———. *Biologische Studien: Studien zur Gastraeatheorie.* Jena: Hermann Dufft, 1877.

———. *Entwicklungsgeschichete einer Jugend: Briefe an die Eltern, 1852–1856.* Edited by Heinrich Schmidt. Leipzig: K. F. Koehler, 1921.

———. *Ernst Haeckel: Biographie in Briefen mit Erläuterungen.* Edited by Georg Uschmann. Gütersloh: Prisma, 1984.

———. *Ernst Haeckel: Gemeinverständliche Werke.* Edited by Heinrich Schmidt. 6 vols. Leipzig: Alfred Kröner, 1924.

———. *Ernst Haeckel, Sein Leben, Denken und Wirken: Eine Schriftenfolge für seine zahlreichen Freunde und Anhänger.* Edited by Victor Franz. 2 vols. Jena: Wilhelm Gronau and W. Agricola, 1943–44.

———. *Ewigkeit: Weltkriegsgedanken über Leben und Tod, Religion und Entwicklungslehre.* Berlin: Georg Reimer, 1915.

———. *Franziska von Altenhausen: Ein Roman aus dem Leben eines berühmten Mannes in Briefen aus den Jahren 1898/1903.* Edited by Johannes Werner. Leipzig: Koehler & Amelang, 1927.

———. *Freedom in Science and Teaching.* Translated and with preface by Thomas Henry Huxley. New York: D. Appleton, 1879.

———. *Freie Wissenschaft und freie Lehre: Eine Entgegnung auf Rudolf Virchow's Münchener Rede über "Die Freiheit der Wissenschaft im modernen Staat."* Stuttgart: E. Schweizerbart'sche Verlagshandlung, 1878.

———. "Die Gastraea-Theorie, die phylogenetische Classification des Thierreichs und die Homologie der Keimblätter." *Jenaische Zeitschrift für Naturwissenschaft* 8 (1874): 1–55.

———. "Die Gastrula und die Eifurchung der Thiere." *Jenaische Zeitschrift für Naturwissenschaft* 9 (1875): 409.

———. *Generelle Morphologie der Organismen.* 2 vols. Berlin: Georg Reimer, 1866.

———. *Gesammelte populäre Vorträge aus dem Gebiete der Entwickelungslehre.* 2 vols. Bonn: Emil Strauss, 1878.

———. *Haeckel e L'Italia: La vita come scienza e come storia.* Edited by Giampiero Bozzolato and Rüdiger Stolz. Brugine: Edizioni Centro Internazionale di Storia dello Spazio e del Tempo, 1993.

———. *Haeckel-Korrespondenz: Übersicht über den Briefbestand des Ernst-Haeckel-Archivs.* Edited by Uwe Hoßfeld and Olaf Breidbach. Berlin: Verlag für Wissenschaft und Bildung, 2005.

———. *Himmelhoch Jauchzend: Erinnerungen und Briefe der Liebe.* Edited by Heinrich Schmidt. Dresden: Carl Reissner, 1927.

———. *The History of Creation.* Translated and revised by E. Ray Lankester. London: H. S. King and Co., 1876.

———. *Italienfahrt: Briefe an die Braut, 1859–1860.* Edited by Heinrich Schmidt. Leipzig: K. F. Koehler, 1921.

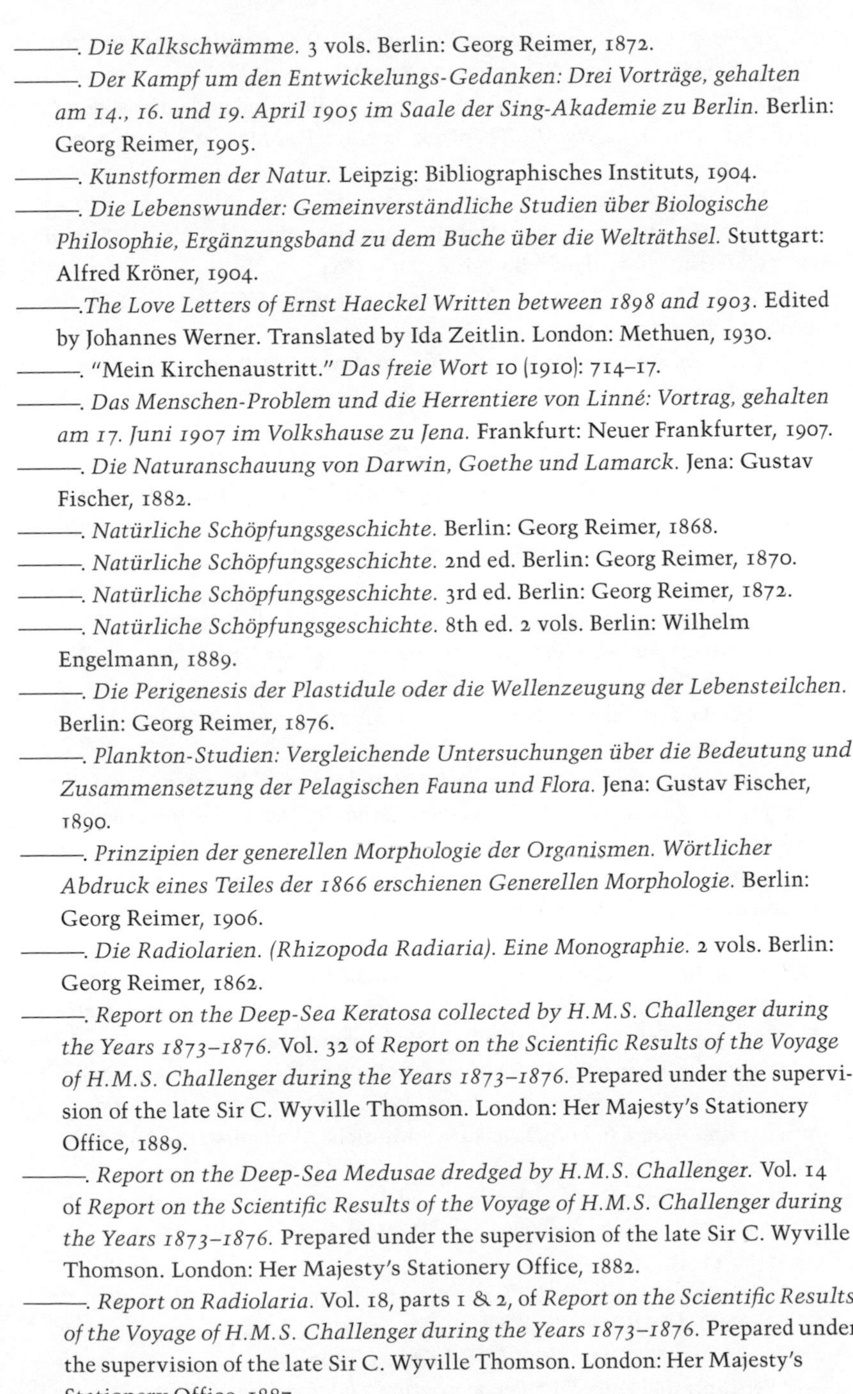

———. *Die Kalkschwämme.* 3 vols. Berlin: Georg Reimer, 1872.

———. *Der Kampf um den Entwickelungs-Gedanken: Drei Vorträge, gehalten am 14., 16. und 19. April 1905 im Saale der Sing-Akademie zu Berlin.* Berlin: Georg Reimer, 1905.

———. *Kunstformen der Natur.* Leipzig: Bibliographisches Instituts, 1904.

———. *Die Lebenswunder: Gemeinverständliche Studien über Biologische Philosophie, Ergänzungsband zu dem Buche über die Welträthsel.* Stuttgart: Alfred Kröner, 1904.

———.*The Love Letters of Ernst Haeckel Written between 1898 and 1903.* Edited by Johannes Werner. Translated by Ida Zeitlin. London: Methuen, 1930.

———. "Mein Kirchenaustritt." *Das freie Wort* 10 (1910): 714–17.

———. *Das Menschen-Problem und die Herrentiere von Linné: Vortrag, gehalten am 17. Juni 1907 im Volkshause zu Jena.* Frankfurt: Neuer Frankfurter, 1907.

———. *Die Naturanschauung von Darwin, Goethe und Lamarck.* Jena: Gustav Fischer, 1882.

———. *Natürliche Schöpfungsgeschichte.* Berlin: Georg Reimer, 1868.

———. *Natürliche Schöpfungsgeschichte.* 2nd ed. Berlin: Georg Reimer, 1870.

———. *Natürliche Schöpfungsgeschichte.* 3rd ed. Berlin: Georg Reimer, 1872.

———. *Natürliche Schöpfungsgeschichte.* 8th ed. 2 vols. Berlin: Wilhelm Engelmann, 1889.

———. *Die Perigenesis der Plastidule oder die Wellenzeugung der Lebensteilchen.* Berlin: Georg Reimer, 1876.

———. *Plankton-Studien: Vergleichende Untersuchungen über die Bedeutung und Zusammensetzung der Pelagischen Fauna und Flora.* Jena: Gustav Fischer, 1890.

———. *Prinzipien der generellen Morphologie der Organismen. Wörtlicher Abdruck eines Teiles der 1866 erschienen Generellen Morphologie.* Berlin: Georg Reimer, 1906.

———. *Die Radiolarien. (Rhizopoda Radiaria). Eine Monographie.* 2 vols. Berlin: Georg Reimer, 1862.

———. *Report on the Deep-Sea Keratosa collected by H.M.S. Challenger during the Years 1873–1876.* Vol. 32 of *Report on the Scientific Results of the Voyage of H.M.S. Challenger during the Years 1873–1876.* Prepared under the supervision of the late Sir C. Wyville Thomson. London: Her Majesty's Stationery Office, 1889.

———. *Report on the Deep-Sea Medusae dredged by H.M.S. Challenger.* Vol. 14 of *Report on the Scientific Results of the Voyage of H.M.S. Challenger during the Years 1873–1876.* Prepared under the supervision of the late Sir C. Wyville Thomson. London: Her Majesty's Stationery Office, 1882.

———. *Report on Radiolaria.* Vol. 18, parts 1 & 2, of *Report on the Scientific Results of the Voyage of H.M.S. Challenger during the Years 1873–1876.* Prepared under the supervision of the late Sir C. Wyville Thomson. London: Her Majesty's Stationery Office, 1887.

———. *Report on the Siphonophorae collected by H.M.S. Challenger during the Years 1873–76.* Vol. 28 of *Report on the Scientific Results of the Voyage of H.M.S. Challenger during the Years 1873–1876.* Prepared under the supervision of the late Sir C. Wyville Thomson. London: Her Majesty's Stationery Office, 1888.

———. "Response to Volger" [in German]. In *Amtlicher Bericht über die acht und dreissigste Versammlung Deutscher Naturforscher und Ärzte in Stettin,* 70–71. Stettin: Hessenland's Buchdruckerei, 1864.

———. *The Riddle of the Universe.* Translated by Joseph McCabe. New York: Harper & Brothers, 1900.

———. *De Rhizopodum finibus et ordinibus.* Berlin: Georg Reimer, 1861.

———. *Sandalion: Eine offene Antwort auf die Fälschungs-Anklagen der Jesuiten.* Frankfurt: Neuer Frankfurther, 1910.

———. *Das System der Medusen.* 2 vols. Jena: Gustav Fischer, 1879.

———. "Systematische Einleitung: Zur Phylogenie der Australischen Fana." In *Zoologische Forschungsreisen in Australien und dem Malayischen Archipel,* edited by Richard Semon, 1:iii–xxiv. 4 vols. Jena: Gustav Fischer, 1893–.

———. *Systematische Phylogenie.* 3 vols. Berlin: Georg Reimer, 1894–96.

———. *De telis quibusdam Astaci fluviatilis.* Berlin: Schade, 1857.

———. *Die Tiefsee-Medusen der Challenger-Reise und der Organismus der Medusen,* with *Atlas.* Jena: Gustav Fischer, 1881.

———. "Über die Eier der Scomberesoces." *Archiv für Anatomie und Physiologie* (1855): 23–32.

———. "Über neue, lebende Radiolarien des Mittelmeeres." *Monatsberichte der Königlichen Preussische Akademie der Wissenschaften zu Berlin* (1860): 794–817, 835–45.

———. "Ueber der Organismus der Schwämme und ihre Verwandtschaft mit den Korallen." *Jenaische Zeitschrift für Naturwissenschaft und Medicin* 5 (1870): 207–35.

———. "Ueber die Entwickelungstheorie Darwins." In *Amtlicher Bericht über die acht und dreissigste Versammlung Deutscher Naturforscher und Ärzte in Stettin,* 17–30. Stettin: Hessenland's Buchdruckerei, 1864.

———. "Ueber die heutige Entwickelungslehre im Verhältnisse zur Gesamtwissenschaft." In *Amtlicher Bericht der 50. Versammlung Deutscher Naturforscher und Aerzte in München,* 14–22. Munich: Akademischen Buchdruckerei von F. Straub, 1877.

———. "Ueber die sexuelle Fortpflanzung und das natürliche System der Schwämme." *Jenaische Zeitschrift für Medicin und Naturwissenschaft* 6 (1871): 642–51.

———. *Ueber unsere gegenwärtige Kenntnis vom Ursprung des Menschen: Vortrag gehalten auf dem Vierten Internationalen Zoologen-Congress in Cambridge, am 26. August 1898.* Stuttgart: Alfred Kröner, 1905.

———. "Vernunft und Krieg." *La Réconciliation* 1 (October 1913): 1–10.

———. *Wanderbilder: Nach eigenen Aquarellen und Oelgemälden.* Gera-Untermhaus: W. Koehler, 1905.

———. *Die Welträthsel, gemeinverständliche Studien über Monistische Philosophie.* Bonn: Emil Strauss, 1899.

———. *Die Welträtsel.* Edited by Olaf Kohr. Berlin: Akademie, 1961.

———. *Zellseelen und Seelenzellen.* 2nd ed. 1878. Reprint, Leipzig: Alfred Kröner, 1923.

———. *Ziele und Wege der heutigen Entwickelungsgeschichte.* Jena: Hermann Dufft, 1875.

———. *Zur Entwickelungsgeschichte der Siphonophoren.* Utrecht: C. Van der Post Jr., 1869.

———. "Eine zoologische Excursion nach den canarischen Inseln." *Jenaische Zeitschrift für Medicin und Naturwissenschaft* 3 (1867): 313–28.

———. "Zwei medizinische Abhandlungen aus Würzburg: I. Über die Beziehungen des Typhus zur Tuberkulose; II. Fibroid des Uterus." *Wiener medizinische Wochenschrift* 6 (1856): 5, 17–20, 97–101.

Haeckel, Ernst, and Hermann Allmers. *Haeckel und Allmers: Die Geschichte einer Freundschaft in Briefen der Freunde.* Edited by Rudolph Koop. Bremen: Arthur Geist, 1941.

Haeckel, Ernst, and Wilhelm Bölsche. *Ernst Haeckel–Wilhelm Bölsche: Briefwechsel 1887–1919.* Edited by Rosemarie Nöthlich. Berlin: Verlag für Wissenschaft und Bildung, 2002.

Haeckel, Ernst, and Eduard von Hartmann. "Metaphysik und Naturphilosophie: Briefwechsel zwischen Eduard von Hartmann und Ernst Haeckel." Edited by Bertha Kern-von Hartmann. *Kant Studien* 48 (1956–57): 3–24.

Haeckel, Ernst, and Agnes Huschke. *Ernst und Agnes Haeckel: Ein Briefwechsel.* Edited by Konrad Huschke. Jena: Urania, 1950.

Haeckel, Ernst, and Thomas Henry Huxley. "Der Briefwechsel zwischen Thomas Henry Huxley und Ernst Haeckel." Edited by Georg Uschmann and Ilse Jahn. *Wissenschaftliche Zeitschrift der Friedrich-Schiller-Universität Jena* (Mathematisch-Naturwissenschaftliche Reihe) 9 (1959–60): 7–33.

Haeckel, Ernst, and Albert von Kölliker. "Über die Beziehungen zwischen Albert Koelliker und Ernst Haeckel." Edited by Georg Uschmann *Wissenschaftliche Zeitschrift der Friedrich-Schiller-Universität Jena* (Mathematisch-Naturwissenschaftliche Reihe) 25 (1976): 125–32.

Haeckel, Ernst, and Frida von Uslar-Gleichen. *Das ungelöste Welträtsel: Frida von Uslar-Gleichen und Ernst Haeckel, Briefe und Tagebücher 1898–1900.* Edited by Norbert Elsner. 3 vols. Göttingen: Wallstein, 2000.

Haeckel, Ernst, and August Weismann. "Der Briefwechsel zwischen Ernst Haeckel und August Weismann." Edited by G. Uschmann and B. Hassenstein. *Jenaer Reden und Schriften* (1965): 7–68.

Hecht, Günther. "Biologie und Nationalsozialismus." *Zeitschrift für die Gesamte Naturwissenschaft* 3 (1937–38): 280–90.

Heikertinger, Franz. "P. Erich Wasmann, S.J.: Ein Nachruf." *Koleopterologische Rundschau* 17 (1931): 88–96.

Heilborn, Adolf. *Die Leartragödie Ernst Haeckels.* Hamburg: Hoffmann & Campe, 1920.

Helmholtz, Hermann von. "Optisches über Malerei." In *Vorträge und Reden,* 2:97–137. 3 vols. Braunschweig: Friedrich Vieweg und Sohn, 1884.

———. *Populäre wissenschaftliche Vorträge.* Braunschweig: Friedrich Vieweg und Sohn, 1865.

———. *Über die Akademische Freiheit der deutschen Universitäten.* Berlin: August Hirschwald, 1878.

Helmhotz, Hermann von, and Emil Du Bois-Reymond. *Dokumente einer Freundschaft: Briefwechsel zwischen Hermann von Helmholtz und Emil du Bois-Reymond, 1846–1894.* Edited by Christa Kirsten. Berlin: Akademie, 1986.

Hensen, Victor. *Die Plankton-Expedition und Haeckel's Darwinismus, ueber einige Aufgaben und Ziele der beschreibenden Naturwissenschaften.* Kiel: Lipsius & Tisher, 1891.

Hertwig, Oscar. "Beiträge zur Kenntniss der Bildung, Befruchtung und Theilung des thierischen Eies." *Morphologisches Jahrbuch* 1 (1876): 347–434.

———. "Die Geschichte der Zellenlehre." *Deutsche Rundschau* 20 (1879): 417–29.

———. *Lehrbuch der Entwicklungsgeschichte des Menschen und der Wirbelthiere.* 2nd ed. Jena: Gustav Fischer, 1888.

Hertwig, Richard. *Zur Histologie der Radiolarien.* Leipzig: Wilhelm Engelmann, 1876.

Heuss, Theodor. *Anton Dohrn: A Life for Science.* Edited by Christiane Groeben. Translated by Liselotte Dieckmann. 1940. Reprint, Berlin: Springer, 1991.

Hildebrandt, Kurt. "Die Bedeutung der Abstammungslehre für die Weltanschauung." *Zeitschrift für die Gesamte Naturwissenschaft* 3 (1937–38): 15–34.

His, Wilhelm. *Anatomie menschlicher Embryonen.* 3 vols. with 3 atlases. Leipzig: F. C. W. Vogel, 1880–85.

———. *Lebenserinnerungen.* Leipzig: Fischer & Wittig, 1903.

———. *Über die Aufgaben und Zielpunkte der wissenschaftlichen Anatomie.* Leipzig: F. C. W. Vogel, 1872.

———. *Ueber die Bedeutung der Entwickelungsgeschichte für die Auffassung der Organischen Natur.* Leipzig: F. C. W. Vogel, 1870.

———. *Unsere Körperform und das physiologische Problem ihrer Entstehung: Briefe an einen befreundeten Naturforscher.* Leipzig: F. C. W. Vogel, 1874. Cover bears 1875 publication date, indicating a delay in publication.

———. *Untersuchungen über die erste Anlage des Wirbelthierleibes.* Leipzig: F. C. W. Vogel, 1868.

Hofsten, Nils. "Obituary Notice: Erik Nordenskiöld (1872–1933)." *Isis* 38 (1947): 103–6.

Holt, Niles. "Monists and Nazis: A Question of Scientific Responsibility." *Hastings Center Report* 5 (1975): 37–43.

Hopwood, Nicholas. *Embryos in Wax: Models from the Ziegler Studio.* Cambridge: Whipple Museum of the History of Science, 2002.

———. "Pictures of Evolution and Charges of Fraud: Ernst Haeckel's Embryological Illustrations." *Isis* 97 (2006): 260–391.

———. "Producing Development: The Anatomy of Human Embryos and the Norms of Wilhelm His." *Bulletin of the History of Medicine* 74 (2000): 29–79.

Horton, R. F. "Ernst Haeckel's 'Riddle of the Universe.'" *Christian World Pulpit* 63 (10 June 1903): 353–56.

Hoßfeld, Uwe. *Geschichte der biologischen Anthropologie in Deutschland von den Anfängen bis in die Nachkriegszeit.* Stuttgart: Franz Steiner, 2005.

———. "Haeckelrezeption im Spannungsfeld von Monismus, Sozialdarwinismus und Nationalsozialismus." *History and Philosophy of the Life Sciences* 21 (1999): 195–213.

———. "Haeckels 'Eckermann': Heinrich Schmidt (1874–1935)." In *Klassische Universität und akademische Provinz: Die Universität Jena von der Mitte des 19. bis in die 30er Jahre des 20. Jahrhunderts,* edited by Matthias Steinbach and Stefan Gerber, 270–88. Jena: Bussert & Stadeler, 2005.

———. "Nationalsozialistische Wissenschaftsinstrumentalisierung: Die Rolle von Karl Astel und Lothar Stengel von Rutkowski bei der Genese des Buches 'Ernst Haeckels Bluts- und Geistes-Erbe' (1936)." In *Der Brief als wissenschaftshistorische Quelle,* edited by Erika Krauße, 171–94. Berlin: Verlag für Wissenschaft und Bildung, 2005.

———. "Staatsbiologie, Rassenkunde und Moderne Synthese in Deutschland während der NS-Zeit." In *Evolutionsbiologie von Darwin bis heute,* edited by R. Brömer, U. Hoßfeld, and N. A. Rupke, 249–305. Berlin: Verlag für Wissenschaft und Bildung, 2000.

Hoßfeld, Uwe, and Olaf Breidbach. "Ernst Haeckels Politisierung der Biologie." *Thüringen: Blätter zur Landeskunde* 54 (2005).

Hoßfeld, Uwe, and Thomas Junker. "Anthropologie und synthetischer Darwinismus im Dritten Reich: Die Evolution der Organismen (1943)." *Anthropologischer Anzeiger* 61 (2003): 85–114.

Hoßfeld, Uwe, Rosemarie Nöthlich, and Lennart Olsson. "Haeckel's Literary Hopes Dashed by Materialism." *Nature* 424 (2003): 875.

Hoßfeld, Uwe, and Lennart Olsson. "The History of Comparative Anatomy in Jena—an Overview." *Theory in Biosciences* 122 (2003): 109–26.

———. "The Road from Haeckel. The Jena Tradition in Evolutionary Morphology and the Origin of 'Evo-Devo.'" *Biology & Philosophy* (2003): 285–307.

Hull, David L. "A Matter of Individuality." *Philosophy of Science* 45 (1978): 335–60.

———. "Are Species Really Individuals?" *Systematic Zoology* 25 (1974): 174–91.

Humboldt, Alexander von. *Ansichten der Natur, mit wissenschaftliche Erläuterungen.* 2nd ed. 2 vols. Stuttgart: J. G. Cotta'schen Buchhandlung, 1826.

———. *Ansichten der Natur.* 3rd ed. Stuttgart: Cotta'schen Buchhandlung, 1871. First published in 1808.

———. *Kosmos: Entwurf einer physischen Weltbeschreibung.* 5 vols. Stuttgart: Gotta'scher, 1845–58.

Humboldt, Alexander von, and Aimé Bonpland. *Voyage aux régions equinoxiales du nouveau continent, fait en 1799–1804.* 29 vols. Paris: F. Schoell, 1807–35.

Humboldt, Wilhelm von. *Über die Kawi-Sprache auf der Insel Java.* 3 vols. Berlin: Königlichen Akademie der Wissenschaften, 1836.

Hunt, James. "On the Negro's Place in Nature." *Memoirs of the Anthropological Society* 1 (1863): 1–64.

Huxley, Thomas Henry. "Criticisms on 'The Origin of Species.'" *Natural History Review* (1864): 566–80.

———. "The Natural History of Creation—by Dr. Ernst Haeckel." *Academy* 1 (1869): 13–14, 40–43.

———. *The Oceanic Hydrozoa; a Description of the Calycophoridae and Physophoridae observed during the Voyage of H.M.S. "Rattlesnake" in the years 1846–1850.* London: Ray Society, 1859.

———. "Zoological Notes and Observations Made on Board H.M.S. Rattlesnake during the Years 1846–50." In *The Scientific Memoirs of Thomas Henry Huxley.* Edited by M. Foster and E. Lankester, 1:86–95. 4 vols. London: Macmillan, 1898. Originally published in the *Annals and Magazine of Natural History,* 1851.

Jahn, Ilse. "Ernst Haeckel und die Berliner Zoologen." *Acta Historica Leopoldina* 16 (1985): 65–109.

Kant, Immanuel. *Kants gesammelte Schriften.* Vol. 5, *Kritik der Urteilskraft.* Edited by the Akademie der Wissenschaften. Berlin: Gruyter, 1902.

———. *Kritik der Urteilskraft.* In *Werke in sechs Bänden.* Edited by Wilhelm Weischedel, V: 549 (A366, B370–71). 6 vols. Wiesbaden: Insel, 1956–64.

Keegan, John. *The First World War.* London: Hutchinson, 1998.

Kelly, Alfred. *The Descent of Darwin: The Popularization of Darwinism in Germany, 1860–1914.* Chapel Hill: University of North Carolina Press, 1981.

Kleeberg, Bernhard. *Theophysis: Ernst Haeckels Philosophie des Naturganzen.* Weimar: Hermann Böhlau, 2005.

Klemm, Peter. *Ernst Haeckel: Der Ketzer von Jena.* Leipzig: Urania-Verlag, 1966.

Kockerbeck, Christoph. *Ernst Haeckel's 'Kunstformen der Natur' und ihr Einfluß auf die deutsche Bildende Kunst der Jahrhunderwende.* Frankfurt: Peter Lang, 1986.

———, ed. *Carl Vogt, Jacob Moleschott, Ludwig Büchner, Ernst Haeckel: Briefwechsel.* Marburg: Basilisken-Presse, 1999.

Kolkenbrock-Netz, Jutta. "Wissenschaft als nationaler Mythos: Anmerkungen zur Haeckel-Virchow-Kontroverse auf der 50. Jahresversammlung deutscher Naturforscher und Ärzte in München (1877)." In *Nationale Mythen und Symbole in der zweiten Hälfte des 19. Jahrhunderts,* edited by Jürgen Link and Wulf Wülfing, 212–36. Stuttgart: Kolett-Cotta, 1991.

Koller, Gottfried. *Das Leben des Biologen Johannes Müller, 1801–1858.* Stuttgart: Wissenschaftliche Verlagsgesellschaft, 1958.

Kölliker, Albert von. *Entwicklungsgeschichte des Menschen und der höheren Thiere.* Leipzig: Wilhelm Engelmann, 1861.

———. *Entwicklungsgeschichte des Menschen und der höheren Tiere.* 2nd ed. Leipzig: Engelmann, 1879.

———. *Die Schwimmpolypen; oder, Siphonophoren von Messina.* Leipzig: Wilhelm Engelmann, 1853.

———. "Ueber die Darwin'sche Schöpfungstheorie." *Zeitschrift für wissenschaftliche Zoologie* 14 (1864): 174–86.

Kowalevsky, Alexander. "Entwicklungsgeschichte des Amphioxus Lanceolatus." *Memoires de l'Academie Imperiale des Sciences de Saint-Petersbourg* 11 (1867): 1–17.

Krauße, Erika. *Ernst Haeckel.* Leipzig: Teubner, 1984.

———. "Pithecanthropus erectus Dubois (1891) in Evolutionsbiologie und Kunst." In *Evolutionsbiologie von Darwin bis heute,* edited by Rainer Brömer, Uwe Hoßfeld, and Nicolaas Rupke, 69–88. Berlin: Verlag für Wissenschaft und Bildung, 2000.

———. "Wege zum Bestseller, Haeckels Werk im Lichte der Verlegerkorrespondenz: Die Korrespondenz mit Emil Strauss." In *Der Brief als wissenschaftshistorische Quelle,* 145–70. Berlin: Verlag für Wissenschaft und Bildung, 2005.

Lange, Albert. *Die Arbeiterfrage in ihrer Bedeutung für Gegenwart und Zukunft.* Duisburg: W. Falk & Volmer, 1865.

Lankester, E. Ray. "On the Primitive Cell-Layers of the Embryo as the Basis of Genealogical Classification of Animals." *Annals and Magazine of Natural History* 11 (1873): 321–28.

Larson, Edward, and Larry Witham. "Leading Scientists Still Reject God." *Nature* 394 (1998): 313.

———. "Scientists Are Still Keeping the Faith." *Nature* 386 (1997): 435–36.

Lea, Henry. "Ethical Values in History." *American Historical Review* 9 (1904): 233–46.

Lerner, Richard M. *Final Solutions: Biology, Prejudice, and Genocide.* University Park: Pennsylvania State University Press, 1992.

Leuba, James. *The Belief in God and Immortality: A Psychological, Anthropological and Statistical Study.* Boston: Sherman, French & Co., 1916.

Leuckart, Rudolf. *Zoologische Untersuchungen, erstes Heft: Siphonophoren.* Giessen: J. Ricker'sche Buchhandlung, 1853.

Levit, George, and Uwe Hoßfeld. "The Forgotten 'Old-Darwinian' Synthesis: The Evolutionary Theory of Ludwig H. Plate (1862–1937)." *Internationale Zeitschrift für Geschichte und Ethik der Naturwissenschaft, Technik und Medizin* (NTM), n.s. 14 (2006): 9–25.

Levs, Sally, and Dafne Eerkes-Medrano. "Gastrulation in Calcareous Sponges: In Search of Haeckel's Gastrea." *Integrative and Comparative Biology* 45 (2005): 342–51.

Lieberkühn, Nathanael. "Neue Beiträge zur Anatomie der Spongien." *Anatomie, Physiologie und wissenschaftliche Medicin* 25 (1859): 353–82, 515–29.

Lifton, Robert Jay. *The Nazi Doctors: Medical Killing and the Psychology of Genocide.* 2nd ed. New York: Basic Books, 2000.

Linnaeus, Carolus. *Systema naturae per regna tria naturae, secundum classes, ordines, genera, species, cum characteribus, differentiis, synonymis, locis.* 3 vols. Hale: Curt, 1760–70.

Loofs, Friedrich. "Offener Brief an Herrn Professor Dr. Ernst Haeckel in Jena." *Die Christliche Welt* 13 (1899): 1067–72.

Lustig, Abigail. "Ants and the Nature of Nature in Auguste Forel, Erich Wasmann, and William Morton Wheeler." In *The Moral Authority of Nature,* edited by Lorraine Daston and Fernando Vidal, 282–307. Chicago: University of Chicago Press, 2004.

———. "Erich Wasmann, Ernst Haeckel and the Limits of Science." *Theory in Biosciences* 121 (2002): 252–59.

Lyell, Charles. *The Geological Evidences of the Antiquity of Man.* London: Murray, 1863.

———. *Principles of Geology.* 3 vols. London: John Murray, 1830–38.

Maillet, Benoît de. *Telliamed; ou, Entretiens d'un philosophe indien avec un missionnaire François.* 2 vols. Amsterdam: L'honore & Fils, 1748.

Marshall, S. L. A. *World War I.* New York: Houghton Mifflin, 2001.

Massin, Benoit. "From Virchow to Fischer: Physical Anthropology and 'Modern Race Theories' in Wilhelmine Germany." *History of Anthropology* 8 (1996): 79–154.

May, Walther. *Ernst Haeckel: Versuch einer Chronik seines Lebens und Wirkens.* Leipzig: Johann Ambrosius Barth, 1909.

Mendel, Gregor. "Versuche über Pflanzen-Hybriden." *Verhandlungen des naturforschenden Vereines in Brünn* 4 (1866): 3–47.

Mikhailov, Alexander, and Scott Gilbert. "From Development to Evolution: the Re-establishment of the 'Alexander Kowalevsky Medal.'" *International Journal of Developmental Biology* 46 (2002): 693–98.

Miklucho, Nikolai. "Beiträge zur Kenntniss der Spongien I." *Jenaische Zeitschrift für Medicin und Naturwissenschaft* 4 (1868): 221–40.

Mikloucho-Maclay, Nikolai. *Nikolai Nikolajewitsch Mikloucho-Maclay Briefwechsel mit Anton Dohrn.* Edited by Irmgard Müller. Norderstedt: Verlag für Ethnologie, 1980.

———. "Ueber einige Schwämme des Nördlichen Stillen Oceans und des Eismeres." *Mémoires de l'Académie impériale des sciences de St.-Pétersbourg,* 7th ser., 15 (1870): part 2.

Mocek, Reinhard. "Ernst Haeckel als Philosoph." In *Leben und Evolution,* edited by Bernd Wilhelmi, 86–95. Jena: Friedrich-Schiller-Universität, 1985.

Moore, James. *The Post-Darwinian Controversies.* Cambridge: Cambridge University Press, 1979.

Moore, Keith L. *Before We Are Born: Basic Embryology and Birth Defects*. 3rd ed. Philadelphia: W. B. Saunders, 1989.

Müller, Fritz. *Facts and Arguments for Darwin*. Translated by W. S. Dallas. London: John Murray, 1869.

———. *Für Darwin*. Leipzig: Wilhelm Engelmann, 1864.

———. *Werke, Briefe und Leben*. Edited by Alfred Möller. 3 vols. in 4. Jena: Gustav Fischer, 1920.

Müller, Johannes. *Handbuch der Physiologie des Menschen für Vorlesungen*. 2 vols. Coblenz: J. Hölscher, 1833–40.

———."Über die im Hafen von Messina beobachteten Polycystinen." *Bericht über die Verhandlungen der Königlichen Preussischen Akademie der Wissenschaften zu Berlin* (1855): 229–54, 671–76.

———."Über die Thalassicollen, Polycystinen und Acanthometren des Mittelmeeres." *Abhandlungen der Königlichen Akademie der Wissenschaften zu Berlin* (1858): 1–62.

Nägeli, Carl von. "Ueber die Schranken der naturwissenschaftlichen Erkenntniss." In *Amtlicher Bericht der 50. Versammlung Deutscher Naturforscher und Aerzte in München*, 25–41. Munich: Akademischen Buchdruckerei von F. Straub, 1877.

National Center for History in the Schools (UCLA). "Standard 5: Historical Issues." http://nchs.ucla.edu/standards/thinking5-12-5.html.

Newman, John Henry. *Certain Difficulties Felt by Anglicans in Catholic Teaching Considered*. New Impression. 2 vols. London: Longmans, Green, 1900.

Nordenskiöld, Erik. *The History of Biology: A Survey*. 2nd ed. Translated by Leonard Bucknall Eyre. New York: Tudor, 1936. Originally published as *Biologens Historia*. Stockholm: Bjorck & Borjesson, 1920–24.

Nöthlich, Rosemarie, Olaf Breidbach, and Uwe Hoßfeld. "'Was ist die Natur?' Einige Aspekte zur Wissenschaftspopularisierung in Deutschland." In *Klassische Universität und akademische Provinz: Die Universität Jena von der Mitte des 19. bis in die 30er Jahre des 20. Jahrhunderts*, edited by Matthias Steinbach and Stefan Gerber, 238–50. Jena: Bussert & Stadeler, 2005.

Novick, Peter. *That Noble Dream. The "Objectivity Question" and the American Historical Profession*. Cambridge: Cambridge University Press, 1988.

Numbers, Ronald. *The Creationists*. New York: Knopf, 1992.

Nyhart, Lynn. *Biology Takes Form: Animal Morphology and the German Universities, 1800–1900*. Chicago: University of Chicago Press, 1995.

———. "The Importance of the 'Gegenbaur School' for German Morphology." *Theory in Biosciences* 122 (2003): 162–73.

Oken, Lorenz. *Ueber die Bedeutung der Schädelknochen*. Bamberg: Göbhardt, 1807.

———. *Die Zeugung*. Bamberg: Goebhardt, 1805.

Oppenheimer, Jane. "Analysis of Development: Problems, Concepts and Their History." 1955. In *Essays in the History of Embryology and Biology*, 117–72. Cambridge, MA: MIT Press, 1967.

———. "An Embryological Enigma in the *Origin of Species*." 1959. In *Essays in the History of Embryology and Biology*, 221–55. Cambridge, MA: MIT Press, 1967.

———. "The Non-Specificity of the Germ Layers." 1955. In *Essays in the History of Embryology and Biology*, 256–94. Cambridge, MA: MIT Press, 1967.

[Owen, Richard]. "Darwin on the *Origin of Species*." *Edinburgh Review* 111 (1860): 487–532.

Owen, Richard. *The Hunterian Lectures in Comparative Anatomy, May and June 1837*. Edited by Phillip R. Sloan. Chicago: University of Chicago Press, 1992.

———. *On the Nature of Limbs*. London: Van Voorst, 1849.

———. "Report on the Archetype and Homologies of the Vertebrate Skeleton." In *Report of the Sixteenth Meeting of the British Association for the Advancement of Science: Held at Southampton in September 1846*, 175–76. London: Murray, 1847.

Paulsen, Friedrich. *Philosophia militans: Gegen Klerikalismus und Naturalismus*. Berlin: Reuther & Reichard, 1901.

Pennisi, Elizabeth. "Haeckel's Embryos: Fraud Rediscovered." *Science* 277 (1997): 1435.

Pick, Daniel. *Faces of Degeneration: A European Disorder, c. 1848–1918*. Cambridge: Cambridge University Press, 1989.

Plate, Ludwig. *Selectionsprinzip und Probleme der Artbildung: Ein Handbuch des Darwinismus*. 3rd ed. Leipzig: Wilhelm Engelmann, 1908.

———. *Ultramontane Weltanschauung und moderne Lebenskunde, Orthodoxie und Monismus*. Jena: Gustav Fischer, 1907.

Proctor, Robert. "Architecture from the Cell-Soul: René Binet and Ernst Haeckel." *Journal of Architecture* 11 (2006): 407–24.

Rabl, Carl. "Die Ontogenie der Süsswasser-Pulmonaten." *Jenaische Zeitschrift für Naturwissenschaften* 9 (1875): 195–240.

Ramaswamy, Sumathi. *The Lost Land of Lemuria: Fabulous Geographies, Catastrophic Histories*. Berkeley: University of California Press, 2004.

Rehkämper, Gerd. "Zur frühen Rezeption von Darwins Selektionstheorie und deren Folgen für die vergleichende Morphologie heute." *Sudhoffs Archiv* 81 (1997): 171–92.

Reil, Johann Christian. "Von der Lebenskraft." *Archiv für die Physiologie* 1 (1796): 8–162.

Remak, Robert. *Untersuchungen über die Entwickelung der Wirbelthiere*. 3 vols. Berlin: Georg Reimer, 1850–55.

Reymond, Moritz von. *Laienbrevier des Häckelismus: Jubiläumsausgabe, 1862–1882–1912*. Munich: Ernst Reinhardt, 1912.

Reynolds, Andrew. "The Theory of the Cell State and the Question of Cell Autonomy in Nineteenth and Early Twentieth Century Biology." *Science in Context* 20 (2007): 71–95.

Richards, Robert J. *Darwin and the Emergence of Evolutionary Theories of Mind and Behavior*. Chicago: University of Chicago Press, 1987.

———. "Ideology and the History of Science." *Biology and Philosophy* 8 (1993): 103–8.

———. "The Linguistic Creation of Man: Charles Darwin, August Schleicher, Ernst Haeckel, and the Missing Link in Nineteenth-Century Evolutionary Theory." In *Experimenting in Tongues: Studies in Science and Language*, edited by Matthias Doerres, 21–48. Stanford, CA: Stanford University Press, 2002.

———. *The Meaning of Evolution: The Morphological Construction and Ideological Reconstruction of Darwin's Theory.* Chicago: University of Chicago Press, 1992.

———. "Neanderthals Need Not Apply." *New York Times Book Review*, August 17, 1997.

———. "Race." In *Oxford Companion to the History of Modern Science*, edited by John Heilbron. Oxford: University of Oxford Press, 2002.

———. "Rhapsodies on a Cat-Piano, or Johann Christian Reil and the Foundation of Romantic Psychiatry." *Critical Inquiry* 24 (1998): 700–36.

———. *The Romantic Conception of Life: Science and Philosophy in the Age of Goethe.* Chicago: University of Chicago Press, 2002.

———. "The Structure of Narrative Explanation in History and Biology." In *History and Evolution*, edited by Matthew Nitecki and Doris Nitecki, 19–53. Albany: State University of New York Press, 1992.

Richardson, Michael, James Hanken, Mayoni L. Gooneratne, Claude Pieau, Albert Paynaud, Lynne Selwood, and Glenda Wright. "There Is No Highly Conserved Embryonic Stage in the Vertabrates: Implications for Current Theories of Evolution and Development." *Anatomy and Embryology* 196 (1997): 91–106.

Richardson, Michael, and Gerhard Keuck. "Haeckel's ABC of Evolution and Development." *Biological Review* 77 (2002): 495–528.

———. "A Question of Intent: When Is a 'Schematic' Illustration a Fraud?" *Nature* 410 (2001): 144.

"Richtilinien für die Bestandsprüfung in den Volksbüchereien Sachsens." *Die Bücherei* 2 (1935): 279–80.

Rickert, Heinrich. *Die Grenzen der naturwissenschaftlichen Begriffsbildung, eine logische Einleitung in die historischen Wissenschaften.* Tübingen: J. C. B. Mohr, 1902.

Rinard, Ruth. "The Problem of the Organic Individual: Ernst Haeckel and the Development of the Biogenetic Law." *Journal of the History of Biology* 14 (1981): 249–75.

Roberts, Jon. *Darwinism and the Divine in America: Protestant Intellectuals and Organic Evolution, 1859–1900.* Madison: University of Wisconsin Press, 1988.

Roll-Hansen, Nils. "E. S. Russell and J. H. Woodger: The Failure of Two Twentieth-Century Opponents of Mechanistic Biology." *Journal of the History of Biology* 17 (1984): 399–428.

Romanes, George. *Darwin and After Darwin.* 3 vols. Chicago: Open Court, 1892–97.

Rothschuh, Karl. *History of Physiology.* Translated by Guenter Risse. Huntington, NY: Krieger, 1973.

Roux, Wilhelm. "Beiträge zur Entwickelungsmechanik des Embryo: Ueber die künstliche Hervorbringung halber Embryonen durch Zerstörung einer der beiden ersten Furchungskugeln, sowie über die Nachentwickelung (Postgeneration) der fehlenden Körperhälfte." *Archiv für pathologische Anatomie und Physiologie und für klinische Medicin* (Virchow's Archive) 94 (1888): 113–53, 246–91.

———. *Der Kampf der Theile im Organismus, ein Beitrag zur Vervollständigung der mechanischen Zweckmässigkeitslehre.* Leipzig: Wilhelm Engelmann, 1881.

———. *Ueber die Leistungsfähigkeit der Principien der Descendenzlehre zur Erklärung der Zweckmässigkeiten des thierischen Organismus.* Breslau: S. Schottänderen, 1880.

———. "Ueber die Verzweigungen der Blutgefässe des Menschen, eine morphologische Studie." *Jenaische Zeitschrift für Naturwissenschaft* 12 (1878): 205–66.

———. *Der züchtende Kampf der Theile; oder, Die "Theilauslese" im Organismus.* In *Gesammelte Abhandlungen über Entwickelungsmechanik der Organismen,* 1:135–422. 2 vols. 1881. Reprint, Leipzig: Wilhelm Engelmann, 1895.

Rupke, Nicolaas. *Alexander von Humboldt: A Metabiography.* Frankfurt: Peter Lang, 2005.

———. "Neither Creation nor Evolution: The Third Way in Mid-Nineteenth-Century Thinking about the Origin of Species." *Annals of the History and Philosophy of Biology* 10 (2005): 143–72.

———. *Richard Owen, Victorian Naturalist.* New Haven, CT: Yale University Press, 1994.

Ruse, Michael. *The Evolution-Creation Struggle.* Cambridge, MA: Harvard University Press, 2005.

———. "Karl Popper and Evolutionary Biology." In *Is Science Sexist? And Other Problems in the Biomedical Sciences,* 65–84. Dordrecht: D. Reidel, 1981.

Russell, E. S. *Form and Function: A Contribution to the History of Animal Morphology.* 1916. Reprint, Chicago: University of Chicago Press, 1982.

———.. *The Study of Living Things: Prolegomena to a Functional Biology.* London: Methuen, 1924.

Rütimeyer, Ludwig. "Referate." Reviews of "Ueber die Entstehung und den Stammbaum des Menschengeslechts" and *Natürliche Schöpfungsgeschichte,* by Ernst Haeckel. *Archiv für Anthropologie* 3 (1868): 301–2.

Sandmann, Jürgen. *Der Bruch mit der humanitären Tradition: Die Biologisierung der Ethik bei Ernst Haeckel und anderen Darwinisten seiner Zeit.* Stuttgart: Gustav Fischer, 1990.

———. "Ernst Haeckels Entwicklungslehre als Teil seiner biologistischen Weltanschauung." In *Die Rezeption von Evolutionstheorien im 19. Jahrhundert,* edited by Eve-Marie Engels, 326–46. Frankfurt: Suhrkamp, 1995.

Sassoon, Siegfried. "Counter-Attack." In *The War Poems*, edited by Rupert Hart-Davis, 33. London: Faber and Faber, 1983.

Schelling, Friedrich. *Schellings Werke*. Münchner Jubiläumsdruck. Edited by Manfred Schröter. 12 vols. Munich: C. H. Beck, 1927–59.

———. *System des transscendentalen Idealismus*. With a new introduction by Walter Schulz. 1800. Reprint, Hamburg: Felix Meiner, 1962.

Schiller, Friedrich. "Was Heisst und zu welchem Ende studiert man Universalgeschichte?" In *Sämtliche Werke*. Edited by Jost Perfahl, 4:705–20. 6th ed. 5 vols. Düsseldorf: Artemis & Winkler, 1997.

Schlegel, August Wilhelm. *Observations sur la langue et la littérature provençales*. Paris: Librarie grecque-latine-allemande, 1818.

Schleicher, August. *Die Darwinsche Theorie und die Sprachwissenschaft*. Weimar: Hermann Böhlau, 1863.

———. "Die Darwin'sche Theorie und die Thier- und Pflanzenzucht." *Zeitschrift für deutsche Landwirthe* 15 (1864): 1–11.

———. *Die Deutsche Sprache*. Stuttgart: Cotta'scher Verlag, 1860.

———. *Über die Bedeutung der Sprache für die Naturgeschichte des Menschen*. Weimar: Hermann Böhlau, 1865.

———. *Zur vergleichenden Sprachengeschichte*. Bonn: H. B. König, 1848.

Schleiden, Matthias Jakob. "Beiträge zur Phytogenesis." *Archiv für Anatomie, Physiologie und wissenschaftliche Medicin* (1838): 137–76.

———. *Die Pflanze und ihr Leben: Populäre Vorträge*. Leipzig: Wilhelm Engelmann, 1848.

———. "Über den Materialismus der neueren deutschen Naturwissenschaft, sein Wesen und seine Geschichte." 1863. In *Wissenschaftsphilosophische Schriften*. Edited by Ulrich Charpa, 265–308. Köln: Jürgen Dinter's Verlag für Philosophie, 1989.

Schleiermacher, Friedrich. *Der christliche Glaube*. 2nd ed. 2 vols. 1835. Reprint, Berlin: De Gruyter, 1960.

Schmidt, Heinrich. *Ernst Haeckel und sein Nachfolger Prof. Dr. Ludwig Plate*. Jena: Volksbuchhandlung, 1921.

———. *Haeckels Embryonenbilder: Dokumente zum Kampf um die Weltanschauung in der Gegenwart*. Frankfurt: Neuer Frankfurter, 1909.

———. *Der Kampf um die "Welträtsel": Ernst Haeckel, die "Welträtsel" und die Kritik*. Bonn: Emil Strauss, 1900.

Schmidt, Heinrich, ed. *Was Wir Ernst Haeckel Verdanken: Ein Buch der Verehrung und Dankbarkeit*. 2 vols. Leipzig: Unesma, 1914.

Schmitz, H. "P. Erich Wasmann S.J." *Tijdschrift voor Entomologie* 75 (1932): 1–57.

Schwann, Theodor. *Mikroskopische Untersuchungen über die Übereinstimmung in der Struktur und dem Wachstum der Tiere und Pflanzen*. Berlin: Sander'schen Buchhandlung, 1839.

Sclater, P. L. "The Mammals of Madagascar." *Quarterly Journal of Science* 1 (1864): 212–19.

Searle, John. "Reply to 'What Is Consciousness?'" *New York Review of Books* 52, no. 11 (23 June 2005): 56–57.

Sebottendorff, Rudolf von. *Bevor Hitler kam*. Munich: Deukula, 1933.

Selenka, Emil. *Menschenaffen (Anthropomorphae): Studien über Entwickelung und Schädelbau*. Vol. 5 of *Zur vergleichenden Keimesgeschichte der Primaten*. Wiesbaden: C. W. Kreidel, 1903.

Semon, Richard, ed. *Zoologische Forschungsreisen in Australien und dem Malayischen Archipel*. 4 vols. Jena: Gustav Fischer, 1893–.

Semper, Carl. *Der Haeckelismus in der Zoologie*. Hamburg: W. Mauke Söhne, 1876.

———. *Offener Brief an Herrn Prof. Haeckel in Jena*. Hamburg: W. Mauke Söhne, 1877.

Shapin, Steven, and Simon Schaffer. *Leviathon and the Air-Pump*. Princeton, NJ: Princeton University Press, 1985.

Sheehan, James J. *German History, 1770–1866*. Oxford: Clarendon Press, 1989.

———. *German Liberalism in the Nineteenth Century*. Chicago: University of Chicago Press, 1978.

Shipman, Pat. *The Evolution of Racism*. Cambridge, MA: Harvard University Press, 1994.

Singer, Charles. *A History of Biology to about the Year 1900: A General Introduction to the Study of Living Things*. London: Abelard-Schuman, 1959.

Spee, Ferdinand Graf von. "Beobachtungen an einer menschlichen Keimscheibe mit offener Medullarrinne und Canalis neurentericus." *Archiv für Anatomie und Entwickelungsgeschichte* (1889): 159–76.

Spilka, Bernard, Ralph Hood, Bruce Hunsberger, and Richard Gorsuch. *The Psychology of Religion: An Empirical Approach*. New York: Guilford Press, 2003.

Stocking, George. *Victorian Anthropology*. New York: Free Press, 1987.

Stölzle, Remigius. *A. Von Köllikers Stellung zur Descendenzlehre: Ein Beitrag zur Geschichte moderner Naturphilosophie*. Munster: Aschendorffschen Buchhandlung, 1901.

Stringer, Christopher, and Robin McKie. *African Exodus: The Origins of Modern Humanity*. New York: Holt, 1996.

Takahashi, Kozo, and Susumu Honjo. *Radiolaria: Flux, Ecology, and Taxonomy in the Pacific and Atlantic*. Ocean Biocoenosis Series, no. 3. Woods Hole, MA: Woods Hole Oceanographic Institution, 1991.

Taub, Liba. "Evolutionary Ideas and Empirical Methods: The Analogy between Language and Species in Works by Lyell and Scleicher." *British Journal for the History of Science* 26 (1993): 171–93.

Teudt, Wilhelm. *Im Interesse der Wissenschaft! Haeckel's "Fälschungen" und die 46 Zoologen*. Schriften des Keplerbundes, Heft 3. Godesberg: Naturwissenschaftlicher, 1909.

Thompson, D'Arcy Wentworth. *On Growth and Form: A New Edition*. Cambridge: Cambridge University Press, 1942.

Thomson, Sir C. Wyville, ed. *Report on the Scientific Results of the Voyage of*

H.M.S. Challenger during the Years 1873–1876. 6 vols. in 50. London: Her Majesty's Stationery Office, 1878–95.

Tiedemann, Friedrich. *Das Hirn des Negers mit dem des Europäers und Orang-outangs verglichen*. Heidelberg: Karl Winter, 1837.

———. *Zoologie, zu seinen Vorlesungen entworfen*. 3 vols. Landshut: Weber, 1808–14.

Trembley, Abraham. *Hydra and the Birth of Experimental Biology—1744: Abraham Trembley's Mémoires Concerning the Polyps*. Translated and edited by Sylvia Lenhoff and Howard Lenhoff. Pacific Grove, CA: Boxwood Press, 1986.

———. *Mémoires pour server à l'histoire d'un genre de polypes d'eau douce, à bras en forme de cornes*. 2 vols. Paris: Durand, 1744.

Unamuno, Miguel de. *The Tragic Sense of Life*. Translated by J. E. Crawford Flitch. London: Macmillan, 1921.

Uschmann, Georg. *Geschichte der Zoologie und der zoologischen Anstalten in Jena 1779–1919*. Jena: Gustav Fischer Verlag, 1959.

Verworn, Max. "Ernst Haeckel." *Zeitschrift für allgemeine Physiologie* 19 (1921): i–xi.

Vesalius. *Vesalius on the Human Brain*. Translated with notes by Charles Singer. London: Oxford University Press, 1952.

Virchow, Rudolf. *Brief an seine Eltern, 1839 bis 1864*. 2nd ed. Edited by Marie Rabl. Leipzig: Wilhelm Engelmann, 1907.

———. *Die Cellularpathologie in ihrer Begründung auf physiologische und pathologische Gewebelehre*. Berlin: Hirschwald, 1858.

———. *Freiheit der Wissenschaft im modernen Staat*. Berlin: Wiegandt, Hempel & Parey, 1877.

———. "Die Freiheit der Wissenschaft im modernen Staatsleben." In *Amtlicher Bericht der 50. Versammlung Deutscher Naturforscher und Aerzte in München*, 65–78. Munich: Akademischen Buchdruckerei von F. Straub, 1877.

———. *Menschen- und Affenschädel*. Berlin: Luderitz'sche, 1870.

———."Über die mechanische Auffassung des Lebens." In *Vier Reden über Leben und Kranksein*. Berlin: Georg Reimer, 1862.

———. "Ueber den vermeintlichen Materialismus der heutigen Naturwissenschaft." In *Amtlicher Bericht über die acht und dreissigste Versammlung Deutscher Naturforscher und Ärzte in Stettin*, 35–42. Stettin: Hessenland's Buchdruckerei, 1864.

———. *Vier Reden über Leben und Kranksein*. Berlin: Georg Reimer, 1862.

———. *Die Vorlesungen Rudolf Virchows über Allgemeine Pathologische Anatomie aus dem Wintersemester 1855/56 in Würzburg*. Transcribed by Emil Kugler. Jena: Gustav Fischer, 1930.

Vogt, Carl. *Vorlesungen über den Menschen, seine Stellung in der Schöpfung und in der Geschichte der Erde*. 2 vols. Gieszen: J. Ricker'sche Buchhandlung, 1863.

Volger, Otto. "Ueber die Darwin'sche Hypothese vom erdwissenschaftlichen Standpunkte aus." In *Amtlicher Bericht über die acht und dreissigste Versammlung*

Deutscher Naturforscher und Ärzte in Stettin, 59–70. Stettin: Hessenland's Buchdruckerei, 1864.

Voltaire, F. *Elémens de la philosophie de Neuton*. Amsterdam: E. Ledet, 1738.

Wagensberg, Jorge. "On the Emergence of Forms in Nature." Lecture. Forma i Funció (6–7 March 2003), Barcelona.

Waldeyer, H. W. "C. E. von Baer und seine Bedeutung für Entwicklungsgeschichte." In *Amtlicher Bericht der 50. Versammlung deutscher Naturforscher und Aerzte in München*, 4–11. Munich: Akademische Buchdruckerei von F. Straub, 1877.

Wallace, Alfred Russel. *Alfred Russel Wallace: Letters and Reminiscences*. Edited by James Marchant. 2 vols. London: Cassell, 1916.

———. *The Geographical Distribution of Animals*. 2 vols. New York: Harper and Brothers, 1876.

———. Review of *Principles of Geology* and *Elements of Geology*, by Charles Lyell. *Quarterly Review* 126 (1869): 359–94.

Wasmann, Erich. *The Berlin Discussion of the Problem of Evolution*. Authorized translation. St. Louis, MO: Herder, 1909.

———. "Die Entstehung der Instincte nach Darwin." *Stimmen aus Maria-Laach* 28 (1885): 333–53.

———. "Gibt es tatsächlich Arten, die heute noch in der Stammesentwicklung begriffen sind?" *Biologisches Zentralblatt* 21 (1901): 685–711, 737–52.

———. *Instinct und Intelligenz im Thierreich*. Freiburg: Herder'sche Verlagshandlung, 1897.

———. "Jugenderinnerungen." *Stimmen der Zeit* 123 (1932): 110–19, 191–99, 259–68, 327–34, 407–13.

———. *Der Kampf um das Entwicklungsproblem in Berlin*. Freiburg: Herdersche Verlagshandlung, 1907.

———. "Konstanztheorie oder Deszendenztheorie?" *Stimmen aus Maria-Laach* 56 (1903): 29–44, 149–63, 544–63.

———. *Modern Biology and the Theory of Evolution*. 3rd ed. Translated by A. M. Buchanan. St. Louis, MO: B. Herder, 1914. Originally published as *Die moderne Biologie und die Entwicklungstheorie*.

———. *Die moderne Biologie und die Entwicklungstheorie*. 2nd ed. Freiburg: Herdersche Verlagshandlung, 1904.

———. "Die monistische Identitätstheorie und die vergleichende Psychologie." *Biologisches Centralblatt* 23 (1903): 545–56.

———. "Offener Brief an Hrn. Professor Haeckel (Jena)." *Kölnische Volkszeitung* 46, no. 358 (2 May 1905): 1–2.

———. *Die psychischen Fähigkeiten der Ameisen*. Stuttgart: E. Schweizerbartsche Verlagshandlung, 1899.

———. *Der Trichterwickler, eine naturwissenschaftliche Studie über den Thierinstinkt*. Münster: Aschendorff'schen Buchhandlung, 1894.

———. *Vergleichende Studien über das Seelenleben der Ameisen und höheren Tiere*. Freiburg: Herder'sche Verlagshandlung, 1897.

———. *Vergleichende Studien über das Seelenleben der Ameisen und höheren Thiere.* 2nd ed. Freiburg: Herder'sche Verlagshandlung, 1900.

———. "Wissenschaftliche Beweisführung oder Intoleranz?" *Biologisches Centralblatt* 25 (1905): 621–24.

———. *Die zusammengesetzten Nester und gemischten Kolonien der Ameisen.* Münster: Aschendorff'schen Buchdruckerei, 1891.

Weber, Heiko. "Der Monismus als Theorie einer einheitlichen Weltanschauung am Beispiel der Positionen von Ernst Haeckel und August Forel." In *Monismus um 1900: Wissenschaftskultur und Weltanschauung,* edited by Paul Ziche, 81–127. Berlin: Verlag für Wissenschaft und Bildung, 2000.

Webster, E. M. *The Moon Man: A Biography of Nikolai Miklouho-Maclay.* Berkeley: University of California Press, 1984.

Weikart, Richard. "Evolutionäre Aufklärung? Zur Geschichte des Monistenbundes." In *Wissenschaft, Politik und Öffentlichkeit,* edited by Mitchell Ash and Christian Stifter, 131–48. Vienna: Universitätsverlag, 2002.

———. *From Darwin to Hitler: Evolutionary Ethics, Eugenics, and Racism in Germany.* New York: Palgrave Macmillan, 2004.

———. "The Origins of Social Darwinism in Germany, 1859–1895." *Journal of the History of Ideas* 54 (1993): 469–80.

Weindling, Paul. *Darwinism and Social Darwinism in Imperial Germany: The Contribution of the Cell Biologist Oscar Hertwig (1849–1922).* Stuttgart: Gustav Fischer, 1991.

———. *Health, Race and German Politics between National Unification and Nazism, 1870–1945.* Cambridge: Cambridge University Press, 1989.

Weingarten, Michael. "Darwinismus und materialistischen Weltbild." In *Darwin und Darwinismus: Eine Ausstellung zur Kultur- und Naturgeschichte,* edied by Bodo-Michael Baumunk and Jürgen Riess, 74–82. Berlin: Akademie, 1994.

Weismann, August. *August Weismann: Ausgewählte Briefe und Dokumente.* Edited by Frederick Churchill and Helmut Risler. 2 vols. Freiburg: Universitätsbibliothek, 1999.

———. *Das Keimplasma: Eine Theorie der Vererbung.* Jena: Gustav Fischer, 1892.

———. *Studien zur Descendenz Theorie.* 2 vols. Leipzig: Wilhelm Engelmann, 1875–76.

———. *Vorträge über Deszendenztheorie.* 3rd ed. 2 vols. Jena: Gustav Fischer, 1913.

Wells, Jonathon. *Icons of Evolution.* Washington, DC: Regnery, 2000.

West, David. *Fritz Müller: A Naturalist in Brazil.* Blackburg,VA: Pocahontas Press, 2003. Based on a translation of *Fritz Müller, Werke, Briefe und Leben,* edited by Alfred Möller.

White, Andrew Dixon. *A History of the Warfare of Science with Theology in Christendom.* 2 vols. 1894. Reprint, New York: D. Appleton, 1896.

———. *A History of the Warfare of Science with Theology in Christendom.* 2 vols. 1894. Reprint, New York: George Braziller, 1955.

Wilson, E. B. *The Cell in Development and Inheritance.* New York: Macmillan, 1897.

Wilson, Edward O. *Insect Societies.* Cambridge, MA: Harvard University Press, 1971.

Wobbermin, Georg. *Ernst Haeckel im Kampf gegen die christliche Weltanschauung.* Leipzig: J. C. Hinrichs'sche Buchhandlung, 1906.

Wolpoff, Milford, and Rachel Caspari. *Race and Human Evolution: A Fatal Attraction.* Boulder, CO: Westview Press, 1998.

Wright, Chauncey. "Limits of Natural Selection." *North American Review* 111 (October 1870): 282–311.

Zigman, Peter. "Ernst Haeckel und Rudolf Virchow: Der Streit um den Charakter der Wissenschaft in der Auseinandersetzung um den Darwinismus." *Medizin-Historisches Journal* 35 (2000): 263–302.

INDEX

Acton, Lord John, 489, 508
Adam's Peak, 349
Adickes, Erich, 7
aesthetics, 2, 9, 22–23, 30, 33–35, 86, 110, 214, 235, 255, 270, 406, 455, 458, 478, 509
Agassiz, Louis, 119, 154–55, 227, 279, 354, 452
allantois, 233, 288–89, 296
Allmers, Hermann, 57–63, 80, 83, 92–93, 104, 106, 113, 117, 156, 217, 221–23, 324, 348, 351, 398, 407
American Indians, 22, 229, 231, 247–48, 272, 469
"An die Kulturwelt!" 434
archetypes, 9, 32, 35–36, 45, 74–75, 134, 151, 304, 310, 312, 341, 460, 470, 472–73, 479, 481, 482, 485
Aristotle, on sponges, 196
Asquith, Herbert, 426
Autenrieth, Johann Heinrich, 149
Aveling, Edward, 11

Baer, Karl Ernst von, 119, 149–51, 227, 281, 298, 302, 312, 341, 354, 440, 444, 478–85
- archetypes, 479
- evolutionary theory, objections to, 479–81
- morphological theory, 478–81
- recapitulation, rejection of, 481
- *Über Entwickelungsgeschichte der Thiere*, 312, 479

Bahr, Hermann, 274
Bailey, John Wendell, 389
Baldwin, James Mark, 128
Balfour, Arthur James, 274
Balfour, Francis, 444
Bartels, Edmund von, 62
Batch, August Carl, 177
Bebel, August, 274, 326–27, 446
Beckmann, Otto, 46
Benet, René, 501
Berlin Academy of Sciences, 65, 80, 345
Bernard, Claude, 282, 312
Bernstein, Eduard, 326, 446
Bethe, Albrecht, 364
Bethmann-Hollweg, Theobald von, 426, 429
Bezold, Albert von, 87
biogenetic law, 4–6, 9, 41, 89, 133, 135, 148–56, 166, 185, 188–90, 202–3, 208, 232, 234, 239, 244, 267, 283, 286, 291–96, 298, 301, 333–34, 336, 338n, 377, 409, 439, 443, 452, 501–2, 510
biogeography, 8, 8n, 21n, 34, 144, 228, 251, 256
Bischoff, Hans, 388
Bischoff, Theodor, 236, 239, 242, 279, 286–87, 301, 333, 388
Bismarck, Otto von, 24, 171–73, 230, 230n, 322, 357–58, 361
Blavatsky, Madame (Helena Hahn), 252n
Blumenbach, Johann Friedrich, 271–72
Bölsche, Wilhelm, 263
Boveri, Theodor, 382
Bowler, Peter, 5n, 5–6, 140, 142, 147–48
Brass, Arnold, 371, 375–82, 386
- failure of, 376n
- fraud, charges against Haeckel, 376–82
- works
  - *Affen-Problem, Das*, 380
  - *Ernst Haeckel als Biologe und die Wahrheit*, 376

Braudel, Fernand, 492n
Braun, Alexander, 27, 39, 128
Bronn, Heinrich Georg, 68, 69, 88, 96, 98, 99, 125, 135, 152, 158, 159, 256, 474, 475, 476, 477, 478, 481, 487
  biogenetic law, 481n
  Darwin's theory, evaluation of, 69, 256n
  evolutionary theory, his version, 476, 478
  morphological theory, 474–78
  Origin of Species, translation of, 98–99, 99n
  phylogenetic tree, 478
  works
    Handbuch einer Geschichte der Natur, 474, 476–77
    Untersuchungen über die Entwickelungs-Gesetze der organischen Welt, 477–78
Brücher, Heinz, 445, 504, 505, 508
Brücke, Ernst, 282
Buch, Leopold von, 177
Büchner, Ludwig, 163, 227, 328
Buden, Alta, 410n
Buffalo Bill, 248
Buffon, Georges-Louis Leclerc, Comte de, 136
Bülow, Bernard Prince von, 426
Burdach, Karl Friedrich, 456, 461, 462, 463, 464, 470, 478
  morphological theory, 461–64
  Ueber die Aufgabe der Morphologie, 462
Burschenschaften, 25
Buttel-Reepen, Hugo von, 387

Camper, Pieter, 467
Canary Islands, 76, 171–78, 199, 213–14, 217
Carl Alexander of Saxe-Weimar-Eisenach, 358
Carl-Zeiss-Stiftung, 425
Carr, Edward Hallett, 497, 506
Carus, Carl Gustav, 53, 74, 105, 227, 470–74, 481–82
  archetypes, 472
  morphological theory, 470–74
  Von den Ur-theilen des Knochen- und Schalengerüstes, 470
Carus, Julius Victor, 53
Carus, Paul, 372, 373n
Catholic Center Party, 46, 357–58
Catholic Church, liberal movements in, 356, 356n
cell nucleus, carrier of heredity, 4, 123, 123n
cell-state, organisms as, 47, 102, 102n, 129
cell theory, 130
Challenger expedition, 75–78, 331–32
Chambers, Robert, 223–24
Chandrasekhar, Subramanian, 265
Chorologie. See biogeography
Churchill, Winston, 426
Collingwood, Robin George, 512
contingency thesis, 14n, 14–15
Craig, Gordon, 357
Cuvier, Georges, 119, 227, 272, 466–67, 479

Darwin, Charles, 1–2, 5–10, 13, 20, 22, 29, 31, 41, 53, 60, 67–72, 74, 82, 86, 88–89, 93–104, 108, 113–17, 119–20, 123, 125, 128, 135–59, 161, 163–64, 166–68, 172–79, 185, 188, 190, 196–97, 202–3, 205–6, 212–13, 215, 218, 220, 223, 225, 227–29, 244, 252, 256, 261–63, 265, 267–71, 277, 279, 282, 284–85, 291–97, 311–15, 328–29, 331, 338, 345, 349–52, 361, 363, 366–67, 375–76, 383–85, 399–400, 409, 414, 420, 439–46, 450, 452–53, 461, 467, 472–78, 484–87, 492, 500, 501, 505–6, 508, 511–12, 529
  archetypes, 485
  brain evolution, 261–63
  gemmules, 146
  Haeckel, Ernst, debt to, 72n, 261–63
  Humboldt, Alexander von, influence of, 31n
  inheritance of acquired characteristics, 146n
  morphological theory, 484–87
  progressive evolution, 98n, 99
  recapitulation, principle of, 41n, 152–55, 484–85
  religion, 350–52
  teleology, 294
  works
    Descent of Man, The, 72, 97, 158, 225, 261, 262, 271, 297, 467
    Naturwissenschaftliche Reisen (*Beagle Voyage*), 20, 22
    Notebooks, 484
    On the Origin of Species, 2, 10, 13, 68, 88, 96, 100, 120, 135–38, 140, 151–53, 156, 158, 190, 244, 255, 268, 291, 383, 385, 443, 474, 478, 485, 501

Über die Entstehung der Arten im Thier- und Pflanzen-Reich (*Origin of Species*), 22, 68, 168
Darwin, Erasmus (grandfather of Charles), 484
Darwin, Henrietta (daughter of Charles), 174
Daston, Lorraine, 303–5, 307, 311n
Dawkins, Richard, 268
De Beer, Gavin, 4
Dennert, Eberhard, 373–76
works
Vom Sterbelager des Darwinismus, 375
Wahrheit über Ernst Haeckel und seine "Welträtsel," 374
Desmond, Adrian, 6
Dewey, John, 125
Die Deutsche Nationalversammlung. *See* Frankfurt Parliament
Di Gregorio, Mario, 260n
Dohrn, Anton, 3, 116, 116n, 211, 221–22, 227, 376
Döllinger, Johann Ignaz von, 357n
Driesch, Hans, 3–4, 167, 186, 188–89, 192–95, 367, 370
experiments on sea urchin eggs, 193
vitalism, 193, 195
Dubois, Eugène, 3, 167, 252–53, 367, 405, 501
Du Bois-Reymond, Emil, 79, 296, 311, 315–18, 323, 327, 345, 359, 399–400
"Über die Grenzen der Naturerkennens," 315
world puzzles, 315, 315n, 399
Duncan, Isadora, 11, 420

Ecker, Alexander, 236–37, 279, 286–87, 301, 333
ecology, 4, 8, 21n, 135, 144, 166, 266, 409, 501
Ehrenberg, Christian, 63n, 65, 68
Eimer, Theodor, 375
embryology as a profession, 282–83, 296, 332
Entwickelungsmechanik, 167, 179, 189, 193–94, 189–95, 285
Escherich, Karl, 382
Eucken, Rudolf, 434
Evangelical Church, 45, 343, 387

Falkenhayn, Erich von, 427–28
Fichte, Johann Gottlieb, 25, 53, 80
Fischer, Kuno, 31, 34, 91, 117, 403
Foch, Ferdinand, 427
Fol, Hermann, 3, 176–77, 179
Fontenelle, Bernard de, 493
Forel, Auguste, 364, 372, 387, 431
Franco-Prussian War, 230n, 230–31, 318, 426
Frankfurt Parliament, 25–26
Franz, Victor, 445
Franz Ferdinand, Archduke of Austro-Hungary, 425
fraud
charges against Haeckel, 234, 243, 278–80, 279n, 283, 286–91, 301, 377–82, 503
Haeckel's response to, 296–303
Frederick William IV, 26
Freud, Sigmund, 282, 450
Friedenthal, Hans, 371
Fürbringer, Max, 83, 89, 252, 382

Galison, Peter, 303–5, 307, 311n
Gandhi, Mohandas, 2, 3n
Gandtner, Otto, 24
Garibaldi, Giuseppe, 61
Gasman, Daniel, 5, 269–70, 273, 448–53, 506–7, 509
gastrulation, 4, 8n, 203, 208, 210, 301, 439, 444, 502
Gegenbaur, Carl, 42, 49–50, 53, 79–91, 104, 106–7, 116, 119, 131, 162, 164, 170, 182, 189, 212–13, 217, 221, 248, 277, 302, 330, 338, 403, 439, 441
anatomy separated from physiology at Jena, 88
biogenetic law, 89, 338n
Catholicism, 85
evolutionary morphology, reticence about, 88–90
friendship, loss of with Haeckel, 403
Gegenbaur school of comparative anatomy, 90
Grundzüge der vergleichenden Anatomie, 89
life and research, early, 84–90
wife, death of, 106
Geoffroy Saint-Hilaire, Étienne, 94
germ layers, 182, 197, 233, 281, 283, 284, 444
Germanies, unification of under Bismarck, 171, 173
Gesellschaft deutscher Naturforscher und Aerzte, 1, 94–104, 312–24, 349
Gibbon, Edward, 499, 500
Gilbert, Scott, 339, 339n, 452–53

God and nature as one, 9, 36, 111, 115, 297, 401, 460, 464
Goethe, Johann Wolfgang von, 6–11, 20–21, 24, 30–32, 35–38, 44, 49, 53, 55–56, 59–60, 62, 71, 74–75, 80, 86–87, 91, 96, 105, 110–11, 119–21, 125–27, 135, 177–78, 215, 227, 266, 268, 297, 304, 311, 314–15, 344, 348, 350–53, 356, 395–96, 404, 454, 456–62, 464, 466, 470, 472, 482, 505
archetype, 458
evolutionist, 119n
monism, 24, 36n
morphological theory, 456–61
Oken, dispute with, 459
Urpflanze, 9, 36, 458
vertebral theory of the skull, 36, 459
works
Faust, 52, 57, 126–27, 454
Metamorphose der Pflanzen, 35, 35n, 458–59, 466
Zur Morphologie, 456, 459
Zwischenkiefer (Intermaxillary bone), 458
Goette, Alexander, 124, 190, 291–93, 296, 298–99, 382–83
Goldschmidt, Richard, 3, 385
Göppert, Ernst, 90
Gordon, Hermann Albert von, 430
Gould, Stephen Jay, 6, 234, 268, 449–53, 506–7, 509
critic of Haeckel, 75n, 449–52
Roux and Driesch, Haeckel's "apostate students," 189n
works
"Abscheulich! (Atrocious!), Haeckel's Distorions Did Not Help Darwin," 452
Ontogeny and Phylogeny, 450
Grant, Robert, 196, 197, 197n, 201, 212
Gray, Asa, 312, 384
Greeff, Richard, 176–77, 179
Gude, Karl, 23
guests of ants, 365–67

Haber, Fritz, 434
Haeckel, Agnes Huschke (second wife), 109, 217, 218, 219, 220, 221, 222, 330, 393, 396, 406, 413, 415, 416, 417, 424, 431
Haeckel, Anna Sethe (first wife), 50, 51, 56, 57, 60, 62, 63, 65, 68, 79, 80, 84, 90, 91, 92, 93, 104, 105, 106, 107, 109, 113, 114, 115, 128, 134, 158, 167, 173, 217, 218, 221, 222, 387, 393, 396, 416, 417, 501
Haeckel, Charlotte Sethe (mother), 20
Haeckel, Ernst
Adam's Peak, climbs, 348
aesthetic education of, 30–37
aesthetics, 9, 33, 75, 346, 346n, 395, 406, 417, 419, 447, 458, 463, 501
Aetna, climbs with Allmers, 62
anti-Catholicism, 46, 275, 356
anti-Semitism, alleged, 273–75, 423, 445, 448, 503n, 506–8
artistic study in Italy, 55
assessments of him
early, 440–42
English-speaking world, 442–44
German, 444–48
today, 448–53
awards, 67, 105, 183, 303, 391
biogenetic law (*see also* main entry), 148–56
Canary Islands, 176–79
cell nucleus, carrier of heredity, 4, 123
cenogenesis, 233–34, 296, 450
Ceylon, travels to, 345–49
correspondence with
Allmers, Hermann, 80, 93, 104, 106, 113, 156, 217, 324, 398
Darwin, Charles, 2, 113, 114, 163, 164, 170, 188, 218, 279n, 328
Duncan, Isadora, 420
Gegenbaur, Carl, 330
Haeckel, Agnes Huschke (second wife), 220, 331, 431
Haeckel, Anna Sethe (first wife), 51–52, 56, 57, 68, 80, 90, 91, 92
Hertwig, Richard, 438
Huxley, Thomas Henry, 172, 329
Kölliker, Albert von, 293
parents, 29, 39, 42, 47, 59, 93, 107
Sethe, Bertha, 48
Uslar-Gleichen, Frida von, 396, 397, 404, 413
Virchow, Rudolf, 172
Weismann, August, 217
Darwin, Charles, visits with, 174
Darwinian, as a, 135–62
death of Anna Sethe Haeckel (first wife), 104–11, 124, 164, 167, 170, 343, 387
Desmonema Annasethe, 109
education, early, 20–24

engagement, 49–53
ethics, theory of, 354
eugenics, 231
evolutionary progress, 148
experiments on siphonophores, 4, 185–88
extraordinary professor at Jena, 91
fraud
charges of, 234, 243, 278–80, 279n, 283, 286–91, 301, 377–82, 503
response to charges of, 296–303
gastraea, theory of, 196, 202, 205, 208, 210, 233, 277n, 301, 331
Gegenbaur, Carl, attitudes about, 90
gymnastics, 83
habilitation, 53, 63, 67, 68, 81
heredity, 123, 123n, 144–46
historical introduction to evolution, 227–28
historical natural science, 285, 296, 313
individual, biological, 128–35
Jena, teaching schedule as *Privatdozent*, 82
Jewish quarter in Morocco, visits, 179, 246n
Jews, relations with, 246, 246n, 274
Kant, reader of, 31n
laws of nature, 120, 121, 123
lectures, 82–83, 93, 117, 117n, 156, 222
life, origin of, 137
Malaya, travel to, 333, 405–13
marriage with Agnes Huschke, 220
medical school, 26–44
medical selection, type of artificial selection, 230
microscope, 55n
military selection, kind of artificial selection, 230–31
Mitrocoma Annae, 109, 180
monera, 137, 138, 328
money, denied for research, 300, 300n, 345, 345n
monism, 11, 24, 84, 111, 114, 115, 124–28, 270, 297, 314, 353, 368, 376, 381, 388, 401, 435, 446, 460
monistic religion, 352–56
Monist League, 419
moral judgment on his character, 502–9
morphology, definition of, 120
Nationalversammlung der Vögel, 26
natural selection, 97, 142–44
natural system of species, 136–42
Natürliche Schöpfungsgeschichte, features of, 269
Nazis
books prohibited by, 447
rejected by, 446–49
use of his work, 505
newspaper articles by or about, 1, 100, 303, 328, 359, 377, 381, 383, 393, 399
Nobel Prize, rumor of, 303n
Ordinarius Professor of Zoology, 116
Origin of Species, first reads, 68
pacifism, 432
palingenesis, 233, 234
photography vs. painting, 410
Pico de Tiede, climbs, 177
plastides, 146, 314
poetry, 38
political judgment, 326–27, 399n, 433, 435–37
popular science, 166
popular science, *Natürliche Schöpfungsgeschichte* as, 263
Privatdozent, 80, 82
progressive evolution, 97, 98, 101, 161
radiolaria, descriptions of, 65
religion
opposition to, 15, 96n, 107, 344, 352, 384
personal struggles with, 46n, 343–44
sexual selection, 158
sued by Hamann, 355–56
suicide, thoughts of, 49n, 218n, 330
techtology, 128
Thule Society, 270n
tragedy of his life, 16, 38, 453–54
travel to Ceylon, 345–49
tree diagrams, 158–62, 245, 503
Tristenspitze, climbs on honeymoon, 220
Uslar-Gleichen, Frida von
death of, 416
love affair with, 391–98, 413–19, 403–5
money for, 415
Vesuvius, climbs with Allmers, 59
Villa Medusa, 110, 349, 425, 438
Virchow, Rudolf
attitudes about, 29–30, 42, 47
response to, 324–29
Wasmann, attack on, 360, 368
wedding to Anna Sethe, 92
Weismann, August, friendship with, 217, 277n
Welträthsel, sales of, 398

Haeckel, Ernst (*continued*)
works
Anthropogenie; oder, Entwickelungsgeschichte des Menschen, 140, 161, 233, 239, 240, 249, 287, 288, 291, 300, 301, 302, 304, 330, 333–34, 338, 379, 384, 409
De Rhizopodum finibus et ordinibus, 81
De telis quibusdam Astaci fluviatilis (medical dissertation), 44
Englands Blutschuld am Weltkriege, 433
Entwickelungsgeschichte der Siphonophoren, 185
"Entwickelungstheorie Darwins," 1n, 94–101, 115, 115n
Ewigkeit: Weltkriegsgedanken über Leben und Tod, Religion und Entwicklungslehre, 435
Franziska von Altenhausen, 391
Freie Wissenschaft und Freie Lehre, 325, 328
Generelle Morphologie der Organismen, 4, 11, 100, 108, 111, 114–15, 117–66, 171, 173, 175, 222, 224, 244, 263, 265–67, 297, 391, 401, 440, 455, 487
Indische Reisebriefe, 346
Kalkschwämme, Die, 203, 205, 330
Kunstformen der Natur, 393, 405–6, 410, 417, 419
Lebenswunder, Die, 231, 401, 504
Love Letters of Ernst Haeckel Written between 1898 and 1903, 392
Malayische Reisebriefe, 410, 413
Menschen-Problem und die Herrentiere von Linné, 377
Monismus als Band zwischen Religion und Wissenschaft, Der, 353–56
Natürliche Schöpfungsgeschichte, 2, 89, 142, 161–62, 165, 223–61, 263, 269, 277–79, 287, 296–97, 330, 332–33, 384, 395, 410, 420, 441, 452, 502–3, 505
Plankton-Studien, 300
Radiolarien, Die, 67–75, 91
Reisebriefe von Ceylon, 395
Reports of the Challenger Expedition, 77, 344, 347, 444
Sandalion: Eine offene Antwort auf die Fälschungs-Anklagen der Jesuiten, 386
System der Medusen, 109, 310, 330
Systematische Phylogenie, 391
Tiefsee-Medusen der Challenger-Reise, 330
"Vernunft und Krieg," 432
Wanderbilder, 419
Welträthsel, Die, 2, 7, 223, 353, 371, 374, 384, 391–98, 400, 402–3, 415, 419, 421, 446
Ziele und Wege der heutigen Entwickelungsgeschichte, 298, 300, 302
Zoological Museum, Jena, 91
Haeckel, Heinrich (nephew), 431
Haeckel, Karl (brother), 20, 50, 80, 106
Haeckel, Karl (father), 20, 25–27
Haeckel, Walter (son), 219, 222, 392, 413, 431, 438
Haig, Douglas, 428–29
Hamann, Otto, 354–55
Hardenberg, Friedrich von (Novalis), 10, 53, 80
Hardenberg, Prince Charles Augustus von, 25
Hartmann, Eduard von, 124n
Hauptmann, Gerhart, 434
Hecht, Günther, 446
Hegel, Georg Friedrich, 53
Helmholtz, Wilhelm von, 128, 265, 329, 345, 439
Hensen, Victor, 300–303, 311
Henslow, John Stevens, 31
Herder, Johann Gottfried von, 25, 350
heredity, 4, 123, 123n, 135, 191, 261, 262, 286, 293, 296, 299, 363, 367, 439
Herodotus, 493
Hertwig, Oscar, 3, 205, 375
Hertwig, Richard, 3, 205, 382, 431, 438
Hildebrandt, Kurt, 446
Hindenburg, Paul von, 427
Hirschfeld, Magnus, 275
Haeckel, friend of, 372n
works
"Ernst Haeckel, ein deutscher Geistesheld," 275
Naturgesetze der Liebe, 275
His, Wilhelm, 236, 238, 241, 243, 244, 280–91, 293, 296, 298–304, 306, 308–11, 334–35, 338, 377, 444
embryogenesis, mechanical principles of, 281n, 283–85, 299
Unsere Körperform, 284, 298

historical responsibility, 491, 492, 505, 506, 509
historiography, 14–15, 489–512
Hitler, Adolf, 270, 276, 445, 449, 496, 505–9, 511
Homo erectus, 253
Hooker, Joseph, 176
Hopwood, Nick, 239n, 310, 310n
human evolution, 156–58
Humboldt, Alexander von, 8, 11, 20–22, 29–37, 44, 60, 86, 110, 173, 177–78, 214–15, 265, 508
  epistemology, 34
  Goethe, Johann Wolfgang von, 35n
  Kant, influence on, 34n
  works
    Ansichten der Natur, 20, 22, 31, 34, 265
    Kosmos, 34, 265
    Voyage aux régions equinoxiales du nouveau continent, fait en 1799–1804, 31, 34
Humboldt, Wilhelm von, 25
Humboldt Stiftung, 345
Humperdinck, Engelbert, 434
Hunt, James, 273
Huschke, Emil, 87, 217
Huschke, Konrad, 431
Huxley, Henrietta, 173
Huxley, Thomas Henry, 53, 108, 108n, 116, 128, 156–57, 164, 167, 169, 172–73, 175, 211, 221, 233–34, 277, 302, 322, 328–29, 345, 380, 439, 452

Ibsen, Henrik, 274
individual, nature of, 128–35
inheritance of acquired characters, 145, 146n, 191, 194, 487
inquilines. *See* guests of ants
intelligent design, 7, 491

jail, student (*Kerker*), 80
James, William, 6, 15, 125
Joffre, Joseph, 427, 428
John Paul II (pope), 386

Kalthoff, Albert, 372
Kammerer, Paul, 431
Kant, Immanuel, 7, 30–36, 110, 117, 121–22, 221, 227, 266, 297, 350, 401, 459–61, 505
  archetype, 121, 472
  intellectus archetypus, 121
  Kritik der Urteilskraft (Critique of Judgment), 32, 35, 459
  laws of nature, 121
  teleology, 32, 33, 121
Kautsky, Karl, 326, 446
Kelly, Alfred, 263–64, 266
Keplerbund, 371, 373, 376, 381–83, 386–87
Kerker. *See* jail, student
Keuck, Gerhard, 334–36, 338
Kielmeyer, Carl Friedrich, 149
Klaatsch, Hermann, 90
Klein, Felix, 434
Kolkenbrock-Netz, Jutta, 6
Kölliker, Albert von, 27, 28, 29, 30, 42, 43, 44, 64, 86, 87, 123, 131, 182, 236, 238, 239, 242, 281, 291, 293, 294, 295, 296, 299, 375, 440
  Darwin's teleology, 294, 295n
  heterogeneous generation, theory of, 295
  works
    Entwicklungsgeschichte des Menschen und der höheren Tiere, 295
    "Ueber die Darwin'sche Schöpfungstheorie," 294
Kovalevsky, Alexander, 208
Kovalevsky, Wladimir, 3
Kraepelin, Karl, 382
Krause, Wilhelm, 289
Krauße, Erika, 447
Kulturkampf, 357, 358, 361

La Valette St. George, Adolph de, 39–40
Lamarck, Jean-Baptiste, 94, 119, 135–36, 142, 149, 197, 227, 351, 440–41, 481, 505
Lamarckism, 13, 150, 228, 260, 466
Lang, Arnold, 3, 382
language and evolution, 84, 96, 125, 157, 256, 257, 258, 260
Lanzarote, island of, 178, 180, 196
Lea, Henry, 489–90
Lebenskraft, 47n
Lemuria, 250–52
Lenin, Vladimir Ilyich, 446
Leo XIII (pope), 357
Leuba, James, 385
Leydig, Franz, 30, 42, 44, 87, 281
Lieberkühn, Nathanial, 197, 199n
Linné (Linnaeus), Carl von, 23, 136, 271, 377
Loder, Johann Christian, 458
Loeb, Jacques, 431

Lowie, Robert, 431
Lubbock, John, 176, 312, 345
Ludendorff, Erich, 427, 436
Lyell, Charles, 157, 173, 225, 227

Maastricht, Natural History Museum of, 388, 390
Mach, Ernst, 11, 125, 128
Maclay, Nikolai Miklucho. *See* Miklucho, Nikolai
Marx, Karl, 11, 315, 319
materialism, 4, 15, 46, 48, 51, 94, 102–3, 124, 126, 270, 297, 353, 401, 440–41, 447–49, 508
Maugham, Somerset, 347
Maurer, Friedrich, 89
Max, Gabriel von, 255, 419
May, Karl, 248
Mayr, Ernst, 3, 268, 385
Meckel, Johann Friedrich, 149, 482
medusae, 3, 40, 63, 77, 87, 109–10, 131, 133, 170, 180–81, 295, 310, 331, 336, 347, 349, 402, 407, 421
Meyer, Elisabeth Haeckel (daughter), 219, 413, 414, 431
Meyer, Else (granddaughter), 219, 424
Meyer, Henriette, 432
Miklucho, Nikolai, 3, 176, 177, 179, 195, 196, 199–212, 222
  rejection of by Haeckel, 210–11
  sponges, work on, 199–202
Milne-Edwards, Henri, 312
missing link, 3, 4n, 157, 167, 252–53, 405, 501
Mommsen, Theodor, 274
monism, 36, 125–28
Monistenbund, 371, 372, 373
Monist League, 384, 419
Moore, James, 6, 385
moral judgment
  historians, 490–92, 500
  principles of, 509–12
Morocco, 179, 246, 426
Morphologie, origin of the term, 456
Müller, Friedrich von, 127
Müller, Fritz, 100, 152n, 152–55, 353n, 374
Müller, Heinrich, 30
Müller, Hermann, 374
Müller, Johannes Peter, 27, 39–40, 42–43, 49–50, 62, 64n, 64–65, 74, 79, 86–88, 116, 120, 122, 129–30, 155, 281, 302, 482
Murray, John, 75, 77
myrmecophile. *See* guests of ants

Nägeli, Carl, 53, 312, 315–18
Nares, George, 75
narrative history, grammar thereof, 492–500
  moral grammar, 497–500
  temporal and causal grammar, 492–97
National Center for History in the Schools, 491
Natural History of Creation. *See* Haeckel, Ernst, *Natürliche Schöpfungsgeschichte*
natural selection, 13, 97–99, 120, 123, 135, 142, 145, 147, 151, 158, 190, 228–32, 261, 291, 295, 354, 366, 400
Naturforschende Gesellschaft des Osterlandes, 353
nauplius, 153, 199, 244
Nazism, 3, 6, 167, 232, 269–70, 275–76, 332, 372, 388, 431, 446–49, 453, 503–4, 506–8, 511–12
Nelson, Garth, 451n
Nernst, Walther, 434
Newman, John Henry (cardinal), 357n
Nietzsche, Friedrich, 354, 401, 508
Nina, Domenico (Haeckel's aide in Sicily), 62
Nordenskiöld, Erik, 2, 223, 440–43, 452, 529
normative judgments in history, 498
Novalis. *See* Hardenberg, Friedrich von
Novick, Peter, 490
Nyhart, Lynn, 89n, 167, 212n

Oken, Lorenz, 23, 53, 80, 94–95, 119, 149, 227, 266, 459, 464–66, 470–71, 481
  Goethe, dispute with, 466
  morphological theory, 464–66
  recapitulation, principle of, 465
  vertebral theory of the skull, 466
  Zeugung, Die, 464
Oppenheimer, Jane, 442, 444
Ostwald, Wilhelm, 372, 431, 434
Owen, Richard, 89, 151, 156, 440, 459, 472, 477, 481–85
  homology, 68, 71, 202–3, 472
  morphological theory, 481–84
  On the Nature of Limbs, 151, 482, 485
  recapitulation, opposition to, 483

Paris Commune, 230n, 318–20
Paris World's Fair, 405, 501
Paulsen, Friedrich, 7

Pershing, John, 430
Peters, Wilhelm, 65, 68, 345
photography, epistemology of, 303–12, 409–10, 410n
Phyletic Museum, 421–25
Pinker, Steven, 268
Pinter, Harold, 493
Pithecanthropus erectus, 4n, 253
Pius IX (pope), 356, 357
Pius X (pope), 358
Planck, Max, 434
plastides, theory of, 146, 314
Plastidulen. See plastides
Plate, Ludwig, 273, 369–71, 376, 382, 421–23
Popper, Karl, 228
popular science, criteria of, 263n, 263–69
Preyer, Wilhelm, 190
primitive streak, 133, 287–89
Princip, Gavrilo, 285, 425–26
progressive evolution, 146–48
Protista, 4, 201

Rabl, Carl, 382–83, 431
race, theory of, 229n, 269–76, 468–69
races, human, 158, 229, 468–69
radiolarians, 1, 3, 63–67, 70, 74–77, 92, 169, 179, 268, 311, 331, 336–47, 444, 487, 501
Ranke, Loepold von, 497
recapitulation. *See* biogenetic law
Rehkämper, Gerd, 6
Reis, Emanuel, 177
religious belief among biologists, 385
Remak, Robert, 281
research trips, motives for, 213–16
Reymond, Moritz von, 328
Richardson, Michael, 304–7, 334–40, 452
Rickert, Heinrich, 490, 508, 510
Roberts, Jon, 385
Romanes, George, 167, 339
Romanticism, 8–13, 19, 25, 38, 44, 80, 95, 100, 121–22, 124, 126, 258, 268, 360, 460, 463, 471, 479, 481
Röntgen, Wilhelm Conrad, 434
Roux, Wilhelm, 3–4, 167, 186, 189–94, 285, 444
  experiments on frog eggs, 191, 192
  Kampf der Thiele im Organismus, Der, 190–91
  mosaic theory of development, 191–94
Ruge, Georg, 89
Rupke, Nicolaas, 475n
Ruse, Michael, 228n
Russell, Bertrand, 125
Russell, Edward Stuart, 4, 440–43
Rütimeyer, Ludwig, 7, 236, 238, 241, 243, 278–80, 284, 286, 291, 296–98, 300–301

Sandmann, Jürgen, 6
Sassoon, Siegfried, 429
Schaffer, Simon, 14n
Schelling, Friedrich Wilhelm Joseph von, 25, 30–34, 53, 80, 121–22, 441, 461, 463–64, 466, 471, 477–79
Schenk, August, 30
Schiller, Friedrich, 20, 38, 46, 53, 80, 87, 425, 456, 529
Schlegel, Friedrich, 80
Schlegel, Wilhelm, 80
Schleicher, August, 83–84, 96, 104, 108, 125–26, 157, 159, 161, 170, 250, 255–62, 314
  languages, types of, 258, 259
  monism, 126
  works
    Darwinsche Theorie und die Sprachwissenschaft, Die, 125, 126, 159, 256
    Deutsche Sprache, 159
    Über die Bedeutung der Sprache für die Naturgeschichte des Menschen, 257, 262
    Zur vergleichenden Sprachengeschichte, 258
Schleiden, Matthias Jakob, 8, 11, 20, 22–23, 26–27, 71, 81, 94, 102, 120, 130, 265
  transformation theory, 22
  works
    Pflanze und ihr Leben, Die, 20, 265
    "Ueber den Materialismus der neueren deutschen Naturwissenschaft," 102
Schleiermacher, Friedrich, 20, 44–45, 48, 343
  nature of religion, 45
  works
    Christliche Glaube, Der, 44
    Über die Religion, 343
Schmidt, Heinrich, 445
schools, evolution in, 11n, 320–24
Schultze, Max, 53
Schwalbe, Gustav, 382
Science for the People, 452

Sclater, Philip, 250, 251, 252
Seebeck, Moritz, 50, 87, 91, 115
Selenka, Emil, 377, 379
Semon, Richard, 3, 335–36, 338, 431
Sethe, Anna. *See* Haeckel, Anna Sethe
Sethe, Bertha (mother's sister), 20, 48
sexual selection, 158
Shapin, Steven, 14n
Singer, Charles, 442–44
siphonophores, 3, 63, 77, 131, 179, 180–89, 192, 194, 212, 268, 336, 402, 407
social Darwinism, 5, 506
Social Democrats, 273, 358, 423
socialism and evolutionary theory, 318–22, 322n, 326
Soemmerring, Samuel Thomas, 467
Sophia (duchess; wife of Archduke Franz Ferdinand), 425
Spee, Ferdinand Graf von, 288
Spencer, Herbert, 128, 227
Spinoza, Baruch, 49, 96, 127, 268, 314, 353, 401
sponges, 3, 77, 179, 196–213, 244, 256, 268, 336, 444, 487
Stammbaum. See tree diagrams
Stein, Baron Karl vom und zum, 25
stem-tree. *See* tree diagrams
Stirner, Max, 401
Stöcker, Adolf, 273
Stöcker, Helene, 431
Strasburger, Eduard, 3
Strauss, David Friedrich, 11
Syllabus errorum (Pius IX), 356

techtology, 128
teleology, 35, 36, 121, 122, 123, 147
theory and fact, epistemology of, 311, 314
Thompson, D'Arcy Wentworth, 474
Thomson, Allen, 239
Thomson, Charles Wyville, 75, 77
Thucydides, 493–98, 510
Tiedemann, Friedrich, 149, 158, 272, 466–69, 471, 481
  morphological theory, 466–69
  recapitulation, principle of, 467
  skull measurements, 467–69
  works
    Hirn des Negers, 272, 468
    Zoologie, 466
Tirpitz, Alfred von, 425
tragic sense of life, 16, 38, 453–54
tree diagrams, 84, 95, 158–62, 137–38, 140, 142, 155, 159, 162, 244, 250, 255, 257, 269
Trembley, Abraham, 181
Treviranus, Gottfried Reinhold, 149, 227

Unamuno, Miguel de, 9, 16, 453–54
Urmensch, 142, 246, 255, 259, 503
Uschmann, Georg, 447
Uslar-Gleichen, Anna von (mother of Frida), 393
Uslar-Gleichen, Bernhard von (father of Frida), 393
Uslar-Gleichen, Frida von, 392–98, 402–8, 413–20

Verworn, Max, 6, 7, 431
Virchow, Rudolf, 27–30, 42–47, 70, 79, 86, 94, 102–4, 116, 120, 124, 129–30, 169, 172, 189, 253, 265, 281, 296, 311, 312, 314, 315, 318–30, 345, 354, 357, 359, 367, 387
  cell-state, organisms as, 47, 102, 129
  cellular basis of life, 28, 129n
  evolution, views of, 70n, 104
  evolutionary theory, objections to, 28, 253, 318–24
  Progressive Party, leader of, 172
  revolution of 1848, 27
  vital forces, 21, 123, 193
Vogt, Carl, 169, 223–24, 265
Volger, Georg Heinrich Otto, 101, 169

Wagner, Cosima, 420
Waldeyer-Hartz, Heinrich Wilhelm Gottfried, 312
Wallace, Alfred Russel, 13, 167, 211, 231, 251–52, 261–62
Wasmann, Erich, S.J., 45, 356, 360–72, 385–90, 422
  amical selection, 366
  evolution, initial opposition to, 361
  guests of ants, 365–67
  Haeckel, response to, 368
  human evolution, 367, 370
  instincts, 361, 364, 364n
  life and education, 360–65
  monists, response to, 371
  Stimmen aus Maria-Laach, editor of, 363

works
Instinct und Intelligenz im Thierreich, 363
Moderne Biologie und die Entwicklungstheorie, Die, 360, 365
Psychischen Fähigkeiten der Ameisen, Die, 363
Trichterwickler, 361
Vergleichende Studien über das Seelenleben der Ameisen und höheren Thiere, 363
Zusammengesetzten Nester und gemischten Kolonien der Ameisen, Die, 362
Weikart, Richard, 270, 273, 449, 491, 506–7, 509, 512
Weismann, August, 146, 162–63, 167, 173, 191–92, 194, 208, 217, 277, 324, 338, 382–83, 439
biogenetic law, 338n
Haeckel, friendship with, 217, 277n
Werner, Johannes, 391
Wheeler, William Morton, 364, 387, 389
White, Andrew Dixon, 383
Wiek, Ferdinand, 24
Wildpret, Hermann, 178
Wilhelm II, 358, 425–26, 430
Wilson, Edward O., 268, 452
Windelband, Wilhelm, 434
World War I, 12, 391, 425–31
Wundt, Wilhelm, 438

Zigman, Peter, 6
Zoological Institute, 86, 396, 420, 422–23, 425, 446